PROCESS-INDUCED CHEMICAL CHANGES IN FOOD

ADVANCES IN EXPERIMENTAL MEDICINE AND BIOLOGY

Recent Volumes in this Series

Volume 429
BRAIN PLASTICITY: Development and Aging
Edited by Guido Filogamo, Antonia Vernadakis, Fulvia Gremo, Alain M. Privat, and Paola S. Timiras

Volume 430
ANALYTICAL AND QUANTITATIVE CARDIOLOGY
Edited by Samuel Sideman and Rafael Beyar

Volume 431
PURINE AND PYRIMIDINE METABOLISM IN MAN IX
Edited by Andrea Griesmacher, Peter Chiba, and Mathias M. Müller

Volume 432
HYPERTENSION AND THE HEART
Edited by Alberto Zanchetti, Richard B. Devereux, Lennart Hansson, and Sergio Gorini

Volume 433
RECENT ADVANCES IN PROSTAGLANDIN, THROMBOXANE, AND LEUKOTRIENE RESEARCH
Edited by Helmut Sinzinger, Bengt Samuelsson, John R. Vane, Rodolfo Paoletti, Peter Ramwell, and Patrick Y-K Wong

Volume 434
PROCESS-INDUCED CHEMICAL CHANGES IN FOOD
Edited by Fereidoon Shahidi, Chi-Tang Ho, and Nguyen van Chuyen

Volume 435
GLYCOIMMUNOLOGY 2
Edited by John S. Axford

Volume 436
ASPARTIC PROTEINASES: Retroviral and Cellular Enzymes
Edited by Michael N. G. James

Volume 437
DRUGS OF ABUSE, IMMUNOMODULATION, AND AIDS
Edited by Herman Friedman, John Madden, and Thomas W. Klein

Volume 438
LACRIMAL GLAND, TEAR FILM, AND DRY EYE SYNDROMES 2: Basic Science and Clinical Relevance
Edited by David A. Sullivan, Darlene A. Dartt, and Michele A. Meneray

PROCESS-INDUCED CHEMICAL CHANGES IN FOOD

Edited by

Fereidoon Shahidi

Memorial University of Newfoundland
St. John, Newfoundland, Canada

Chi-Tang Ho

Rutgers University
New Brunswick, New Jersey

and

Nguyen van Chuyen

Japan Women's University
Tokyo, Japan

Springer Science+Business Media, LLC

Library of Congress Cataloging in Publication Data

Process-induced chemical changes in food / edited by Fereidoon Shahidi, Chi-Tang Ho, and Nguyen van Chuyen

p. cm.—(Advances in experimental medicine and biology; v. 434)

Includes bibliographical references and index.

1. Food—Analysis—Congresses. 2. Food industry and trade—Quality control—Congresses. 3. Food—Quality. I. Shahidi, Fereidoon, 1951- . II. Ho, Chi-Tang, 1944- . III. Chuyen, Nguyen van. IV. Series.

TP372.5.P78 1998

664—dc21 98-15604

CIP

Based, in part, on proceedings of Pacifichem 95, held December 18 – 22, 1995, in Honolulu, Hawaii

DOI 10.1007/978-1-4899-1925-0

Originally published by Plenum Press, New York in 1998

MyCopy version of the original edition 1998

http://www.plenum.com

10 9 8 7 6 5 4 3 2 1

PREFACE

Chemical changes that occur in foods during processing and storage are manifold and might be both desirable and undesirable in nature. While many of the processes are carried out intentionally, there are also certain unwanted changes that naturally occur in food and might have to be controlled. Therefore, efforts are made to devise processing technologies in which desirable attributes of foods are retained and their deleterious effects are minimized. While proteins, lipids and carbohydrates are the main nutrients of food that are affected by processing, it is their interaction with one another, as well as involvement of low-molecular-weight constituents that affects their flavor, color and overall acceptability. Thus, generation of aroma via thermal processing and bioconversion is of utmost importance in food preparation. Furthermore, processing operations must be optimized in order to eliminate or reduce the content of antinutrients that are present in foods and retain their bioactive components. Therefore, while novel processing technologies such as freezing, irradiation, microwaving, high pressure treatment and fermentation might be employed, control process conditions in a manner that both the desirable sensory attributes and wholesomeness of foods are safeguarded is essential. Obviously, methodologies should also be established to quantitate the changes that occur in foods as a result of processing.

This volume was developed from contributions provided by a group of internationally-recognized lead scientists. It serves as a resource book with extensive bibliography for chemists, biochemists, food scientists and nutritionists working in the industry, academic institutions and government laboratories. It may also be used as a complementary text for graduate students in food chemistry.

Fereidoon Shahidi
Chi-Tang Ho
and Nguyen Van Chuyen

CONTENTS

1. Process-Induced Chemical Changes in Foods: An Overview 1
   Fereidoon Shahidi and Chi-Tang Ho

2. Methods to Monitor Process-Induced Changes in Food Proteins: An Overview 5
   E. C. Y. Li-Chan

3. Proteolysis and Gelation of Fish Proteins under Ohmic Heating 25
   Jae W. Park, Jirawat Yongsawatdigul, and Ed Kolbe

4. Effect of Maturity and Curing on Peanut Proteins: Changes in Protein Surface Hydrophobicity 35
   Si-Yin Chung, John R. Vercellotti, and Timothy H. Sanders

5. High Pressure Processing Effects on Fish Proteins 45
   T. C. Lanier

6. Effect of High Hydrostatic Pressure on Pacific Whiting Surimi 57
   Michael T. Morrissey, Yildiz Karaibrahimoglu, and Jovi Sandhu

7. High Pressure Processing of Fresh Seafoods 67
   Benjamin K. Simpson

8. High Pressure and Heat Treatments Effects on Pectic Substances in Guava Juice 81
   Gow-Chin Yen and Hsin-Tang Lin

9. Chemometric Applications of Thermally Produced Compounds as Time-Temperature Integrators in Aseptic Processing of Particulate Foods 91
   H.-J. Kim and Y.-M. Choi

10. Heating Rate of Egg Albumin Solution and Its Change during Ohmic Heating .. 101
    T. Imai, K. Uemura, and A. Noguchi

11. Chemical Changes during Extrusion Cooking: Recent Advances 109
    Mary Ellen Camire

12. Sucrose Loss and Color Formation in Sugar Manufacture 123
Les A. Edye and Margaret A. Clarke

13. Process-Induced Changes in Edible Oils 135
P. K. J. P. D. Wanasundara and F. Shahidi

14. Effects of Processing Steps on the Contents of Minor Compounds and Oxidation of Soybean Oil 161
David B. Min, Tsung-Lin Li, and Hyung-Ok Lee

15. Antioxidizing Potentials of BHA, BHT, TBHQ, Tocopherol, and Oxygen Absorber Incorporated in a Ghanaian Fermented Fish Product 181
Toshiaki Ohshima, Vivienne V. Yankah, Hideki Ushio, and Chiaki Kiozumi

16. Minimizing Process Induced Prooxidant Stresses 189
R. J. Evans and T. S. Jones

17. Antioxidative Properties of Products from Amino Acids or Peptides in the Reaction with Glucose 201
N. V. Chuyen, K. Ijichi, H. Umetsu, and K. Moteki

18. Maillard Reaction and Food Processing: Application Aspects 213
N. V. Chuyen

19. Generation and the Fate of C_2, C_3, and C_4 Reactive Fragments Formed in Maillard Model Systems of [^{13}C]Glucose and [^{13}C]Glycine or Proline 237
Varoujan A. Yaylayan, Anahita Keyhani, and Alexis Huygues-Despointes

20. Metal Chelating and Antioxidant Activity of Model Maillard Reaction Products 245
Arosha N. Wijewickreme and David D. Kitts

21. Volatile Components Formed from Reaction of Sugar and β-Alanine as a Model System of Cookie Processing 255
S. Nishibori, R. A. Berhnard, T. Osawa, and S. Kawakishi

22. Amino-Reductones: Formation Mechanisms and Structural Characteristics 269
T. Kurata and Y. Otsuka

23. Effects of Gamma Irradiation on the Flavor Composition of Food Commodities 277
Jui-Sen Yang

24. Flavor Deterioration in Yogurt 285
Naomi Harasawa, Hideki Tateba, Nobuko Ishizuka, Toshiyuki Wakayama, Katsumi Kishino, and Mitio Ono

25. Flavor Generation during Extrusion Cooking 297
William E. Riha, III and Chi -Tang Ho

26. Process-Induced Compositional Changes of Flaxseed 307
P. K. J. P. D. Wanasundara and F. Shahidi

27. Effect of Processing on Phenolics of Wines 327
V. Z. Blanco, J. M. Auw, C. A. Sims, and S. F. O'Keefe

28. Photochemical Reactions of Flavor Compounds 341
Chung-Wen Chen and Chi-Tang Ho

Index 357

1

PROCESS-INDUCED CHEMICAL CHANGES IN FOODS

An Overview

Fereidoon Shahidi[1] and Chi-Tang Ho[2]

[1]Department of Biochemistry
Memorial University of Newfoundland
St. John's, NF, A1B 3X9, Canada
[2]Department of Food Science
Rutgers University
New Brunswick, New Jersey 08903

Processing of foods induces changes in their physical, chemical and sensory characteristics. Many researchers have shown the chemical consequences of food processing on acceptability and sensory attributes, nutritive value and wholesomeness of foods. A cursory account of these changes is provided.

INTRODUCTION

Chemical changes that occur in food during harvesting, processing and storage affect its overall quality and may influence both its major and minor components and sensory attributes. These changes might be advantageous or deleterious. Advantages that might be attained due to processing of food might relate to reduction in its content of hazardous antinutrients such as enzyme inhibitors, cyanogens, glucosinolates, among others, as well as control of microbial and enzymatic spoilage. In addition, they may improve the color, flavor, texture and other quality characteristics of foods and enhance availability of perishable products and convenience foods. However, there are also disadvantages that might be experienced during food processing. Deleterious effects on food quality relate to a decrease in its nutritive value and sensory attributes which may lead to the production of toxicologically undesirable compounds.

The changes occurring in foods as a result of processing and storage are both chemical and physical in nature. However, often chemical changes also affect the physical na-

Process-Induced Chemical Changes in Food
edited by Shahidi *et al.* Plenum Press, New York, 1998

ture of food and vice-versa. Proteins, lipids and carbohydrates are the major food components that may be affected by processing, but minor constituents such as pigments, minerals, vitamins, free amino acids, nucleotides as well as enzyme and toxicants might serve as important culprits in process-induced changes in sensory quality and acceptability of food. The following provides a cursory account of some important changes in foods induced by processing and storage.

CHANGES IN MAJOR AND MINOR CONSTITUENTS OF FOODS

Changes induced by processing in food proteins relate mainly to their denaturation, binding of flavor-active and lipid oxidation products to them and modifications brought about intentionally by derivatization and enzymatic treatment. Therefore, functional properties of proteins and their role in foods, other than nutritional attributes, may be dictated by the system in which they are present as well as presence of other ingredients in the food matrix and storage conditions. Structural effects on properties of proteins from different source materials is also quite important and therefore protein type and its molecular features and hydrophilicity/hydrophobicity characteristics influence the degree which processing induces changes in them. It is also worth noting that enzymes, enzyme inhibitors, allergenic proteins and proteins carrying prosthetic groups are all affected by processing and caution should be exercised in order to optimize process conditions in such a way that their potential beneficial attributes are retained and deleterious effects eliminated.

Among lipid constituents of foods, both polar and non-polar components are affected by processing. While many of the changes that occur in food lipids are experienced during their production, many of these effects are system-dependant. In addition to changes that occur in the constituents of edible oils during primary production processes, they may also undergo major changes during heating as they serve as a heat transfer medium for processing of foods. Stability of lipids during processing and storage is affected by their chemical nature, degree of unsaturation, storage condition and presence of minor constituents such as tocopherols, other phenolics, carotenoids, chlorophylls and phospholipids. Interactions of lipids with carbohydrates, flavor defects in foods due to oxidation of lipids, free radial reactions in lipids and their influence on protein oxidation and characteristics of oils in bulk, emulsions and in complex matrices are also important. Lipids might also be present in the form of conjugates with carbohydrates and proteins and the extent to which glycolipid and lipoproteins are influenced by processing is varied.

Carbohydrates present in food are either in the polymeric, oligomeric or simple form. Many of the characteristics of carbohydrates in foods and their physio-chemical changes during processing are dictated by structural characteristics. Complex carbohydrate constituents of foods and the source material from which they are derived also affect many of their properties. Meanwhile, simple carbohydrates are important taste-active components of foods and these also undergo many chemical changes during food preparation. However, polymeric carbohydrates such as amylose, amylopectin and glycogen are often affected by the presence of other food ingredients during processing.

Simple carbohydrates might be conjugated with food phenolics, thus may affect oxidative stability of foods. In addition, reducing sugars and lipid oxidation products interact with free amino acids in foods to produce a wide-rage of flavor active components. Thus, low-molecular-weight components of foods and their degradation products have a major effect on aroma generation in foods. Meanwhile, pigments, vitamins and minerals in foods are also affected by processing such as those encountered during production of raw material, preparation of food and subsequent storage.

The following chapters of this monograph (Chapters 2–28) provide a concise account of changes that are brought about in foods during processing. Major and minor constituents of foods, flavor characteristics and methodologies involved for their evaluation are thoroughly discussed.

2

METHODS TO MONITOR PROCESS-INDUCED CHANGES IN FOOD PROTEINS

An Overview

E. C. Y. Li-Chan

Department of Food Science
The University of British Columbia,
6650 N.W. Marine Drive, Vancouver, BC, Canada, V6T 1Z4

Proteins in food systems may undergo various changes in their structural properties as a consequence of processing. Whether these changes are beneficial or detrimental in terms of the nutritional, biological or functional properties of the processed system, it is important to apply analytical methods which can monitor the course of protein structural changes, in order to elucidate the underlying mechanism behind the results of different processes. Proteins are usually found in high concentrations in foods; furthermore, these proteins frequently may either initially be part of a solid food or may become insoluble due to processing. As a result, many of the traditional biochemical methods for analysis of protein structural properties in dilute solution cannot be applied directly to study food proteins. This chapter gives an overview of some potential methods which may be used to monitor the changes in quaternary, tertiary, secondary and primary structure of proteins in food systems.

INTRODUCTION

Proteins in food systems may undergo various reactions as a result of processing, intentional modification or storage. The effects of thermal treatment has been the most widely studied process, but recent investigations have also focussed on alternative processes such as high pressure or ohmic heating. On the other end of the temperature spectrum, effects of freezing and frozen storage have also been the subject of many investigations, with recent emphasis turning to the importance of water and glass transitions on protein structure. Furthermore, the sensitivity of some proteins to drying proc-

Process-Induced Chemical Changes in Food
edited by Shahidi *et al.* Plenum Press, New York, 1998

esses such as lyophilization or freeze-drying has also been related to the effects of freezing, in contrast to the destructive effects of high temperatures as in the case of drying processes such as drum or spray drying.

Common to all of these processes, is the concept of protein denaturation, the definition of which has come under scrutiny (Stanley and Yada, 1992). The term "denaturation" has been defined in terms of the effect on the protein structure, as "simply a major change from the original native structure, without alteration of the amino acid sequence, i.e., without severance of any of the primary chemical bonds which join one amino acid to another" (Tanford, 1968). One criticism of this definition lies in its exclusion of oxidation-reduction or interchange reactions of covalent disulfide bonds, which may be crucial in the denaturation process for some proteins. Denaturation has also been defined with respect to its effect on protein functionality, as "any nonproteolytic modification of the unique structure of a native protein giving rise to definite changes in chemical, physical or biological properties" (Neurath and colleagues, as cited by Colvin, 1964).

For scientists studying the effects of process-induced changes in food proteins, assessment of denaturation has frequently been based on the one hand by a measured loss of solubility, i.e., the functional definition of denaturation, and on the other hand, by changes from the "native structure" using techniques requiring non-turbid solutions of proteins at low concentrations. Neither of these criteria is completely satisfactory for food systems in which proteins may be present at high concentrations, and in which solubility or loss thereof may not be directly related to other functional properties of interest. In fact, many foods are not solutions.

Processing may have adverse as well as beneficial effects on the properties of food proteins, as a result of physical and chemical changes to one or more of the four hierarchical levels of protein structure. These can include changes in quaternary structure reflected by the state of aggregation or dissociation of protein molecules, in tertiary structure as indicated by increased exposure of hydrophobic groups previously buried in the core, in secondary structure as reflected by changes in the proportion of helical, sheet, turn and random coil or unordered structures, and finally, in primary structure as evidenced by hydrolytic cleavage of peptide bonds or by the production of modified side chains of amino acid residues.

If the basic assumption that the interactions and functions of proteins are ultimately controlled by their molecular structure holds true, then it is critical to be able to monitor structural properties of proteins in food systems *in situ*, both in the so-called native state as well as after changes induced by processing. The objective of this chapter is to present an overview of some recent techniques used to study structural changes induced in proteins by processing or intentional modification, primarily from the point of view of methodologies which might be of relevance to studying food systems. Not included in the scope of this overview are a range of techniques yielding complementary information, such as separation methodologies based on electrophoresis and chromatography, thermoanalytical methodologies such as differential scanning calorimetry, or techniques to monitor rheological properties and biological functions. An overview of the methodologies to be discussed is shown in Table 1.

MONITORING QUATERNARY STRUCTURE

Quaternary structure of proteins in food systems is important for a number of reasons. Firstly, many food proteins are large and oligomeric in nature, composed of either identical or heterogeneous subunits. Examples include β-lactoglobulin (dimer), avidin

Table 1. An overview of some potential methods for monitoring process-induced changes in food proteins

	Protein sample		Information on structure			
Method	Concentration	Type	4°	3°	2°	1°
Analytical ultracentrifugation	low	solution	yes	some	no	no
Light scattering	moderate	solution/suspension	yes	some	no	no
Neutron scattering	high	solution	yes	some	some	no
UV absorption	low	solution	no	aromatic	no	no
Fluorescence, intrinsic	low	solution	no	aromatic	no	no
Fluorescence, probes	low	solution	no	various	no	no
Nuclear magnetic resonance	moderate to high	solution or solid	some	yes	yes	no
Vibrational spectroscopy (FTIR, Raman)	moderate to high	solution or solid	no	yes	yes	no
Circular dichroism	moderate to low	solution	no	yes (near-UV)	yes (far-UV)	no
Mass spectrometry	variable	may be on matrix	possible	not usually	not usually	yes (MW)
X-ray diffraction	high	crystals required	yes	yes	yes	yes

(tetramer), myosin (hexamer) and 11S soy protein (dodecamer) (Stanley and Yada, 1992). Secondly, the influence of heat and other processes on quaternary structures and subsequent structure formation is of major importance in food systems. The formation of soluble aggregates is an example of interactions at the quaternary structural level, an intermediate stage induced by processing en route to formation of larger networks including gels or precipitates.

Sedimentation or Ultracentrifugation

Analytical ultracentrifugation is considered a classical technique, having been used for many years to study the molecular weight distribution of proteins and their mixtures. However, it has been reported that this technique is still the best method for quantitative studies of the interactions between macromolecules (Schachman, 1989), and recent developments of computerized and simpler instrumentation may lead to a renewed interest in application to food systems. By performing either sedimentation velocity or sedimentation equilibrium experiments, the molecular weight, sedimentation coefficient, frictional coefficient and general shape factor of a protein can be determined. For quantitative characterization of reversible associations, including the interaction stoichiometry and association or dissociation constants, the non-destructive nature of analytical ultracentrifugation gives it advantages over methods such as size exclusion chromatography or electrophoresis. When used in conjunction with simplex optimization for the best fit of the equilibrium sedimentation patterns, accurate estimates were computed for the molecular weight and composition of the components in mixtures containing up to six components (Nakai and Nonaka, 1992a and b).

Light Scattering

Large aggregates of proteins which exhibit an opalescent or hazy appearance may be easily measured by light scattering phenomena. Protein pharmaceutical products are commonly evaluated visually for clarity and degree of opalescence; a relatively simple turbidimetric method was proposed for more accurate, reproducible and objective categorization of products (Eckhardt *et al.*, 1994). A similar approach could be useful to monitor food protein suspensions.

In many cases, it is desired to measure the formation of smaller aggregates, which may or may not appear turbid. Dynamic or quasi-elastic light scattering (DLS or QELS) has recently emerged as a useful tool to assess particle size distribution and therefore to monitor intermolecular interactions, or quaternary structural changes, as well as unfolding or tertiary structural changes of proteins. Since most of these small aggregates are much smaller than the wavelength of light used in scattering studies, the dependence of the scattered intensity upon scattering angle is not strong and simple right angle scattering studies may be useful for molecular weight determination. Two modes of analysis are possible: optical mixing or spectrum analyzer spectroscopy, which employs a frequency domain experiment, and intensity fluctuation or photon correlation spectroscopy, which employs a time domain experiment. An introduction to the basics and important considerations in use of these techniques for size distribution analysis is given by Yada *et al.* (1996). A recent example was reported by Yang *et al.* (1994), who performed a dynamic light scattering study of changes in protein particle size of β-galactosidase from *Aspergillus oryzae*, under various salt, pH and temperature conditions. Changes in activity and stability of the enzyme were correlated with size changes of the protein particles due to unfolding and aggregation.

Photon correlation spectroscopy is usually applicable only to dilute solutions to avoid multiple scattering, and therefore has not been useful to study concentrated or gelling food protein systems. However, optically opaque solutions may be studied by a technique known as diffusing wave spectroscopy, by measuring the scattered light at angles close to 180°, enabling the study of the dynamics of interacting colloids and gelling systems (Horne, 1991). This technique was applied to study different coagulation and gelation modes of bovine casein micelles (Dalgliesh and Horne, 1992).

Other Methods

Other tools have recently been used to study the quaternary structure of proteins. Pulsed neutron scattering using a recently developed scattering camera LOQ was applied to investigate the 12-domain structures of bovine IgG_1 and IgG_2; relative displacements and coordinates of the various domains of the Fab and Fc fragments were examined to elucidate the structural basis of the large differences in effector functions between these two isotypes (Mayans *et al.*, 1995). High-resolution transmission electron microscopy coupled with computer image analysis techniques has been proposed for food protein imaging and was used to demonstrate the dodecameric structure of the seed globulin from *Amaranthus hypochondriacus* K343 (Marcone *et al.*, 1994, Yada *et al.*, 1995). Techniques such as electrospray ionization mass spectrometry can be used to detect non-covalent complexes if solution conditions and instrumental parameters are appropriately adjusted. Observation of the oligomeric complexes requires a compromise between adequate heating and activation for ion desolvation and gentle enough interface conditions to avoid disruption of the complex (Smith *et al.*, 1996).

TERTIARY STRUCTURAL CHANGES

At present only two techniques can elucidate the three-dimensional structure of a protein to high resolution: X-ray or neutron diffraction analysis of single crystals and

NMR analysis of small proteins in solution (Creighton, 1993). Unfortunately, these techniques are not always applicable to food proteins, due to their large size or difficulties in obtaining suitable crystals; this is further complicated by process-induced changes. Much interest has recently centred on an intermediate state called the "molten globule" state (Ptitsyn, 1995), in which the protein molecule characteristically has its polypeptide chain with the native-like secondary structure, in contrast to amino acid side chains in a denatured or partly denatured state. Involvement of the molten globule state in various functional properties of food proteins, including gelling, emulsifying and foaming properties, has been suggested (Hirose, 1993). Techniques to monitor tertiary structural changes including exposure of hydrophobic groups and enhancement in dynamic accessibility of peptide N-H protons have been proposed to monitor this intermediate state.

Ultraviolet (UV) Spectroscopy

The UV absorption spectra of the aromatic amino acids (Phe, Tyr and Trp) are sensitive to their environment. The hydrophobic nature of these residues frequently results in the majority being buried in the interior of the protein molecules in the native state. Upon processing, some of these hydrophobic residues become exposed to a more polar environment, with resultant changes in their chromophoric properties. However, complexity of the spectra, arising from overlapping contributions of various residues in proteins, makes interpretation a difficult task.

Derivative spectrophotometry, including second-derivative and fourth-derivative spectroscopy, has been used to reveal the underlying changes in ultraviolet absorbance spectra of proteins (e.g. Padrós *et al.*, 1984; Mozo-Villarias *et al.*, 1991; Mach *et al.*, 1995). A method for simultaneous monitoring of the environment of the aromatic residues in proteins by near-UV second derivative spectroscopy was suggested based on numerically shifting the spectra of tryptophan and tyrosine model compounds to create a set of reference spectra corresponding to the expected peak positions in protein environments of different polarity (Mach and Middaugh, 1994). The maximum spectral shift observed between solvent-exposed model compounds and side chains completely buried in apolar protein core was found to be 5 nm for tyrosine, 4 nm for tryptophan and 2 nm for phenylalanine. However, for all three aromatic residues in proteins, there was no consistent correlation between absolute spectral band positions and average solvent accessibility, suggesting significant influence of other local effects such as electrostatic interactions, on the near-UV spectra of proteins.

Derivative analysis favors narrow bands to the detriment of broader ones, and therefore, theoretically, the higher the order of derivative, the greater the resolution. However, a difficulty arises in the form of a decrease in the signal-to-noise ratio (Padrós *et al.*, 1984). Yesilada *et al.* (1992) generated even-order derivatives up to eighth order of spectra for bovine and porcine insulin, and concluded that information in the derivative spectra above fourth order were of limited value, probably due to the nature of the complex overlapping satellite bands thereby generated and the increased noise level. An advantage of even-numbered derivatives lies in the fact that the maxima of the derivative spectra are at the same wavelengths as the original absorption spectra. Moderately turbid samples can be analyzed by derivative spectroscopy, since a horizontal baseline is obtained even with appreciable non-selective light-scattering (Padrós *et al.*, 1984). For example, air-stressed insulin solutions exhibiting visual opalescence and increased background absorption due to polymerization could still be analyzed by derivative spectroscopy (Yesilada *et al.*, 1992). In contrast, techniques such as near-UV circular dichroism spectroscopy, which are also sensitive to tertiary structure in the environment of aromatic chromophores, require non-turbid solutions.

Fluorescence Spectroscopy

The fluorescent properties of aromatic residues, particularly tryptophan residues, are also sensitive to their environment, but fluorescence spectroscopy has the advantage of 100–1000 fold higher sensitivity than spectrophotometric techniques. Shifts in the fluorescence emission peak of tryptophan from ~350 nm in highly polar media to ~315 nm in apolar media can be used to monitor the environment around these residues in proteins. In addition to the shifts in wavelength of maximum emission, the fluorescence spectra of proteins often exhibit a decrease in intensity upon denaturation, as a result of collisional quenching as the chromophores become exposed to the solvent. Changes in the intrinsic anisotropy may be used to monitor changes in shape upon unfolding, in mass or size from aggregation, or in conformation from alterations in a particular protein domain (Strasburg and Ludescher, 1995). Variations in the technique have been reported. For example, Andrade *et al.* (1984) monitored the conformational changes of human plasma fibronectin on hydrophobic and hydrophilic silica by a technique known as intrinsic ultraviolet total internal reflection fluorescence (UV TIRF). These authors reported a red shift in the emission maximum, from 321 nm in solution or when adsorbed to hydrophilic surfaces, to 326 nm when adsorbed to hydrophobic surfaces; the latter was suggestive of partial denaturation on interaction with a hydrophobic surface. In contrast bovine serum albumin showed a 9 nm blue shift from 342 nm in solution to 333 nm upon adsorption on hydrophilic quartz, suggesting a more hydrophobic environment for the Trp residue as a result of adsorption.

Unfortunately, the extreme sensitivity to environmental effects makes detailed interpretation of fluorescence changes difficult, particularly when several tyrosine and tryptophan residues are present. For example, Pjura *et al.* (1993) reported that large changes in fluorescence and phosphorescence caused by perturbation of a tryptophanyl residue were not always correlated with the three-dimensional structure, stability and solvation properties of mutants of T4 lysozyme. Increases in polar relaxation about the excited state of tryptophan could result from only small increases in local dynamics or solvent exposure.

Intrinsic fluorescence spectroscopy is limited to yielding information on the aromatic chromophores of proteins, whereas fluorescence spectroscopy based on extrinsic probes may be more versatile in yielding information about different residues or domains. Although fluorescent probes for hydrophobic sites on proteins are the most commonly used extrinsic probes, the technique can also give useful information about other functional groups. For example, 5-(iodoacetamidoethyl)aminonaphthalene-1-sulfonic acid (1,5-IAEDANS) is a specific probe which covalently labels free sulfhydryl groups or methionine residues of proteins. This probe has been used to detect conformational changes in troponin C upon binding of calcium ions (Grabarek *et al.*, 1986).

Hydrophobic probes have been widely used to monitor changes in the exposure of aliphatic and aromatic residues of food proteins (Nakai and Li-Chan, 1988). Two of the more popular probes are of the anionic type, 1-anilinonaphthalene-8-sulfonic acid (ANS) and *cis*-parinaric acid (CPA). Table 2 shows the hydrophobicity of food proteins monitored by these two probes (Li-Chan, 1991). These probes have high quantum yields of fluorescence in nonpolar environment, and are therefore useful to monitor the accessible or surface hydrophobicity of proteins. However, fluorescence of the ANS probe is influenced by solvents which favor the rigid planar configuration, including aqueous $MgCl_2$ so-

Table 2. Hydrophobicity values of some food proteins measured by two fluorescence probes. (Adapted from Li-Chan, 1991)

Protein[1]	Hydrophobicity index[2], S_o	
	ANS	CPA
Albumin, bovine serum	1600	2750
Canola isolate	110	370
Casein, bovine	150	400
Egg albumen	40	135
Gelatin	0	0
Lactalbumen	145	1700
Lactalbumin, α-	90	280
Lactoglobulin, β-	40	7000
Lysozyme, hen egg white	1	15
Muscle, chicken breast salt soluble proteins	75	350
Myosin, chicken breast	100	—
Ovalbumin, hen egg white	10	40
Pea protein isolate	280	825
Soy protein isolate	250	1000
Sunflower protein isolate	155	910
Whey protein concentrate	70	3600
Zein	410	390

[1]Proteins were dissolved or dispersed in 0.01M sodium phosphate buffer at pH 7. The buffers for the salt-soluble proteins of chicken breast muscle and for isolated chicken myosin also included 0.6 M NaCl and 0.3M NaCl, respectively.

[2]The hydrophobicity index, S_0, was calculated as the initial slope of the relative fluorescence intensity versus protein concentration (%), as described by Kato and Nakai (1980). ANS, 1-anilinonaphthalene-8-sulfonic acid; and CPA, *cis*-parinaric acid.

lution and, therefore, the enhanced fluorescence in protein solutions needs to be interpreted with caution. Since the structure of CPA is similar to naturally occurring fatty acids, interaction of CPA with food proteins has been suggested to be useful to probe for hydrophobic regions which may be important in protein-lipid interactions in food systems.

Since both ANS and CPA are anionic probes, the possibility of charged interactions with the protein binding sites has been raised. Thus, neutral or uncharged molecules have been proposed as alternative fluorescent probes for protein hydrophobicity. One such probe is diphenylhexatriene (DPH), which has been primarily used to estimate membrane fluidity, but has been applied to study food proteins (Tsutsui *et al.*, 1986). Nile Red is an uncharged phenoxazone dye which was first introduced as a fluorescent stain for lipids, but has also been reported to be useful as a polarity-sensitive probe of hydrophobic protein surfaces (Sackett and Wolff, 1987). One difficulty in the use of the uncharged types of probes is their low solubility in the aqueous systems typically used for studying proteins. Nevertheless, their application may yield useful additional information, especially when interfering electrostatic contributions exclude clear interpretation of results based on charged extrinsic probes. For example, Gatti *et al.* (1995) studied the interactions of ANS and Nile red with bovine casein micelles by fluorescence spectroscopy. Although both probes showed blue shifts of their fluorescence emission peaks and enhancement of fluorescence intensity, indicative of binding in low-polarity regions of the casein micelles, the

ANS binding site was suggested to involve a weak interaction of high binding capacity, probably involving both hydrophobic and electrostatic components, while the Nile red binding site was clearly hydrophobic and of lower binding capacity.

Parameters which have been used to quantify the probe-protein interactions include the wavelength of maximum emission, the quantum yield, the affinity constant and the number of binding sites (Cardamone and Puri, 1992). Titration of protein solutions with increasing amounts of the fluorescence probe can provide information on both the number and affinity of binding sites, which can yield information on whether high fluorescence intensity arises from the presence of many binding sites of moderate hydrophobic character, or from one or two high affinity sites with high hydrophobicity. Alternatively, the initial slope of a plot of fluorescence intensity versus protein concentration has been proposed as an index for protein surface hydrophobicity (Kato and Nakai, 1980).

Some limitations to the applications and interpretations of the results of fluorescence from extrinsic probes should be noted. Extrinsic probe fluorescence has frequently been used to monitor protein hydrophobicity as a function of varying pH, temperature or in the presence of denaturing agents. However, the effects of these conditions on the fluoresence properties of the probes must be considered. For example, fluorescence experiments using CPA at pH below 5 are not reliable due to the poor solubility of the probe in the undissociated acid form predominating at low pH. Preliminary experiments in our laboratory have shown that very low fluorescence values were obtained when ANS or CPA was added to protein solutions containing varying concentrations of the denaturant urea, necessitating a modification of the fluorescence probe method (Arteaga, 1994). Some researchers have interpreted decreases in fluorescence intensity upon proteolytic action as a decrease in hydrophobicity. This interpretation may be valid during the initial stages of proteolysis or with only limited proteolysis, resulting from unfolding of the native protein structure and increasing exposure of hydrophobic residues from the core to the aqueous environment. On the other hand, a sharp decrease in fluorescence intensity which may be observed upon extensive proteolysis cannot be interpreted as evidence of production of hydrophilic peptides, since the fluorescence probes may have low quantum yields of fluorescence even when bound to small hydrophobic peptides.

A final limitation in the use of fluorescence spectroscopy in general is related to the high sensitivity of this technique. To avoid self-absorption and quenching effects on fluorescence, only dilute protein solutions can be studied. However, many phenomena which may be induced by processing are concentration-dependent, and it may not be relevant to investigate structural changes in dilute solution. For example, Iametti *et al.* (1995) studied concentration dependence of the modifications resulting from thermal treatment of β-lactoglobulin using a combination of techniques. The required temperature for occurrence of protein swelling, the first step in formation of associated forms of the protein, increased with protein concentration. Heating of dilute (<3.8 mg/mL) protein solutions induced an increase in the number of ANS binding sites; this heat-induced increase was less evident in concentrated solutions and was completely reversible upon cooling.

Nuclear Magnetic Resonance Spectroscopy

Nuclear magnetic resonance or NMR spectroscopy has a long history of use, but the advent of three- and four-dimensional spectroscopy, together with very high magnetic fields, has expanded the applications for determination of solution and solid state structure, with detailed dynamical information (Belton, 1993). Hydrogen-deuterium exchange rates, number and intensity of nuclear Overhauser enhancements, coupling constants, line widths and intensities may be measured and correlated to mobility and exposure. However, with simple or one-dimensional proton NMR, even a small protein of 50 residues will yield an NMR spectrum containing 300–400 individual signals, and assignment of these signals to yield a fingerprint is a major task. Larger proteins will yield highly complicated and congested spectra even using two-dimensional techniques, such as COSY (correlated spectroscopy) which can be used to determine connectivities between neighbouring amino acids. Three- and four-dimensional methods have made it possible to address the problem of moderately large proteins, but at the expense of time and significant cost.

As an alternative, relaxation time methods can be more useful to give information on overall dynamics and solvation effects of moderately large proteins. For example, solid-state relaxation methods have been applied to study wheat gluten and cereals, bovine serum albumin and whey protein concentrates (Belton, 1993). A spin diffusion NMR experiment based on the cross saturation method reported by Akasaka (1983) was used in our laboratory to estimate the exposure of hydrophobic residues in several proteins from egg white and milk (Arteaga, 1994; Arteaga *et al.*, 1995). The return to equilibrium depends on the mobility and electron environment of the proton, and is characterized by spin-lattice or longitudinal relaxation (T_1) and spin-spin or transverse relaxation (T_2). In large molecules such as proteins, T_2 is influenced by the exchange of spin magnetization or energy between neighbouring protons. This effect, called cross-relaxation or spin diffusion, depends on inter-proton distances and molecular motions of the macromolecules; a higher degree of spin diffusion usually results from relatively immobile or rigid protons. By applying a cross-saturation pulse at a given frequency before the actual NMR pulse, the rigid protons become saturated and their NMR signals are decreased significantly. On the other hand, the signals of mobile protons are slightly or not affected by the cross-saturation pulse. Thus the ratio of the peak areas for a normal NMR spectrum (no pre-saturation pulse) and that for a spectrum after cross saturation pulse may be used to monitor the degree of exposure of groups.

Examples of typical NMR spectra obtained before and after cross-saturation are shown in Figures 1 and 2. Proteins with flexible or less ordered structures, such as the milk caseins, showed less changes in the area or intensity of the NMR signals in these cross-saturation experiments, when compared to proteins with globular, compact three-di-

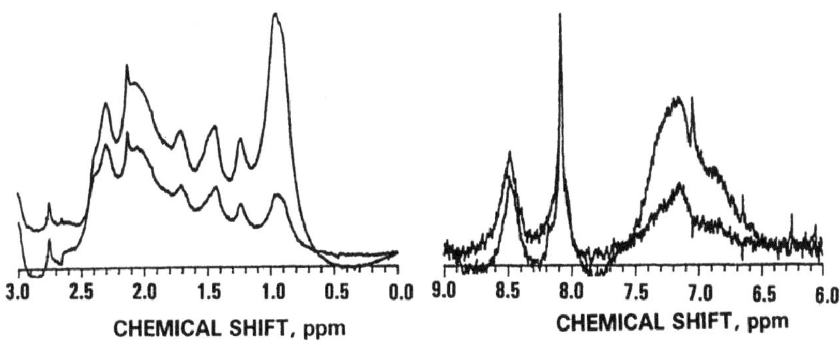

Figure 1. Normal and cross-saturated ^{1}H-NMR spectra of 5% α_{s1}-casein in 0.1M deuterated phosphate buffer (pD_{app} = 7.5) in the aliphatic (left) and aromatic (right) regions. The spectra with lower intensities in each region was obtained after cross saturation. (Adapted from Arteaga, 1994).

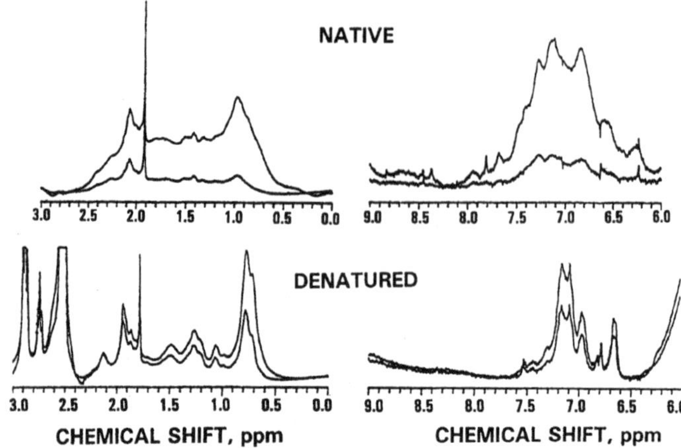

Figure 2. Normal and cross-saturated ^{1}H-NMR spectra of 5% native (top) and denatured (bottom) ovalbumin in the aliphatic (left) and aromatic (right) regions. Native ovalbumin was dissolved in 0.1M deuterated phosphate buffer (pD_{app} = 7.5), while denatured ovalbumin was prepared in the same buffer containing 8M deuterated urea and 0.1M β-mercaptoethanol. The spectra with lower intensities in each set correspond to the spectra obtained after cross-saturation. (Adapted from Arteaga, 1994).

mensional structures, such as the whey and egg white proteins. Denaturation of the globular proteins with 8M urea and β-mercaptoethanol also resulted in a decreased sensitivity to the cross saturation pulse. The ratio of the areas under the NMR spectrum before and after cross-saturation were calculated for two regions, corresponding to protons from aliphatic residues (0.00–3.00 ppm) and from aromatic residues (6.50–8.00 ppm). These ratios were expressed as the changes in areas in the aliphatic (CHAL) and aromatic (CHAR) regions. An inverse relationship was observed between CHAL and hydrophobicity determined by the aliphatic fluorescence probe CPA, and between CHAR and hydrophobicity determined by the aromatic fluorescence probe ANS, suggesting the usefulness of this type of NMR experiment to study surface hydrophobicity of proteins (Arteaga, 1994).

Raman Spectroscopy

Raman spectroscopy is a vibrational spectroscopic technique which can be a useful probe of protein structure, since both intensity and frequency of vibrational motions of the amino acid side chains or polypeptide backbone are sensitive to chemical changes and the microenvironment around the functional groups. Thus, it can monitor changes related to tertiary structure as well as secondary structure of proteins. An important advantage of this technique is its versatility in application to samples which may be in solution or solid, clear or turbid, in aqueous or organic solvent. Since the concentration of proteins typically found in food systems is high, the classical dispersive method based on visible laser Raman spectroscopy, as well as the newer technique known as Fourier-transform Raman spectroscopy which utilizes near-infrared excitation, are more suitable to study food proteins (Li-Chan *et al.*, 1994). In contrast the technique based on ultraviolet excitation, known as resonance Raman spectroscopy, is more commonly used to study dilute protein solutions.

The applications of visible laser Raman spectroscopy to monitor process-induced structural changes are illustrated in a study of the formation of transparent gels of whey

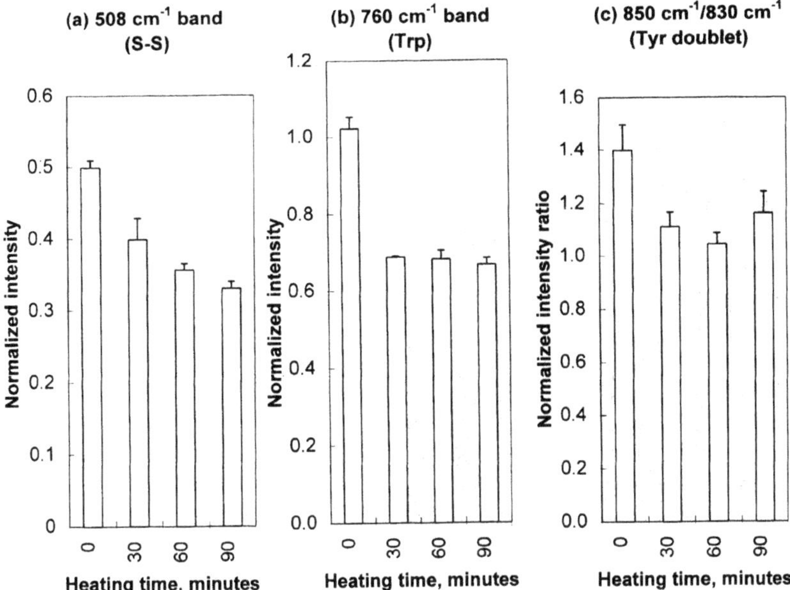

Figure 3. Changes in the normalized intensity of bands in the Raman spectra of 15% α-lactalbumin in deuterium oxide containing 20 mM NaCl (pD_{app} = 6.8) upon formation of transparent gels as a function of heating time at 90°C. The bands are assigned to vibrational motions of amino acid side chains as follows: (a) 508 cm^{-1} (S-S); (b) 760 cm^{-1} band (tryptophan); (c) 850 cm^{-1}/830 cm^{-1} doublet (tyrosine). Error bars represent the standard deviation for three replicate samples, with 8 spectral scans averaged per replicate. (Adapted from Nonaka *et al.*, 1993).

proteins as a function of thermal processing (Nonaka *et al.*, 1993). Some of the changes in the spectra obtained upon heating bovine α-lactalbumin at 90°C are shown in Figure 3. Decrease in intensity of the 508 cm^{-1} band assigned to S-S stretching vibrations of disulfide bonds in an all-*gauche* conformation suggested a change in the environment of cystine residues after heating. Exposure of buried tryptophan residues was indicated by the decrease in intensity of a band at 760 cm^{-1}, while a drop in the intensity ratio (I_{850}/I_{830}) of the doublet assigned to tyrosine residues suggested either an increase in buriedness or an increase in involvement of some tyrosine residues as strong hydrogen bond donors.

Other Methods

The above-mentioned techniques are based on the interaction of electromagnetic radiation with the protein molecules. With the exception of high-resolution NMR, these methods cannot yield specific information on process-induced changes in particular localized regions of the protein molecule. However, specific affinity probes, such as antibodies against known epitopic regions, may be used for this purpose. For example, Hattori *et al.* (1993) used four monoclonal antibodies against bovine β-lactoglobulin to study the completeness of refolding of the protein after denaturation and renaturation. Binding studies with two monoclonal antibodies which recognized epitope regions only in the native protein were used to demonstrate that complete refolding was not attained for some specific moieties in the renatured molecules. These small local structural differences did not affect the biological function of ligand (retinol) binding, and only very subtle differences were noted between native and renatured proteins in their circular dichroism and intrinsic fluorescence spectra.

Three-dimensional molecular modelling of proteins has become feasible with the advent of powerful computers. For purified proteins whose primary sequence is known, and whose X-ray crystallographic structure is unknown, these techniques can be useful for elucidating the tertiary structure. For example, Kumosinski *et al.* (1991) have proposed 3-D molecular models for the bovine caseins. Especially when used in conjunction with other techniques such as spectroscopic results, this approach can yield useful clues to the relationships between structure and function.

Global analysis of multidimensional spectroscopic data was suggested to monitor unfolding of proteins (Ramsay and Eftink, 1994). This approach uses multiple data sets, and involves simultaneous nonlinear analysis to obtain a single set of parameters to describe all sets of data. These authors collected far- and near-UV CD, tryptophan fluorescence, and absorbance data using a modified Aviv 62 DS circular dichroism spectrophotometer, and then applied global analysis of the weighted multiple data sets to test the applicability of different denaturation models. Hydrogen-deuterium exchange as a function of temperature-induced structural changes in various proteins, including albumin, imunoglobulin G, fibrinogen, lysozyme and α-lactalbumin, was monitored by FTIR spectroscopy and globally fitted with a two-state thermodynamic model (Van Stokkum *et al.*, 1995). Enhanced exchange rates occurred at temperatures well below unfolding of secondary structure, and were interpreted as changes in tertiary structure leading to enhanced solvent accessibility.

CHANGES IN SECONDARY STRUCTURE FRACTIONS

The regular conformations observed in polypeptides include the antiparallel and parallel β-sheet, the right handed α-helix, 3_{10}-helix, π-helix, and the polyproline and polyglycine forms (Creighton, 1993). Only the first four of these conformations are typically found in proteins, and together with β-turns and random coil or unordered structure, are commonly known as the secondary structure of proteins. As reported by Sreerama and Woody (1994), the secondary structure elements in X-ray-derived structures deviate from the ideal geometry, and algorithms to identify the secondary structure elements in globular proteins have been developed. Up to 25% of residues in globular protein structures remain unassigned, and short segments of the less common structures, such as poly(L-proline)-type conformations, have been suggested in some proteins. As mentioned before, X-ray diffraction and multi-dimensional NMR can provide a detailed description of protein structure but is not always applicable. Methods based on circular dichroism (CD), infrared and Raman spectroscopy can be used to monitor changes in the estimated fractions of residues in the different secondary structure types upon processing.

Circular Dichroism

The far-UV (170–250 nm) CD region resulting from peptide groups is characteristic for various types of secondary structure, and forms the basis for estimation of secondary structure fractions in proteins using various algorithms, including the convex constraint analysis of Perczel *et al.* (1992), the variable selection method of Johnson and colleagues (Tournadje *et al.*, 1992) and the self-consistent method of Woody and colleagues

(Sreerama and Woody, 1993). In many cases, especially when the objective is to follow the transitions between helical and unordered conformations as a function of heating or denaturing conditions, comparisons of the molar ellipticity values at 220 nm may suffice. In contrast to the extensive studies of helix-coil transitions, measurements of β-coil transitions have been much less common, possibly due to the limited solubility of many β-sheet models (Jardine, 1990).

Many examples of application of CD to analysis of secondary structural analysis of food proteins can be found in the literature. However, a major limitation lies in the necessity for clear samples. Careful selection of the solvating medium is also required, to avoid interference in the specttrum. Absorbance of various salt and buffer substances in the far-UV region has been reported by Stanley and Yada (1996). Wang and Damodaran (1991) conducted CD analysis of the fluid expressed from protein gels by centrifugation, to indirectly elucidate the structural conformation of the proteins in the gels. Tani *et al.* (1995) measured the CD spectra of soluble linear aggregates, an intermediate stage to the formation of transparent gels. Matsuura and Manning (1994) reported that they were able to use far-UV and near-UV (250–350 nm) CD analysis to monitor the secondary and tertiary structural changes due to heat-induced gel formation of β-lactoglobulin, but were limited to conditions which produced only clear or transparent gels. Analysis of the CD spectra demonstrated a dependence of secondary and tertiary structures on protein concentration, both before and after heating.

Vibrational Spectroscopy

The different secondary structures of polypeptides and proteins give characteristic bands in vibrational spectra, including both infrared (IR) and Raman spectra. The amide I band primarily represents C=O stretching vibrations, with a minor contribution from C-N stretching vibrations. The amide II band arises from N-H bending and C-N stretching, while the amide III band arises predominantly from C-N stretching vibrations coupled to N-H in-plane bending vibrations, with weak contributions from C-C stretching and C=O in-plane bending. The latter is the least well- characterized with respect to the underlying nature of the vibration, as well as due to interference from CH_2 wagging vibrations.

Typical assignment of the components of the amide I band in IR spectra are shown in Table 3; corresponding assignments of the band in the Raman spectra have also been reported (e.g. Susi and Byler, 1988a). Quantitative estimation of the proportions of secondary structure types is usually based on Fourier self-deconvolution, least squares analysis or other mathematical algorithms (Jackson and Mantsch, 1995; Li-Chan *et al.*, 1994). However, as noted by Wilder *et al.* (1992), IR alone is not sufficient to unequivocally establish secondary protein structure without verification by other analytical methods, including CD, X-ray diffraction and NMR spectroscopy. These authors reported that two recombinant murine interleukin proteins exhibited an IR absorption band at 1656 cm^{-1}, which is usually assigned to α-helical or random structures. However, no evidence of helical structures was supported by any other analytical methods, and the IR band was assigned instead to the presence of large loops. An IR and CD combined approach to the analysis of protein secondary structure, using factor analysis methods, has been suggested (Sarver and Krueger, 1991).

The potential of FTIR spectroscopy to monitor changes in food proteins has been explored in an increasing number of publications (e.g. Susi and Byler, 1988b; Kumosinski and Farrell, 1993; Boye *et al.*, 1996). As Boye *et al.* (1996) noted, many studies have re-

Table 3. Assignment of the Amide I frequency to common protein structures. (Adapted from Jackson and Mantsch, 1995)

Structure	Frequency of the amide I band, cm^{-1}
Anti-parallel β-sheet or aggregated strands	1675–1695
3_{10}-helix	1660–1670
α-helix	1648–1660
Unordered	1640–1648
β-sheet	1625–1640
Aggregated strands	1610–1628

ported the peak temperature of the thermogram from differential scanning calorimetry (DSC) experiments as the denaturation temperature of β-lactoglobulin. However, these authors noted that the starting rather than peak temperatures of the DSC thermograms correlated with denaturation temperatures based on the amide I band of FTIR spectra (loss of 1648, 1636, and 1624 cm^{-1} bands). The peak temperatures obtained from the DSC thermogram, on the other hand, were similar to the temperatures at which distinct aggregation bands (1618 and 1684 cm^{-1} bands) could be observed in the FTIR spectra. These results suggested that unfolding of β-lactoglobulin actually occurred at an earlier temperature than that of the DSC peak temperature, and that the latter appeared to be more indicative of aggregation than denaturation.

Despite the popularity of FTIR to investigate protein secondary structure, many potential pitfalls exist for the unwary investigator (Jackson and Mantsch, 1992 and 1995). Water-soluble proteins have frequently been analyzed as films by attentuated total reflectance (ATR) IR. Possible artifacts which should be considered in this technique are the potential interactions between the ATR crystal and the protein, and lack of relevance of spectra obtained for proteins in the solid or hydrated film state compared to the protein solution. The strong O-H bending absorption of water in the IR spectrum obscures the amide I band used for protein secondary structural analysis. To study aqueous protein solutions, therefore, very short pathlengths (10 μm or less) need to be used, and the water signal is commonly digitally subtracted using a water or buffer baseline spectrum. Alternatively, deuterium oxide may be used instead of water as the solvent.

Since water yields only weak Raman scattering, Raman spectroscopy has advantages over the IR technique to study food proteins *in situ* (Li-Chan *et al.*, 1994). Estimation of the secondary structural fractions based on analysis of the Amide I band in Raman spectra has been reported for various food proteins, such as the lyophilized proteins from milk and other foods (Susi and Byler, 1988a), casein in cheese (Fontecha *et al.*, 1993), and aqueous solutions and gels of hen egg white lysozyme (Li-Chan and Nakai, 1991). An example of the application of Raman spectroscopy to monitor protein structural changes induced by processing is shown in Table 4. Changes in secondary structure fractions of proteins in Pacific whiting were monitored *in situ* upon conversion from the raw surimi to the cooked kamaboko products.

Although fluorescence can be a serious problem in Raman studies of some biological systems, the availability of FT-Raman instrumentation which use near-infrared excitation has been suggested to eliminate this problem for many cases (Colaianni *et al.*, 1995).

Table 4. Changes in the secondary structure fractions of proteins during processing of Pacific whiting surimi to kamaboko. (Adapted from Bouraoui, 1995)

	Secondary structure fractions[2]			
Treatment[1]	α-helix	Anti-parallel β-sheet	Parallel β-sheet	Unordered
Raw surimi	0.44	0.24	0.00	0.32
Salted	0.37	0.26	0.00	0.37
Set	0.36	0.41	0.00	0.23
Cooked	0.09	0.31	0.20	0.40
Set-cooked	0.20	0.49	0.00	0.31

[1]The surimi was salted (ground with 3% NaCl in a vacuum silent cutter), set (32 °C for 19 min), cooked (86°C for 12 min) or set then cooked (32°C for 19 min followed by 86°C for 12 min).
[2]Secondary structure fractions were estimated by the Raman spectral analysis program of Przybycien and Bailey (1989), based on the algorithms of Williams (1983) for least-squares analysis of the Amide I band.

Other Methods

Vibrational CD (VCD) is a very promising technique for providing secondary structure estimates for proteins, with advantages over far-UV CD of involving more localized transitions and much lower sensitivity to side chain contributions, especially those of aromatic residues (Woody, 1995). Its main disadvantage is the present limited availability of VCD instrumentation. One other cited limitation, namely the much higher concentrations needed compared to far-UV CD, may be of advantage for food systems if clear, non-particulate solutions are maintained at these high protein concentrations.

An excellent overview of methods for determining and predicting secondary structure in proteins, including a critical evaluation of the applicability of some of these methods to the major milk proteins, was written by Sawyer and Holt (1993). Experimental methods described include CD, IR, NMR two-dimensional spectroscopic techniques, and X-ray crystallography. At present, NMR spectroscopy cannot be readily applied to the study of proteins having a molecular weight of 20,000 daltons or to proteins in membrane environment due to line-broadening effects resulting from motional restriction; turbid solutions or membrane systems are not suitable for CD spectroscopy, due to artifacts associated with light scattering; X-ray diffraction requires suitable high quality crystals (Jackson and Mantsch, 1995).

With regard to theoretical methods, several approaches based on statistical, hydrophobic and pattern recognition methods have been proposed (Sawyer and Holt, 1993). Cumulative or joint prediction methods, with supplementary information from spectroscopic methods and the use of templates and sequence information from related proteins, were shown to improve the confidence of prediction, as assessed by comparison to X-ray crystallographic structures. Despite the great interest and advances in research in these areas, the accuracy of these secondary structure predictions (i.e. theoretical methods) still remains at only about 60%. Even when the structure of structurally related or homologous proteins is known, the accuracy of prediction is only 70.9% (Mehta *et al.*, 1995). Furthermore, these methods cannot easily be applied to monitor changes in protein secondary structure induced by processing.

CHANGES IN PRIMARY STRUCTURE

Ultimately, the functional and nutritional properties of proteins depends on their primary structure. Methods for evaluation of protein quality, including both *in vivo* and *in vitro* approaches to assess the effects of processing on nutritional availability or quality, are under continual review (FAO/WHO, 1990). Advances in instrumentation and protocols for amino acid composition and sequence analysis have taken great strides in the past decade, especially with techniques involving DNA cloning and sequencing for prediction of primary amino acid sequence. Furthermore, recent developments in mass spectrometry, such as electrospray ionization and matrix-assisted laser desorption/ionization mass spectrometry (ESI-MS and MALDI-MS, respectively), have allowed its application to protein structural analysis (e.g. Jardine, 1990; Stults, 1995; Siuzdak, 1996). The higher mass resolution, sensitivity and accuracy enable identification and characterization of even fairly large proteins. Although not yet widely used in the study of proteins in food systems, these techniques may have potential as a sensitive method to monitor interactions or changes in molecular mass with processing, including changes of single amino acid residues. Deconvolution of the electrospray molecular mass profile data of a sample of ovalbumin (estimated molecular weight of 45,000 daltons) was reported to indicate the presence of at least two components at 44,308 and 44,559 daltons, possibly due to carbohydrate heterogeneity (Jardine, 1990). The use of gentle electrospray ionization mass spectrometer interface conditions was used to study the oligomeric state of the chaperone protein SecB from *E. coli*; location of the site of proteolytic cleavage was also identified by analysis of the proteolyzed monomer (Smith *et al.*, 1996). However, at high mass, even very-high resolution mass analysis will be insufficient to detect small changes, such as deamidation of asparaginyl residues or oxidation of sulfur-containing residues, necessitating preliminary sample treatment or even analysis by complementary techniques (Jardine, 1990).

REFERENCES

Akasaka, K. Spin diffusion and the dynamic structure of a protein *Streptomyces subtilisin* inhibitor. *J. Magnet. Res.* **1983,** *36,* 135–140.

Andrade, J.D.; Hlady, V.L.; Van Wagenen, R.A. Effects of plasma protein adsorption on protein conformation and activity. *Pure Appl. Chem.* **1984**, *56*, 1345–1350.

Arteaga, G.E. Assessment of protein surface hydrophobicity by spectroscopic methods and its relation to emulsifying properties of proteins. Ph.D. dissertation, University of British Columbia, Vancouver, BC, Canada, 1994.

Arteaga, G.E.; Li-Chan, E.; Nakai, S. Assessment of protein surface hydrophobicity by proton nuclear magnetic resonance spectroscopy. Poster 68A-39, presented at Annual Meeting, Institute of Food Technologists, Anaheim, CA, June 3–7, 1995.

Belton, P.S. New methods for monitoring changes in proteins. Food *Rev. Intern.* **1993,** *9,* 551–573.

Bouraoui, M. Surimi-based product development and viscous properties of surimi paste. Ph. D. Thesis, Department of Chemical Engineering, The University of British Columbia, Vancouver, BC, Canada, 1995.

Boye, J.I.; Ismail, A.A.; Alli, I. Effects of physicochemical factors on the secondary structure of β-lactoglobulin. *J. Dairy Res.* **1996,** *63,* 97–109.

Cardamone, M.; Puri, N.K. Spectrofluorimetric assessment of the surface hydrophobicity of proteins. *Biochemical J.* **1992,** *282,* 589–593.

Colaianni, S.E.M.; Aubard, J.; Hansen, S.H.; Nielsen, O.F. Raman spectroscopic studies of some biochemically relevant molecules. *Vibrational Spectroscopy* **1995,** *9,* 111–120.

Colvin, J.R. Denaturation: A Requiem. In *Symposium on Foods: Proteins and Their Reactions*; Schultz , H.W.; Anglemier, A.F., Eds.; AVI Publishing Co., Westport, CT, 1964; pp 69–83.

Creighton, T. E. *Proteins. Structures and Molecular Properties*. Second Edition. W.H. Freeman and Company, New York, 1993.

Dalgliesh, D.G.; Horne, D.S. Different coagulation and gelation modes of casein micelles followed by diffusing wave spectroscopy. In *Protein Interactions*; Visser. H., ed.; VCH Verlagsgesellschaft, Weinheim, Germany 1992, pp 87–101.

Eckhardt, B.M.; Oeswein, J.Q.; Yeung, D.A.; Milby, T.D.; Bewley, T.A. A turbidimetric method to determine visual appearance of protein solutions. *PDA J. Pharm. Sci. & Technol.* **1994,** *48,* 64–70.

FAO/WHO. Protein Quality Evaluation. Report of a joint FAO/WHO Expert Consultation. Food and Agriculture Organization of the United Nations, Rome, 1990.

Fontecha, J.; Bellanato, J.; Juarez, M. Infrared and Raman spectroscopic study of casein in cheese: effect of freezing and frozen storage. *J. Dairy Sci.* **1993,** *76,* 3303–3309.

Gatti, C.A.; Risso, P.H.; Pires, M.S. Spectrofluorometric study on surface hydrophobicity of bovine casein micelles in suspension and during enzymic coagulation. *J. Agric. Food Chem.* **1995**, *43,* 2339–2344.

Grabarek, Z.; Leavis, P.C.; Gergely, J. Calcium binding to the low affinity sites in troponin C induces conformational changes in the high affinity domain. *J. Biol. Chem.* **1986,** *261*, 608–613.

Hattori, M.; Ametani, A.; Katakura, Y.; Shimizu, M.; Kaminogawa, S. Unfolding/refolding studies on bovine β-lactoglobulin with monoclonal antibodies as probes. Does a renatured protein completely refold? *J. Biol. Chem.* **1993,** *268,* 22414–22419.

Hirose, M. Molten globule state of food proteins. *Trends Food Sci. Technol.* **1993,** *4*, 48–51.

Horne, D.S. Diffusing wave spectroscopy studies of gelling systems. In *Photon Correlation Spectroscopy: Multicomponent Systems*; Schmitz, K.S., Ed., SPIE International Society for Optical Engineering, Bellingham, WA, 1991, pp 166–180.

Iametti, S.; Cairoli, S.; De Gregori, B.; Bonomi, F. Modifications of high-order structures upon heating of β-lactoglobulin: dependence on the protein concentration. *J. Agric. Food Chem.* **1995,** *43,* 53–58.

Jackson, M.; Mantsch, H.H. The use and misuse of FTIR spectroscopy in the determination of protein structure. *Crit. Rev. Biochem. Molec. Biol.* **1995,** *30,* 95–120.

Jackson, M.; Mantsch, H.H. Artifacts associated with the determination of protein secondary structure by ATR-IR spectroscopy. *Appl. Spectrosc.* **1992**, *46*, 699–701.

Jardine, I. Molecular weight analysis of proteins. *Methods in Enzymol.* **1990,** *193,* 441–455.

Kato, A.; Nakai, S. Hydrophobicity determined by fluorescence probe method and its correlation with surface properties of proteins. *Biochim. Biophys. Acta* **1980,** *624,* 13–20.

Kumosinski, T.; Brown, E.M.; Farrell, H.M., Jr. Three-dimensional molecular modeling of bovine caseins: α_{s1}-casein. *J. Dairy Sci.* **1991,** *74,* 2889–2895.

Kumosinski, T.F.; Farrell, H.M., Jr. Determination of the global secondary structure of proteins by Fourier transform infrared (FTIR) spectroscopy. *Trends Food Sci. Technol.* **1993,** *4,* 169–175.

Li-Chan, E. Hydrophobicity in food protein systems. In *Encyclopedia of Food Science and Technology;* Hui, Y.H., Ed.; John Wiley and Sons, Inc., New York, NY, 1991; pp 1429–1439, .

Li-Chan, E.; Nakai, S. Raman spectroscopic study of thermally and/or dithiothreitol induced gelation of lysozyme. *J. Agric. Food Chem.* **1991,** *39,* 1238–1245.

Li-Chan, E.; Nakai, S.; Hirotsuka, M. Raman spectroscopy as a probe of protein structure in food systems. In *Protein Structure-Function Relationships in Foods*; Yada, R.Y.; Jackman, R.L.; Smith ; J.L., Eds.; Blackie Acadmic & Professional, Chapman & Hall Inc., London, England, 1994; pp 163–197.

Mach, H.; Middaugh, C.R. Simultaneous monitoring of the environment of tryptophan, tyrosine and phenylalanine residues in proteins by near-ultraviolet second-derivative spectroscopy. *Anal. Biochem.* **1994,** *222,* 323–331.

Mach, H.; Volkin, D.B.; Burke, C.J.; Middaugh, C.R. Ultraviolet absorption spectroscopy. *Methods in Molecular Biol.* **1995,** *40,* 91–114.

Marcone, M.F.; Beniac, D.R.; Harauz, G.; Yada, R.Y. Quaternary structure and model for the oligomeric seed globulin from *Amaranthus hypochondriacus* K343. *J. Agric. Food Chem.* **1994,** *42,* 2675–2678.

Matsuura, J.E.; and Manning, M.C. Heat-induced gel formation of β-lactoglobulin: A study on the secondary and tertiary structure as followed by circular dichroism spectroscopy. *J. Agric. Food Chem.* **1994,** *42,* 1650–1656.

Mayans, M.O.; Coadwell, W.J.; Beale, D.; Symons, D.B.; Perkins, S.J. Demonstration by pulsed neutron scattering that the arrangement of the Fab and Fc fragments in the overall structures of bovine IgG_1 and IgG_2 in solution is similar. *Biochemical J.* **1995,** *311 (Pt. 1),* 283–291.

Mehta, P.K.; Heringa, J.; Argos, P. A simple and fast approach to prediction of protein secondary structure from multiply aligned sequences with accuracy above 70%. *Protein Sci.* **1995,** *4,* 2517–2525.

Mozo-Villarías, A.; Morros, A.; Andreu, J.M. Thermal transitions in the structure of tubulin: environments of aromatic amino acids. *Eur. Biophys. J.* **1991,** *10,* 295–300.

Nakai, S.; Nonaka, M. Computation of molecular weight and weight fraction of five and six components in mixtures from model equilibrium ultracentrifugation data. *J. Agric. Food Chem.* **1992a,** *40,* 824–829.

Nakai, S.; Nonaka, M. Protein-protein interaction: an ultracentrifugal approach. In *Protein Interactions*; Visser, H., Ed.; VCH Verlagsgesellschaft, Weinheim, Germany, 1992b; pp 57–72,.

Nakai, S.; Li-Chan, E. Hydrophobic Interactions In Food Systems. CRC Press, Boca Raton, FL, 1988

Nonaka, M.; Li-Chan, E.; Nakai, S. Raman spectroscopic study of thermally induced gelation of whey proeins. *J. Agric. Food Chem.* **1993,** *41,* 1176–1181.

Ogawa, M.; Kanamaru, J.; Miyashita, H.; Tamiya, T.; Tsuchiya, T. Alpha-helical structure of fish actomyosin: changes during setting. *J. Food Sci.* **1995,** *60,* 297.

Padrós, E.; Dunach, M.; Morros, A.; Sabés, M.; Manosa, J. Fourth-derivative spectrophotometry of proteins. *Trends Biochem. Sci.* **1984,** *9,* 508–510.

Perczel, A.; Park, K.; Fasman, G.D. Analysis of the circular dichroism spectrum of proteins using the convex constraint algorithm: a practical guide. *Anal. Biochem.* **1992,** *203,* 83–93.

Pjura, P.; McIntosh, L.P.; Wozniak, J.A.; Matthews, B.W. Perturbation of Trp 138 in T4 lysozyme by mutations at Gln 105 used to correlate changes in structure, stability, solvation and spectroscopic properties. *Proteins* **1993,** *15,* 401–412.

Przybycien, T.M.; Bailey, J.E. Structure-function relationships in the inorganic salt-induced precipitation of α-chymotrypsin. *Biochim. Biophys. Acta* **1989,** *995,* 231–245.

Ptitsyn, O.B. Molten globule and protein folding. *Adv. Protein Chem.* **1995,** *47,* 83–229.

Ramsay, G.D.; Eftink, M.R. Analysis of multidimensional spectroscopic data to monitor unfolding of proteins. *Methods in Enzymol.* **1994,** *240,* 615–645.

Sackett, D.L.; Wolff, J. Nile red as a polarity-sensitive fluorescent probe of hydrophobic protein surfaces. *Anal. Biochem.* **1987,** *167,* 228–234.

Sarver, R. W., Jr.; Krueger, W.C. An infrared and circular dichroism combined approach to the analysis of protein secondary structure. *Anal. Biochem.* **1991,** *199,* 61–67.

Sawyer, L.; Holt, C. The secondary structure of milk proteins and their biological function. *J. Dairy Sci.* **1993,** *76,* 3062–3078.

Schachman, H. K. Analytical ultracentrifugation reborn. *Nature* **1989,** *341,* 259–260.

Siuzdak, G. *Mass Spectrometry for Biotechnology*. Academic Press, San Diego, CA, 1996.

Smith, V.F.; Schwartz, B.L.; Randall, L.L.; Smith, R.D. Electrospray mass spectrometric investigation of the chaperone SecB. *Protein Sci.* **1996,** *5,* 488–494.

Sreerama, N.; Woody, R.W. A self-consistent method for analysis of protein secondary structure from circular dichroism. *Anal. Biochem.* **1993,** *209,* 32–44.

Sreerama, N.; Woody, R.W. Poly(Pro)II helices in globular proteins: identification and circular dichroic analysis. *Biochemistry* **1994,** *33,* 10022–10025.

Stanley, D.W.; Yada, R.Y. Physical consequences of thermal reactions in food protein systems. In *Physical Chemistry of Foods*; Schwartzberg, H.G.; Hartel, R.W., Eds.;Marcel Dekker, Inc., New York, NY, 1992; pp 669–733.

Strasburg, G.M.; Ludescher, R.D. Theory and applications of fluorescence spectroscopy in food research. *Trends Food Sci. Technol.* **1995,** *6,* 69–75.

Stults, J.T. Matrix-assisted laser desorption/ionization mass spectrometry (MALDI-MS). *Current Opinion Structural Biol.* **1995,** *5,* 691–698.

Susi, H.; Byler, D.M. Fourier deconvolution of the Amide I Raman band of proteins as related to conformation. *Appl. Spectrosc.* **1988a,** *42,* 819–826.

Susi, H.; Byler, D.M. Fourier transform infrared spectroscopy in protein conformation studies. In *Methods for Protein Analysis*; Cherry, J.P.; Barford, R.A., Eds.; American Oil Chemists' Society, Champaign, IL, 1988b; pp 235–255.

Tanford, C. Protein denaturation. Part B. *Adv. Protein Chem.* **1968,** *23,* 121–275.

Tani, F.; Murata, M.; Higasa, T.; Goto, M.; Kitaatake, N.; Doi, E. Molten globule state of protein molecules in heat-induced transparent food gels. *J. Agric. Food Chem.* **1995,** *43,* 2325–2331.

Tournadje, A.; Alcorn, S.W.; Johnson, W.C., Jr. Extending CD spectra of proteins to 168 nm improves the analysis for secondary structures. *Anal. Biochem.* **1992,** *200,* 321–331.

Tsutsui, T.; Li-Chan, E.; Nakai, S. A simple fluorometric method for fat-binding capacity as an index of hydrophobicity of proteins. *J. Food Sci.* **1986,** *51,* 1268–1272.

VanStokkum, I.H.M.; Linsdell, H.; Hadden, J.M.; Haris, P.I.; Chapman, D.; Bloemendal, M. Temperature-induced changes in protein structures studied by Fourier transform infrared spectroscopy and global analysis. *Biochemistry* **1995,** *34,* 10508–10518.

Wang, C.-H.; Damodaran, S. Thermal gelation of globular proteins: influence of protein conformation on gel strength. *J. Agric. Food Chem.* **1991,** *39,* 433–438.

Wilder, C.L.; Friedrich, A.D.; Potts R.O., Daumy, G.O.; Francoeur, M.L. Secondary structural analysis of two recombinant murine proteins, interleukins 1α and 1β: Is infrared spectroscopy sufficient to assign structure? *Biochemistry* **1992,** *31,* 27–31.

Williams, R.W. Estimation of protein secondary structure from the laser Raman Amide I spectrum. *J. Mol. Biol. 1983, 166,* 581–603.

Woody, R.W. Circular dichroism. *Methods in Enzymol.* **1995,** *246,* 34–71.

Yada, R. Y.; Harauz, G.; Marcone, M.F.; Beniac, D.R.; Ottensmeyer, F.P. Visions in the mist: The *Zeitgeist* of food protein imaging by electron microscopy. *Trends Food Sci. Technol. 1995, 6,* 265–270.

Yada, R.Y.; Jackman, R.L.; Smith, J.L.; Marangoni, A.G. Analysis: Quantitation and physical characterization. In *Food Proteins. Properties and Characterization*; Nakai , S.; Modler , H.W., Eds.; VCH Publishers, Inc. , New York, NY, 1996; pp 71–165.

Yang, S.T.; Marchio, J.L.; Yen, J.W. A dynamic light scattering study of β-galactosidase: environmental effects on protein conformation and enzyme activity. *Biotechnol. Prog.* **1994,** *10,* 525–531.

Yesilada, A.; Theobald, A.; Hider, R.C. Discrimination of bovine and porcine insulin by higher-order derivative UV-spectroscopy. *J. Pharm. Biomed. Anal.* **1992**, *10,* 699–703.

3

PROTEOLYSIS AND GELATION OF FISH PROTEINS UNDER OHMIC HEATING

Jae W. Park, Jirawat Yongsawatdigul, and Ed Kolbe

OSU Seafood Laboratory and
Department of Food Science and Technology
Oregon State University
250 36th Street, Astoria, Oregon 97103

Pacific whiting surimi gels heated slowly in a water bath exhibited poor gel quality, while the ohmically heated gels without holding at 55°C showed more than a twofold increase in shear stress and shear strain over conventionally heated gels. Degradation of myosin and actin was minimized by ohmic heating, resulting in a continuous network structure. Ohmic heating with a rapid heating rate was an effective method for maximizing gel functionality of Pacific whiting surimi without enzyme inhibitors. In linear heating, slow heating rates increased proteolysis in Pacific whiting surimi as shown by degradation of myosin heavy chain and low shear stress and shear strain. Proteolysis of whiting surimi was decreased by the presence of beef plasma protein (BPP) to a greater extent at rapid heating rates (20 and 30°C/min) than at slow heating rates (1 and 5°C/min). Shear stress of Alaska pollock surimi gels with or without BPP increased as heating rate decreased, but shear strain was unaffected. An increase in shear stress was accompanied by an increase of cross-linked myosin heavy chain.

INTRODUCTION

Pacific whiting *(Merluccius productus)* surimi without enzyme inhibitors normally undergoes texture degradation during slow heating (Chang-Lee *et al.*, 1989; Morrissey *et al.*, 1993; Yongsawatdigul *et al.*, 1995). This is due to the presence of heat stable endogenous protease(s) that exhibit their high hydrolytic activity on muscle myosin (Erickson *et al.*, 1983; Patashnik *et al.*, 1982). In conventional heating methods, heat is transferred within the solid sample by means of conduction. Rate of conductive heat transfer is typically slow and depends on various factors such as temperature of heating medium, geometry, and thermal prop-

Process-Induced Chemical Changes in Food
edited by Shahidi *et al.* Plenum Press, New York, 1998

erties of the sample. Temperature at geometric center of whiting surimi paste in stainless steel tube (i.d. = 1.9 cm) increased from 10 to 80°C within 8 min when heated in a 90°C water bath (Yongsawatdigul *et al.*, 1995). Such a slow heating process allowed the protease(s) to be active for a relatively long period of time. As a result, substantial loss of myosin heavy chain was evident in concomitant with very low shear stress and shear strain of surimi gel. In commercial processing of surimi from Pacific whiting, the use of food grade inhibitors was inevitable. Beef plasma proteins, egg white, and other food grade enzyme inhibitors have been used to inhibit proteolytic activity (Nagahisa *et al.*, 1981; Chang-Lee *et al.*, 1989; Hamann *et al.*, 1990; Porter, 1992; Morrissey *et al.*, 1993). However, these enzyme inhibitors used commercially have some negative effects such as off-odor, off-color, high cost, and labeling concerns from marketing groups. Consequently, a processing alternative to overcome the textural deterioration of Pacific whiting surimi was suggested by Yongsawatdigul *et al.* (1 995) and Yongsawatdigul (1996).

Ohmic heating is a method in which alternating electrical current is passed through an electrically conducting food product (Biss *et al.*, 1989). Heat is internally generated, resulting in a rapid heating rate. Because heat is simultaneously generated in liquid and solid phases, the temperature increase in the product is uniform, compared with the conventional process in which heat is applied at the external boundary (de Alwis and Fryer, 1990; Parrot, 1992). Rapid heating methods, such as microwave and ohmic heating, have been reported to effectively minimize textural degradation of fish muscle caused by endogenous heat stable proteases (Greene and Babbitt, 1990; Yongsawatdigul *et al.*, 1995; Yongsawatdigul and Park, 1996). This is because the rapid increase of temperature causes protease to be inactivated before it could substantially hydrolyze myofibrillar proteins (de Alwis and Fryer, 1990).

Protein gels are commonly defined as three-dimensional matrices with polymerpolymer and polymer-solvent interactions in an ordered manner. Therefore, fish protein gels immobilize large amounts of water by a relatively small portion of protein. Studies of gelation have been conducted quite extensively since Ferry (1948) proposed the formation of proteins gels is a two-stage process of denaturation and aggregation. Denaturation is a process in which native proteins undergo conformational changes, including alterations of hydrogen bondings, hydrophobic interactions, and ionic linkages (Mulvihill and Kinsella, 1987). Denaturation is a prerequisite for protein gelation (Hermansson, 1979). The following process is aggregation in which denatured protein molecules align themselves and interact each other at specific points to form a threedimensional network. Heating modes certainly influence gelation of surimi which stabilize fish myofibrillar proteins. A two-step heating process improves textural properties of surimi made from a variety of fish species such as Alaska pollock

(Montejano *et al.*, 1984; Numakura *et al.*, 1985), sardine (Roussel and Cheftel, 1988), Atlantic croaker (Kim, 1987; Kamath *et al.*, 1992), Southern blue whiting, hoki (MacDonald *et al.*, 1994), and Pacific whiting with beef plasma protein (Park *et al.*, 1994). Gel elasticity increased when surimi pastes were subjected to setting at 5–40°C prior to heating to 90°C. In pollock surimi, improved gel quality is associated with increased ε-(γ-glutamyl)-lysine contents, resulting from polymerization of myosin heavy chains. Endogenous transglutaminase was reported to catalyze such reactions (Seki *et al.*, 1990; Kimura *et al.*, 1991).

PROTEOLYTIC DEGRADATION DUE TO DIFFERENT HEATING METHODS

Textural degradation of Pacific whiting at 50–65°C has been extensively studied. However, it is still unclear whether the protease is produced from the parasite or from im-

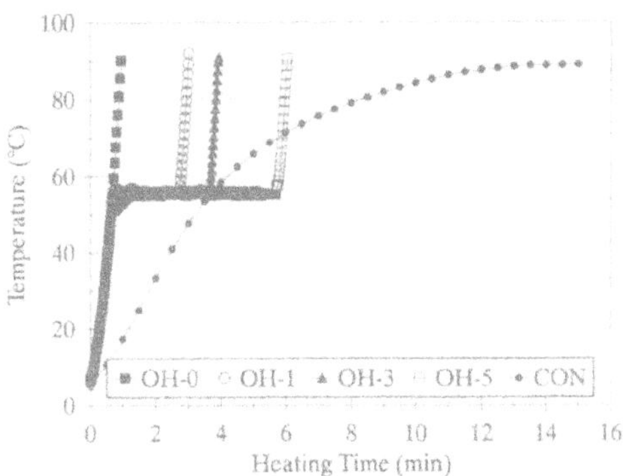

Figure 1. Temperature profiles of surimi gels heated by various heating methods. Adapted from Yongsawatdigul *et al.* (1995). **CON:** conventional heating (heated in a 90°C water bath for 15 min). **OH-0:** heated ohmically with voltage gradient of 13.3 V/cm (applied voltage = 200 V) to reach 90°C. **OH-1:** heated ohmically with voltage gradient of 13.3 V/cm and held for 1 min at 55°C before heating to 90°C. **OH-3:** heated ohmically with voltage gradient of 13.3 V/cm and held for 3 min at 55°C before heating to 90°C. **OH-5:** heated ohmically with voltage gradient of 13.3 V/cm and held for 5 min at 55°C before heating to 90°C.

munological responses of the fish. Nevertheless, most researchers have reported the role of cathepsins in proteolysis of whiting. The purified cathepsin L hydrolyzes myofibrillar proteins, myosin, and native as well as denatured collagen (Toyohara *et al.*, 1993; An *et al.*, 1994). The degradation pattern of myofibrillar proteins induced by the purified cathepsin L at its optimum pH (5.5) is consistent to that of proteolysis of surimi at 55°C, at pH 6.8–7.0, suggesting that textural degradation of whiting surimi at elevated temperatures is attributed to proteolytic activity of cathepsin L (An *et al.*, 1994). Temperature histories of surimi gels heated conventionally (CON) in a water bath and ohmically are clearly differentiated in Figure 1 (Yongsawatdigul *et al.*, 1995). It took less than 1 min to reach the internal temperature of 90°C using ohmic heating, while approximately 13 min was required when using water bath. Shear stress and shear strain of gels, indicating hardness and cohesiveness, respectively (Hamann and MacDonald, 1992), are compared by various heating methods (Figure 2). Hardness of gels (OH-0) heated continuously by ohmic heating and gels (OH-1) heated to 55°C and held for 1 min before heating to 90°C was significantly higher than that of CON, OH-3 (3 min holding at 55°C), and OH-5 (5 min holding at 55°C) gels. No significant differences occurred in cohesiveness between CON and OH-3 gels. However, cohesiveness of these two gels was significantly lower than that of OH-0 and OH-1, and higher than that of OH-5 gels. Since an acceptable surimi gel should have a strain value of at least 2.0 (Hamann *et al.*, 1990), textural properties of surimi gels heated in a 90°C water bath and those held ohmically at 55°C for 3 and 5 min

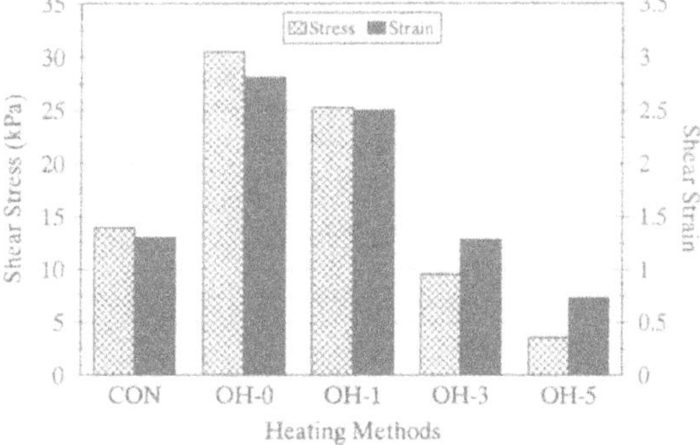

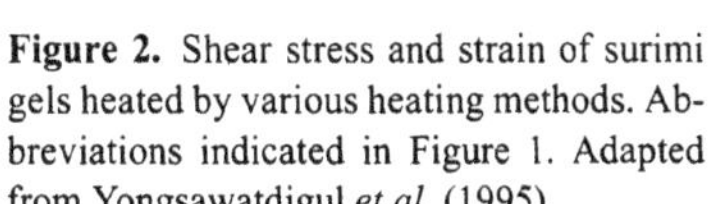
Figure 2. Shear stress and strain of surimi gels heated by various heating methods. Abbreviations indicated in Figure 1. Adapted from Yongsawatdigul *et al.* (1995).

were very poor. Low shear stress and shear strain in OH-3 and OH-5 illustrated the effect of protease activity at 55°C on textural characteristics of Pacific whiting surimi gel. Those gels heated ohmically to 90°C with a voltage gradient of 13.3 V/cm (200 V) and those held in ohmic heating at 55°C for 1 min showed good textural properties with strain values of 2.81 and 2.50, respectively. These were more than twice the value of conventionally heated surimi gels. Based on torsion test results, Pacific whiting gels without enzyme inhibitors heated ohmically without holding at 55°C were slightly better than those heated conventionally with incorporation of enzyme inhibitors such as beef plasma proteins, egg white, and potato extract. These had a maximum strain value of around 2.5 (Morrissey *et al.*, 1993).

Seymour *et al.* (1994) illustrated that activity of the protease purified from Pacific whiting gradually increased as temperature increased until peaking at 55°C. It then started decreasing and reached a minimum at 70°C. Temperature profiles of conventional heating (Figure 1), indicate that surimi was slowly heated in a 90°C water bath and was exposed to a temperature range of 40–60°C for >2 min. Moreover, it took about 6 min to reach 70°C at which the protease could be inactivated. Due to the slow heating rate, the enzyme was activated and started degrading the myosin before thermal inactivation occurred. Since the myosin plays an important role in gel network formation (Niwa, 1992), severe degradation of the myosin heavy chain (Figure 3) resulted in low gel strength (Figure 2). When holding time at 55°C in ohmic heating was prolonged, integrity of the myosin heavy chain and actin were diminished (Figure 3). The OH-0 and OH-1 samples exhibited higher intensity of myosin and actin than the CON sample, while less intensity was ob-

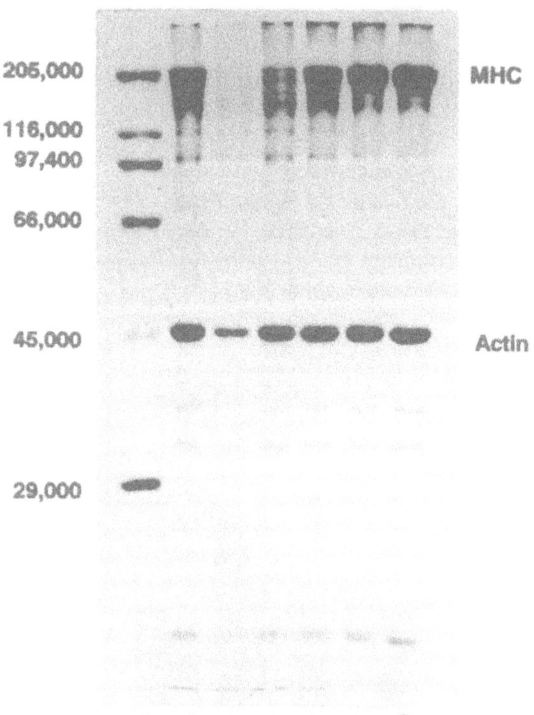

Figure 3. SDS-PAGE separation pattern of surimi proteins under various heating methods: (R) unheated (raw) surimi paste; (0) surimi heated ohmically with 13.3 V/cm to 90°C; (1) surimi heated ohmically with 13.3 V/cm and held at 55°C for 1 min before heating to 90°C; (3) surimi heated ohmically with 13.3 V/cm and held at 55°C for 3 min before heating to 90°C; (5) surimi heated ohmically with 13.3 V/cm and held at 55°C for 5 min before heating to 90°C; (C) surimi heated conventionally; (S) High molecular weight standard; NMC: Myosin heavy chain.

served in the OH-3 and OH-5. As surimi was subjected to optimum temperature of the enzyme for a longer period of time, more degradation of myosin and actin occurred. For the samples held at 55°C for 5 min, the myosin heavy chain and actin almost disappeared. These results suggest that gel-weakening associated with breakdown of the myosin heavy chain and actin in Pacific whiting surimi was due to proteolytic activity (Yongsawatdigul *et al.*, 1995). The highest gel strength and the most intense myosin heavy chain and actin bands were observed in OH-0 gels that were heated to 90°C within 1 min. The myosin and actin bands of the OH-0 were comparable to those of unheated surimi paste (Figure 3). Probably rapid thermal inactivation minimized degradation of the myosin and actin.

Although the intensity of the myosin heavy chain band of CON gels was visually greater than that of OH-3 gels (Figure 3), no differences occurred in shear strain and shear stress between those two gels (Figure 2). Toyohara and Shimizu (1988) also found a similar discrepancy. These authors indicated that the intensity of myosin did not correspond to their gel strength. Yongsawatdigul *et al.* (1995) hypothesized that once the myosin was degraded by the enzyme to a specific level, the gel network could not be properly formed regardless of degree of degradation. This would agree with the fact that intact myosin is required for gel formation (Ziegler and Acton, 1984). When the myosin is severely hydrolyzed to smaller protein fragments (OH-3 and OH-5 in Figure 3), a gel network was hardly developed, resulting in extremely low gel strength and a very mushy texture.

It is very important to note that actual heating processes conducted on the primary cooking machine of surimi seafood are not like those used for gel preparation using plastic casings or stainless tubes, but rather similar to ohmic heating. Extruded thin surimi sheet is approximately 2 mm thick so that heat penetration on the cooking machine using steam and gas is very rapid. As a result, enzyme inhibitors may not be required for this specific process because enzyme can be easily inactivated in a rapid heating regime. However, the addition of protein additives could increase gel strength, not primarily because they work well as enzyme inhibitors, but because they contain their own gel enhancing ability.

GELATION OF FISH PROTEINS UNDER LINEAR HEATING

Surimi is stabilized fish myofibrillar proteins which provide important textural characteristics. Heating modes influence gelation of surimi. Yongsawatdigul *et al.* (1995) demonstrated that Pacific whiting surimi exhibited high shear stress and shear strain when heated ohmically from 5 to 90°C within 1 min, while rheological properties of surimi heated in a 90°C water bath were unacceptably low. This was attributed to a rapid inactivation of endogenous protease with ohmic heating; consequently, degradation of myosin was greatly reduced.

The effect of linear heating rate on surimi gelation has not been extensively studied. Since water bath heating provides non-linear temperature profiles (Yongsawatdigul *et al.*, 1995), the influence of such heating rates on gelation cannot be validated. Linear heating rates may be achieved using a programmable water bath only when very slow heating rates are applied (Foegeding *et al.*, 1987a, b; Camou *et al.*, 1989; Arntfield and Murray, 1992). Yongsawatdigul and Park (1996) applied ohmic heating to obtain linear heating rates using appropriate voltage gradients and temperature controllers. Heating rates obtained ohmically were referred to as linear since temperature increased equally in each time interval (Figure 4). Temperature was assumed to be uniformly distributed throughout the sample because of the homogeneity of the surimi paste and internal heat generation (Yongsawatdigul and Park, 1996). As a result, the protease was thermally inactivated in a

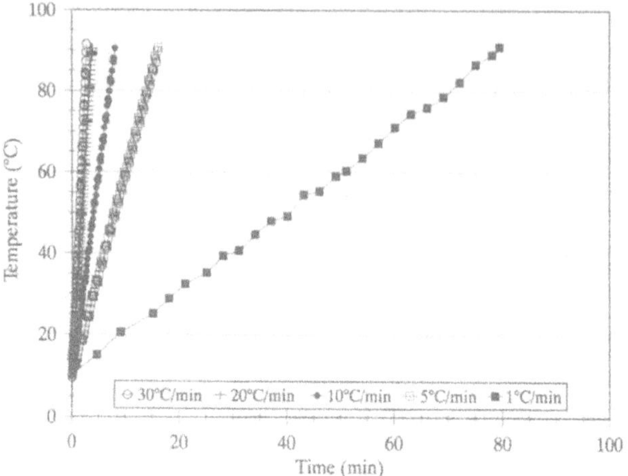

Figure 4. Temperature profiles of various linear heating treatments. Adapted from Yongsawatdigul and Park (1996).

rapid manner. High quality whiting surimi gel (shear stress and shear strain of 30 kPa and 2.8, respectively) was obtained when heated ohmically from 10 to 90°C within 1 min (Yongsawatdigul *et al.*, 1995). On the other hand, as the time of ohmic heating was prolonged to 80 min at the same temperature range, very soft and mushy gels were obtained, accompanying complete disappearance of myosin heavy chain and slightly reduced actin band (Yongsawatdigul and Park, 1996). It is clear that heating time and temperature are critical factors controlling textural properties of whiting surimi.

The whiting samples heated at 1 and 5°C/min were very soft and mushy (Figure 5). At 10°C/min, the whiting gel was strong enough to measure its rheological properties. Shear stress and shear strain of the whiting gels increased when heating rates were raised. However, shear stress of samples heated at 30°C/min was not different from those heated at 20°C/min. Textural properties of whiting surimi were increased at all heating rates when BPP was added (Yongsawatdigul and Park, 1996). This implied that textural degradation due to proteolysis also occurred in the whiting samples heated at 20 and 30°C/min.

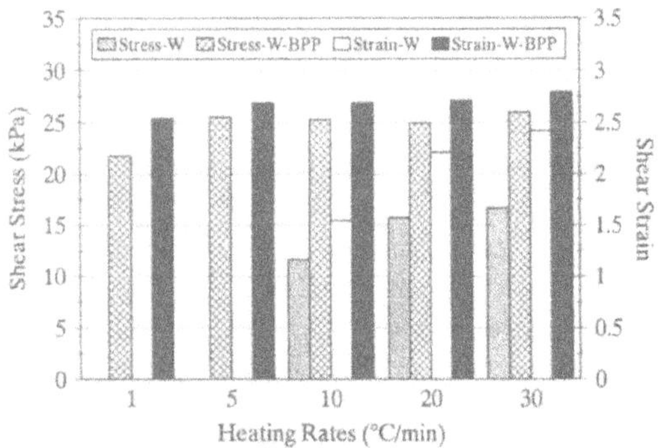

Figure 5. Shear stress and shear strain of whiting surimi gels heated at various rates. W denotes whiting surimi and BPP indicates the inclusion of beef plasma protein.

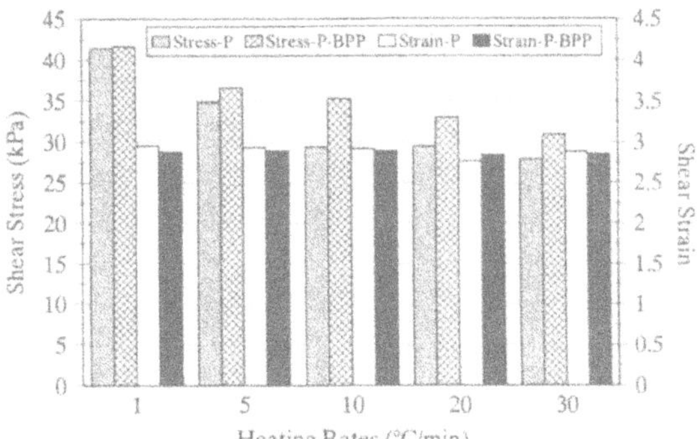

Figure 6. Shear stress and shear strain of pollock surimi gels heated at various rates. P denotes pollock surimi and BPP indicates the inclusion of beef plasma protein.

Yongsawatdigul *et al.* (1995) reported that the gel-weakening of whiting surimi was effectively minimized when the surimi was heated ohmically at approximately 80°C/min. Note that shear stress and shear strain of the whiting gels with BPP heated at 1°C/min were lower than those heated at 5, 10, 20, and 30°C/min.

Shear stress of pollock gels heated at 1°C/min was higher than those heated at faster rates (Figure 6). Lowest shear stress was observed in the sample heated at 30°C/min. Heating rates did not appear to affect shear strain of these samples (Figure 6). Changes of shear stress and shear strain of pollock gels with BPP showed similar trends to those of pollock alone (Figure 6). The heating rate of 1°C/min provided the highest shear stress, and the effect of heating rate on shear strain was insignificant. Yonsawatdigul and Park (1996) reported no statistical difference in shear stress and shear strain between the pooled data of pollock surimi with and without 1.4% BPP. This indicated that proteolysis of pollock myosin did not occur or was minimal during heating. Moreover, BPP at that level did not significantly enhance gel-forming ability of pollock surimi. Montejano *et al.* (1984) and Kamath *et al.* (1992) reported setting treatments had a greater effect on shear stress than on shear strain of pollock surimi, which was similar to the effect of heating rate observed by Yongsawatdigul and Park (1996).

Myosin heavy chain of whiting decreased as heating rate decreased and primary hydrolysis products (66,000 < Mr < 116,000) appeared in samples heated at 5, 10, 20, and 30°C/min (Yongsawatdigul and Park, 1996). Disappearance of myosin heavy chain and the extensive loss of actin were observed in the sample heated at 1°C/min. Consequently, the number of smaller molecular weight proteins (29,000 < Mr < 45,000) increased. A decreased myosin heavy chain at lower heating rates coincided with changes in shear stress and shear strain (Figure 5). This emphasized the importance of myosin as a major component responsible for gel-forming ability of surimi. Although a considerable amount of actin remained in the surimi gel heated at 5°C/min, the sample was very soft and mushy. This indicated that actin affected gel functionality to a lesser extent. The endogenous protease in Pacific whiting surimi was purified and characterized as cathepsin L, exhibiting its maximum activity at 55°C with no activity at temperatures higher than 70°C (Seymour *et al.*, 1994). Proteolysis could occur in whiting surimi heated at 1°C/min for approximately 1 hr (Figure 4) before the enzyme was thermally inactivated at >70°C. This was about 30 times longer than the samples heated at 30°C/min (2 min). Therefore, the extent of proteolysis was greater in the slow heating treatment than in the rapid heating treatment.

In pollock samples, as observed on the 10% polyacrylamide, myosin heavy chain markedly decreased at the heating rate of 1°C/min. Small fragments of degraded myosin

Table 1. Degradation and gelation of myosin heavy chain at various heating rates

	Heating Rates, °C/min					
	Raw	1	5	10	20	30
1) Comparison of myosin heavy chain content						
Pollock—10%[1]	48.15[2]	19.17	44.07	49.35	44.46	49.72
Pollock—3%	59.29	33.40	48.50	48.61	46.44	53.71
Whiting—10%	45.78	0	0	2.91	23.68	42.66
Whiting—3%	56.99	0	0	8.85	38.29	47.25
2) Relative contents of cross-linked proteins (Mr>205,000)						
Pollock—3%	0.63	6.3	3.37	2.95	2.96	2.26
Whiting—3%	0	0	0	0	0	0

[1]Indicates concentration of polyacrylamide used for SDS-PAGE.
[2]Represents relative density of protein band measured using a densitometer.

heavy chain were not noted. However, further analysis on 3% polyacrylamide elucidated that proteins with higher molecular weight than myosin heavy chain were formed; the highest intensity appeared on the gel heated at 1 °C/min (Table 1). Although myosin heavy chains decreased as high molecular weight proteins increased, the changes in actin band were not pronounced. This suggested that high molecular weight proteins were formed by cross-linking of myosin heavy chain (Yongsawatdigul and Park, 1996). Since an increase in cross-linked myosin was accompanied by an increase in shear stress, myosin cross-linking could be responsible for a change in rheological properties of pollock surimi gels heated at slow heating rates. Cross-linking of pollock myosin was possibly catalyzed by endogenous transglutaminase (Seki *et al.*, 1990; Kimura *et al.*, 1991; Yongsawatdigul and Park, 1996). Probably slow heating enhanced enzymatic cross-linking of pollock myosin by prolonging the reaction period. Since surimi gels were solubilized using urea, SDS, and β-mercaptoethanol buffer prior to electrophoresis, polymerization of myosin heavy chain involved nondisulfide covalent bonds (Yongsawatdigi and Park, 1996). However, no cross-linking of whiting myosin chain with Mr > 205 kDa were found at all heating rates (Table 1). It indicated that nondisulfide covalent bonds were not involved in gelation of whiting surimi regardless of the inclusion of BPP. Chan *et al.* (1992) also reported that polymerization of silver hake myosin during thermal gelation was noncovalent. Thus, low gel-forming ability of whiting surimi could be the result of both the presence of an endogenous protease and lack of covalent cross-linking of whiting myosin (Yongsawatdigul and Park, 1996).

CONCLUSION

Pacific whiting gels rapidly heated to 90 °C by ohmic heating with a voltage gradient of 13.3 V/cm demonstrated superior shear stress and shear strain. Under these conditions, degradation of the myosin heavy chain and actin was significantly minimized. A decreased gel strength associated with poor gel network and degradation of the myosin heavy chain and actin was noticed as holding time at optimum temperature of 55 °C of the protease was prolonged. Whiting surimi gels were susceptible to proteolysis under conventional heating since a slow heating rate allowed the enzyme to degrade more myosin heavy chain and actin.

Linear heating rates were achievable using ohmic heating. The effect of heating rate on gelation was different, depending on inherent characteristics of the fish. Textural properties of pollock surimi gels could be improved through slow heating regimes. At rapid heating rates, degradation of whiting myosin heavy chain was greatly reduced. Rapid inactivation of endogenous protease resulted in increased shear stress and shear strain of whiting surimi gels.

REFERENCES

An, H.; Weerasinghe, V.; Seymour, T.A.; Morrissey, M.T. Cathepsin degradation of Pacific whiting proteins. *J. Food Sci.* **1994,** *59,* 1013–1017, 1033.

Arntfield, S.D.; Murray, E.D. Heating rate affects thermal properties and network formation for vicilin and ovalbumin at various pH values. *J. Food Sci.* **1992,** *57,* 640-646.

Biss, C.H.; Coombes, S.A.; Skudder, P.J. The development and application of ohmic heating for the continuous processing of particulate food stuffs. In *Processing Engineering in the Food Industry;* Field, R.W.; Howell J.A., Eds.; Elsevier Applied Science Publishers, Essex, U.K., 1989; pp. 17–27.

Camou, J.P.; Sebranek, J.G.; Olson, D.G. Effect of heating rate and protein concentration on gel strength and water loss of muscle protein gels. *J. Food Sci.* **1989,** *54,* 850-854.

Chan, J.K.; Gill, T.A.; Paulson, A. Cross-linking of myosin heavy chains from cod, herring and silver hake during thermal setting. *J. Food Sci.* **1992,** *57,* 906–912.

Chang-Lee, M.V.; Pacheo-Aguilar, R.; Crawford, D.L.; Lampila, L. Proteolytic activity of surimi from Pacific whiting *(Merluccius productus)* and heat-set gel texture. *J. Food Sci.* **1989,** *54,* 1116–1119, 1124.

de Alwis, A.A.P.; Fryer, P.J. The use of direct resistance heating in the food industry. *J. Food Eng.* **1990,** *11,* 3–27.

Erickson, M.C.; Gordon, D.T.; Anglemier, A.F. Proteolytic activity in the sarcoplasmic fluid of parasitized Pacific whiting *(Merluccius productus)* and unparasitized true cod *(Gadus macrocephalus). J. Food Sci.* **1983,** *48,* 1315–1319.

Ferry, J.D. Protein gels. *Adv. Protein Chem.* **1948,** *4,* 1–78.

Foegeding, E.A.; Dayton, W.R.; Allen, C.E. Effect of heating rate on thermally formed myosin, fibrinogen, and albumin gels. *J. Food Sci.* **1986a,** *51,* 104–108, 112.

Foegeding, E.A.; Dayton, W.R.; Allen, C.E. Interaction of myosin-albumin and myosin-fibrinogen to form protein gels. *J. Food Sci.* **1986b,** *51,* 109–112.

Greene, D.H.; Babbitt, J. Control of muscle softening and protease parasite interactions in arrowtooth flounder *(Atheresthes stomias). J. Food Sci.* **1990,** *55,* 579–580.

Hamann, D.D.; MacDonald, G.A. Rheology and texture properties of surimi and surimi-based foods. In *Surimi Technology;* Lanier, T.C.; Lee, C.M., Eds., Marcel Dekker, Inc., New York, 1992; pp. 429–501.

Hamann, D.D.; Amato, P.M.; Wu, M.C.; Foegeding, E.A. Inhibition of modori (gel weakening) in surimi by plasma hydrolysate and egg white. *J. Food Sci.* **1990,** *55,* 665–669.

Hermansson, A.M. Aggregation and denaturation involved in gel formation. In *Functionality and Protein Structure;* Pour-El, A., Ed., ACS Symposium Series 92, Amer. Chem. Soc., Washington, DC, 1979; p. 81.

Kamath, G.G.; Lanier, T.C.; Foegeding, E.A.; Hamann, D.D. Nondisulfide covalent cross-linking of myosin heavy chain in "setting" of Alaska pollock and Atlantic croaker surimi. *J. Food Biochem.* **1992,** *16,* 151–172

Kim, B.Y. *Rheological Investigation of Gel Structure Formation by Fish Proteins during Setting and Heat Processing,* Ph.D. thesis, North Carolina State University, Raleigh, NC, 1987.

Kimura, I.; Sugimoto, M.; Toyoda, K.; Seki, N.; Arai, K.; Fujita, T. A study on the cross-linking reaction of myosin in kamaboko "suwari" gels. *Nippon Suisan Gakkaishi* **1991,** *57,* 1389–1396.

MacDonald, G.A.; Stevens, J.; Lanier, T.C. Characterization of New Zealand hoki and southern blue whiting surimi compared to Alaska pollock surimi. *J. Aquatic Food Product Technol.* **1994,** *3,* 19–38.

Montejano, J.G.; Hamann, D.D.; Lanier, T.C. Thermally induced gelation of selected comminuted muscle systems: Rheological changes during processing, final strengths and microstructure. *J. Food Sci.* **1984,** *49,* 1496–1505.

Morrissey, M.T.; Wu, J.W.; Lin, D.; An, H. Protease inhibitor effects on torsion measurements and autolysis of Pacific whiting surimi. *J. Food Sci.* **1993,** *58,* 1050–1054.

Mulvihill, D.M.; Kinsella, J.E. Gelation characteristics of whey proteins and beta-lactoglobulin. *Food Technol.* **1987,** *41(9),* 102–111.

Nagahisa, E.; Nishimuro, S.; Fujita, T. Kamaboko-forming ability of the jellied meat of Pacific hake. *Bull. Jap. Soc. Sci. Fish.* **1981,** *49,* 901–906.

Niwa, E. Chemistry of surimi gelation. In *Surimi Technology;* Lanier, T.C.; Lee, C.M., Ed.; Marcel Dekker, Inc., New York, 1992; pp. 429–501.

Numakura, T.; Seki, N.; Kimura, I.; Toyoda, K.; Fujita, T.; Takama, K.; Arai, K. Cross-linking reaction of myosin in the fish paste during setting (suwari). *Nippon Suisan Gakkaishi* **1985,** *53,* 633–639.

Park J.W.; Yongsawatdigul, J.; Lin, T.M. Rheological behavior and potential cross-linking of Pacific whiting *(Merliccius productus)* surimi gel. *J. Food Sci.* **1994,** *59,* 773-776.

Patashnik, M.; Groninger, H.S.,Jr.; Barnett, H.; Kudo, G.; Koury, B. Pacific whiting *(Merluccius productus): I.* Abnormal muscle texture caused by myxosporidianinduced proteolysis. *Mar. Fish. Rev.* **1982,** *44(5),* 1–12.

Parrot, D.L. Use of ohmic heating for aseptic processing of food particulates. *Food Technol.* **1992**, *46(12)* ,68–72.

Porter, R.W. Use of potato inhibitor in Pacific whiting surimi. In *Pacific Whiting: Harvesting, Processing, Marketing and Quality;* Sylvia, G.; Morrissey, M.T., Eds.; Oregon Sea Grant Publication, Corvallis, OR, 1992; pp. 33–35.

Roussel, H.; Cheftel, J.C. Characteristics of surimi and kamaboko from sardines. *Int. J. Food. Sci. Technol.* **1988,** *23,* 607–623.

Seki, N.; Uno, H.; Lee, N.H.; Kimura, I.; Toyoda, K.; Fujita, T.; Arai, K. Transglutaminase activity in Alaska pollock muscle and surimi, and its reaction with myosin B. *Nippon Suisan Gakkaishi* **1990,** *56,* 125–132.

Seymour, T.A.; Morrissey, M.T.; Peters, M.Y.; An, H. Purification and characterization of Pacific whiting proteases. *J. Agric. Food Chem.* **1994,** *42,* 2421–2427.

Toyohara, H. and Shimizu, Y. Relation between the modori phenomenon and myosin heavy chain breakdown in threadfin-bream gel. *Agric. Biol. Chem.* **1988,** *52,* 255–257.

Toyohara, H.; Kineshita, M.; Kimura, I.; Satake, M.; Sakaguchi, M. Cathepsin LO like protease in Pacific hake muscle infected by myxosporidian parasites. *Nippon Suisan Gakkaishi* **1993,** *59,* 1101–1107.

Yongsawatdigul, J. *Textural and Electrical Properties of Pacific Whiting Surimi under Ohmic Heating,* Ph.D. Thesis, Oregon State University, Corvallis, OR, 1996.

Yongsawatdigul, J.; Park, J.W. Linear heating rate affects gelation of Alaska pollock and Pacific whiting surimi. *J. Food Sci.* **1996,** *61,*149–153.

Yongsawatdigul, J.; Park, J.W.; Kolbe, E.; AbuDagga, Y.; Morrissey, M.T. Ohmic heating maximizes gel functionality of Pacific whiting surimi. *J. Food Sci.* **1995,** *60,* 10-14.

Ziegler, G.R.; Acton, J.C. Mechanisms of gel formation by proteins of muscle tissue. *Food Technol.* **1984,** *38(5),*77–82.

4

EFFECT OF MATURITY AND CURING ON PEANUT PROTEINS

Changes in Protein Surface Hydrophobicity

Si-Yin Chung,[1] John R. Vercellotti,[1] and Timothy H. Sanders[2]

[1]USDA, ARS
Southern Regional Research Center
1100 Robert E. Lee Boulevard, New Orleans, Louisiana 70179
[2]USDA, ARS
Marketing Quality Handling Research, North Carolina State University
Raleigh, North Carolina 27695-7624

A hydrophobic fluorescence probe, 1,8-anilinonaphthalene sulfonate (ANS), was used to study the changes in protein surface hydrophobicity (PSH) occurring during peanut maturation and curing. PSH increased with the degree of maturity and during curing (windrow drying). The increase of PSH during curing or heating was more pronounced in immature peanuts than their mature counterparts, suggesting that more hydrophobic sites are hidden in the former proteins. PSH decreased when proteins were chemically modified with phenylglyoxal (an arginine-modifying agent), suggesting that arginine might play a role in hydrophobicity. The findings indicate that maturation and curing affect PSH, and that there is a relationship between PSH and peanut maturity. Possible factors contributing to the increase of PSH are discussed.

INTRODUCTION

Proteins may bind a variety of hydrophobic molecules or flavor compounds because protein molecules possess a number of hydrophobic sites on their surfaces (i.e., protein surface hydrophobicity, PSH). Proteins such as bovine serum albumin (Beyeler and Solms, 1974; Cardamone and Puri, 1992; Fujii *et al.*, 1992; Richieri *et al.*, 1993), soy proteins (Beyeler and Solms, 1974; Petruccelli and Anon, 1994) and b-lactoglobulin (Akita and Nakai, 1980) are known for their binding properties or hydrophobic traits. Because

Process-Induced Chemical Changes in Food
edited by Shahidi *et al.* Plenum Press, New York, 1998

the binding of flavors to a protein is dependent on the protein structure (Damodaran and Kinsella, 1981; O'Keefe *et al.*, 1991), a change in protein structure is potentially indicative of changes in the binding capacity and PSH. Several studies (Fujimaki *et al.*, 1968; Arai *et al.*, 1970; Noguchi *et al.*, 1970) have shown that treatment of soy proteins with proteases aids in the reduction of off-flavor compounds. A possible mechanism for this is that upon cleavage by proteases, proteins change structurally (e.g., they break down into polypeptides) and as a result, the number of hydrophobic sites is reduced and/or the affinity of proteins for flavor compounds is weakened. Proteins cleaved by chymotrypsin have also been shown to exhibit lower binding capacity or PSH (Mahmoud *et al.*, 1992; Ploug *et al.*, 1994). This is because chymotrypsin disrupts the protein structure by specifically cleaving at sites containing such hydrophobic amino acids as phenylalanine, tyrosine, tryptophan, and leucine. Modifications of proteins by heat (Sorgentini *et al.* 1995; Tani *et al.*, 1995), methylation (Dufour *et al.*, 1992), and succinylation (Paulson and Tung, 1987; Lakkis and Villota, 1992) also change PSH due to their effects on protein structure. Because the protein structure (Chung *et al.*, 1994) and binding capacity (Chung *et al.*, 1995) have been shown to change during peanut maturation and curing, PSH is expected to change as well. Changes in PSH during peanut maturation and curing using a fluorescence probe 1,8-anilinonaphthalene sulfonate (ANS) which binds to accessible hydrophobic sites on the protein surfaces (Semisotnov *et al.*, 1991; Cardamone and Puri, 1992). The investigation of changes in PSH may aid in understanding its potential role were studied in this work in peanut flavor development.

EXPERIMENTAL

Materials

The compound 1,8-anilinonaphthalene sulfonate (ANS) was purchased from Sigma Co. (St. Louis, MO). Bicinchoninic acid (BCA)-protein assay kit, microdialyzer System 500, and framed dialysis membrane with a molecular weight cut-off of 8000 Da were purchased from Pierce Chemical Co. (Rockford, IL). Reversed-phase HPLC C_{18} column (5 μ, 4.6 x 150 mm) was purchased from Vydac (Hesperia, CA). A Hewlett Packard (Willmington, IL) HPLC 1050 system was used. Peanuts (*Arachis hypogaea* L., var. Florunner) were planted at the USDA-ARS National Peanut Research Laboratory (Dawson, GA), dug 120 days after planting and subjected to windrow drying where samples were taken 0, 1, 2, 3, and 4 days after windrow drying, and day-4 samples were further dried to a 10% moisture content with heated air and used for PSH comparison after curing. Day-0 samples were used for corresponding PSH comparison before curing. At each sample date, peanuts were subjected to gentle abrasion to remove the exocarp, sorted by pod color, hand shelled, and stored at -80° C. Peanut maturity [defined as yellow (least), orange, brown, and black (most)] was based on the visual hull-scrape/color method (Williams and Drexler, 1981).

Preparation of Protein Extract from Peanuts

Protein extracts were prepared with some modification, as described by Chung *et al.* (1995). Briefly, 0.2 g of defatted peanut meal were suspensed in 1.5 mL of 0.05 M sodium phosphate buffer pH 8, containing 0.5 M NaCl. The mixture was stirred for 2 hr at 4° C and then centrifuged at 5000xg for 15 min. The supernatant (i.e., extract) was recovered and the fat layer (on top of supernatant) removed. The resultant protein extract was used for determination of the hydrophobicity assay.

Determination of Protein Surface Hydrophobicity

Protein surface hydrophobicity was determined as described by Tsutsui *et al.* (1986) with minor modifications. Protein extract (250 μL) was diluted with 0.05 M phosphate buffer (2 mL) pH 8, containing 0.5 M NaCl, to a protein concentration of approximately 2.5 mg/mL. To the diluted protein extract (1 mL) was added 1, 8-anilinonaphthalene sulfonate (ANS) (5 μL). Fluorescence measurements were carried out in triplicate at an excitation wavelength of 390 nm and an emission wavelength of 470 nm using a Gilford Fluoro IV^{TM} spectrofluorometer. The measurement was standardized using a mixture of methanol (1 mL) and 2 mM ANS (5 μL) along with a blank containing the buffer (1 mL) and 2 mM ANS (5 μL). Protein surface hydrophobicity was measured as the fluorescence intensity (FI) divided by the amount (mg) of protein. Protein concentration was determined using the BCA-protein assay kit. To determine possible reaction of other free components in the extract reacted with the fluorescent probe ANS, the extract was treated with trichloroacetic acid (TCA) (to remove proteins) at a final concentration of 5% and centrifuged. The resultant supernatant was adjusted to pH 8 and analyzed using ANS as described above. Results showed that little fluorescence was observed from the supernatant, indicating that the high fluorescence obtained in subsequent measurement of the crude protein extract was mostly due to the binding of ANS to the proteins.

Modification of Peanut Proteins with Phenylglyoxal

Protein modification was carried out according to Mukherjee and Dekker (1992) with some modifications. Briefly, 5 μl of a solution of phenylglyoxal (50 mM) in ethanol was added to 150 μL of a protein extract. The mixture was then incubated at 40° C for 1 hr and dialyzed using a microdialyzer (System 500) containing a framed dialysis membrane with a molecular weight cut-off of 8000 Da. After dialysis, the samples were diluted with buffer to a final volume of approximately 1.1 mL, 1 mL of which was used for protein surface hydrophobicity measurement and the remainder for protein determination. A control without modification was prepared in the same way. To determine protein modification, modified and unmodified proteins were digested with trypsin and analyzed using a Vydac reversed-phase HPLC using a C_{18} column. For HPLC analyses, 10 μL sample was injected and a gradient of 0–50% acetonitrile with 0.1% trifluoroacetic acid (TFA) over a 50 min period at a flow rate of 1 mL/min at 40° C was used. Results showed that the peptide profiles of modified and unmodified proteins were different, thus indicating that modification was successful.

RESULTS AND DISCUSSIONS

Effect of Peanut Maturity on PSH

An hydrophobic fluorescent probe such as 1, 8-anilinonaphthalene sulfonate (ANS) gives a very low quantum yield of fluorescence in water, but becomes highly fluorescent upon binding to protein hydrophobic sites. The excitation and emission wavelengths for the protein-ANS conjugate vary slightly with different proteins (Cardamone and Puri,

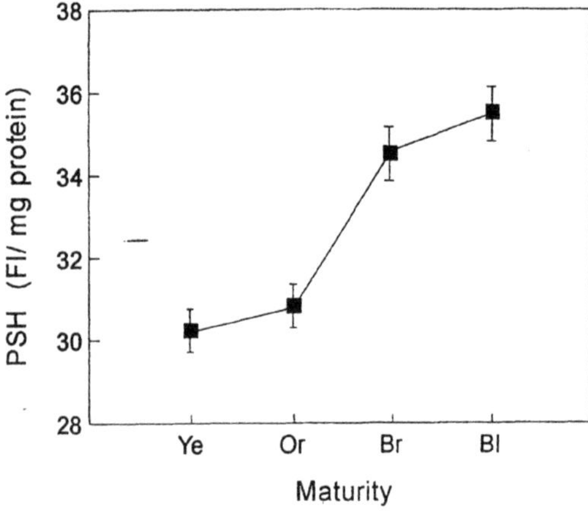

Figure 1. Effect of peanut maturity on protein surface hydrophobicity (PSH). Ye = yellow; Or = orange; Br = brown; and Bl = black. Each data point is the mean of three determinations (n=3).

1992). For peanut protein-ANS conjugate, the optimal excitation and emission wavelengths were found to be 390 and 470 nm, respectively.

The effect of peanut maturity on protein surface hydrophobicity (PSH) is shown in Figure 1. PSH is shown to increase with the degree of maturity (i.e., PSH is higher in mature peanuts). The increase of PSH reflects an exposure of additional hydrophobic sites on the protein surface during peanut maturation. The exposure may be caused by changes in peanut protein structure (Chung *et al.*, 1994), changes in protein conformation (Semisotnov *et al.*, 1991; Albani, 1994), or progressive actions by enzymes (Iametti *et al.*, 1991; Peri *et al.*, 1990). Enzymes such as chymotrypsins, on the contrary, lead to a loss of hydrophobic sites and subsequent decrease of PSH (Mahmoud *et al.*, 1992; Ploug *et al.*, 1994). Another possible factor contributing to the increase of PSH is water reduction or stress. Reduction of moisture content in peanuts is known to occur during maturation and, therefore, may potentially aid in the development of a hydrophobic environment in which hydrophobic sites on proteins are exposed. Under the condition of water stress, the activity of alcohol dehydrogenase (ADH) has been shown to increase (Chung *et al.*, in press), and its increase coincides with the increase of PSH. For this reason, ADH is probably related to PSH since ADH also binds ANS ((Brand *et al.*, 1967). Sugars are also a potential factor because sugars such as glucose and sucrose are known to enhance PSH at high concentrations (Fujii *et al.*, 1992). In this study, sugars were inversely related to PSH because their total as well as individual contents have been shown to decrease during peanut maturation and curing (Vercellotti *et al.*, 1995). Although Vercellotti *et al.* (1995) indicated some decrease in concentration of individual sugars, they probably have a role in hydrophobicity because the metabolism of sugars (i.e., glycolysis) potentially leads to the production of energy/heat and ADH (Tajima and LaRue, 1982; Russell *et al.*, 1990), both of which are related to PSH (see below).

Effect of Curing on PSH

Figure 2 shows the changes in PSH between the mature and immature peanuts during curing. Both groups of peanuts exhibited an increase in PSH at the curing stages indicated except for day 1. The increase was more pronounced in immature peanuts than in

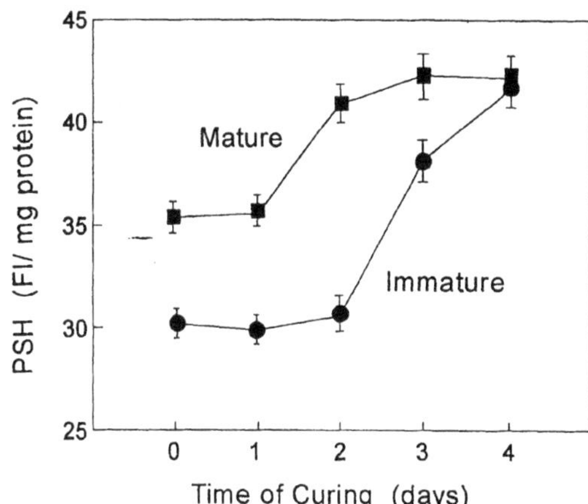

Figure 2. Effect of curing (windrow drying) on PSH from mature and immature peanuts. Mature and immature refer to seeds from "black" and "yellow" peanut pods.

their mature counterparts; as a result, PSH in immature peanuts was almost the same as that in mature ones at the final curing stage (i.e., Day 4). The marked increase in PSH indicates continued exposure of additional hydrophobic sites on proteins during curing. As curing is a drying process, the increase of PSH may be attributed to the reduction of moisture content (or water stress) during curing. However, in one study (Iametti *et al.*, 1991), dehydration was found to lead to a decrease rather than an increase in PSH. Iametti *et al.* (1991) explained that on dehydration proteins rearrange their structures and form a more compact and impermeable network, which consequently prevents permeation of the fluorescent probe into the binding sites of proteins.

Heat might be another possible factor for the increase of PSH during curing. Heat could be generated as a result of temperature increase during windrow/heated-air drying or carbohydrate metabolism (i.e., glycolysis) or curing (Vercellotti *et al.*, 1995). Heat is thought to be responsible because it has been known to cause a change in protein conformation and, consequently, an increase in PSH due to exposure of additional hydrophobic sites on proteins (Bertazzon *et al.*, 1990; Eynard *et al.*, 1992; Sorgentini *et al.*, 1995; Tani *et al.*, 1995). However, when too many hydrophobic sites are exposed (this occurs especially on prolonged heating at a high temperature), hydrophobic interactions between the exposed sites may occur, and subsequently lead to protein aggregation and a decrease in PSH (Bonomi and Iametti, 1991; Eynard *et al.*, 1992; Zheng *et al.*, 1993; Petruccelli and Anon, 1994). In this study, no decrease in PSH during curing was observed (Figure 2), indicating that the curing or heating process was mild and did not lead to the formation of protein aggregrates.

The postulate that heat is responsible for the increase of PSH was demonstrated by incubating the protein extracts overnight at 40°C and 4°C, respectively, and comparing the PSH of the treated proteins. Figure 3 shows a typical profile of PSH of uncured immature and mature peanut proteins incubated at 40°C and 4°C, respectively. As shown in both cases, PSH at 40 °C was greater than that at 4°C. However, when mature and immature peanut proteins were compared with respect to the percent increase of PSH due to heating at 40°C, the latter exhibited a greater increase (approximately 25% increase in immature and 15% increase in mature), suggesting that more hydrophobic sites are buried in the interior of immature peanut proteins than their mature counterparts.

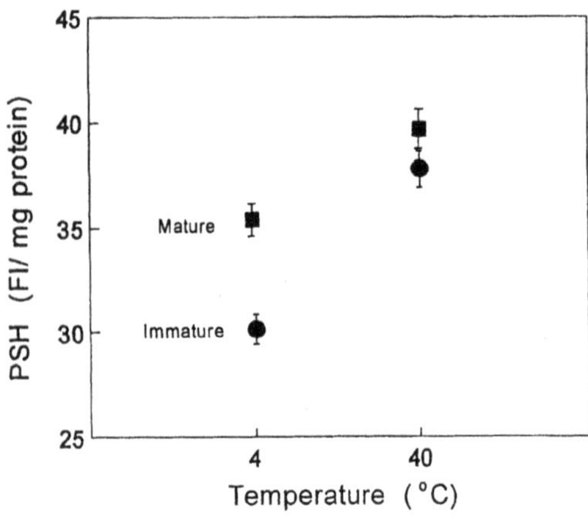

Figure 3. Effect of temperature on PSH from mature and immature peanut proteins. Protein extracts were incubated overnight at the temperature indicated prior to the measurement of PSH.

Other factors that might be associated with the change of PSH include pH (Farahbakhsh *et al.*, 1987; Goto and Fink, 1989; Petruccelli and Anon, 1994), surface charges (Lakkis and Villota, 1992; LeBlanc and LeBlanc, 1992), and sugars (Fujii *et al.*, 1992). A pH change could occur if there is carbohydrate metabolism (glycolysis); the latter leads potentially to the production of lactic acid and release of carbon dioxide from pyruvate (a reaction catalyzed by pyruvic decarboxylase) (Tajima and LaRue, 1982; Laszlo and Lawrence, 1983; Mucke *et al.*, 1995), which in turn forms carbonic acid. In this case, induction of PSH by acidification or change of pH is possible because carbohydrate metabolism is known to occur during peanut maturation and curing (Vercellotti *et al.*, 1995).

Effect of Protein Modification on PSH

Modification of proteins by methylation (Dufour *et al.*, 1992) or succinylation (Damodaran and Kinsella, 1981; Paulson and Tung, 1987; Lakkis and Villota, 1992) is known to increase or decrease PSH due to the changes in net charges caused by modification. In this study, peanut proteins were modified with an arginine-specific modifying agent, phenylglyoxal. Modification of the arginine residue in proteins is of interest because arginine has been found in many hydrophobic peptides (Otagiri *et al.*, 1985) and shown to be an active and ANS-binding site of alcohol dehydrogenase (ADH) (Lange *et al.*, 1974; Brand *et al.*, 1967). ADH could be related to PSH because like PSH, ADH increases during peanut maturation and curing (Chung *et al.*, in press). Thus, the increase of PSH observed in this study could be due to a high level of ADH and its affinity for ANS (Brand *et al.*, 1967). Whether or not there is a relationship between ADH and PSH was determined in this modification study. As shown in Figure 4, modification with phenylglyoxal led to a decrease in PSH in both mature and immature peanuts. The latter showed a greater decrease in PSH (a decrease by approximately 12% as compared to 3% in mature peanuts). The decrease in PSH could be a response to the change of net charges of proteins due to modification of arginine by phenylglyoxal. Change of net charges has been shown to lead to polymerization or formation of soluble aggregates, and consequently, a decrease in hydrophobicity (Lakkis and Villota, 1992). This finding suggests that arginine might have a role in the surface hydrophobicity of peanut proteins. Although arginine is

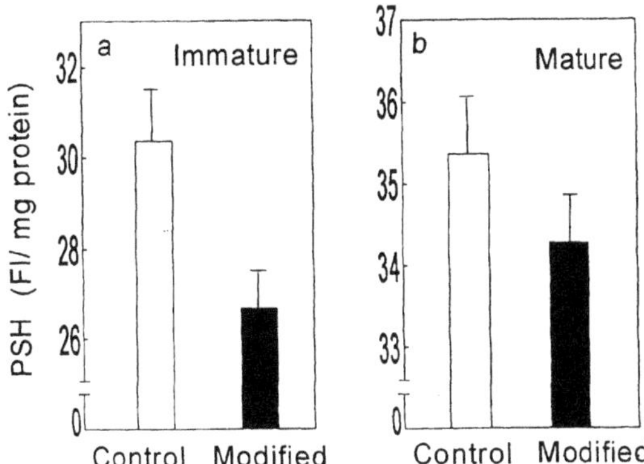

Figure 4. Effect of modification with phenylglyoxal on PSH.

an active site of ADH (Lange *et al.*, 1974) which increases during peanut maturation/curing (Chung *et al.*, in press) and binds ANS (Brand *et al.*, 1967), it appears that ADH is not the sole factor contributing to the increase of PSH during maturation and curing. If it were ADH, this experiment (i.e., modification) should have shown a greater decrease in PSH in mature peanuts due to the availability (for modification) of more arginine residues in ADH which exists in high levels in mature peanuts (Chung *et al.*, in press). This study indicates that ADH may not be totally responsible for the increase of PSH.

CONCLUSIONS

PSH as measured by the fluorescence probe ANS was found to increase during peanut maturation and curing. The marked increase in PSH during curing or heating of immature peanuts suggests that there are more hydrophobic sites hidden in the interior of immature peanut proteins than in their mature counterparts. The increase of PSH is probably due to the temperature increase/heat or water reduction (a stress) which causes the exposure of additional hydrophobic sites on proteins. Modification of peanut proteins with phenylglyoxal decreases PSH, suggesting that arginine might play a role in hydrophobicity. However, the finding implies that alcohol dehydrogenase (ADH), the active site of which is arginine and binds to ANS, may not be totally responsible for the increase of PSH as observed in this study. Correlation between PSH and foaming/emulsifying properties has been reported (Matsudomi *et al.*, 1985; Petruccelli and Anon, 1994) and this study indicated the existence of a relationship between PSH and peanut maturity. Whether or not PSH is related to the flavor characteristics of mature and immature peanuts is still unclear. However, hydrophobicity has been related to the bitterness of peptides (Lau *et al.*, 1991).

ACKNOWLEDGMENTS

The authors thank Maurice Brett and Lisa L. Bothman for assisting in the preparation of peanut samples and fluorescence assays.

REFERENCES

Arai, S.; Noguchi, M.; Yamashita, M.; Kato, H.; Fujimaki, M. Applying proteolytic enzymes on soybean. Part VII. *Agric. Biol. Chem.* **1970**, *34*, 1338–1345.

Akita, A.M.; Nakai, S. Lipophylization of b-lactoglobulin. Effect on hydrophobicity, conformation and surface functional properties. *J. Food Sci.* **1980**, *55*, 711.

Albani, J.R. Effects of sialic acids and the drug adrenergic blocker, propranolol, on the dynamics of human acid glycoprotein: a fluorescence study. *J. Biochem.* **1994**, *116*, 625–630.

Bertazzon, A.; Tian, G.H.; Lamblin, A.; Tsong, T.Y. Enthalpic and entropic contributions to actin stability: Calorimetry, circular dichroism, and fluorescence study and effects of calcium. *Biochem.* **1990**, *29*, 291–298.

Beyeler, M.; Solms, J. Interaction of flavor model compounds with soy protein and bovine serum albumin. *Lebensm-Wiss. u. Technol.* **1974**, *7*, 217–219.

Bonomi, F.; Iametti, S. Real-time monitoring of the surface hydrophobicity changes associated with isothermal treatment of milk and milk protein fractions. *Milchwissenschaft* **1991**, *46*, 71–74.

Brand, L.; Gohlke, J.R.; Rao, D.S. Evidence for binding of rose bengal and anilino-naphthalenesulfonates at the active site regions of liver alcohol dehydrogenase. *Biochem.* **1967**, *6*, 3510–3518.

Cardamone, M.; Puri, N.K. Spectrofluorimetric assessment of the surface hydrophobicity of proteins. Biochem. J. **1992**, *282*, 589–593.

Chung, S.Y., Vercellotti, J.R., Sanders, T.H. Rapid test for alcohol dehydrogenase during peanut maturation and curing. In *Chemical Markers for the Quality of Processed and Stored Foods*, ACS Symposium Series 631, American Chemical Society, Washington, D.C., pp 197–188.

Chung, S.Y.; Ullah, A.H.; Sanders, T.H. Peptide mapping of peanut proteins: Identification of peptides as potential indicators of peanut maturity. *J. Agric. Food Chem.* **1994**, *42*, 623–628.

Chung, S.Y.; Vercellotti, J.R.; Sanders, T.H. An enzyme-amplified microtiter plate assay for ethanol: Application to the detection of apparent ethanol in peanuts. *J. Agric. Food Chem.* **1995**, *43*, 1545–1548.

Damodaran, S.; Kinsella, J.E. Interaction of carbonyls with soy protein: Conformational effects. *J. Argic. Food Chem.* 1981, **29,** 1253–1257.

Dufour, E.; Roger, P.; Haertle, T. Binding of benzopyrene, ellipticine, and cis-parinaric acid to β-lactoglobulin: Influence of protein modifications. **1992.**

Eynard, L.; Iametti, S.; Relkin, P.; Bonomi, F. Surface hydrophobicity changes and heat-induced modifications of lactalbumin. *J. Agric. Food Chem. 40*, **1992,** 1731–1736.

Farahbakhsh, Z.; Baldwin, R. L.; Winsnieski, B.J. Effect of low pH on the conformation of Pseudomonas Exotoxin A. *J. Biol. Chem.* **1987,** *262*, 2256–2261.

Fujimaki, M.; Kato, H.; Arai, S.; Tamaki, E. Applying proteolytic enzymes on soybean. Part 1. *Food Technol.* **1968,** *22*, 889–893.

Fujii, N.; Hamano, M.; Hashimoto, H.; Ono, F. Solubilization of lipohilic compound in highly concentrated saccharide solutions containing protein. *Biosci. Biotech. Biochem.* **1992,** *56*, 118–121.

Goto, Y.; Fink, A. Conformational states of lactamase: Molten-globule states at acidic and alkaline pH with high salt. *Biochem.* **1989,** *28*, 945–952.

Iametti, S.; Negri, E.; Bonomi, F.; Giangiacomo, R. A spectrofluorimetric approach to the stimation of changes in protein surface hydrophobicity during cheese ripening. *Neth. Milk Dairy J.* **1991**, *45*, 183–191.

Lakkis, J.; Villota, R. Effect of acylation on substructural properties of proteins: A study using fluorescence and circular dichroism. *J. Agric. Food Chem.* **1992,** *40*, 553–560.

Lange, L.G.; Riordan, J.F.; Vallee, B.L. Functional arginyl residues as NADH binding sites of alcohol dehydrogenases. *Biochem.* **1974,** *13*, 4361–4370.

Laszlo, A.; Lawrence, P. S. Parallel induction and synthesis of PDC and ADH in anoxic maize roots. *Mol. Gen. Genet.* **1983,** *192*, 110–117.

Lau, K.Y.; Barvano, D.M; Rasmussen, R.R. Influence of Pasteurization of milk on protein breakdown in chesddar cheese during aging. *J. Dairy Sci.* **1991**, *74*, 727–740.

LeBlanc E.; LeBlanc, R. Determination of hydrophobicity and reactive groups in proteins of cod (*Gadus morhua*) muscle during frozen storage. *Food Chem.* **1992**, *43*, 3–11.

Mahmoud, M.I.; Malone, W.T.; Cordle, C.T. Enzymatic hydrolysis of casein: Effect of degree of hydrolysis on antigenicity and physical properties. *J. Food Sci.* **1992**, *57*, 1223–1228.

Matsudomi, N., Mori, H., Kato, A.,and Kobayashi, K. Emulsifying and foaming properties of heat-denatured soybean 11S globulins in relation to their surface hydrophobicity. *Agric. Biol. Chem.* **1985**, *49*, 915–919.

Mucke,U.; Konig, S.; Hubner, G. Purification and characterization of pyruvate decarboxylase from pea seeds (*Pisum sativum* cv. Miko). *Biol. Chem. Hoppe-Seyler* **1995**, *376*, 111–117.

Mukherjee J.J.; Dekker, E.E. Inactivation of E. Coli 2-amino-3-ketobutyrate CoA ligase by phenylglyoxal and identification of an active-site arginine peptide. *Arch. Biochem. Biophys.* **1992**, *299*, 147–153.

Noguchi, M.; Arai, S.; Fujimaki, M. Applying proteolytic enzymes on soybean. Part 2. *J. Food Sci.* **1970**, *35*, 211–214.

Ohnishi, M., Sugawara, R., and Kusano, T. Structure-activity relationship between the hydrophobicity of alkali metal salts of warfarin [3-(acetonyl-benzyl)-4-hydroxycoumarin] and the effectiveness of the taste response to these salts in mice. *Biosci. Biotech. Biochem.* **1995**, *59*, 995–1001.

O'Keefe, S.F.; Wilson, L.A.; Resurreccion, A.P.; Murphy, P. Determination of the binding of hexanal to soy glycinin and conglycinin in an aqueous model system using a headspace technique. *J. Agric. Food Chem.* **1991**, *39*, 1022–1028.

Otagiri, K.; Nosho, Y.; Shinoda, I.; Fukui, H.; Okai, H. Studies on a model of bitter peptides including arginine, proline and phenylalanine residues. *Agric. Biol. Chem.* **1985**, *49*, 1019–1026.

Paulson, A.T.; Tung, M.A. Solubility, hydrophobicity and net charge of succinylated canola protein isolate. *J. Food Sci.* **1987**, *52*, 1557–1561.

Peri, C.; Pagliarini, E.; Iametti, S.; Bonomi, F. A study of surface hydrophobicity of milk proteins during enzymic coagulation and curd hardening. *J. Dairy Res.* **1990**, *57*, 101–108.

Petruccelli, S.; Anon, M.C. Relationship between the method of obtention and the structural and functional properties of soy protein isolates. 2. Surface properties. *J. Agric. Food Chem.* **1994**, *42*, 2170–2176.

Ploug, M.; Ellis, V.; Dano, K. Ligand interaction between urokinase-type plasminogen activator and its receptor probed with 8-anilino-1-naphthalenesulfonate,. Evidence for a hydrophobic binding site exposed only on the intact receptor. *Biochem.* **1994**, *33*, 8991–8997.

Richieri, G.V.; Anel, A.; Kleinfeld, A.M. Interactions of long-chain fatty acids and albumin: Determination of free fatty acid levels using the fluorescent probe of ADIFAB. *Biochem.* **1993**, *32*, 7574–7580.

Russell, D.A.; Wong, D.M.L.; Sachs, M.M. The anaerobic response of soybean. *Plant Physiol.* **1990**, *92*, 401–407.

Semisotnov, G.V.; Rodionova, N.A.; Razgulyaev, O.I.; Uversky, V.N.; Gripas, A.F.; Gilmanshin, R.I. Study of the Molten globule intermediate state in protein folding by a hydrophobic fluorescence probe. Chemtracts-Biochem. *Molecular. Biology* **1991**, *2*, 393–397.

Sorgentini, D.A.; Wagner, J.R.; Anon, M.C. Effects of thermal treatment of soy protein isolate on the characteristics and structure-function relationship of soluble and insoluble fractions. *J. Agric. Food Chem.* **1995**, *43*, 2471–2479.

Tajima, S.; LaRue, T.A. Enzymes for acetaldehyde and ethanol formation in legume nodules. *Plant_Physiol.* **1982**, *70*, 388–392.

Tani, F.; Murata, M.; Higasa, T.; Goto, M.; Kitabatake, N.; Doi, E. Molten globule state of protein molecules in heat-induced transparent food gels. J. Agric. Food Chem. **1995**, *43*, 2325–2331.

Tsutsui, T.; Li-Chan, E.; Nakai, S. A simple fluorometric method for fat-binding capacity as an index of hydrophobicity of proteins. *J. Food Sci.* **1986**, *51*, 1268–1272.

Vercellotti, J.R.; Sanders, T.H.; Chung, S.Y.; Bett; K.L.; Vineyard, B.T. Carbohydrate metabolism in peanuts during postharvest curing and maturation. In *Food Flavors: Generation, Analysis and Process Influence*. G. Charalambous (Ed.). Elsevier Science Publishers, Amsterdam, The Netherlands, 1995; pp. 1547–1578.

Williams, E.J.; Drexler, J.S. A non-destructive method for determining peanut pod maturity. *Peanut Sci.* **1981**, *8*, 134.

Zheng, B.A.; Matsumura, Y.; Mori, T. Conformational changes and surface properties of legumin from broad beans in relation to its thermal aggregation. *Biosci. Biotech. Biochem.* **1993**, *57*, 1366–1368.

5

HIGH PRESSURE PROCESSING EFFECTS ON FISH PROTEINS

T. C. Lanier

Food Science Department
North Carolina State University
Raleigh North Carolina 27695-7624

Salted pastes of surimi, a myofibrillar concentrate of fish muscle, gel at pressures near 300 MPa. High pressure processing has been thought to induce denaturation and gelation of myofibrillar proteins mainly by disruption of protein intramolecular hydrophobic interactions which subsequently reform intennolecularly. We have shown that pressure-induced surimi gels evidence disulfide bonding as well. Endogenous transglutaminase (TGase) evidently survives the pressure treatment, and subsequent TGase-mediated setting of Alaska pollock surimi pastes at 25°C results in very strong gels as compared to those prepared without prior pressurization. High pressure during freezing or thawing greatly accelerates these operations and can reduce ice crystal size and associated tissue damage. Yet pressure treatment can destabilize proteins which might lower fish quality. Infusion of certain carbohydrates into muscle prior to pressure-assisted freezing/thawing can achieve both baroprotection and cryoprotection of the muscle proteins. Pressure treatment has not proven useful for inactivation of proteolytic enzymes that degrade fish quality.

INTRODUCTION

Muscle myosin is more labile to denaturing forces than most plant structural proteins, and fish muscle is even more so compared to meat from homeotherms (Connell, 1961; Hashimoto *et al.*, 1982). Fish proteins thus possess unique gelling properties (Lanier, 1986), and are more subject to changes during freezing, frozen storage, and thawing than other meats (Shenouda, 1980). High pressure affects water structure and changes the equilibrium of forces that stabilize proteins (Cheftel, 1992). Aside from applications in pasteurization of seafoods, pressure may be a useful adjunct to heat-induced gelation or

Process-Induced Chemical Changes in Food
edited by Shahidi *et al.* Plenum Press, New York, 1998

freezing/thawing of fish muscle proteins. Pressure effects on muscle proteases that affect fish meat quality and functionality have also been studied.

PRESSURE-INDUCED GELATION OF FISH MUSCLE PROTEIN

Fish muscle proteins are typically heat-gelled in the manufacture of crab analog products. High temperatures (>40–60°C, depending on species) are normally required to denature fish myosin, leading to intermolecular covalent and non-covalent interactions including disulfide bond formation and hydrophobic group interactions that result in gelation (Lee and Lanier, 1995).

Proteins are destabilized and may also be induced to gel at low temperature by pressure treatment ranging from 100–1000 MPa (Ohshima *et al.*, 1993). A typical temperature-pressure denaturation boundary for proteins is shown in Figure 1 (Heremans, 1995). Gels formed by pressure treatment generally possess increased glossiness and deformability, as well as a more natural (uncooked) flavor as compared to heat-induced gels (Okamoto *et al.*, 1990). Fish protein gels can be formed at ambient or lower temperatures by pressures ranging from 200 to 500 MPa (Shoji, 1990).

Heremans and Heremans (1989) proposed that pressure-induced protein denaturation proceeded via a cascade effect. Initially, hydrophobic interactions which stabilize the native structure of the protein are disrupted under the influence of pressure. This causes an opening of the protein structure, allowing hydrophobic groups to be exposed to the aqueous environment. Changes result from pressure-induced destabilization of the protein and produce further volume decreases due to electrostriction around charged groups, water structuring around exposed apolar groups, and salvation of polar groups through hydrogen bonding. Upon release of pressure, the hydrophobic groups interact once again to minimize exposure to the aqueous environment. Along with other protein-protein interactions such as possibly disulfide bonds formed under pressure and hydrogen bonds that form upon release of pressure, the intermolecular hydrophobic interactions result in the formation of a gel structure when protein concentration is sufficient. More recent information suggests that it is mainly the myosin heads that aggregate by hydrophobic interactions, as

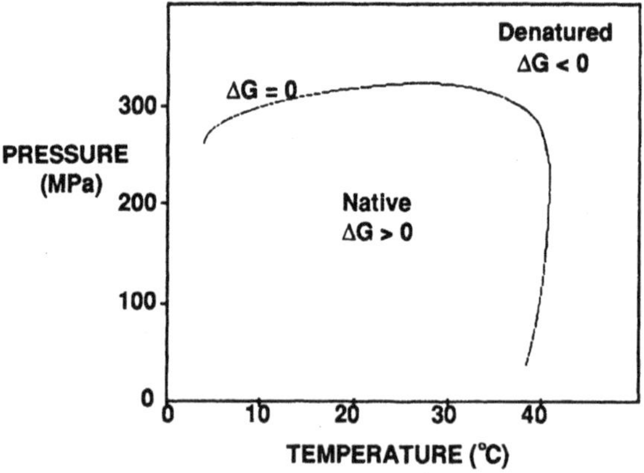

Figure 1. Typical pressure-temperature denaturation boundary for proteins. Adapted from Hermans, 1995.

pressure had little effect upon the a helical tail portion (Yamoto, 1995). Changes in actin induced by pressure may also alter the gel-forming properties of fish muscle (Ikeuchi *et al.*, 1995)

Even at atmospheric pressure muscle pastes from most species of fish can gel at low temperatures (0–40 °C), but the time required for gelation is greater than by heat or pressure (Kamath *et al.*, 1992). Endogenous transglutaminase (TGase) is thought to induce this low temperature gelation (termed "setting"), which also imparts added strength to the gel upon subsequent cooking at higher temperatures. TGase forms intermolecular covalent F--(,y-glutamyl) lysine bonds between myosin heavy chains (MHC), the polymerization of which can be measured by SDS-PAGE (Joseph *et al.*, 1994). The endogenous TGase requires Ca 2+ to be active, and thus can be inhibited with EDTA (Kumazawa *et al.*, 1993). Since pressure causes denaturation and gelation of proteins at low temperatures, it may also affect the activity of TGase (Low and Somero, 1975).

Gilleland and Lanier (1997) subjected salted pastes of Alaska pollock *(Theragra chalcogramma)* surimi (refined myofibrillar protein containing 4% sucrose, 4% sorbitol and 0.3% sodium tripolyphosphate) to cooking (90 °C for 30 min.), setting (25 °C for 2 hr), pressure treatment (300 MPa isostatic pressure for 30 min at 5 °C), or a combination of these treatments, setting or cooking always being carried out at atmospheric pressure. Stress (strength) and strain (deformability) of gels were determined at the tensile failure point, measured at 25 °C. It is apparent from Figure 2 that the various methods of processing the paste into a gel had dramatic effects on tensile strength, but little effect on tensile deformability. The setting treatment strengthened the gel, particularly when preceded by a pressure treatment.

This enhancement of the setting effect by a prior pressure treatment raises the question of whether the endogenous TGase presumed to be responsible for gelation during setting is also active during the pressure treatment, and survives pressure-induced denaturation. Since the enzyme is known to be calcium activated, we can inhibit its action by addition of EDTA (Figure 3, setting only treatment) (Low and Somero, 1975). EDTA addition had no effect on the pressure-only treatment; thus it is logical to conclude that TGase is not involved in pressure-induced gelation.

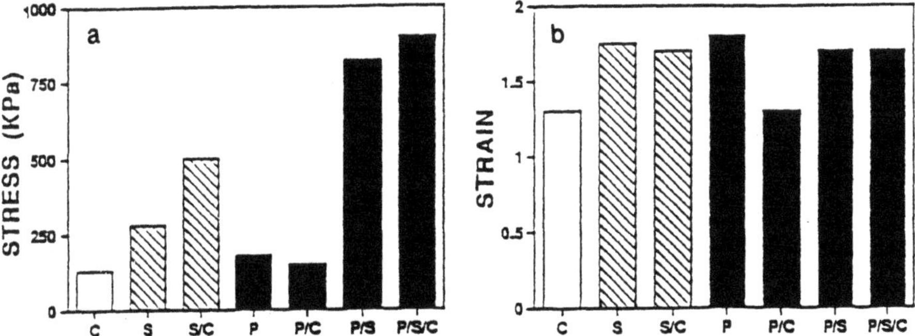

Figure 2. Effect of gelling treatment on tensile stress and strain at failure of surimi gels (Shoji *et al.*, 1990). Treatments: P = 30 Mpa isostatic pressure, 30 min, 5°C; S = 25°C, 2 hr, atmospheric pressure; C = 90°C 30 min, atmospheric pressure. From (Gilleland and Lanier, 1997).

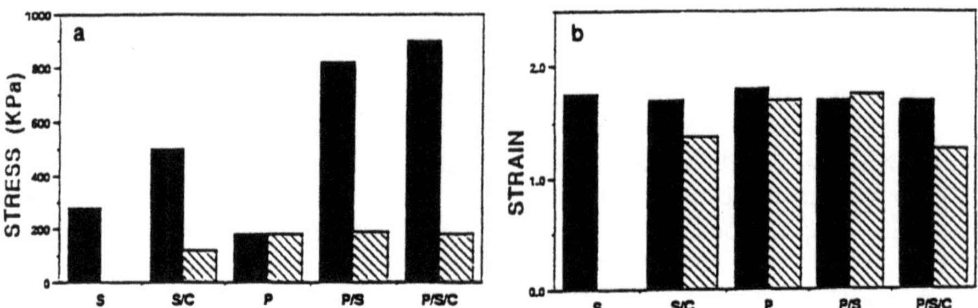

Figure 3. Effect of EDTA and gelling treatment on tensile stress and strain at failure of surimi gels. Treatments are same as in Fig. 2; solid: no EDTA, crosshatched: with EDTA. From (Gilleland and Lanier, 1997).

Shoji *et al.* (Shoji *et al.* 1994) gave data that would indicate TGase may be inactivated by such a pressure treatment (Figure 4). Yet the dramatic effect of EDTA on gel stress in the pressure+setting treatments (Figure 3) indicates that TGase must in fact survive a pressure treatment of this magnitude.

The data of Figure 2 and 3 were corroborated by SDS-polyacrylamide gel electrophoresis of these gels, which indicated non-disulfide polymerization of myosin heavy chain as a result of including a setting step as part of the treatment. Loss of myosin heavy chain monomer (due to polymerization) was approximately the same for setting with or without a prior pressure treatment. (Gilleland and Lanier, 1997).

The activity of the endogenous TGase is dependent upon denaturation of the substrate myosin to expose available binding sites (Joseph *et al.*, 1994). Thus it is reasonable to believe that the greater strength of gels prepared by the pressure+setting+cook treatment as compared to those prepared by setting + cooking (Figure 2) might be attributed to a more available substrate, which facilitates more active crosslinking of myosin, in the

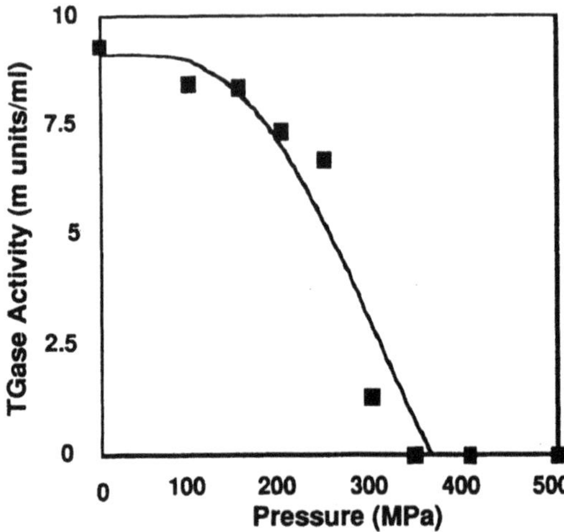

Figure 4. Effect of pressure on activity of TGase endogenous to surimi. From (Shoji *et al.*, 1990).

former treatment. However, we did not quantitate numbers of P--(y-glutamyl) lysine bonds directly to verify this hypothesis, and it should be noted that we have previously seen instances where the rate of bond formation, rather than total bonds formed, seemed to correlate better with ultimate gel strength (Lee *et al.*, 1997).

The pressure treatment, having induced gelation of the fish proteins, certainly introduced intermolecular bondings, most likely hydrophobic and possibly even covalent/disulfide (Heremans and Heremans, 1989). Solubility of the gels for which fracture data are presented in Figure 2 were determined in the manner of Buttkus (Buttkus, 1971) in 2% SDS-8M urea, with or without added 2% P-mercaptoethanol (0-Nffi) to disrupt disulfide bondings. These data indicated that all treatments decreased solubility in SDS-urea in the absence of P-ME, but this loss was restored upon inclusion of P-ME for all treatments except those which included a setting step. This remaining insoluble fraction in gels subjected to setting can be attributed to formation of non-disulfide P--(,y-glutamyl) lysine bonds by TGase. No such remaining insolubility occurred in samples subjected to setting which contained added EDTA.

A remarkable loss of solubility in SDS-urea for pressure (only)-treated gels indicated that substantial disulfide bonding likely occurred due to the pressure treatment, in excess of that which resulted from subsequent cooking at 900°C. EDTA addition had little or no effect on solubility of these gels. Berg *et al.* (Berg *et al.*, 1965) had noted a dramatic increase in the numbers of readily reacting SH groups of myosin at pressures of 300 MPa.

This work indicated that gelation of surimi pastes by high pressure does likely involve formation of intermolecular disulfide bonds, though probably the gel structure is mainly stabilized by intermolecular hydrophobic bondings. TGase-mediated covalent crosslinking proceeded unabated after a 300 MPa/30 min treatment when samples were warmed to 25°C, and the effects of this subsequent setting treatment on gel strength were quite synergistic with the prior pressure treatment. Conformational changes in the proteins must have occurred that, after pressure release, increased the effectiveness of TGase in catalyzing covalent crosslinking of MHC. This opens the possibility of enhancing the gelling effects of added microbial transglutaminase in protein foods (Neilsen, 1995) by a prior pressure treatment, a hypothesis we are presently investigating.

PRESSURE EFFECTS ON FISH PROTEINS DURING FREEZING OR THAWING

The bulk of the world's seafood supply today is handled frozen at some point in the distribution chain. Freezing and frozen storage offer protection for seafoods against bacterial attack and therefore the potential to safely, economically, and efficiently distribute seafoods to consumers in a convenient manner. However, seafoods often suffer substantial losses in quality during frozen storage (Love, 1988). These are normally manifest as water weepage from the thawed muscle, and a concurrent toughening and dryness/fibrousness development in the muscle texture upon cooking. Fish proteins in the minced state are even more reactive in that components of the muscle are no longer compartmentalized as in the intact muscle. In leached minces the myofibrillar proteins have been refined from the water-soluble fraction, which is comprised largely of sarcoplasmic proteins known to have a stabilizing effect on myofibrillar proteins during frozen storage (Jiang *et al.*, 1987a,b; Loomis *et al.*, 1989). To compensate for this loss, cryoprotective additives (commonly sugars or polyalcohols, plus phosphate) are added to stabilize leached minces in surimi manufacture.

Quality loss occurs during freezing, thawing, and upon prolonged frozen storage of foods, particularly when the storage temperature is above -30°C and fluctuates, as commonly occurs during shipping or in many cold-store facilities. During slow freezing, ice forms predominantly in the extracellular solution of seafood muscle. To equilibrate the difference in chemical potential between the intracellular solution and the extracellular solution, water leaves the cell through the cell membrane to freeze in the extracellular space. Consequently the cell shrinks and the intracellular solution becomes hypertonic (Rubinsky *et al.*, 1994). The intracelluar myofibrillar proteins are quite prone to denaturation due to this hypertonic condition (Love, 1962).

During rapid freezing of muscle, or freezing of prerigor seafoods at any freezing rate, ice crystals form predominantly within the muscle cells, their size decreasing with increasing freezing rate. The amount of water which initially freezes depends on the freezing rate and the temperature attained, though this tends to equilibrate upon storage to a level dependent upon the composition of the sarcoplasmic fluid. Typically this amount is near 90% at conventional cold stores temperatures (near -25O°C; Powrie, 1973). Thus there is considerable concentration of muscle salts, and the remaining intracellular fluid is again quite hypertonic.

Slow freezing and the recrystallization of water that occurs during normal temperature fluctuations of frozen storage lead to large ice crystal size. This, along with variation in osmotic pressure across cell membranes, can cause disruption of the sarcolemma and thus textural damage to the seafood (Rubinsky *et al.*, 1994; Martino and Zaritzky, 1988). Though rapid freezing induces ice crystals of smaller size, recrystallization leading to ice growth during normal frozen storage can result in greater damage to the muscle cells than when slowly frozen initially, since the ice crystals are confined by the cell membrane (Powrie, 1973).

Thus, while freezing prevents microbial deterioration of seafoods, damage to the textural quality due to freezing/thawing and frozen storage can be considerable. Structural damage, and protein denaturation induced by the concentration of salts combined with their lesser stability at temperatures below 40C, may inflict considerable quality deterioration (Fennema, 1973). Enzymes and substrates as well as oxygen are also concentrated by the freezing out of water, such that lipid oxidation and enzymically linked deterioration processes (such as formation of formaldehyde by TMAO demethylase in gadoids) can be accelerated (Shenouda, 1980; Sikorski, 1976).

To date the best advice has been to freeze seafoods quickly and hold them at a very low 300°C), constantly maintained temperature, with an adequate moisture barrier applied to surfaces in contact with air to prevent lyophilization. Thawing should be equally fast to prevent ice recrystallization and tissue damage. Yet these conditions are difficult to achieve economically and logistically, and consequently the public image of the quality of frozen seafood has justifiably been a good notch below that of fresh seafood.

Pressure treatment offers an extremely rapid means of freezing and thawing that results in smaller ice crystal formation and lesser tissue damage (Kalichevsky *et al.*, 1995). For example, thaw drip was reduced up to six fold by pressure thawing of tuna back muscle, while thawing time was reduced to only 30–60 min at O°C using 50–150 MPa pressure, versus 15 hours required thawing time under atmospheric pressure at 5°C (Murakami *et al.*, 1992). Rapid freezing of tofu (soybean curd) was demonstrated by first pressurizing to 200 MPa, then rapid cooling to -20°C, followed by release of pressure. The product exhibited much smaller ice crystal formation, and better texture upon thawing, than product frozen under atmospheric pressure (Deuchi and Hayashi, 1992).

These benefits of pressure-assisted freezing and thawing must, however, be balanced with the denaturing effects of pressure on fish proteins discussed in the previous section.

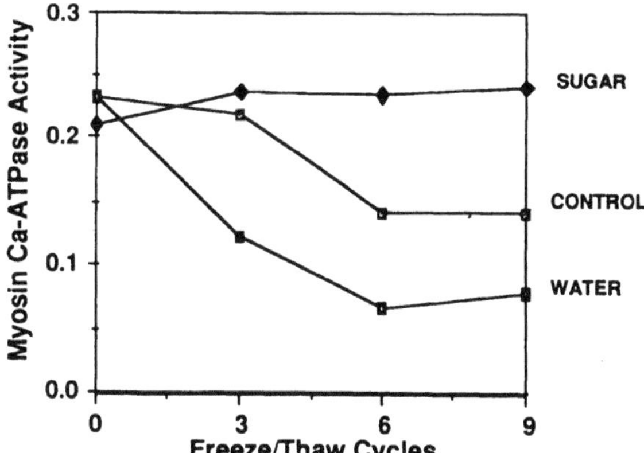

Figure 5. Effect of vacuum infusion of sucrose on stability of intact fish meat during freeze/thaw cycling. From (Carvajal and Lanier, 1997).

Greater acceleration of the freezing process requires pressures in excess of 150 MPa, but reportedly causes discoloration, which is evidence of pressure-induced protein denaturation, in meats at pressures in excess of 50 MPa (Deuchi and Hayashi, 1992). Dumay *et al.* (1994) reported the baroprotective effects of sucrose during pressure treatment of P-lactoglobulin. Since sucrose and other low molecular weight (MW) sugars and polyols are known to also stabilize proteins to freeze-induced denaturation, it is conceivable that they could be used as additives to both baro- and cryo-protect fish proteins during pressure assisted freezing/thawing and during frozen storage.

Carvajal *et al.* (1997) recently demonstrated that vacuum infusion of low NM sugars and polyols into intact seafood muscle dramatically reduced myosin denaturation during subsequent freeze/thaw abuse (Figure 5) as compared to untreated (control) meat or a water-infused sample. Treated muscle samples were also frozen both intact and finely minced after sugar infusion as a means of determining whether more intimate contact of the infused sugar was achieved by prior mincing. The data revealed no difference in stability between minced or intact samples within any treatment, indicating that intimate association of sugar with muscle proteins was achieved by infusion alone. Note the water-infused sample was least stable, indicating that soluble muscle constituents exert a stabilizing effect on myofibrillar proteins.

Subsequent to this work, Ashie and Lanier (Ashie and Lanier, 1996) showed that sucrose and other low MW carbohydrates baroprotect fish myosin during pressure treatment of surimi (Figure 6).

These studies show the potential of infusing low MW carbohydrates into intact fish muscle to achieve both cryoprotection and baroprotection of fish muscle. Cryoprotection of muscle proteins inhibits protein denaturation and aggregation that can manifest as poor water-holding and textural properties of previously frozen fish meat. The baroprotective properties allow pressure-assisted freezing or thawing to be used, which results in more rapid freezing and thawing, thus smaller ice crystals and less tissue damage. Ice crystal damage can likewise contribute to poor water-holding and textural damage to previously-frozen fish.

We have also considered the use of higher MW carbohydrates such as starch hydrolysate products (SHP) to infuse fish meats for purposes of cryo- and baro-stabilization. These offer several advantages such as lower cost, lower sweetness (especially the lower

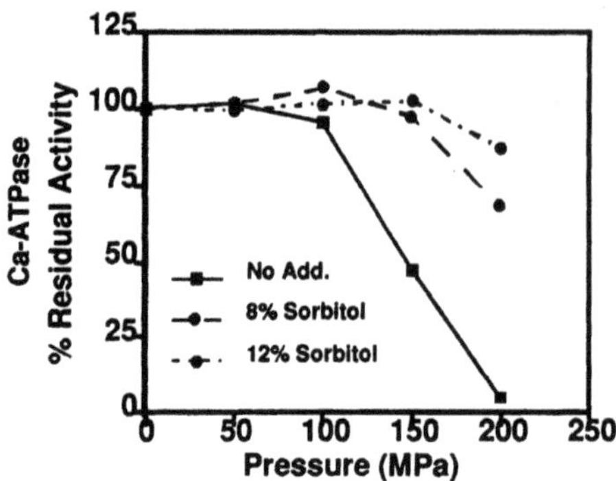

Figure 6. Baroprotective effect of sorbitol on washed Alaska pollock mince. From (Ashie and Lanier, 1996).

D.E., higher MW maltodextrins), and a greater ability to reduce the mobility of water and thereby inhibit many diffusion-dependent deteriorative processes that occur during freezing (Levine and Slade, 1986, 1988; Fennema, 1996, Parker and Ring, 1995). This latter property derives from the ability of higher polymers to greatly increase the viscosity of the freeze concentrated unfrozen solution in a food. Carvajal *et al.* (1997) explored the cryoprotective properties of maltodextrins in Alaska pollock surimi. They found that higher MW maltodextrins were relatively ineffective in protecting proteins during freeze/thaw abuse, but were effective during isothermal storage at temperatures near -20°C. Maltodextrins of lower mean MW were effective at higher frozen storage temperatures and during freeze/thaw abuse, but not to the same extent as sucrose or sorbitol. This result was explained as evidence that the higher MW maltodextrins are effective water immobilizers near the glass transition temperature (Tg') of the system (which is effectively raised by addition of high MW SHP), while low Nff carbohydrates are known to stabilize proteins at any temperature through a solute exclusion mechanism (Carvajal *et al.*, 1997).

Ashie and Lanier (1996) also found that 20 D.E. maltodextrin evidenced baroprotection of washed Alaska pollock mince at pressures of 150 - 200 MPa, but to a lesser extent than sucrose, sorbitol or lacfitol. Besides the lesser effectiveness of SHP as protein stabilizers when compared to low MW carbohydrates, a further potential problem is the difficulty of infusing these large polymers through muscle membranes to effectively immobilize intracellular water or stabilize myofibrillar proteins.

PRESSURE EFFECTS ON FISH MEAT PROTEASES

Proteolytic degradation is helpful in the tenderization of mammalian meats postmortem, and pressurization has been shown to enhance the tenderization process by disrupting lysomal membranes, thereby releasing proteases into the muscle tissue. Proteolysis is not advantageous to fish texture, however, and degradation of myosin can considerably reduce the gel-forming Functionality of mince or surimi. Chung *et al.* (1994) evaluated pressurization of surimi as a means of inhibiting the heat-stable proteolytic activity (attributed to cathepsin L) that degrades its gel forming potential. They found that pressures up to 250 Mpa had no inhibitory effect on this degradative activity. Ashie (1995) determined that pressure up to 300 MPa inhibited isolated fish cathepsin C and chymotrypsin-like activities up to 80%, yet had little effect on total proteolytic activity at 25°C of fish tissue homogenates. Similar pressure treatments had little effect on the bovine-derived chymotrypsin or cathepsin C. Homma *et al.* (1994) also reported that bovine cathepsins B, D and L were stable up to 400 MPa. Both Ohmori *et al.* (1991) and Homma *et al.* (1994) measured an increase in proteolytic activity in beef muscle up to 300 Mpa which they attributed to lysomal disruption.

Thus it is apparent that high pressure treatment is not a feasible means of reducing proteolytic activity in **fish** muscle.

OUTLOOK FOR USE OF HIGH PRESSURE IN PROCESSING OF FISH MUSCLE

The high cost of pressure-treatment equipment, concerns over safety of operation, and the relatively low volume of through-put have hindered use of high hydrostatic pressure as a unit operation in food processing thus far. However, since 1989 there has been concentrated research activity in Japan to develop food applications of pressure (Farr, 1990). With the expansion of this work in Europe and America, several equipment companies now exhibit pressure-treatment equipment at national and international food processing shows. Both Demazeau (1992) and Deplace and Mertens (1992) have reviewed the development of equipment for high-pressure processing of foods, and express optimism for its widespread adoption by the food industry. Ohshima *et al.* (1993), in discussing the future of pressure processing of fishery products, note that equipment costs can be decreased, and throughput increased, dramatically by lowering the pressure requirements of the equipment. For promising seafood applications of high pressure processing, such as pressure induced gelation and pressure-assisted freezing/thawing, the pressure requirements are likely to be much less than are required for pasteurization, making pressure treatment more commercially feasible.

ACKNOWLEDGMENT

The author gratefully acknowledges support from the National Fisheries Institute, Arlington, VA, for a portion of the work reported herein.

REFERENCES

Ashie, I.N.A. Application of high hydrostatic pressure and oc2-macroglobulin to control postharvest seafood texture deterioration. Ph.D. thesis, McGill Univ. , Montreal Canada, 1995, 239 pp.

Ashie, I.; Lanier, T.C. Baroprotection of fish proteins by sugars and polyols. *Abst. Intl. Conf. Food Sci. Technol., Fine Particle Soc., Chicago, IL,* 1996.

Berg Y.N.; Lebedeva, N.A.; Markina, E.A.; Ivanov, I.I. The effect of high pressure on some properties of myosin. *Biokhimiya* **1965,** *30,* 277–281.

Buttkus, H. The sulfhydryl content of rabbit and trout myosins in relation to protein stability. *Can. J. Biochem.* **1971,** *49,* 97–107.

Carvajal, P.A.; Lanier, T.C. Cryoprotection of intact meats by carbohydrate infusion. *J. Food Sci.* **1997** (in press).

Carvajal, P.A.; MacDonald, G.A.; Lanier, T.C. Cryostabilization of muscle proteins. *Cryobiology* **1997** (in press)

Cheftel, J.-C. Effects of high hydrostatic pressure on food constituents: an overview. In *High Pressure and Biotechnology;* Balny, C.; Hayashi, R.; Heremans, K.; Masson, P., Eds.; Colloque INSERM, John Libbey Eurotext Ltd., 1992; Vol. 224, pp 195–209.

Chung, Y.C.; Gebrehiwot, A.; Farkas, D.F.; Morrissey, M.T. Gelation of surimi by high hydrostatic pressure. *J. Food Sci.* **1994,** *59,* 523–524&543.

Connell, J. J. The relative stabilities of the skeletal muscle myosins of some animals. *Biochem. J.* **1961,** *80,* 503–509.

Demazeau, G. The demystification of the pressure parameter for industrial applications. In *High Pressure and Biotechnology;* Balny, C.; Hayashi, R.; Heremans, K.; Masson, P., Eds.; Colloque INSERM, John Libbey Eurotext Ltd., 1992; Vol. 224, pp 481–491.

Deplace, G.- Mertens, B. The commercial application of high pressure technology in the food processing industry. In *High Pressure and Biotechnology;* Balny, C.; Hayashi, R.; Heremans, K.; Masson, P., Eds.; Colloque INSERM, John Libbey Eurotext Ltd., 1992, Vol. 224, pp 469–479.

Deuchi, T.; Hayashi, R. High pressure treatments at subzero temperature: application to preservation, rapid freezing and rapid thawing of foods. In *Hgh Pressure and* Biotechnology; Balny, C.; Hayashi, R.; Heremans, K.; Masson, P., Eds.; Colloque INSERM, John Libbey Eurotext Ltd., 1992; Vol. 224, pp 353–355.

Dumay, E.M.; Kalichevsky, M.T.; Cheftel, J.C. High-pressure unfolding and aggregation of lactoglobulin and the baroprotective effects of sucrose. *J. Agric. Food Chem.* **1994,** *42,* 1861-1868.

Farr, D. High pressure technology in the food industry. *Trends in Food Sci. Technol.* 1990, *1,* 14–16.

Fennema, O.R. Water and ice. In *Low-Temperature Preservation of Foods and Living Matter;* Fennema, O.R.; Powrie, W.D.; Marth, E.H., Eds.; Marcel Dekker, Inc., N.Y., 1973; pp 3–78.

Fennema, O.R. Water and ice. In *Food Chemistry, 3rd edition;* Fennema, O.R., Ed.; Marcel Dekker, Inc., N.Y., 1996; pp 17–94.

Gilleland, M.G.; Lanier, T.C. Investigation into the mechanism of gelation of surimi pastes treated by high isostatic pressures. *Food Sci.* **1997,** *4,* 1–5

Hashimoto, A.; Kobayashi, A; Arai, K. Thermostability of fish myofibrillar Ca-ATPase and adaption to environmental temperature. *Bull. Japan. Soc. Sci. Fish.* **1982,** *48,* 671–684.

Heremans, L.; Heremans, K. Raman spectroscopic study of the changes in secondary structure of chymotrypsin: Effect of pH and pressure on the salt bridge. *Biochim. Biophys. Acta.* **1989,** *999,* 192–197.

Heremans, K. Pressure behavior of proteins: infrared studies in the diamond anvil cell. *Abst. Int. Conf. High Press. Biosci. Biotechnol., Kyoto, Japan,* 1995.

Homma, N.; Ikeuchi, Y.; Suzuki, A. Effects of high pressure treatment on the proteolytic enzymes in meat. *Meat Sci.* **1994,** *38,* 219–228.

Ikeuchi, Y.; Tanji, H.; Kim, K.; Takeda, N.; Kakimoto, T.; Suzuki, A. Dynamic rheological behavior and biochemical properties of pressurized actomyosin. *Abst. Int. Conf High Press. Biosci. Biotechnol. Kyoto, Japan,* 1995.

Jiang, S.T.; Hwang, B.O.; Tsao, CT. Protein denaturation and changes in nucleotides of fish muscle during frozen storage. *J. Agric. Food Chem.* **1987a,** *35,* 22–27.

Jiang, S.T.; Hwang, B.O.; Tsao, C.T. Effect of adenosine-nucleotides and their derivatives on denaturation of myofibrillar proteins *in vitro* during frozen storage at -20°C. J. *Food Sci.* **1987b,** *52,* 117–123.

Joseph, D.; Lanier, T.; Hamann, D. Temperature and pH affect transglutaminase-catalyzed setting of crude fish actomyosin. *J. Food Sci.* **1994,** *59(5),* 1019–1027.

Kalichevsky, M.T.; Knorr, D.; Lilliford, P.J. Potential food applications of high-pressure effects on ice-water transitions. *Trends in Food Sci. & Technol.* **1995,** *6,* 253–259.

Kamath, G.; Lanier, T.; Foegedin, E.; Hamann, D. Nondisulfide covalent cross-linking of myosin heavy chain in "setting" of Alaska pollock and Atlantic croaker surimi. *J. Food Biochem.* **1992,** *16,* 151–172.

Kumazawa, Y.; Numazawa, T.; Motoki, M.; Takamura, M. Participation of transglutaminase in the manufacturing process of "kamaboko". *Abstr. Ann. Meet., Inst. Food Technol., Chicago, IL,* 1993.

Lanier, T. C. Functional properties of surimi. *Food Technol.* **1986,** *40,* 107–124.

Lee, H. Lanier, T. The role of covalent crosslinking in the texturizing of muscle protein sols. *J. Muscle Foods* **1995, 6,**125–138.

Lee, H.G.; Lanier, T.C.; Hamann, D.D.; Knopp, J.A. Investigation of the role of transglutaminase in the low temperature gelation of fish protein sols. *J. Food Sci.* **1997,** *62,* 20–24.

Levine, H.; Slade, L. A polymer physico-chemical approach to the study of commercial starch hydrolysis products (SEPs). *Carbohydr. Polym.* **1986,** *6,* 213.

Levine, H.; Slade, L. A food polymer science approach to the practice of cryostabilization technology. *Comments Agric. Food Chem.* **1988,** *1,* 315–395.

Loomis, S.H.; Carpenter, T.J.; Anchordoguy, T.J.; Crowe, J.H.- Branchini, B.R. Cryoprotective capacity of end products of anaerobic metabolism. *J. Exp. Zoo.* **1989,** *252,* 9- .

Love, R. M. New factors involved in the denaturation of frozen cod muscle protein. *J. Food Sci.* **1962,** *27,* 544–550.

Love, R. M. The Food Fishes: Their Intrinsic Variation and Practical Implications. Van Nostrand Reinhold, New York, 1988; pp. 130–140.

Low, P.; Somero, G. Pressure effects on enzyme structure and function in vitro and under simulated in vivo conditions. *Comp. Biochem. Physiol.* **1975,** *52B,* 67–74.

Martino, M.N.; Zaritzky, N.E. Ice crystal size modifications during frozen beef storage. *J. Food Sci.* **1988,** *53,* 1631–1637.

Murakami, T.; Kimura, I.; Yamagishi, T.; Yamashita, M.; Sugimoto, M.; Satake, M. Thawing of frozen **fish** by hydrostatic pressure. In *High Pressure and Biotechnology;* Balny, C.; Hayashi, R.; Heremans K.; Masson, P.; Eds.; Colloque INSERM, John Libbey Eurotext Ltd., 1992; Vol. 224, pp 329–331.

Nielsen, P.M. Reactions and potential industrial applications of transglutaminase. Review of literature and patents. *Food Biotechnology* **1995,** *9,* 119–156.

Ohmori, T.; Shigehiza, T.; Taji, S.; Hayashi, R. Effect of high pressure on the protease activities in meat. *Agric. Biol. Chem.* **1991,** *55,* 357–361.

Ohshima, T.; Ushio, H.; Koizumi, C. High-pressure processing of fish and fish products. *Trends Food Sci. Technol.* **1993,** *4(11),* 1370–375.

Okamoto, M.; Kawamura, Y.; Hayashi, R. Application of high pressure to food processing: Textural comparison of pressure and heat induced gels of food proteins. *Agric. Biol. Chem.* **1990,** *54(l),* 183–189.

Parker, R.; Ring, S.G. A theoretical analysis of diffusion-controlled reactions in frozen solutions. *Cryo-Letters* **1995,** *16,* 197–208.

Powrie, W.D. Characteristics of food myosystems and their behavior during freezepreservation. In *Low-Temperature Preservation of Foods and Living Matter;* Fennema, O.R.; Powrie, W.D.; Marth, E.H., Eds.; Marcel Dekker, Inc., N.Y., 1973; pp 282–353.

Rubinsky, B.; Arav, A.; Hong, J.-S.; Lee, C.Y. Freezing of mammalian livers with glycerol and antifreeze proteins. *Biochem. Biophys. Res. Comm.* **1994,** *200,* 732–741.

Shenouda, S.Y. Theories of protein denaturation during frozen storage of fish flesh. *Adv. Food Res.* **1980,** *26,* 275–311.

Shoji, T.; Saeki, H.; Wakarneda, A.; Nakamura, M.; Nonaka, M. Gelation of salted paste of Alaska pollock by high hydrostatic pressure and change in myofibrillar protein in it. *Nippon Suisan Gakkaishi* **1990,** *56(12),* 2069–2076.

Shoji, T.; Saeki, H. Wakameda, A.; Nonaka, M. Muence of ammonium salt on the formation of pressure-induced gel from Walleye pollock suiimi. *Nippon Suisan Gakkaishi* **1994,** *60(l),* 101–109.

Sikorski, Z.; Olley, J.; Kostuch, S. Protein changes in frozen fish. *Crit. Rev. Food Sci. Nutri.* **1976,** *8,* 97.

Yamamoto, K. Changes in myosin molecule and its proteolytic subfragments induced by high hydrostatic pressure. *Abst. Int. Conf High Press. Biosci. Biotechnol. Kyoto, Japan*, 1995.

6

EFFECT OF HIGH HYDROSTATIC PRESSURE ON PACIFIC WHITING SURIMI

Michael T. Morrissey, Yildiz Karaibrahimoglu, and Jovi Sandhu

Oregon State University Seafood Laboratory
Astoria, Oregon 97331

The effects of high hydrostatic pressure (HHP) on gel strength, microbial numbers, proteolytic activity, color and pH of Pacific whiting surimi gels were investigated. Strong gels without the use of protease inhibitors were formed with HHP treatments from 1–4 kBars. Gel strength was not affected by holding time. Total plate counts showed that destruction of vegetative cells was accomplished with 4 kBar. Proteolytic activity was diminished but not eliminated with pressure treatments. The color of the surimi gels treated were translucent while the heat-treated gels were opaque. The pH was increased slightly with HHP treatment. HHP was an effective processing method for making high quality surimi gels from Pacific whiting.

INTRODUCTION

High hydrostatic pressure (HHP) has become a processing method of increasing interest to food scientists. The main areas of research with HHP have been destruction of microorganisms, activation and inactivation of enzymes, and formation of protein gels. Typically, these treatments are carried out in batch processes at pressures from 0.5–4 kBar (493–3,948 atmospheres). HHP is currently used in the food industry in Japan, mainly for pasteurization of jams and jellies (Farr, 1990). The advantages have been the retention of natural flavor and inherent food characteristics while extending the shelf-life of these foods. The main drawbacks to the food industry have been the need to use a batch-process operation and high cost of the equipment. Continuing efforts in this field may provide industry with the necessary research and engineering knowledge to build equipment with increased capacity and lower operating costs (Ledward, 1995).

HHP has a significant impact on microorganisms in food systems. A reduction of vegetative cells occurs at approximately 2 kBar at room temperature with complete de-

Process-Induced Chemical Changes in Food
edited by Shahidi *et al.* Plenum Press, New York, 1998

struction at 4–5 kBar (Hoover, 1993). Spores are considerably more resistant to pressure, however, and a combination of pressure and heat treatment is necessary to significantly reduce these microorganisms (Ludwig *et al.*, 1992). Some multi-step processes to increase lethality of microorganisms include pH, addition of additives, as well as temperature and pressure cycling (Cheftel 1992). HHP as low as 2 kBar have been effective in killing parasites such as nematodes (Morrissey *et al.*, 1995).

The activation and inactivation of enzymes has been of particular interest to food scientists. Knorr *et al.* (1992) reported on the inactivation of polyphenol oxidase in real food systems in response to moderate temperatures and pressures. Ohmori *et al.* (1991, 1992) showed that neutral and alkaline proteases are more sensitive than acid proteases in beef muscle. The rupture of lysosomal membranes, releasing catheptic proteases, occurs at 1–2 kBar pressure. In some cases, HHP can catalyze reactions such as the synthesis of oligosaccharides (Pelenc *et al.*, 1992) and affect the activity of cellulases (Murao *et al.*, 1992).

Protein modification of a number of food systems include meat tenderization, cold temperature gelation, texture modification, shaping and molding of food products, and increased binding of ligands to proteins (Cheftel 1992). The use of pressure to tenderize beef is well documented even at pressures as low as 0.5 kBar (Beilken *et al.*, 1990). Gelation of Pacific whiting and Alaska pollock surimi at pressures as low as 1 kBar at room temperature for 1 hr has been demonstrated (Chung *et al.*, 1994). Surimi is a washed fish mince that has a high concentration of myofibrillar proteins. The gelation is believed to be the result of cross-linking of myosin heavy chains. Okazaki and Nakamura (1992) have shown that fish sarcoplasmic proteins can also form gels after HHP treatment. Ohshima *et al.* (1993) have provided a review of the effects of high pressure treatment on fish and fish products.

The present contribution intended to study the effect of HHP on Pacific whiting surimi. Pacific whiting has high levels of proteases in the muscle tissue. These enzymes rapidly breakdown myofibrillar protein during the heat-setting of surimi, thus causing poor gel formation (Morrissey *et al.*, 1993). Therefore, effects of HHP on proteases in surimi as well as time/pressure effects on gelation as well as total plate count, pH and color were also studied.

EXPERIMENTAL

Sample Preparation

Commercially frozen Pacific whiting surimi was obtained from American Seafoods Co. (Seattle, WA). The surimi contained 4% sorbitol, 4% sucrose, 0.30% tripolyphosphate, and 0.12% mono- and diacylglycerols. Surimi blocks were stored at -20°C at the Oregon State University, Seafood Laboratory, Astoria, OR. Fish paste was prepared as described in the surimi testing manual (Lanier *et al.*, 1991). All samples were standardized at 2% NaCl and 78% moisture. The ingredients were blended in a Stephan vacuum mixer (Model UM5, Stephan Machinery Corporation, Columbus, OH) for approximately 4 min. The mixed paste was transferred to a sausage stuffer (5 lb capacity, The Sausage Maker, Buffalo, NY) and extruded into stainless steel cooking tubes (17.8 × 2.2 cm i.d.). Control samples were cooked in a water bath at 90°C for 15 min. After cooking, the gels were transferred to an iced water bath for 15 min. The gels were removed from the tubes, placed in plastic bags, and stored at refrigerated temperatures for testing the following day.

Pressure Treatments

Samples were extruded into stainless steel tubes as described above. The tubes were sealed with a fold-down rubber stopper on one side and a metal screw top on the other side. Both ends of the cooking tube were sealed tight to prevent moisture penetration during pressure treatment. The tubes were placed into polyethylene bags filled with distilled water, and vacuum sealed. The tubes were placed in an isostatic press (Model L 400–2S, Autoclave Engineers Inc., Erie, PA) with a pressure chamber of 55.88 x 7.62 cm. Samples were submerged in water containing 2% hydraulic fluid (Hydrolubric 120-B, E. F. Houghton and Co., Valley Forge, PA), which acted as the hydrostatic fluid medium in the press. All pressure treatments were run at 18°C. The pressure/time settings for this study were: (1) 1.0 kBar/15, 30, 45, 60 min; (2) 2.0 kBar/15, 30, 45, 60, min; (3) 3.0 kBar/15, 30, 45, 60 min and (4) 4.0 kBar/15, 30, 45, 60 min. The pressure-treated gels were removed from the tubes, and stored in plastic bags at refrigerated temperatures to be analyzed the following day.

Torsion Measurements

Gel texture properties for both heat-treated and pressure-treated gels were determined by torsion. The gels were cut into 2.8 cm lengths and formed into an hourglass shape with a 1 cm die on a lathe-type apparatus (Gel Consultants Inc., Raleigh, NC). The samples were subjected to torsional strain in a modified Brookfield viscometer (Gel Consultants Inc., Raleigh NC). Shear stress and shear strain, at failure, were calculated using equations developed by Hamann (1983).

Autolytic Assays

Autolytic assays were undertaken according to the method described by Morrissey *et al.* (1993) with the following changes. A 3 g surimi gel sample was finely chopped with a razor blade and incubated at 55°C for 1 hr. Autolysis was stopped by adding 27 mL 5% cold trichloroacetic acid (TCA), incubating the mixture at 4°C and centrifuging at 6100 x g for 15 min. The supernatant was analyzed for oligopeptide content by the Lowry assay (Lowry *et al.*, 1951) and expressed as mmoles of tyrosine released.

Color Measurement

The CIE Lab color scale was used to measure Hunter **L*** (Lightness), Hunter **+a*** (redness), Hunter **-a*** (greenness), Hunter **+b*** (yellowness), and Hunter **-b*** (blueness) values. Color values of gels were measured using a Chroma Meter (Model CR-310, Minolta Corporation, Ramsey, NJ). The instrument was standardized for surimi gel measurement by using a Minolta calibration plate (No. 18133009 Y_{CIE} = 94.5, X_{CIE}= 0.3160, Y_{CIE}= 0.3330) and a Hunter lab standard plate (Standard No. D44C-1618, **L*** = 82.13, **a*** = -5.24, **b*** = -0.55) with D65 illuminant and 2° observer.

Microbial Count

Total plate counts were run on dehydrated plate count agar (Difco, Detroit, MI). Ten grams of surimi gels were homogenized with 90 mL distilled water using an Osterizer pulsematic 10 blender (Oster Corporation, Milwakuee, WI) at frappe speed for 1 min. Dilutions were made from 10^{-2} to 10^{-6}. All plates were run in duplicate. Plates were incubated for 48 hr at 36°C.

pH Determination

A slurry consisting of 10 g of surimi and 90 mL of distilled water was measured for pH using a standard meter (Corning Ciba Diagnostics Co., Corning, NY).

Statistical Analysis

A sub-set of four samples from each treatment was used for each analysis. Statistical analysis of data was carried out using one way analysis of variance (ANOVA). Differences among mean values were established using the least significant difference (LSD) multiple range test (Steel and Torrie, 1980). Values were considered significant when $p<0.05$.

RESULTS AND DISCUSSION

Gel Strength

Shear stress is indicative of gel hardness, while strain measures the cohesiveness or elasticity of surimi gels (Hamann and Lanier, 1987). The shear stress and shear strain values of Pacific whiting surimi increased significantly for all pressure treatments when compared to the heat-set control gels (Tables 1 and 2). The strain for Pacific whiting surimi cooked at 90°C for 15 min was 0.90, and the shear stress value was 10.4 kPa. These low values are a direct result of protease activity during the cooking stage (Morrissey *et al.*, 1993). Pressures as low as 1 kBar doubled the strain values of gels at all times tested while stress values increased slightly. At pressure treatments of 2 kBar, both stress and strain values increased significantly. Stress values increased to a range of 26–34 kPa at this pressure while strain increased to a value of 2.5. At 2 kBar, stress values increased with holding times while strain remained constant. Higher pressure increased both stress and strain of the surimi gels and maximum values were shown at 4 kBar. The maximum strain value was 3.1 (4 kBar, 15 min) and the maximum stress value was 73.5 kPa (4 kBar, 60 min). There were no significant differences in strain values at different holding times at the same pressure level. The increase in strain was related to increases in pressure. There were few differences among stress values at different holding times of gels held at the

Table 1. Stress values (kPa) for pacific whiting surimi at different pressure and time treatments

	Time (min)			
Pressure(kBar)	15	30	45	60
Control—0.001/90°C	10.5			
1.0	11.9	13.1	13.0	12.3
2.0	26.1	28.1	31.7	34.4
3.0	40.0	44.1	46.6	44.1
4.0	57.2	68.9	67.8	73.5

Table 2. Strain values for Pacific whiting surimi at different pressure and time treatments

	Time (min)			
Pressure(kBar)	15	30	45	60
Control—0.001/90°C	0.90			
1.0	1.97	2.05	2.48	2.08
2.0	2.59	2.49	2.52	2.57
3.0	2.63	2.65	2.72	2.57
4.0	3.10	2.94	3.08	2.90

same pressure. Shoji *et al.* (1990) reported that strong gels were formed from surimi paste by treatment at 2.0–4.0 kBar. Chung *et al.* (1994) reported high stress and strain values for Pacific whiting surimi gels with pressure treatments of 1 kBar at 28°C for 1 hr. They found that stress increased with higher pressures while strain decreased.

Autolysis Assays

Proteolytic activity was measured in Pacific whiting surimi treated at different pressures and time (Table 3). Overall, there was a decrease in enzymatic activity from the control sample (43.95 nmol Tyr). Enzyme activity between the pressures fluctuated at different levels. The lowest enzymatic activity recorded was at 2.0 kBar/15 min compared to the highest activity of 45.04 nmol Tyr at 3.0 kBar/60 min. There were few differences in proteolytic activity between time intervals at the 1 and 4 kBar treatments. At 2 and 3 kBar there were increases in enzymatic activity with holding time.

The majority of muscle degradation that occurs in Pacific whiting is caused by cathepsin (An *et al.*, 1994). Seymour *et al.* (1994) identified cathepsin L, a lysosomal protease, as the protease responsible for gel softening in Pacific whiting surimi. Ohmori *et al.* (1992) noted that pressures of 2.0 kBar caused the rupture of the lysosomal membrane in bovine liver, releasing enzymes and increasing proteolytic activity. For whiting surimi, HHP treatment at 3.0 kBar showed the highest proteolytic activity. Proteolysis has its greatest affect on texture and gel strength. Although stress increased with increased pressure, it is worthy to note that there was no significant difference in the strain values between 2 and 3 kBar. A likely scenario is that the lysosomal membranes break down at pressures above 2 kBar resulting in increased proteolytic activity. However, this is somewhat counteracted by the partial inactivation of proteases as the pressure is increased. At

Table 3. The effect or protease activity as measured by autolysis for Pacific whiting surimi under different pressure treatments

	Time (min)			
Pressure(kBar)	15	30	45	60
Control—0.001/90°C	43.95			
1.0	33.30	30.03	34.59	31.78
2.0	18.62	22.80	25.35	20.61
3.0	35.98	38.53	42.18	45.04
4.0	32.94	29.68	32.26	35.01

the highest pressure (4 kBar), there is further denaturation of all proteins and a decrease in enzymatic activity.

The autolytic assay cannot differentiate between proteolysis that occurred during gel formation (releasing tyrosine reactive oligopeptides) and activity that remains after the gels have been formed. The heat-set gels are cooked at high enough temperatures (90°C, 15 min) to completely denature and inactivate all protease activity. Tyrosine is formed during the heating process especially as the surimi paste is heated at the optimum protease temperature range (50–60°C). For the pressure treated gels, proteolytic activity can be diminished by several methods. One is denaturation of the enzyme due to pressure, while another is cross-linking of proteins to each other during gel formation. This would render potential substrates less reactive to available protease and/or inactivate the protease as it is cross-linked to other macromolecules. The net results are strong gels that are resistant to proteolysis.

Microbial Analysis

Total plate counts of pressurized samples showed increased microorganism inactivation with pressure and time (Table 4). All vegetative cells were killed at 4 kBars for all holding times. There was a significant reduction of total plate counts for 2 and 3 kBars of pressure. A similar result showed by Miyao *et al.* (1993) was that most microorganisms in surimi were inactivated by the treatment of high pressure between 3 and 4 kBars. Styles *et al.* (1991) found inactivation of *L. monocytogenes*, and *V. parahemolyticus* under similar pressures in milk products. The destruction of vegetative cells extended the shelf-life of pressure treated products. However, spores of lethal bacteria can survive the process and care must be taken to keep the product frozen or at refrigerated temperatures below 3.3°C to inhibit the growth of pathogenic *Clostridia spp.*

Color Analysis

Conventional cooking (90°C/15 min) of Pacific whiting surimi produces gels with an opaque appearance. When surimi is treated by HHP, it becomes transparent. The differences in color are shown in Table 5. The Hunter **L*** value of the heat-set gel was 78.05 and the HHP-treated gels ranged from values of 57.91–71.82. The Hunter **L*** values increased significantly with pressure and the gels were less transparent at the higher values. The Hunter **a*** value for the heat-set gel was 2.90 while the pressure treated values ranged from -4.18 to -4.72. The Hunter **b*** values for the heat-set surimi gel was positive (2.95) while the pressure treated gels had negative readings. The positive Hunter **b*** value re-

Table 4. Total plate counts (CFU/mL) of Pacific whiting surimi gels treated at different pressure and holding times

	Time (min)			
Pressure(kBar)	1.0	2.0	3.0	4.0
Control—0.001/90°C	1.9×10^4			
15	8.5×10^4	1.5×10^4	1.5×10^4	0
30	8.5×10^4	3.0×10^4	0.8×10^4	0
45	6.0×10^4	4.5×10^4	0.5×10^2	0
60	7.0×10^3	3.4×10^4	0.5×10^2	0

Table 5. Hunter color values of heat-set and pressure treated Pacific whiting surimi gels

Hunter values	Pressure (kBar)	Time (min)			
		15	30	45	60
L*	Control—0.001/90°C	78.05			
	1.0	58.41	58.16	58.07	57.91
	2.0	59.92	61.42	63.31	63.74
	3.0	68.34	69.89	69.92	70.03
	4.0	71.02	65.93	71.82	71.64
a*	Control—0.001/90°C	−2.90			
	1.0	−4.51	−4.50	−4.45	−4.47
	2.0	−4.72	−4.71	−4.55	−4.47
	3.0	−4.30	−4.28	−4.33	−4.34
	4.0	−4.17	−4.20	−4.18	−4.21
b*	Control—0.001/90°C	2.95			
	1.0	−2.28	−2.29	−2.26	−2.28
	2.0	−2.35	−2.24	−1.86	−1.67
	3.0	−1.10	−0.74	−0.61	−6.50
	4.0	−0.21	−0.29	0.07	−0.05

flects a slight yellow hue while a negative Hunter **b*** represents a slight blueness to the gel reflecting its transparency. Both pressure and holding time affected Hunter **b*** values. Gels treated at 1 kBar for 15 min had a Hunter **b*** value of -2.28, while gels formed at the highest pressure and longest time (4 kBar, 60 min) showed a Hunter **a*** value of -0.05. Similar changes in appearance occurred when Alaska pollock surimi was treated at up to 500 MPa (Shoji *et al.*, 1990). The most striking difference is the transparent nature of the HHP treated gels. This was probably due to the alignment of protein molecules in the gel network. Highly modified starch gels will be more transparent than the corresponding native starch gel (Park, 1995).

pH Analysis

The pH of samples treated by HHP ranged from 7.33 to 7.42 and were all higher than control at 7.02 (Table 6). This is in contrast to work done by MacFarlane (1985) with pressured meat. In his findings, the pH of pressurized meat samples decreased 0.6–0.8 pH units with samples treated at 1 kBar. The texture of a surimi gel is greatly affected by the pH. Chung *et al.* (1993) showed that Pacific whiting gels can increase as much as 1.0 kPa in stress

Table 6. pH of heat-set and pressure treated Pacific whiting surimi gels

Pressure (kBar)	Time (min)			
	15	30	45	60
Control—0.001/90°C	7.02			
1.0	7.35	7.33	7.34	7.34
2.0	7.34	7.39	7.42	7.43
3.0	7.44	7.42	7.47	7.46
4.0	7.34	7.43	7.33	7.42

for each 0.1 increase in pH. They speculated that the increase in pH causes a greater unfolding of the protein opening additional binding sites and facilitating gel formation.

CONCLUSIONS

HHP was very effective in forming good quality surimi gels from Pacific whiting. There was a positive correlation between increased treatment (pressure and time), and increased stress and strain values of the surimi gels. Heat-set gels of Pacific whiting surimi, without protease inhibitors, were soft and had low stress and strain values due to proteolytic activity. Gels formed by HHP circumvented the protease problem and formed strong gels. The HHP treated gels were transparent and very elastic indicating that a different gelation mechanism may occur through HHP.

REFERENCES

An, H.; Weerasinghe, V.; Seymour, T.; Morrissey, M.T. Degradation of Pacific whiting proteins by cathepsin. *J. Food Sci.* **1994,** *59,* 1013–1017, 1033.

Beilken, S.L.; MacFarlane, J.J.; Jones, P.N. Effect of high pressure during heat treatment on the Warner-Bratzler shear force values of selected beef muscles. *J. Food Sci.* **1990,** *55,* 15–19, 42.

Cheftel, J.C. Effects of high hydrostatic pressure on food constituents: an overview. In *High Pressure and Biotechnology*; Balny, C.; Hayashi, R.; Heremans, K; Masson P., Eds.; John Libbey Eurotext, London, 1992; pp. 195–210.

Chung, Y.C.; Richardson, L.; Morrissey, M.T. Effects of pH and NaCl on gel strength of Pacific whiting surimi. *J. Aquatic Food Product Technol.* **1993,** *2(3),* 19–35.

Chung, Y.C.; Gebrehiwot, A.; Farkas, D.F.; Morrissey, M.T. Gelation of surimi by high hydrostatic pressure. *J. Food Sci.* **1994,** *59,* 523–524,543.

Farr, D. High pressure technology in the food industry. *Trends Food Sci. Technol.* **1990,** *1(7),* 14–16.

Hamann, D.D. Structural failure in solid foods. In Physical Properties of Foods; Peleg, M.; Bagely, E.B., Eds.; Avi Pub. Co., Inc., Westport, CT, 1983; pp. 351–383.

Hamann, D.D. and Lanier, T.C. Instrumental methods for predicting seafood sensory texture quality. In *Seafood Quality Determination*; Kramer, D.E.; Liston J., Eds.; Proceedings of an international symposium coordinated by the University of Alaska Sea Grant Program, Elsevier Science Publishers, New York, 1987.

Hoover, D.G.Pressure effects on biological systems. *Food Technol.* **1993,** *47(6), 150–155.*

Knorr, D.; Bottcher, A.; Dornenburg, H.; Eshtiaghi, M.; Oxen, P.; Richwin, K.; Seyderhelm, I. High pressure effects on microorganisms, enzymes activity and food functionality. In *High Pressure and Biotechnology*; Balny, C.; Hayashi, R.; Heremans, K.; Masson, P., Eds.; John Libbey Eurotext, London, 1992; pp. 211–218.

Lanier, T.C.; Hart, K.; Martin, R.E., Eds. *A Manual of Standard Methods for Measuring and Specifying the Properties of Surimi*; University of North Carolina Sea Grant College Program, Raleigh, NC, 1991.

Ledward, D.A. High pressure processing - the potential. In *High Pressure Processing of Foods;* Ledward, D.A.; Johnson, D.E.; Earnshaw, R.G.; Hasting, A.P.M., Eds.; Nottingham University Press, Loughborough, England, 1995; pp. 1–6.

Lowry, O.H., Rosebrough, N.J., Farr, A.L., and Randall, R.J. Protein measurement with Folin phenol reagent. *J. Biol. Chem.* **1951,** *193,* 256–275.

Ludwig, H.; Bieler, C.; Hallbauer, K.; Scigalla, W. Inactivation of microorganisms by hydrostatic pressure. In *High Pressure and Biotechnology*; Balny, C.; Hayashi, R.; Heremans, K,; Masson, P., Eds.; John Libbey Eurotext, London, 1992; pp. 25–32

MacFarlane, J.J. High pressure technology and meat quality. In *Developments in Meat Science-3*; Lawrie, P., Ed.; Elsevier Applied Science Publishers Ltd., London, 1985; p. 135.

Miyao, S.; Shindoh, T.; Miyamori, K.; Arita, T.Effects of high pressurization on the growth of bacteria derived from surimi (fish paste). *Nippon Shokuhin Kogyo Gakkaishi* **1993,** *40,* 478–484.

Morrissey, M.T.; Karaibrahimoglu, Y. The effects of high hydrostatic pressure on nematodes in fish. Presented at the World Food Congress, Budapest, Hungary, August 4–10, 1995.

Morrissey, M.T.; Wu, J.W.; Lin, D.D.; An, H. Effect of food grade protease inhibitor on autolysis and gel strength of surimi. *J. Food Sci.* **1993,** *58,* 1050–1054.

Murao, S.; Nomura, Y.; Yoshikawa, M.; Shin, T.; Oyama, H.; Arai, M. Enhancemment of activity of cellulases under high hydrostatic pressure. *Biosci. Biotech. Biochem.* **1992,** *56,* 1366–1367.

Ohmori, T.; Shigehisa, T.; Taji, S.; Hayashi, R. Effect of high pressure on the protease activities in meat. *Agric. Biol. Chem.* **1991, 55,** 357–361.

Ohmori, T.; Shigehisa, T.; Taji, S.; Hayashi, R.Biochemical effects of high pressure on the lysosome and proteases involved in it. *Biosci. Biotech. Biochem.* **1992,** *56,* 1285–1288.

Ohshima, T.; Ushio, H.; Koizumi, C. High-pressure processing of fish and fish products. *Trends Food Sci. & Technol.* **1993,** *4,* 370–375.

Okazaki, E.; Nakamura, K. Factors affecting texturization of sarcoplasmic proteins of fish by high pressure treatment. *Bull. Jpn. Soc. Fish.* **1992,** *58,* 2197–2206.

Park, J. Personal communication. Oregon State University Seafood Laboratory, Astoria, OR, 1995.

Pelenc, V.; Paul, F.; Monsan, P.; Masson, P. Enzymatic synthesis of oligosaccharides under pressure. In *High Pressure and Biotechnology*; Balny, C.; Hayashi, R.; Heremans, K.; Masson, P., Eds.; John Libbey Eurotext, London, 1992; pp. 349–352.

Seymour, T.A.; Morrissey, M.T.; Peters, M.Y.; An, H. Purification and characterization of Pacific whiting proteases. *J. Agric. Food Chem.* **1994,** *42,* 2421–2427.

Shoji, T.; Saeki, H.; Wakameda, A.; Nakamura, M.; Nonaka, M. Gelation of salted paste of Alaska pollock by high pressure and change in myofibrillar protein in it. *Bull. Jpn. Soc. Fish.* **1990,** *56,* 2069–2076.

Steel, R.G.; Torrie, J.H. *Principles and Procedures of Statistics*, 2nd ed.; Mcgraw-Hill Book Co., NY, 1980.

Styles, M.F.; Hoover, D.G.; Farkas, D.F. Response of *Listeria monocytogenes* and *Vibrio parahaemolyticus* to high hydrostatic pressure. *J. Food Sci.* **1991,** *56,* 1404–1407.

7

HIGH PRESSURE PROCESSING OF FRESH SEAFOODS

Benjamin K. Simpson

Department of Food Science and Agricultural Chemistry
Macdonald Campus of McGill University
Ste. Anne de Bellevue, PQ, Canada, H9X 3V9

Crude proteolytic enzyme extracts were prepared from the muscle tissues of two fish species, bluefish and sheephead, and subjected to high hydrostatic pressure treatments (from 1,000–3,000 atm), and monitored for residual activity for cathepsin C, collagenase, chymotrypsin-like and trypsin-like enzymes versus homologous enzymes from bovine. The fish enzymes were more sensitive to hydrostatic pressure than the mammalian enzymes. The extent of enzyme inactivation achieved depended on both the amount of pressure applied, the duration of pressurization, and on the source material. Pressure treatment of fresh fish flesh formed products whose color deteriorated (cooked appearance) with increasing pressure as well as holding time. Application of pressure also improved tissue firmness or strength of fresh fish up to 2,000 atm and a holding time of 10 min, beyond which texture generally deteriorated. The combined use of pressure in combination with the broad spectrum protease inhibitor, α_2-macroglobulin, enhanced the capacity of the hydrostatic pressure technology to achieve a more lasting inactivation of endogenous enzymes to form stable fish gels.

INTRODUCTION

Concerns about declining fish stocks have increased research to develop innovative procedures to enhance the quality, safety, and shelflife of traditional fish species, and also to put non-traditional fish species to better economic use.

Seafoods generally comprise of about 18–35% total solids, with 14–20% protein, 0.2–20% fat, and the balance as ash and carbohydrates. The major biological molecules (i.e., carbohydrates, proteins, and lipids) all serve as substrates for microorganisms and endogenous enzymes.

Process-Induced Chemical Changes in Food
edited by Shahidi *et al.* Plenum Press, New York, 1998

Microorganisms are diverse in their biochemical activities. Some ferment carbohydrates to generate acid and/or gas, others degrade proteins to liberate ammonia and other amino compounds. Some others convert nitrates or nitrites into nitrogen or its oxides. Microbial spoilage, therefore, may vary from gross and stinking putrefaction, to acid and/or gas production. In addition, certain toxin-producing bacteria may release their toxins into food to render them harmful for human consumption. Certain strains of *Bacillus* and *Clostridium perfringens* may induce severe diarrhea, while others like *C. botulinum* produce potent toxins in foods with fatal consequences to humans.

Important groups of endogenous enzymes that promote autolysis in postmortem seafoods include those originating from the gut, present in muscle tissues or synthesized and secreted in the extracellular matrix. The major autolytic reactions that cause deterioration of fresh seafood quality include proteolysis, glycolysis, nucleic acid (ATP) breakdown, and lipid hydrolysis/oxidation.

Fish muscle proteins are degraded by endogenous proteases to result in quality defects such as "belly burst" in fish species like capelin, herring, and mackerel (Haard *et al.*, 1979), and mushiness in crustaceans (Nip *et al.*, 1985). Endogenous proteases have been implicated in salmon flesh softening (Konagaya, 1985), in the deterioration of connective tissue matrix of cod muscle leading to losses in surimi yield, and mushiness in ice-stored and/or cooked prawn (Lindner *et al.*, 1988).

As a result of glycolysis, muscle glycogen is converted to lactate leading to a decline in muscle pH, and eventually, to severe exudation conditions, opaque or cooked appearance, and poor texture.

The nucleotide, adenosine triphosphate (ATP) is enzymatically degraded in seafoods through various intermediates to hypoxanthine (Figure 1). ATP breakdown leads to rigor mortis and a decline in muscle pH, resulting in poor quality meat.

Fish muscle contains high levels of ω3 fatty acids, and thus is highly susceptible to autooxidation (Lakidos and Lougovois, 1990). Lipolytic enzymes promote hydrolytic rancidity in raw seafood, and are particularly important in high fat fish species resulting in off flavors during storage. Lipoxygenases in fish promote lipid peroxidation to form TBA-reactive substances (Hultin *et al.*, 1982), and bleaching of carotenoid pigments in the flesh and skin of salmonids (Tsukuda, 1970).

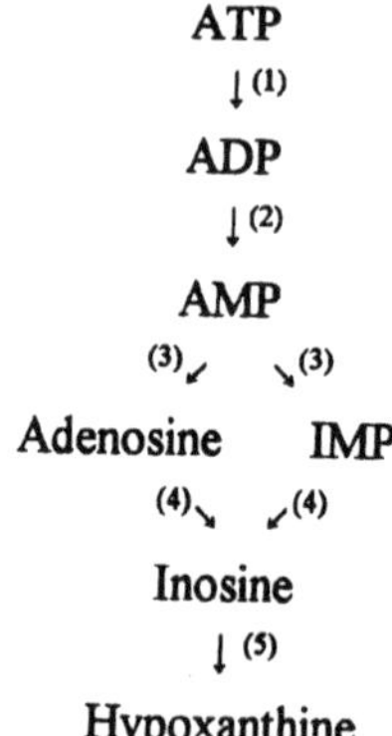

Figure 1. Scheme for enzymatic degradation of ATP in fresh fish (Flick and Lovell, 1970). The numbers in parentheses denote the enzymes catalyzing steps in the sequence, viz., : (1) ATPase; (2) myokinase; (3) AMP deaminase; (4) IMP phosphorylase; and (5) nucleoside phosphorylase.

Such microbial and enzymatic changes in the postmortem animal adversely impact the prime quality attributes of raw fish during handling and iced storage. Of all the muscle foods used for human consumption, fresh seafoods are most susceptible to postmortem texture deterioration. In general, consumers prefer fresh seafoods over their frozen counterparts. This fact coupled with the growing consumer wariness about the use of chemicals in foods, and the ample demonstration of the importance of endogenous enzymes in seafood spoilage, suggest that novel preservation techniques are needed to control both the deleterious effects of microorganisms, and enzyme-induced autolysis in raw seafood during storage.

Current methods used to preserve fresh seafood quality include rapid chilling, storage in melting ice or in refrigerated sea water, gentle handling, bleeding and gutting. Other methods are superchilling, use of modified atmospheres, low dose ionizing radiation, and treatment with chemicals (Ashie *et al.*, 1996a). Seafood preservation by icing retards both autolysis and bacteria growth, and is the most popular way of extending the market life of fresh seafood. However, even with proper handling, premium quality of iced fish is lost within a few days. The disadvantages with the refrigerated seawater technique for preserving fresh seafoods include high capital and operating costs, requirement for skilled personnel, potential for growth of psychrophilic bacteria, and salt uptake from seawater to alter product flavor. With regards to superchilling, it is found to be most effective at -3°C, however, a slightly lower temperature of about -5°C promotes enzymatic hydrolysis of phospholipids and protein denaturation, resulting in poor quality products (Toyomizu *et al.*, 1981). Low dose radiation does not entirely inhibit enzymatic reactions, and may promote lipid peroxidation and other autolytic reactions. Modified atmosphere treatments result in increased drip loss, darkening of red muscle, and rapid conversion of inosine to hypoxanthine, and the use of chemicals is not appealing to consumers (Simpson and Haard, 1987a).

Studies have shown that the shelflife of sterile fresh fish is similar to that of normal fish, and although exclusion of bacteria reduces gross spoilage/putrefaction of seafood, it does not alter the rate at which the product becomes unacceptable to sensory panelists (Fletcher and Satham, 1988). This revelation underscores the argument that novel strategies for enhancing the quality and shelflife of fresh fish must also aim at controlling the deleterious effects of endogenous enzymes in fish tissues.

One such novel approach which is gaining popularity as a food processing tool is high hydrostatic pressure technology. It must be pointed out that high pressure technology is not entirely new, and interest in the application of high hydrostatic pressure technology in food processing was demonstrated over nine decades ago (Bridgman, 1914; Hite *et al.*, 1914). Furthermore, studies were carried out as far back as the early 1880's on the effects of hydrostatic pressure on biological molecules and/or intact organisms on the *Talisman* dredging expedition of 1882–1883. These initial studies on high pressure effects on biological systems established certain basic relationships on the pressure effects on biological systems, some of which have been confirmed and extended by more precise studies in recent years. The studies revealed that living organisms could subsist and maintain normal biological function at depths of 6000 m below sea level, where hydrostatic pressure is in excess of 600 atm. Studies with small aquatic animals (cyclops and daphnia), fish, and invertebrates, revealed that with brief periods of compression, experimental animals recovered their activity after release of pressure, while longer periods of compression induced coma or deaths in the animals.

The effects and mechanism of action of hydrostatic pressure on biological materials at moderate pressures are quite different from those at high pressures. Hochachka *et al.*,

(1975) demonstrated that a pressure of 200 atm activated citrate synthase from *Antimora* gill; the catalytic activity of the enzyme was optimum at 2000–2500 atm; but beyond 4000 atm, the enzyme was dramatically and irreversibly inactivated. Curl and Jansen (1950) inactivated trypsin, chymotrypsin, and pepsin at 25°C with pressures from 3,000–9,000 atm, while a 25% decrease in peptic activity was achieved with 5,000 atm in 5 h at 0°C versus a 60% loss in activity at 35.5°C. It is now well known that hydrostatic pressure achieves various degrees of inactivation of both the spore and vegetative forms of microorganisms; and enzymes (Ashie and Simpson, 1995; Sareevoravitkul *et al.*, 1996; Hayakawa et al, 1994; Ogawa et al, 1990).

The idea of using hydrostatic pressure in combination with other barriers is based on the advantages of high pressure technology. For example, pressure treatments result in glossier and smoother products which show little shrinkage, and have a large extensibility. Products formulated by high hydrostatic pressure technology also retain the nutrients, color, flavor, and freshness of the raw material better (Sareevoravitkul *et al.*, 1996; Farr, 1990). Furthermore, since the freezing point of H_20 decreases with pressure, the use of hydrostatic pressure in combination with sub-zero temperatures, has been proposed as a means of storing fresh food products without formation of ice crystals (Farr, 1990). Lanier (1995) demonstrated that freezing under high pressure conditions can reduce ice formation, and facilitate freezing and thawing operations. Thus, the perceived advantages of the high hydrostatic pressure approach for preserving seafood freshness include controlled enzymatic activity, less protein denaturation, and inactivation of microorganisms. Farkas (1987) has suggested that combination of hydrostatic pressure technology with refrigeration, etc., has the potential to extend the shelflife of various products from several days to several weeks.

The approach we have taken in this work is that the contribution of endogenous enzymes to the postmortem fresh fish spoilage phenomenon has been grossly overlooked in developing methods for preserving the premium quality of fresh fish. Thus, high hydrostatic pressure technology was used in combination with other barriers to evaluate its efficacy in controlling texture degradation of fresh fish and related products. This contribution describes the hydrostatic pressure effects on four seafood proteases, i.e., chymotrypsin, trypsin, cathepsin, and collagenase, which may be involved in postharvest seafood texture deterioration (Wasson, 1992; Lindner *et al.*, 1988). While it is true that trypsin and chymotrypsin are digestive proteases which may not be expected to participate in fish texture deterioration, in conditions of "belly burst", these enzymes may seep out of the gut and come into contact with the muscle (Haard *et al.*, 1979). Also, trypsin-like and chymotrypsin-like enzymes occur in fish flesh (Guizani *et al.*, 1992).

EXPERIMENTAL

Materials

Trypsin, chymotrypsin, and cathepsin C (all from bovine), and the synthetic substrates glycyl-L-phenylalanine β-naphtylamide (GPNA), 4-phenylazo benzyloxy-carbonyl-pro-leu-gly-pro-D-arg (PZ-peptide), N-benzoyl-L-tyrosine ethyl ester (BTEE), and Nα-benzoyl-DL-arginine p-nitroanilide (BAPNA), were purchased from Sigma Chemical Co (St. Louis, MO). Fresh bluefish *(Pomatomus saltatrix)* and sheephead samples were purchased from a local fish market (Waldman Plus, Montreal, PQ) and kept in ice until ready for use.

Methods

The crude fish enzyme extracts were prepared as per the method of Baranowski *et al.* (1984), and stored in ice for use in the pressure treatments and subsequent enzyme assays. The spectrophotometric methods of Hummel (1959) and Erlanger *et al.* (1961) were used to assay for chymotrypsin-like and trypsin-like enzyme activities using BTEE and BAPNA as substrates, respectively. Cathepsin C activity was assayed using gly-phe-NA as substrate (Lee *et al.*, 1971), and collagenase activity in the fish extracts were assayed as per the method of Wunsch and Heidrich (1963). Protein content of the crude enzyme extracts from fish was determined by the method of Hartree (1972).

For the pressurization studies, enzymes (fish extract or commercial enzyme solutions) were sealed in 2.5"/5" whirl-pak bags (VWR Scientific, Montreal, PQ) and pressurized in an isostatic unit (Autoclave Engineers, Columbus, OH) from 1,000 - 3,000 atm. Fish samples were also placed in 4.5"/9" whirl-pak bags, vacuum sealed, and similarly pressurized.

The textural properties of the pressure-treated and control samples were measured with a Texture Testing instrument (Lloyd Instruments - Omnitronix, Mississauga, ON) at about 25°C using a compressive probe at a speed of 20 mm/min. The fish samples were taken out of the whirl-pak bags and cut into cubes of dimensions (mm) 30 x 20 x 15, and then subjected to a steadily increasing load up to a maximum of 45 Newtons. The corresponding deformations in the samples were measured, and the values of textural parameters were expressed as the means of three measurements.

The color changes in the samples were monitored by measuring the Hunter L, a, b values of the pressurized and control fish samples using a Minolta Chromameter II reflectance equipment. The differences between the mean values were analyzed using the least significant difference (LSD) multiple comparison test of Steel and Torrie, (1960). The general linear model (GLM) procedure of the Statistical Analysis System (SAS) on McGill University mainframe was used.

RESULTS AND DISCUSSION

The data presented in Figures 2a-c indicate that the fish enzymes were generally more sensitive to pressure than their bovine counterparts. For example, pressurization at 3,000 atm for 30 min had little or no effect on bovine cathepsin C activity, while cathepsins C from the two species of fish investigated, i.e., sheephead and bluefish, lost as much as 80 and 91% of their original activities, respectively, after a similar treatment (Figure 2a). The behavior of bovine chymotrypsin versus the fish trypsins under pressure (Figure 2b), was similar to that described for the cathepsins (Figure 2a). However, bovine trypsin was more pressure-sensitive than either bovine cathepsin or bovine chymotrypsin. Nevertheless, the degree of inactivation of bovine trypsin (50%) achieved with the same level of pressure treatment and holding time was lesser than that for the trypsins from sheephead (65%) and bluefish (75%) - Figure 2c. The greater susceptibility of seafood enzymes to pressure inactivation versus their bovine counterparts correlates well with the adaptive differences in the nature of enzymes from warm- and cold-adapted organisms (Simpson and Haard, 1987b).

When the enzymes from the two fish species were pressurized at 3,000 atm pressure for 30 min, and then stored at refrigerated temperatures (4–7°C), they all showed various degrees of reactivation over a 21 day period (Figures 3a and 3b). For the three sheephead

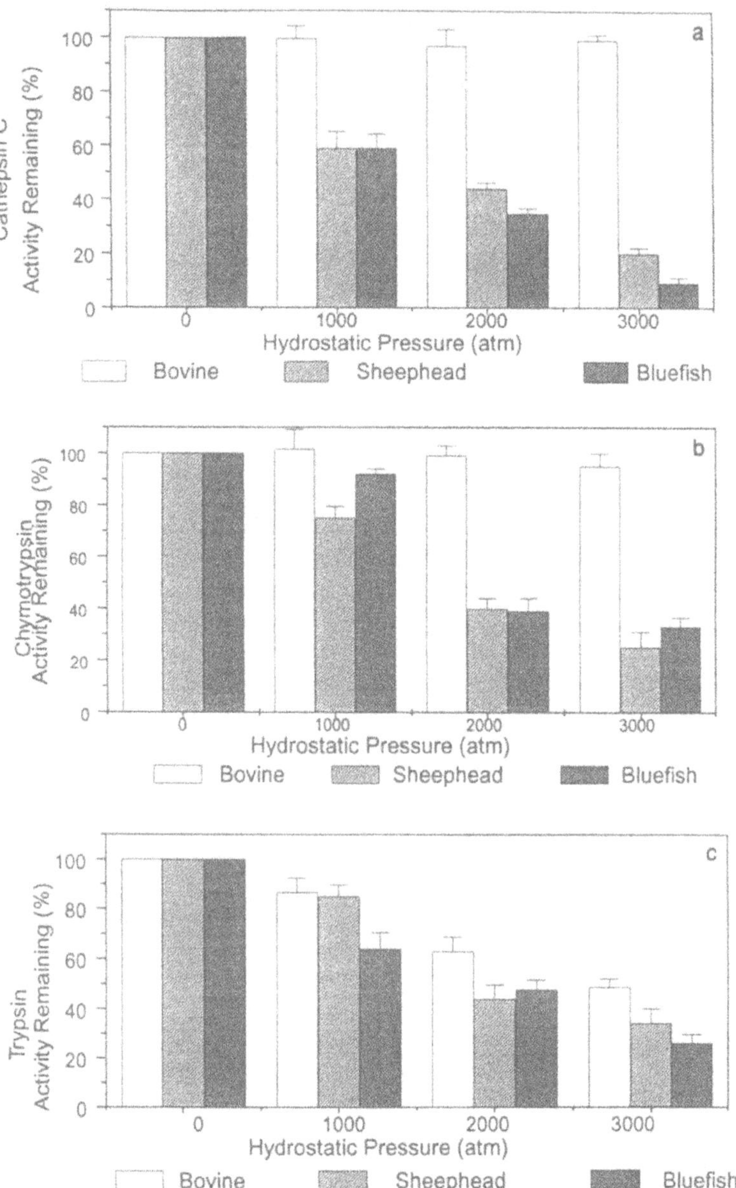

Figure 2. Hydrostatic pressure inactivation of proteases from fish and mammalian sources: (a) => cathepsin C; (b) => chymotrypsin; and (c) => trypsin.

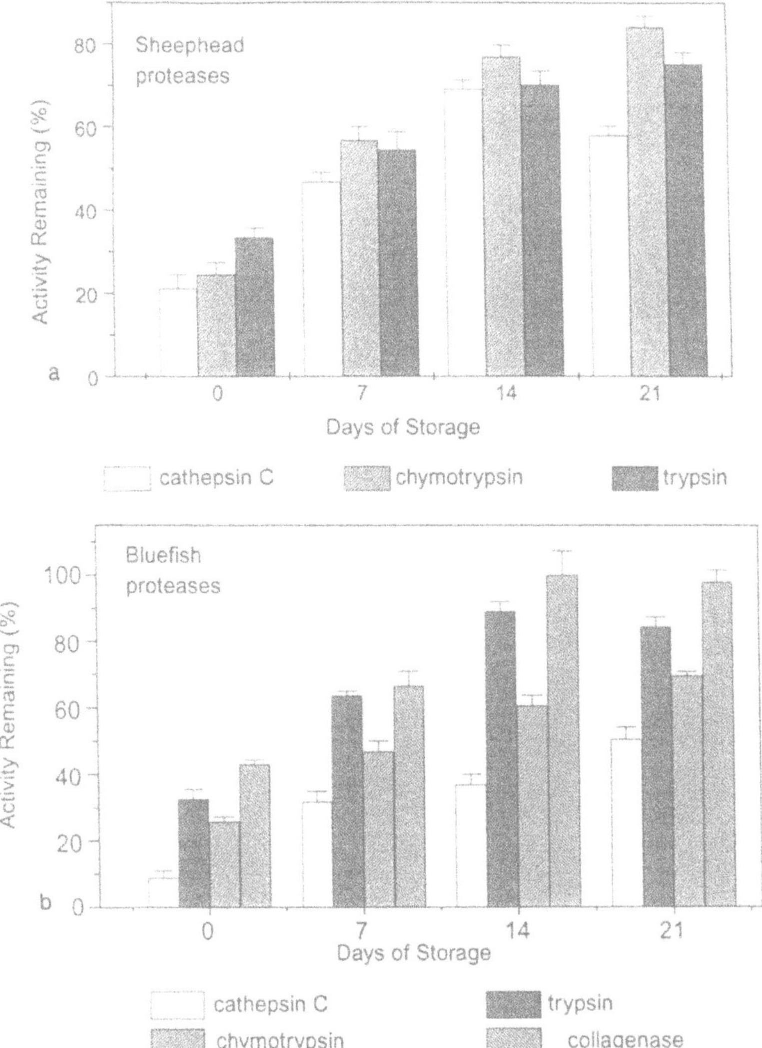

Figure 3. Reactivation of pressurized enzymes during storage at 4–7°C. Enzymes were pressurized at various levels for 30 min. Enzyme sources: (a) => sheephead; and (b) bluefish.

proteases studied, the order of sensitivity to pressure was: cathepsin C > chymotrypsin > trypsin (Figure 3a). After 7 days of storage, the residual activities of all three enzymes had virtually doubled as compared with the residual activity levels at day 0 (Figure 3a). For the four proteases from bluefish tissue investigated, the order of sensitivity to pressure was: cathepsin C > trypsin > chymotrypsin > collagenase (Figure 3b). There was also a reactivation of all the enzymes during storage.

Bluefish muscle tissues were used for the studies of pressure effects on texture. The data for textural properties, presented in Figures 4a and 4b, indicate that the sensitivity of

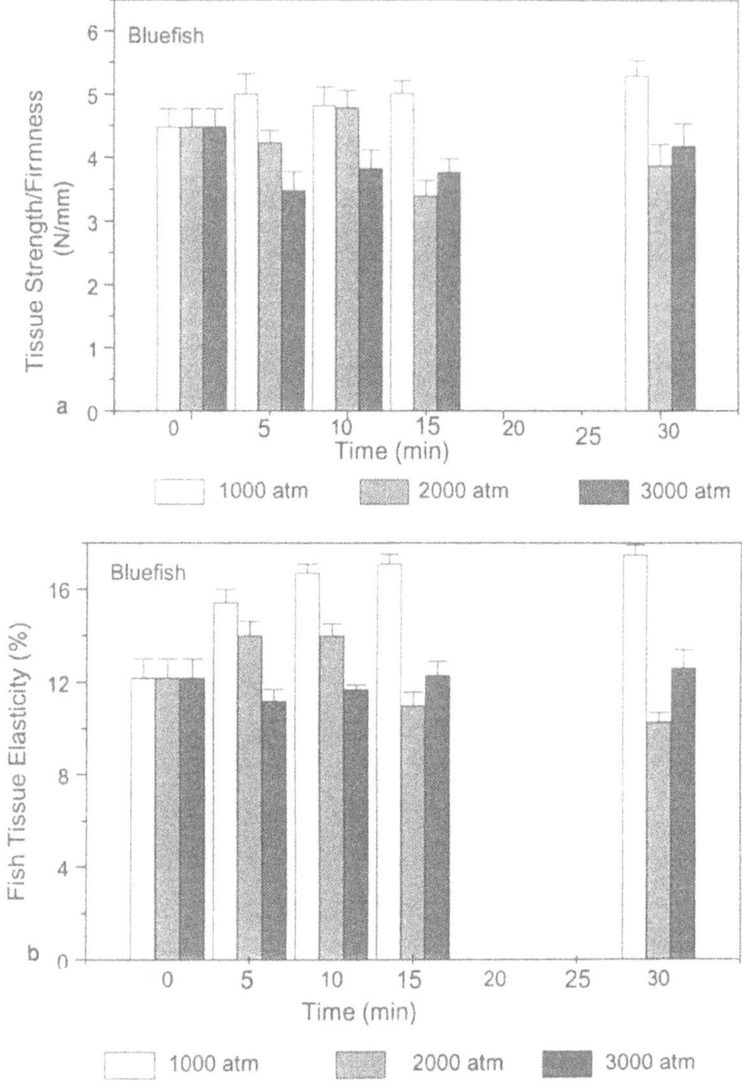

Figure 4. Effect of pressurization on (a) tissue strength, and (b) elasticity of bluefish meat.

fresh fish flesh to hydrostatic pressure depended on both the amount of pressure applied as well as the holding time. For example, fish tissue strength generally increased with the application of 1,000 atm pressure for the entire period used in this study (5–30 min). At 2,000 atm, however, there was no significant change in strength up to 10 min of holding (Figure 4a). Beyond this point, there was a general reduction in strength. Pressurization at 3,000 atm even for 5 min resulted in products with a softer texture than the controls (Figure 4a). The data on fish tissue elasticities presented in Figure 4b show some similarities with the responses of tissue strength after pressure treatment. For example, when fish tissues were pressurized at 1000 atm for up to 30 min, fish tissue elasticities generally increased with holding time of up to the 30 min; at 2000 atm fish tissue elasticity increased with holding time of up to 10 min and then declined; at 3000 atm, tissue elasticity generally declined with holding time. The "apparent" drop in tissue strength or firmness at higher pressures is possibly due to the loss of elasticity as a result of disruption of cohesive forces within the tissue (Bourne, 1976). It is suggested that the increase in fish tissue firmness at 1000 atm for up to 30 min, and/or at 2000 atm for up to 10 min of pressurization, was due to a decrease in volume as a result of compaction which facilitated protein-protein interactions, similar to other observations made with some proteins in cod and mackerel (Ohshima *et al.*, 1993). Such interactions would lead to an increase in bond formation causing an increase in tissue elasticity and firmness. However, excessive pressure beyond a certain critical point (2,000 atm for 10 min in this case), may result in the disruption of these and other bonds to cause a reduction in tissue elasticity and/or firmness.

The results reported here indicate that the upper limit of pressure application for maintaining or enhancing the firmness of fresh bluefish meat should not exceed 2,000 atm for 10 min, since higher pressures would lead to loss of tissue elasticity and firmness (flesh softening). These conditions may vary for the treatment of other fish species since tissue toughness varies among fish species.

In other studies, crude enzyme extracts were obtained from pressurized fish tissue and compared with pressurized "tissue-free" crude enzyme extracts for residual activity and was found that the extent of inactivation of the proteases was generally higher for the enzyme extracts from the pressurized tissue (Figures 5a and 5b). For example, the data presented in Figures 5a and 5b indicate that pressurization of intact fish tissue at 1000 atm for 30 min resulted in loss of about 66 and 77% of trypsin-like and chymotrypsin-like activities, respectively. In contrast, only 36 and 7% of their respective activities were destroyed for the same treatment on "tissue-free" crude enzyme extract. The reason for the difference in extent of inactivation of the proteases from pressure-treated intact fish tissue and the crude enzyme extract must await further studies. For now, it is speculated that other components naturally present in the more complex intact fish tissue may have acted in concert with pressure to achieve greater inactivation of the enzymes. It must also be borne in mind that the mechanisms of enzyme inactivation in solution versus enzyme inactivation in the intact tissue may be quite different. While intramolecular changes in enzyme structure/conformation is the main cause of inactivation in solution, possible enhancement of protein-protein interactions in the whole fish tissue could also result in intermolecular changes being involved in the process of inactivation.

Certain time-pressure combination treatments of the fish muscle resulted in significant ($p < 0.05$) changes in color as shown by the tristimulus Hunter color values in Table 1. Fish tissue color generally became lighter with increasing pressure presenting an increasingly cooked appearance to the flesh. The Hunter L values (indices of brightness or lightness of the sample) showed progressive increase with increase in pressure and duration of pressure application. The indicators of yellowness or blueness color (Hunter b values) also showed a similar trend with both the amount of pressure applied and holding time. The Hunter a values (indicators of greenness or redness) on the other hand, decreased with increase in pressure and holding time. Statistical analysis of the total color difference (i.e., ΔE values, where $\Delta E^2 = \Delta L^2 + \Delta a^2 + \Delta b^2$; Francis and Clydesdale, 1975, showed that there was no significant difference ($p > 0.05$) in color between sample treatments up to 2,000 atm for 10 min. Beyond this limit, however, the changes were quite significant (Table 1). The color changes noted here are similar to those reported for

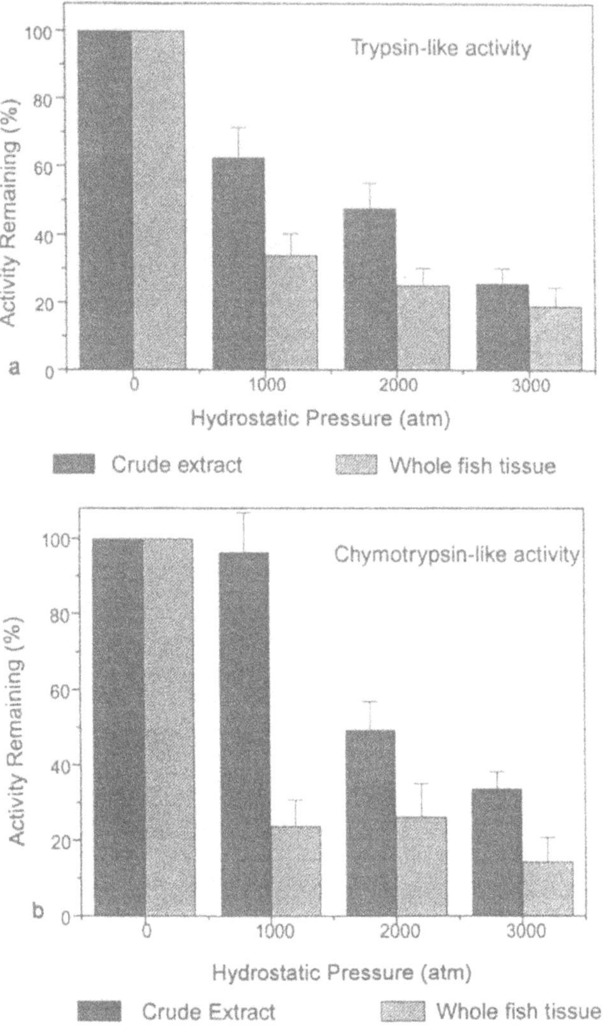

Figure 5. Comparison of residual activities of (a) trypsin-like, and (b) chymotrypsin-like enzymes in pressurized intact fish tissues versus pressurized "tissue-free" enzyme extracts.

pressure-treated squid mantle, cod and mackerel muscle, as well as Alaska pollock surimi (Nagashima *et al.*, 1993; Ohshima *et al.*, 1993; Shoji *et al.*, 1990).

The pressure levels used in the treatment of fish tissues, in studies presented in Figures 3a and 3b, were not adequate to ensure complete and/or permanent inactivation of fish proteases. All four enzymes investigated showed various degrees of reactivation during refrigerated storage. In order to achieve a more lasting inactivation of the enzymes from the fish tissues, the pressure treatments were combined with the addition of the broad spectrum protease inhibitor, α_2-macroglobulin. Previous studies on the effects of pressure on α_2-macroglobulin activity indicated that the activity of this inhibitor was not affected by increasing pressures up to 1,500 atm. However, increases in pressure beyond this level inactivated the inhibitor to various extents (Ashie and Simpson, 1995). Endogenous protease activity in minced fish tissue was reduced with increasing pressure and/or α_2-macroglobulin concentrations (Ashie *et al.*, 1996b). Based on these observations, the inhibitor was incorporated in raw bluefish meat paste and used to formulate fish gels by pressure or heat treatment. The pressure-induced gels and heat-induced gels formulated with added α_2-macroglobulin were evaluated for their textural properties (Figures 6a and 6b). The heat-induced gels were significantly firmer than their pressure-induced counterparts ($p < 0.01$), while the pressure-induced gels were more elastic than those of the heat-induced gels ($p < 0.01$). Firmness of the heat-treated samples generally increased up to 7 days ($p <$

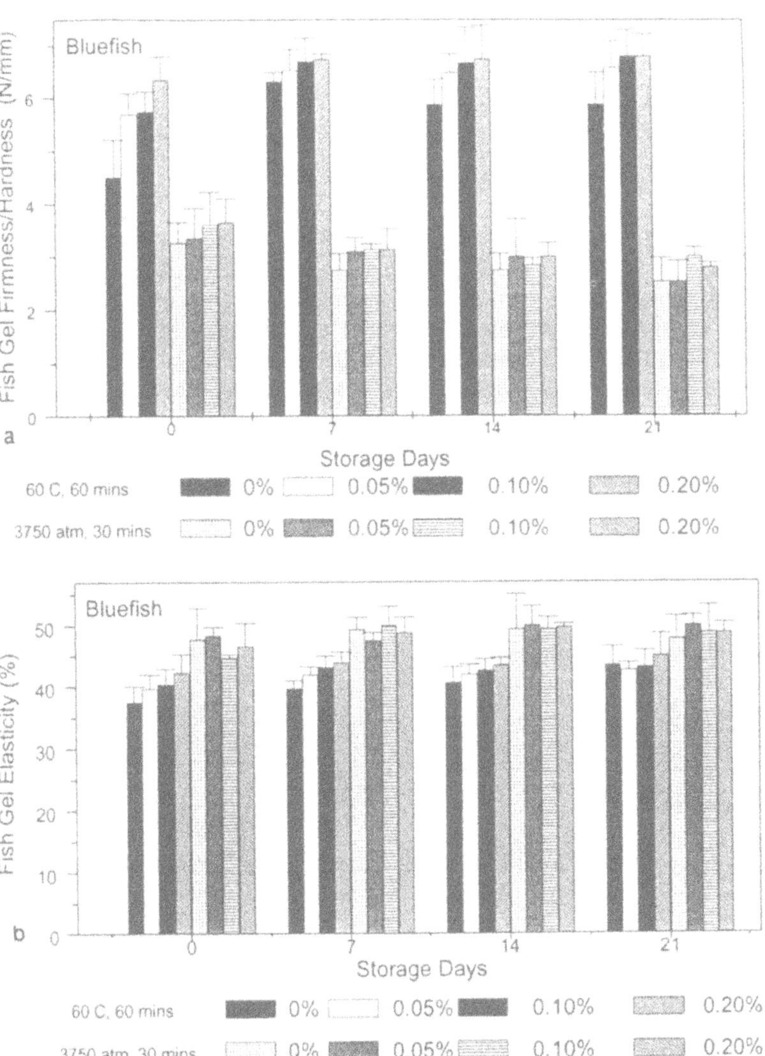

Figure 6. Comparison of textural properties (a => tissue strength; b => elasticity) of pressure-induced versus heat-induced gels formulated from bluefish meat paste with added α_2-macroglobulin.

Table 1. Tristimulus Hunter color-values of pressure-treated fish tissue

Pressure applied	Holding time	L	a	b	ΔE
1	0	40.5	1.5	−2.6	40.562[f]
1000	5	43.3	1.5	−2.6	43.405[f,e]
	10	46.3	1.5	−2.7	46.407[d,e]
	15	46.9	1.6	−2.3	46.988[d,e]
	30	49.7	1.4	−2.5	49.784[d,c]
2000	5	48.7	1.8	−2.2	48.784[d,c]
	10	52.3	1.5	−1.4	52.290[c]
	15	56.2	1.3	−1.2	56.179[b]
	30	59.3	0.6	−0.8	59.259[b]
3000	5	67.7	1.1	−0.5	67.661[a]
	10	68.8	0.9	0.2	68.807[a]
	15	69.0	0.6	−0.3	69.004[a]
	30	69.1	0.2	0.8	69.105[a]

Mean ΔE values with the same superscripts are not significantly different ($p > 0.05$). Least significant difference (LSD) = 3.6341.

0.01) irrespective of α_2-macroglobulin concentrations, then remained constant throughout the rest of the 21 day storage period. On the other hand, there were no significant changes ($p > 0.05$) in firmness of the pressure-treated samples during storage. The increase in hardness of the heat-induced gels versus the pressure-induced gels found in the early stages of the storage period could have resulted from a greater decrease in water-holding capacity coupled with an increased crosslinking in protein molecules of the heat-treated gels due to pH changes. As suggested by Farr (1990) and Okamoto *et al.* (1990), the mechanisms of gel formation or protein denaturation caused by heat and pressure are quite different.

Furthermore, both firmness and elasticity of the heat-induced gels in the early stages of storage (< 7 days) were enhanced significantly ($p < 0.05$) over the control samples as the levels of incorporated α_2-macroglobulin was increased, suggesting that α_2-macroglobulin could control texture degradation of gels due to proteolysis during the storage period. In contrast, addition of α_2-macroglobulin did not appear to have a significant effect ($p < 0.05$) on the texture of the pressure-induced gels even though gel hardness increased slightly with increasing concentration of the inhibitor, suggesting that α_2-macroglobulin was more effective in controlling the activity of heat stable/active endogenous proteases involved in texture degradation of bluefish gels formulated by heat at 60°C for 60 min.

CONCLUSION

This study has demonstrated that fish enzymes are generally more susceptible to hydrostatic pressure inactivation than their mammalian counterparts. However, pressurization of fish enzymes up to 3000 atm was inadequate to achieve complete inactivation of the enzymes. The pressurized enzymes also showed various degrees of reactivation during storage at 4–7°C. Pressurization resulted in tissue color becoming lighter presenting a cooked appearance to the fresh fish meat. Both tissue strength and elasticity showed a biphasic response on pressurization. Pressurization, up to 2,000 atm for 10 min, resulted in the enhancement of both textural parameters whereas pressures beyond these conditions generally caused weakening of the tissue. Furthermore, application of hydrostatic pressure and incorporation of protease inhibitor (α_2-macroglobulin) were both effective in controlling texture softening in fish gels. The results presented give further support for the potential of hydrostatic pressure technology as a food processing tool. Thus hydrostatic pressure technology may be used in combination with other barriers, like enzyme inhibitors, to control texture deterioration of seafoods and other muscle foods due to its ability to inactivate the endogenous proteases.

REFERENCES

Ashie, I.N.A.; Simpson, B.K. Effects of hydrostatic pressure on alpha-Macroglobulin and selected proteases. *J. Food Biochem.* **1995**, *18*, 377–391.

Ashie, I.N.A.; Simpson, B.K.; Smith, J.P. Spoilage and shelf-life extension of fresh fish and shellfish. *CRC Crit. Rev. Fd. Sci. Nutr.* **1996a**, *36*, 1–30.

Ashie, I.N.A.; Simpson, B.K.; Ramaswamy, H.S. Control of endogenous enzyme activity by inhibitors and hydrostatic pressure using RSM. *J. Food Sci.* **1996b**, *61*, 350–356.

Baranowski, J.D.; Nip, W.K.; Moy, J.H. Partial characterization of a crude enzyme extract from the freshwater prawn, (*Marcobrachium rosenbergii*), *J. Food Sci.* **1984**, *49*, 1494–1495.

Bourne, M.C. Interpretation of force curves from instrumental texture measurements. In *Rheology and Texture in Food Quality;* Deman, *et al.*, Eds; AVI Publishing Company Inc. Wesport, CN, 1976; pp 244–274.

Bridgman, P.W. The coagulation of albumen by pressure. *J. Biol. Chem.* **1914**, *19*, 511–512.

Curl, L.A.; Jansen, E.F. Effect of high pressures on trypsin and chymotrypsin. *J. Biol. Chem.* **1950**, *184*, 45–54.

Erlanger, B.F.; Kokowsky, N.; Cohen, W. The preparation and properties of two new chromogenic substrates of trypsin. *Archs. Biochem. Biophys.* **1961**, *95*, 271–278.

Farkas, D.F. Novel Processes - Ultra High Pressure Processing. In *Food Protection Technology;* Felix, C.W., Ed.; Lewis Pub., Inc. Chelsea, Michigan, 1987.

Farr, D. High pressure technology in the food industry. *Trends Food Sci. Technol.* **1990**, *1*, 14–16.

Fletcher, G.C.; Statham, J.A. Shelf-life of sterile yellow-eyed mullet (*Aldrichetta forsteri*) at 4°C. *J. Food Sci.* **1988**, *53*, 1030–1035.

Flick, G.J.; Lovell R. T. Postmortem degradation of nucleotides and glycogen in Gulf shrimp. *Dissert. Abst. Int. Section B. Science and Engineering.* **1970**, *30*, 1743.

Francis , F. J.; Clydesdale, F. M. Food Colorimetry: Theory and Applications. AVI Publishing Co., Inc. Westport, CT, 1975.

Guizani, N.; Marshall, M.R.; Wei, C.I. Purification and characterization of a trypsin-like enzyme from the hepatopancreas of crayfish (*Procambarus clarkii*). *Comp. Biochem. Physiol.* **1992**, *103B*, 809–815.

Haard, N.F.; Martins, I.; Newbury, R.; Botta, R. Hypobaric storage of Atlantic herring and cod. *Can. Inst. Food Sci. Technol. J.* **1979**, *12*, 84–87.

Hartree, E.F. Determination of protein: a modification of the Lowry method that gives a linear photometric response. *Anal. Biochem.* **1972**, *48*, 422–427.

Hayakawa, I.; Kanno, T.; Tomita, M.; Fujio, Y. Application of high pressure for spore inactivation and protein denaturation. *J. Food Sci.* **1994**, *59*, 159–163.

Hite, B.H.; Giddings, N.J.; Weakly, C.E. The effects of pressure on certain microorganisms encountered in the preservation of fruits and vegetables. Bull. 146, W. Va. Univ. Agric. Exp. Sta. Morgantown, USA, 1994.

Hochachka, P.W.; Storey, K.B.; Baldwin, J. Gill citrate synthase from an abyssal fish. *Comp. Biochem. Physiol.* **1975**, *52B*, 43–49.

Hultin, H.; McDonald, R.; Kelleher, S. Lipid oxidation in fish muscle microsome. In *Chemistry & Biochemistry of Marine Food Products*; Martin *et al.*, Eds.; AVI, Westport, Connecticut, 1982; pp 1–9.

Hummel, B.C.W. A modified spectrophotometric determination of chymotrypsin, trypsin, and thrombin. *Can. J. Biochem. Physiol.* **1950**, *37*, 1393–1399.

Konagaya, S. Proteases responsible for softening or lysing of meat of chum salmon caught during spawning migration. *Bull. Tokai Reg. Fish. Res. Lab.* **1985**, *116*, 39–47.

Lakidos, D.; Lougovois, V. Lipid oxidation in muscle foods. *Food Chem.* **1990**, *35*, 295–314.

Lanier, T.C.. High pressure processing effects on fish products. In *Int. Chem. Congress of Pacific Basin Societies*; Dec. 1995, 17–22.

Lee, H-J.; LaRue, J. N.; Wilson, I. B. A simple spectrophotometric assay for amino acyl arylamidases (naphtylamidases, aminopeptidases). *Anal. Biochem.* **1971**, *41*, 397–401.

Lindner, P.; Angel, S.; Weinberg, Z.G.; Granit, R. Factors inducing mushiness in stored prawns. *Food Chemistry.* **1988**, *29*, 119–132.

Nagashima, Y.; Ebina, H.; Tanaka, M.; Taguchi, T. Effect of high hydrostatic pressure on the thermal gelation of squid mantle meat. *Food Res. Int.* **1993**, *26*, 119–123.

Nip, W.K.; Lan, C. Y.; Moy, J.H. Partial characterization of a collagenolytic enzyme fraction from the hepatopancreas of the freshwater prawn, (*Macrobrachium rosenbergii.*) *J. Food Sci.* **1985**, *50*, 1187–1188.

Ogawa, H.; Fukuhisa, K.; Kubo, Y.; Fukumoto, H. Pressure inactivation of yeasts, molds, and pectinesterase in Satsuma Mandarin juice: effects of juice concentration, pH, and organic acids, and comparison with heat sanitation. *Agric. Biol. Chem.* **1990**, *54*, 1219–1225.

Okamoto, M.; Kawamura, Y.; Hayashi, R. Application of high pressure to food processing: textural comparison of pressure- and heat-induced gels of meat proteins. *Agric. Biol. Chem.* **1990**, *54,* 183–189.

Ohshima, T.; Ushio, H.; Koizumi, C. High pressure processing of fish and fish products. *Trends Food Sci. Technol.* **1993**, *4*, 370–375.

Sareevoravitkul, R.; Simpson, B.; Ramaswamy, H. Comparative properties of bluefish (*Pomatomus saltatrix*) gels formulated by high hydrostatic pressure and heat. *J. Aquat. Food Prod. Tech.* **1996**, (In press).

Shoji, I.; Saeki, H.; Wakameda, A.; Nakamura, M.; Nonaka, M. Gelation of salted paste of Alaska pollock by high hydrostatic pressure and change in myofibrillar protein in it. *Nippon Suisan Gakkaishi* **1990**, *56*: 2069–2076

Simpson, M.V.; Haard, N.F. Temperature acclimation of Atlantic cod (*Gadus morhua*) and its influence on freezing point and biochemical damage of postmortem muscle during storage at 0°C and -3°C. *J. Food Biochem.* **1987a**, *11*, 69–93

Simpson, B.K.; Haard, N.F. Cold-adapted enzymes from fish. In *Food Biotechnology;* Knorr, D., Ed.; Marcel Dekker, Inc., 1987b; pp 495–527.

Steel, R.J.D.; Torrie, J.H. Principles and Procedures of Statistics - A Biometric Approach. (Steel, R.J.D., and Torrie, J.H., eds.). McGraw Hill Pub. Co., 1980, pp 173–175.

Toyomizu, M.; Hanaoka, K.; Yamaguchi, K. Effect of release of free fatty acids by enzymatic hydrolysis of phospholipids on lipid oxidation during storage of fish at -5°C. *Nippon Suisan Gakkaishi.* **1988**, *47*, 615–620.

Tsukuda, N. Studies on the discoloration of red fishes. VI. *Bull. Jpn. Soc. Sci. Fish.* **1970**, *36*, 725–730.

Wasson, D.H. Fish muscle proteases and heat-induced myofibrillar degradation: A Review. *J. Aquat. Food Prod. Technol.* **1992**, *1(2)*, 23–41.

Wunsch, E.; Heidrich, H-G. Quantitative determination of collagenase. *Biochem. J.* **1963**, *333*, 149–151.

HIGH PRESSURE AND HEAT TREATMENTS EFFECTS ON PECTIC SUBSTANCES IN GUAVA JUICE

Gow-Chin Yen and Hsin-Tang Lin

Department of Food Science
National Chung Hsing University
250 Kuokuang Road, Taichung, Taiwan, Republic of China

Effects of high pressure treatment on changes in pectic substances in guava juice were investigated and compared with those of heat treated samples. The viscosity and turbidity of guava juice pressurized at 6000 atm and 25°C for 10 min increased slightly, whereas the viscosity of juice heated at 95°C for 5 min decreased from 362 to 285 cps while turbidity increased from 0.87 to 1.15 (OD 600 nm). There were no apparent changes in water soluble, oxalate soluble and alkali soluble pectins in the pressurized juice. However, heat treated juice exhibited a decrease in its water and alkali soluble pectins and a slight increase in oxalate soluble pectin. The DEAE-cellulose profiles of pectic substances in guava juice were apparently unchanged after high pressure treatment while they were markedly changed by heat treatment, due to coagulation or degradation. During thermal processing, the degradation of pectin in guava juice caused a decrease in viscosity while the coagulation of pectin resulted in an increase in turbidity and cloud content. High pressure treatment showed no marked changes in pectic substances and cloud content in guava juice and maintained its natural viscous properties.

INTRODUCTION

Depending on the maturity of the fruit, Guava fruit contains a high quantity of pectin (Burns, 1991). Ferro *et al.* (1969) reported that pectin from guava has a high methoxy index. Pectin is a polysaccharide and is believed to impart a viscous property to the guava puree or juice. In some fruit juices such as orange or guava juices, the pectic substances would combine with protein or polyphenolic compounds to form a suspension phenomenon, termed as "cloud" (Binning, 1992). Therefore, highly viscous and cloudiness are special characteristics of guava juice. Guava juice is one of the most popular fruit juices in Taiwan, and cannot be substituted by imported fruit juices. However, some problems still exist regarding the color and flavor quality of the juice arising during processing and storage (Yen *et al.*, 1992a, 1994). Yen *et al.* (1992b) reported that the decrease in viscosity and turbidity of guava puree during storage is due to the degradation of pectin by pectin esterase. The quality of guava puree should hence be judged by the stability of the pectin that it contains.

It has been suggested that high hydrostatic pressure may be applied to food during its processing rather than heating (Hayashi, 1987). This technique has high potential for applications in the food industry (Knorr, 1993). Several studies have been carried out in order to investigate the sterilization and inactivation of enzymes in fruit juices using high pressure treatment, which would reduce the undesired change in quality as a result of heat treatment (Ogawa *et al.*, 1990; Takahashi *et al.*, 1993; Lin and Yen, 1995). The high pressure treatment has gained recognition because it does not cause the formation and destruction of covalent bonds but affects hydrogen, ionic and hydrophobic bonds (Heremans, 1982). However, this treatment causes changes in denaturation and gelation of protein and pectin in the foods.

The viscosity and turbidity characteristics of guava juice are due to its high content of pectin and cloud. However, these characteristics could be changed during thermal processing and storage, thus resulting in a loss of quality. The objective of this work was to investigate the effect of high pressure treatment on changes in pectic and cloud substances in guava juice and to compare the results with those of heat-treated samples.

EXPERIMENTAL

Preparation of Guava Juice

Guava puree was prepared from fully ripe guava fruit (*Psidium guajava* L. cv. Chung shan) in accordance with the method described by Boyle *et al.* (1957). Guava juice (30 %) was prepared from guava puree by diluting with distilled water. The guava juice (originally having pH 4.7, 3°Brix) was adjusted to pH 3.8 and 12°Brix as in commercial products and stored at 4°C until utilized for heat and high pressure treatments.

High Pressure Equipment

The hydrostatic pressurization equipment (Pmax 6000 atm, Mitsubishi Heavy Industries, Ltd., Type 471–046, Hiroshima, Japan) used in the present study consisted of a pressure container and a pressure cell, with approximately 120 mL capacity. The pressure cell was maintained at a constant temperature with a thermo-controlled water bath (Hakke GH, Germany). Pressure was applied by operating a piston with an oil hydraulic motor, that is used in the high pressure equipment. The time required to build up pressures of up to 4000 and 6000 atm was about 30 and 40 sec, respectively. The time required to return to atmospheric pressure was about 5 sec.

High Pressure Treatment

The pH adjusted guava juice was poured into a plastic container (volume of about 100 mL) and treated under 1000, 3000, 5000 and 6000 atm pressure for 10 min. The treatment temperature was controlled at 25°C.

Thermal Pasteurization Treatment

The pH adjusted guava juice was pasteurized by thermal treatment at 95°C for 5 min and was then chilled with ice water to reduce the temperature to below 20°C. The cooled juice was then stored at 4°C until use.

Analysis of Soluble and Insoluble Solids

The content of soluble solids was determined by centrifuging a sample at 7500 x *g* for 10 min and measuring the°Brix with a refractometer at 25°C. The insoluble solid content was estimated by centrifuging a sample at 1270 x g for 10 min and calculating the volume ratio (v/v %) of the precipitate to juice.

Turbidity Determination

Guava juice (20 mL) was centrifuged at 800 x *g* for 10 min. The turbidity of the upper layer solution was estimated by measuring the absorbance at 600 nm using a spectrophotometer, as described by Chandler and Robertson (1983).

Viscosity Measurement

The viscosity (cps) of samples was measured with a digital viscometer (Brookfield, USA) at the rotation speed of 100 rpm with a No. 21 spindle.

Pectin Fraction and Determination

The alcohol insoluble solids (AIS) of the heated, pressurized and untreated guava juice samples were obtained by the procedure proposed by Plat *et al.* (1991). The AIS were fractionated into water soluble pectin (WSP), oxalate soluble pectin (OSP) and alkali soluble pectin (ASP) fractions, according to the procedure given by Dietz and Rouse (1953) and quantified using the carbazole colorimetric method given by Ting and Rouseff (1986).

Degree of Esterification

The degree of esterification of samples was determined using the method developed by McCready (1970).

Separation of the WSP, OSP and ASP on DEAE-Cellulose

The DEAE-cellulose column chromatography of pectic substances was carried out by the method of Plat *et al.* (1991) with a slight modification. Pectic substances (in 50 mM sodium phosphate buffer, pH 6.3) were added to a DEAE cellulose column (2 x 40 cm) equilibrated with the same phosphate buffer. The solution was then eluted with a 50 mM sodium phosphate buffer (pH 6.3) with a flow rate of 30 mL/h. The fraction (2 mL) was collected and monitored by the carbazole colorimetric method and expressed as the absorbance at 525 nm.

Cloud Determination

The cloud content of guava juice was determined by the method of Klavons *et al.* (1991). Pulp was removed from all samples by low-speed centrifugation (360 x g for 10 min). The pulpless juice (15 mL) was then centrifuged at 27000 x g for 15 min, to produce a supernatant, the optical density (OD) of which was less than 0.05 at 600 nm. The supernatant was then decanted and the cloud pellet was redissolved in 10 mL of deionized water by vortexing. This process was repeated and the total cloud weight of the sample was obtained by freeze-drying to a constant weight.

Preparation for Scanning Electron Microscopy (SEM)

The cloud was dehydrated overnight in a Labconco freezer-dryer (USA). All samples were mounted on aluminum stubs using a double-sided tape, sputter-coated with gold-palladium and investigated using a Topcon ABT-150S SEM (Japan) at 15 kV.

Statistical Analysis

All data were analyzed using ANOVA and Duncan's multiple range tests by means of SAS (SAS Institute Inc., 1985) statistical system.

RESULTS AND DISCUSSION

Changes in Viscosity and Turbidity

The viscosity and turbidity are important indices of the viscous properties of guava juice. Table 1 shows the changes in viscosity and turbidity of guava juice after subjecting it to heat and high pressure treatments. The viscosity of guava juice was found to be reduced from 362 to 285 cps after heating at 95°C for 5 min. This reduction in viscosity might be due to the decomposition of pectic substances by heat treatment which results in reducing the interaction between pectin molecules and/or pectin with other components. The viscosity of guava juice did not change significantly (P>0.05) under 1000 and 3000 atm pressure at 25°C for 10 min, but it was increased slightly from 362 to 385 cps under 5000 atm and to 405 cps under 6000 atm. The high pressure treatment did not destroy the components of interest but resulted in either formation or destruction of non-covalent bonds (hydrogen, ionic and hydrophobic bonds) (Heremans, 1982). This would influence

Table 1. Effect of high pressure and heat treatments on changes in the viscosity and turbidity of guava juice[1]

			High pressure treatment (25°C, 10 min)			
	Untreated	Heating 95°C, 5 min	1,000 atm	3,000 atm	5,000 atm	6,000 atm
Viscosity (cps)	362[2]	285[c]	357[a]	366[a]	385[b]	405[b]
Turbidity (Abs. at 660 nm)	0.87[a]	1.15[c]	0.83[a]	0.86[a]	0.89[ab]	0.96[b]

[1]Guava juice (30%) was adjusted to pH 3.8 and 12° Brix. Means within a row with different letters are significantly different at 5% level.

the structure of large molecules and the intermolecular interaction forces. Therefore, the viscosity was increased when a pressure higher than 5000 atm was applied, perhaps due to the interaction between pectin molecules and/or with other components in juice. Horie (1991) has reported that gelation of pectin by pressure treatment is similar to that caused by heat treatment at certain pectin contents, and brix and pH values. The content of pectin in guava juice was not high enough to cause gelation by pressure treatment, though it caused slight changes in viscosity and turbidity after undergoing high pressure treatment.

The turbidity of heated guava juice was found to increase significantly ($P<0.05$). Although the pectic substance in guava juice was degraded by heat and hence reduced the viscosity, the degraded pectin could probably combine with protein and increase the turbidity of guava juice. Klavons *et al.* (1992) have observed that the turbidity of orange juice increases after heat treatment due to the combination of pectin and proteins. In the present study, the turbidity of guava juice did not show any significant change ($P>0.05$) under 1000 to 5000 atm pressure for 10 min, but it was increased ($P<0.05$) under 6000 atm pressure. The increase in viscosity might be due to the interaction of pectin with proteins as a result of high pressure treatment. From these results, it can be concluded that the viscosity and turbidity of guava juice changed appreciably after undergoing heat treatment, whereas these properties showed only a slight increase after high pressure treatment, particularly beyond 5000 atm.

CHANGES IN SOLUBLE AND INSOLUBLE SOLID CONTENTS

The soluble solids in juice include sugars, organic acids and other soluble components. Meanwhile, the insoluble components in the juice may dissolve after heat treatment, hence increasing the soluble solid content. However, the soluble components may combine with other insoluble substances resulting in a decrease in soluble solid contents. The soluble solid content in guava juice exhibited no significant change ($P>0.05$) after heating, when compared to the unheated sample (Table 2). Similarly, high pressure treatment of guava juice under 1000 to 6000 atm for 10 min exhibited no significant ($P>0.05$) change in its soluble solid content. The insoluble solid content of guava juice, on the other hand, showed a significant increase from 5.2 to 10.6 % ($P<0.05$) after heating. Such an increase in the content of insoluble solids might be due to the conversion of some suspended components in juice to insoluble solids after heat treatment. The heat treatment also causes degradation of pectin in the cloud which allows the proteins in the cloud to combine with other components and form precipitates (Binnig, 1992). However, the insoluble solid contents of pressurized juice did not change significantly ($P>0.05$) when treated under 1000 to 6000 atm for 10 min. Takahashi *et al.* (1993) have also reported similar results for high pressure treated orange juice.

Table 2. Effect of high pressure and heat treatments on changes in the solid content of guava juice[1]

			High pressure treatment (25°C, 10 min)			
Solids	Untreated	Heating 95°C, 5 min	1,000 atm	3,000 atm	5,000 atm	6,000 atm
Soluble (°Brix)	11.8[a]	11.6[a]	11.7[a]	11.8[a]	11.7[a]	11.6[a]
Insoluble (%)	5.20[a]	10.6[b]	5.10[a]	5.2[a]	5.2[a]	5.2[a]

[1]Guava juice (30%) was adjusted to pH 3.8 and 12° Brix. Means within a row with different letters are significantly different at 5% level.

Table 3. Effect of high pressure and heat treatments on changes in the pectin content of guava juice[1]

Pectins	Untreated	Heating 95°C, 5 min	High pressure treatment (25°C, 10 min) 1,000 atm	3,000 atm	5,000 atm	6,000 atm
Water-soluble (mg/100 mL)	95.35[a]	82.49[b]	94.46[a]	96.84[a]	97.01[a]	93.88[a]
Oxalate-soluble (mg/100 mL)	102.42[a]	100.36[b]	104.12[a]	99.79[a]	101.22[a]	103.93[a]
Alkali-soluble (mg/100 mL)	67.20[a]	47.40[b]	68.57[a]	65.73[a]	57.86[a]	66.01[a]

[1]Guava juice (30%) was adjusted to pH 3.8 and 12° Brix. Means within a row with different letters are significantly different at 5% level.

CHANGE IN PECTIN CONTENT

The effects of processing conditions on pectin content of guava juice are presented in Table 3. The pectic substances are classified into water soluble, oxalate soluble and alkali soluble pectin according to the extraction procedure employed. Guava juice contains high amounts of oxalate pectin (102.42 mg/100 mL) and low amounts of water soluble pectin (67.2 mg/100 mL). The contents of water soluble and alkali soluble pectins were observed to decrease after heating at 95°C for 5 min while the content of oxalate soluble pectin was found to increase from 102.42 to 110.36 mg/100 mL. This difference between water soluble, oxalate and alkali soluble pectin is due to methoxy content and the binding sugars. Burns (1991) reported that heating results in conversion of propectin to pectin in juice but over cooking causes the degradation of pectin. This result confirms the decrease in viscosity after heat treatment as shown in Table 1. However, these three types of pectin did not change significantly after high pressure treatment under 6,000 atm and 25°C for 10 min.

CHANGES IN THE DEGREE OF ESTERIFICATION

The degree of pectin esterification affects the solubility of pectin and its interaction with other components in juice. The effects of processing conditions on the degree of esterification of pectin in guava juice are presented in Table 4. The alkali soluble pectin in untreated guava juice had the highest degree of esterification (47.2 %), whereas oxalate soluble pectin had the lowest degree of esterification (27.4 %). The degree of pectin esterification in heated juice however, decreased significantly ($P<0.05$). This result im-

Table 4. Effect of high pressure and heat treatments on changes in the degree of esterification of pectin in guava juice[1]

Pectin esterification %	Untreated	Heating 95°C, 5 min	High pressure treatment (25°C, 10 min) 1,000 atm	3,000 atm	5,000 atm	6,000 atm
Water-soluble	40.3[a]	32.6[b]	41.2[a]	42.4[a]	40.7[a]	42.1[a]
Oxalate-soluble	27.4[a]	20.5[b]	26.5[a]	26.3[a]	27.0[a]	27.6[a]
Alkali-soluble	47.2[a]	23.8[b]	46.0[a]	46.3[a]	47.4[a]	45.7[a]

[1]Guava juice (30%) was adjusted to pH 3.8 and 12° Brix. Means within a row with different letters are significantly different at 5% level.

Table 5. Effect of high pressure and heat treatments on changes in the total cloud weight of guava juice[1]

Untreated	Heating 95°C, 5 min	High pressure treatment (25°C, 10 min) 1,000 atm	3,000 atm	5,000 atm	6,000 atm
42.30[a]	63.5[b]	42.6[a]	42.0[a]	42.5[a]	42.7[a]

[1]Unit: mg% guava juice (30%) was adjusted to pH 3.8 and 12° Brix. Means within a rowwith different letters are significantly different at 5% level.

plies the occurrence of demethylation by heat treatment. However, it was observed that the degree of esterification of these three types of pectin in guava juice remained unchanged (P>0.05) after treatment under 1000 to 6000 atm for 10 min.

CHANGE IN CLOUD CONTENT

The changes in cloud content of guava juice under different processing conditions are presented in Table 5. The cloud content of guava juice was found to increase from 42.3 to 63.5 mg % after heating at 95°C for 5 min. This increase might be due to coagulation between components in juice by heat treatment (Shomer, 1991a, b). This result is similar to the increasing trend in turbidity as shown in Table 1. This result thus indicates a significant correlation between turbidity and cloud content in juice. The cloud content of pressurized juice was around 42.0–42.7 mg% when treated under 1000–6000 atm and 25°C for 10 min and showed no significant change (P>0.05) when compared with the untreated sample. It is thus clear that the cloud (Table 5) and pectin (Table 3) contents in juice are not affected by high pressure treatment, since the high pressure treatments does not cause formation or destruction of covalent bonds.

SEM OF CLOUD SUBSTANCE

Figure 1 depicts the SEM of cloud substance from guava juice under various processing conditions. The cloud surface of the fresh sample (unheated and not pressurized) was irregular and with unique particle size (Fig. 1, A & B). The particle distribution and appearance of pressurized juice (6,000 atm, 25°C, 15 min) were similar to the fresh sample (Fig. 1, C & D). Takahashi *et al.* (1993) have reported that the soluble solid particle distribution in citrus juice does not change after high pressure treatment. However, the cloud surface of pasteurized juice was observed to be greatly different from the pressurized juice (Fig. 1, E & F) due to coagulation of the small particles.

CHANGES IN MOLECULAR CHARACTERISTICS OF PECTIN

The pectic substances in guava juice, and as affected by changes by heat or high pressure treatment, were separated by DEAE cellulose column. The changes of water soluble, oxalate soluble and alkali soluble pectin in guava juice after heat or high pressure treatment are presented in Fig. 2. The results indicate that the observed molecular distribution of these three types of pectin of high pressure treated guava juice were similar to un-

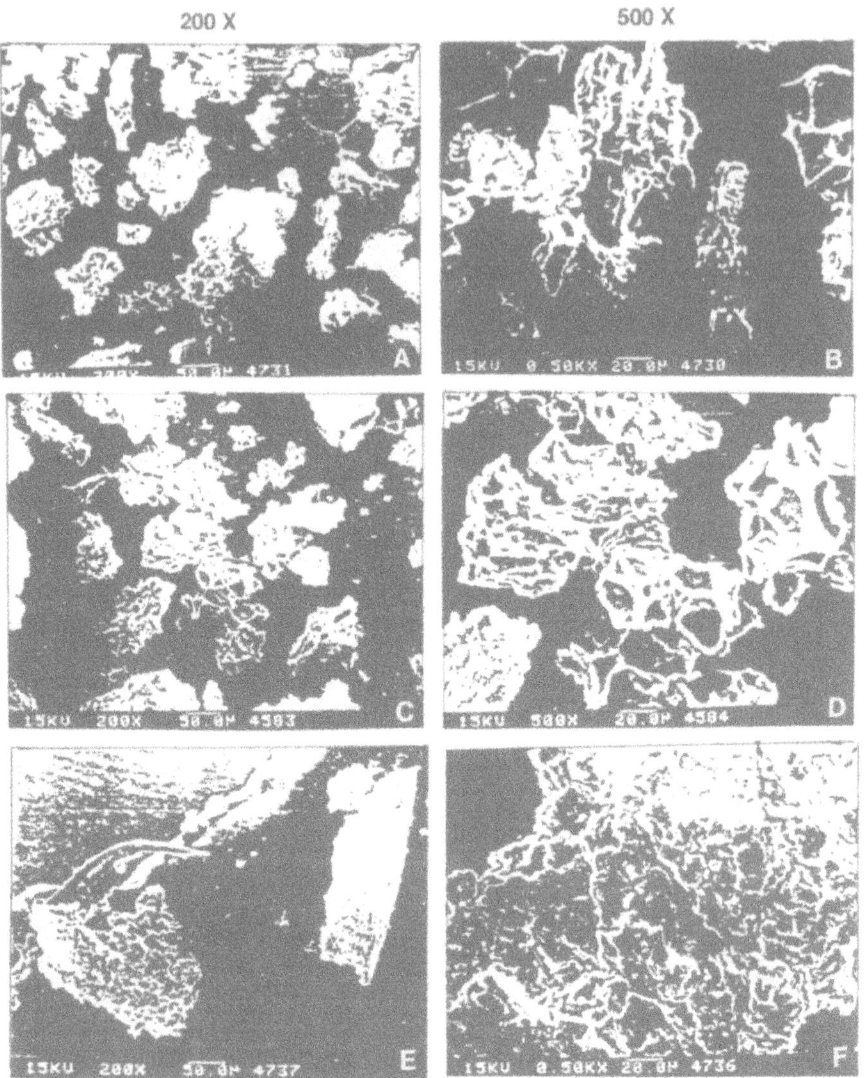

Figure 1. SEM pictures of cloud substance is untreated, pressurized and heated guava juice. Guava juice (30%) was adjusted to pH 3.8 and 12°Brix. **A, B:** untreated guava juice. **C, D:** guava juice was pressurized at 6,000 atm and 25°C for 10 min. **E, F:** guava juice was heated at 95°C for 5 min.

treated juice (fresh). This result further confirms that the pectin in guava juice was not effected by high pressure treatment. However, the molecular characteristic of pectin in guava juice was changed markedly after heat treatment. These three soluble pectins degraded to smaller molecules, and then some of these smaller molecules coagulated to form larger molecules in alkali pectin. This implies that the pectic substance was degraded by heat treatment. The degraded pectic substances then coagulated to form insoluble components which resulted in an increase of the clouding substance and the turbidity of guava

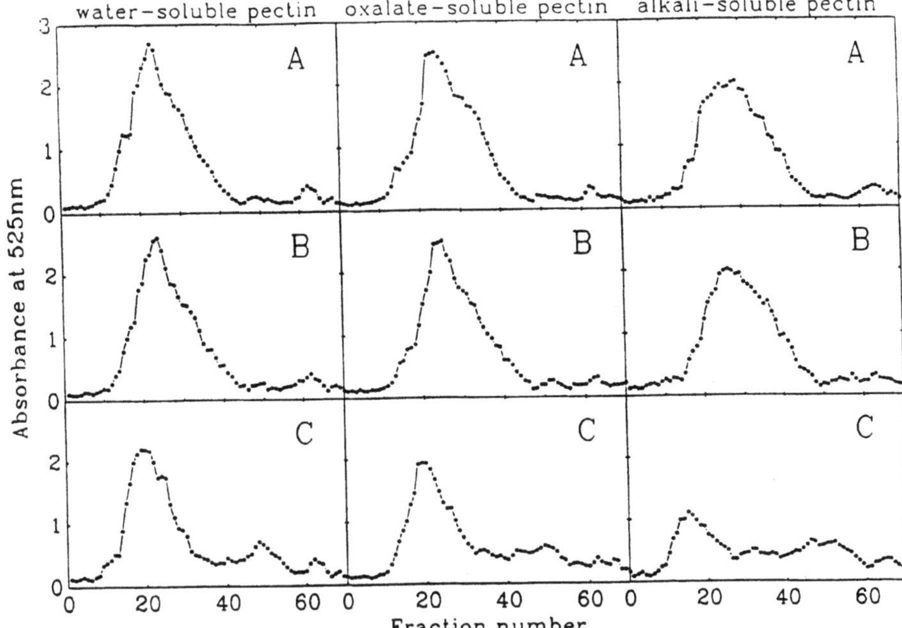

Figure 2. Separation of soluble pectin extracted from the alcohol-insoluble solids of guava juice on DEAE-cellulose. Guava juice (30%) was adjusted to pH 3.8 and 12°Brix. A, untreated guava juice; B, guava juice was pressurized at 6,000 atm and 25°C for 10 min.; C, Guava juice was heated at 95°C for 5 min. Chromatographic conditions: Column, 2.0 x 40 cm (packed with DEAE-cellulose); Elution solvent, 50 nm potassium phosphate buffer (pH 6.3); Flow rate, 30 mL/h; Fraction volume, 2.0 mL/fraction; Pectin extraction sequence, water-soluble pectin → oxalate-soluble pectin → alkali-soluble pectin.

juice. However, this phenomenon was not observed for high pressure treatment, in which the pectic substances in guava juice retained their original characteristics.

CONCLUSION

The viscosity and turbidity of guava juice changed markedly by heat treatment (95°C, 5 min) but not by high pressure treatment under 5000 atm. The contents of water soluble, oxalate soluble and alkali soluble pectin in guava juice remained unchanged after high pressure treatment, whereas heat treatment resulted in the loss of water soluble and alkali soluble pectin and a slight increase in the oxalate soluble pectin. The decrease of viscosity in heat treated juice was due to the degradation of pectic substances. The coagulation of degraded pectin combined with other components then resulted in an increase in turbidity and cloud content. The SEM result confirmed the coagulation of clouding substances in the heat treated sample but not in its high pressure treated counterpart. The profile of pectin in guava juice did not change after high pressure treatment but it showed coagulation and degradation in heat treated sample as separated by DEAE cellulose. Since the pectic and cloud substances in guava juice are not affected by high pressure treatment, the high pressure treated guava juice could retain its high quality.

ACKNOWLEDGMENT

The authors thank the National Science Council, Republic of China for financial support under Grant No. NSC 83-0406-E005-003 and NSC84-2214-E005-006.

REFERENCES

Binnig, R. Light color-stable cloudy apple juices. *Fruit Process.* **1992**, *2*, 133–136.

Burns, J. *The Chemistry and Technology of Pectin.* Walter, R.H. (Ed.), Academic Press: San Diego, CA, 1991, pp. 165–174.

Chandler, B.U.; Robertson, G.L. 1983. Effect of pectic enzymes on cloud stability and soluble limonin concentration in stored orange juice. *J. Sci. Food Agric.* **1983**, *34*, 599–611.

Dietz, J, H.; Rouse, A. H. 1953. A rapid method for estimating pectin substances in citrus juices. *Food Res.* **1953**, *18*, 169–177.

Ferro, R.; Martha, L.; Castelblanco, R. Extraction and characterization of pectin from two varieties of guava (*psidium guajava* L.). In *Chemical Abstracts*, Vol. 71, 1969, 48482b.

Hayashi, R. Possibility of high pressure technology for cooking, sterilization, processing and storage of foods. *Shokuhin to Kaihatau (Food & Development)* **1987**, *22*, 55–62.

Heremans, K. High pressure effects on proteins and other biomolecules. *Ann. Rev. Biophys. Bioeng.* **1982**, *11*, 1–21.

Horie, Y.; Kimura, K.; Ida, M.; Yoshida, Y.; Aohki, K. Jam preparation by pressurization. *Nippon Nogeikaguku Kaishi* **1991**, *65*, 975–980.

Klavons, J.A.; Bennett, R.D.; Vannier, S.H. Nature of the protein constituent of commercial orange juice cloud. *J. Agric. Food Chem.* **1991**, *39*, 1545–1548.

Klavons, J. A.; Bennett, R.D.; Vannier, S. H. Stable clouding agent from isolated soy protein. *J. Food Sci.* **1992**, *57*, 945–947.

Knorr, D. Effects of high-hydrostatic-pressure processes on food safety and quality. *Food Technol.* **1993**, *47(6)*, 156–161.

Lin, H.T.; Yen, G.C. Effect of high hydrostatic pressure on the inactivation of enzymes and sterilization of guava juice. *J. Chin. Agric. Chem. Soc.* **1995**, *33*: 18–29.

McCready, R.M. In *Methods in Food Analysis*, Maynard, A.J. (Ed.). Academic Press: New York, 1970, pp. 565–595.

Ogawa, H.; Fukuhisa, K.; Kubo, Y.; Fukumoto, H. Pressure inactivation of yeasts, molds and pectinesterase in satsuma mandarin juice: effects of juice concentration, pH and organic acids and comparison with heat sanitation. *Agric. Biol. Chem.* **1990**, *54*, 1219–1225.

Plat, D.; Ben-shalom, N.; Levi, A. Changes in pectic substances in carrots during dehydration with and without blanching. *Food Chem.* **1991**, *39*, 1–12.

SAS. *SAS User's Guide*. SAS Institute, Inc., Cary, NC, 1985.

Shomer, I. Protein coagulation cloud in citrus fruit extract. 1. formation of coagulates and their bound pectin and neutral sugars. *J. Agric. Food Chem.* **1991a**, *39*, 2263–2266.

Shomer, I. Protein coagulation cloud in citrus fruit extract. 2. Structural characterization of coagulates. *J. Agric. Food Chem.* **1991b**, *39*, 2267–2273.

Takahashi, Y.; Ohta, H.; Yonei, H.; Ifuku, Y. Microbicidal effect of hydrostatic pressure on satsuma mandarin juice. *Internat. J. Food Sci. Technol.* **1993**, *28*, 95–102.

Ting, S.V.; Rouseff, R.L. Chemical constituents affecting quality characteristics of citrus products. In *Citrus Fruits and their Products*. Marcel Dekker, Inc., New York, 1986, pp. 99–103.

Yen, G.C.; Lin, H.T.; Yang, P. Changes in volatile flavor components of guava puree during processing and frozen storage. *J. Food Sci.* **1992a**, *57*, 679–681 & 685.

Yen, G.C.; Lin, H.T.; Yang, P. Effects of enzymatic activities on changes in quality of guava puree during low temperature storage. *J. Chin. Agric. Chem. Soc.* **1992b**, *30*, 253–263.

Yen, G.C.; Lin, H.T.; Yang, P. Effects of heating process and frozen storage on changes in quality of guava puree. *J. Chin. Agric. Chem. Soc.* **1994**, *32*, 156–160.

CHEMOMETRIC APPLICATIONS OF THERMALLY PRODUCED COMPOUNDS AS TIME-TEMPERATURE INTEGRATORS IN ASEPTIC PROCESSING OF PARTICULATE FOODS

H.-J. Kim[1] and Y.-M. Choi[2]

[1]Department of Chemistry
College of Natural Sciences, Seoul National University
Seoul, Korea
[2]Institute of Biotechnology
Korea University
Seoul, Korea

The chemometric principle was used to derive a guideline for obtaining a simple "yes or no" answer about the sterility of food particulates heated at aseptic processing temperatures. A quadratic temperature pulse model was used to estimate bacterial destruction from the fractional yield of thermally produced chemical marker compounds (2,3-dihydro-3,5-dihydroxy-6-methyl-4(H)-pyran-4-one, M-1, and 4-hydroxy-5-methyl-3(2H)-furanone, M-2) and the rate constants and the activation energies of the chemical and bacterial systems. The model yielded a conservative estimate of lethality at the center of meatballs heated under different time-temperature conditions. A scheme for determining the minimum marker yield for a designated F_0-value is provided.

INTRODUCTION

The objective of chemometrics is to extract useful information from complex analytical data. Hence, valid analytical data and an appropriate model linking the data to the desired information are two key elements in chemometrics. Recent examples include applications to spectroscopy (Oka *et al.*, 1991) and chromatography (Baiocchi *et al.*, 1993).

Process-Induced Chemical Changes in Food
edited by Shahidi *et al.* Plenum Press, New York, 1998

One of the key potential applications of chemometrics in food science is to determine whether or not the food particulates undergoing high temperature/short time (HTST) processing are sterile from post process chemical measurements. Such a determination is possible in principle because both the yield of a chemical reaction and bacterial destruction depend on the temperature/time history.

Even though conceptually straightforward, the development of time-temperature integrators (TTI) in food processing and preservation has been a challenge for many years (Mulley *et al.*, 1975; Sastry *et al.*, 1988; Berry *et al.*, 1990; Taoukis *et al.*, 1991; Weng *et al.*, 1991; Awuah *et al.*, 1993, Maesmans *et al.*, 1994) and continues to be an active research subject (Van Loey *et al.*, 1996; Kim *et al.*, 1996a,b). A major challenge in the use of TTI in thermal processing is to determine process lethality (F_0) from the TTI measurements under variable time-temperature conditions.

Several chemical marker compounds formed at sterilizing temperatures from precursor compounds inherently present in the foods have been reported (Kim and Taub, 1993; Kim *et al.*, 1994). In order to apply these chemical markers to real processing conditions, where lethality accumulates under nonisothermal conditions, accurate informations about the reaction rate constants and their temperature dependence (activation energy or Z-value) are needed. The variability of the precursor concentration within the particulates and among different particulates also needs to be considered carefully.

As a first step in addressing this challenging problem, we evaluated the possibility of determining the minimum lethality delivered to the center of a particulate, with known initial precursor concentrations, from the marker yields and the kinetic data obtained in the laboratory under an isothermal condition. The mathematical model used (Ross, 1993) yields bacterial destruction from the fractional marker yield as long as the peak temperature of the temperature pulse is provided.

EXPERIMENTAL

Sample Preparation and Chemical/Microbiological Measurements

Meatballs weighing approximately 5g each were made from ground beef containing 10% D-glucose (precursor of M-1), 1% D-ribose (precursor of M-2), and 0.2% $CaCl_2$, and coated with alginate (Kim *et al.*, 1996a). For rate measurements, a meatball was placed in a stainless cup (2.1 cm inside diameter, 3.6 cm height) and 7 mL water was added to the cup to avoid direct steam heating. The sainple was heated at 110, 114, 118, 121, 126, and 131±1C for different time periods with Biological Indicator Evaluator Resistometer (BIER, Joslyn Valve, Macadon, NY), which uses pressurized steam for rapid heating. Temperature at the center of the meatball was continuously monitored with a thermocouple.

For test samples, a small alginate bead containing approximately 6×10^7 spores of *Bacillus stearothermophilus* was placed at the center of the meatball and the meatball was similarly heated until the center temperature increases from room temperature to a preset target temperature. The particulate center temperature dropped abruptly as the system was evacuated at the end the temperature pulse- after cooling about 0. 15g portion of the meat at the center of the meatball was used for extraction, with 10-fold excess water, of the chemical markers. The chemical markers, M-1 and M-2, were extracted with water and determined by ion exclusion chromatography following centrifugation and filtration. The peak absorbance values at 285 nm were measured for both M-1 and M-2 using a photodiode array detector. A variable wavelength detector set at 285 nm will work as long as the absorbance of a reference compound such as 5-hydroxymethylfurfural is measured under the same experimental conditions (Kim *et al.*, 1996a).

Rate Constants and Activation Energies

Results from isothermal experiments were used to determine the rate constants and activation energies for both M-1 and M-2 formation. The chemical marker yields, in absorbance at 285 mn, as a function of time, t, at 121 °C were fitted in the following equation to yield M_∞, the limiting value of the marker concentration, and the rate constant, k_m:

$$M = M_\infty\left(1 - e^{-km(t-a)}\right) \quad (1)$$

A parameter, a, was needed in fitting to account for the come-up time. The rate constants at other temperatures were obtained by fitting the marker data in Eq. (1) using the temperature data.

Determination of the Molar Concentration of the Markers

It is convenient to present the chemical marker results as absorbance at 285 nm after correction for the dilution factor. However, the measured absorbance is influenced by experimental conditions such as the injection volume and the size of the analytical column.

To convert absorbance to molar concentration of the markers in the extract, authentic 5-hydroxymethy furfural (HMF) with known extinction coefficient (ε = 16,700 M^{-1} cm^{-1} at 284 mn; Singh *et al.*, 1948) was used as a reference. A 10 ppm solution of authentic HMF (0.0794 mM) gave peak absorbance of 0.21 at 285 nm when the same HPLC system was used. The extraction coefficient of M-1 at 298 nm, the absorption maximum is 6,300 M^{-1} cm^{-1} (Shaw *et al.*, 1967). For simultaneous quantitation of M-1 and M-2, absorbance was routinely measured at 285 nm. From the absorption spectrum of M-1, obtained with the photodiode array detector, the ratio of absorbance at 285 nm (Abs_{285}) to that at 298 mn was determined as 0.837. Hence, ε = 5,270 M^{-1} cm^{-1} at 285 mn for M-1. Thus, absorbance of M-1 can be converted to molar concentration by Eq. (2):

$$\begin{aligned}[M-1] &= 0.0794 \times (16{,}700 / 5{,}270) \times (Abs_{285} / 0.21) \\ &= 1.2 \times Abs_{285} \quad (mM) \end{aligned} \quad (2)$$

Similarly, absorbance of M-2 (ε = 9,600 at 284 nm; Hicks *et al.*, 1974) can be converted to molar concentration by Eq. (3):

$$\begin{aligned}[M-2] &= 0.0794 \times (16{,}700 / 9{,}600) \times (Abs_{285} / 0.21) \\ &= 0.66 \times Abs_{285} \quad (mM) \end{aligned} \quad (3)$$

Application of the Quadratic Model

In the mathematical model of Ross (1993), the decadic reduction (DR) in bacterial spore population, $_{log}$(No/N), can be calculated from Eq. (4):

$$\log(N_o/N) = -[\log(1 - M/M_\infty)](k_n^r/k_m^r)(E_m/E_n)^{1/2}\exp[(E_m - E_n)x_p] \quad (4)$$

where N_0 and N are spore population before and after heat treatment, respectively.

k^r_n are k^r_m are rate constants for bacterial destruction and the marker formation, respectively, at the reference temperature, 121.1°C. E_m is a dimensionless activation energy for the chemical marker obtained by Eq. (5):

$$E_m = E_a/RT_r \tag{5}$$

where R is the gas constant (1.987 kcal $°K^{-1}$) and T_r the reference temperature in Kelvin. E_n for bacterial destruction is calculated similarly, x_p is a dimensionless peak temperature transformation defined by Eq. (6):

$$X_p = (T_r/T_p) - 1 \tag{6}$$

where T_p is the peak temperature of the pulse in Kelvin. It is important to note that, according to Eq. (4), the initial temperature and the duration of the pulse are not needed to convert the marker yield information to the bacterial destruction information.

RESULTS AND DISCUSSION

Rate Constants and Activation Energies

Figure 1 shows the result of M-1 and M-2 measurements at 12 1°C. In fact, the temperature at the center of the meatball increased from room temperature to 121°C in 8 min. After 8 min, the temperature was constant within ±0.4°C. The M-1 yield was almost zero until 8 min. Therefore, the rate constant at 121°C was determined as 0.0302 min^{-1} by fitting results between 8 and 70 min with Eq. (1). Clearly, a series of reactions leading to the formation of M-1 from glucose and amines (Kim *et al.*, 1996b) takes place during the come-up time, which is a very nice feature of these marker reactions.

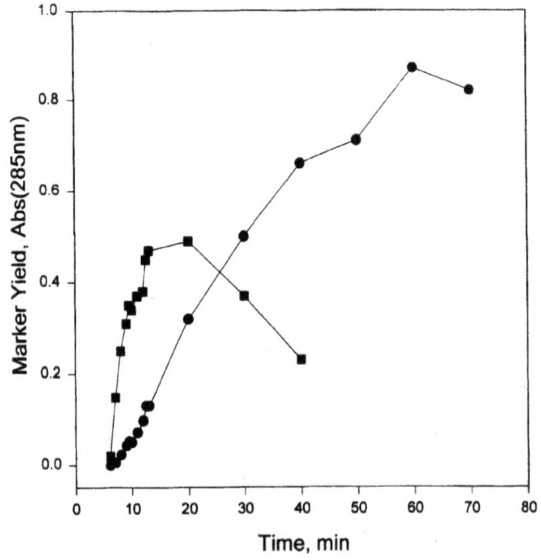

Figure 1. The yield of chemical markers, M-1 (-●-) and M-2 (-■-) at 12 1°C as a function of time used for determination of the rate constants. Each point represents an average of duplicate measurements.

The asymptotic value of M-1 (M_∞) was determined as 1.02 absorbance (285nm), which corresponds to 12 mM according to Eq. (2) after correction for 10-fold dilution during extraction. The initial concentration of glucose in the meatball (10%) corresponds to 0.56M. Therefore, the limiting M-1 yield is equivalent to 2.1% conversion of glucose to M-1. After 80 min, the M-1 concentration started to decrease indicating that thermal destruction of M-1 (Kim and Baltes, 1996) is overtaking its formation. The same limiting value of 1.02 absorbance was used for fitting data at other temperatures.

Figure 1 shows that the rate constant for M-2 formation is higher than that for M-1. The plateau value of M-2 is reached in 20 min and the M-2 concentration decreases while the M-1 concentration is still increasing. The M-2 concentration was 0.15 after 7 min heating, when the center temperature was 120.2°C. The temperature remained 121±0.4°C after 8 min. The rate constant for M-2 formation was obtained by fitting the data for 7–13 min with Eq. (1). The asymptotic value of 0.502, obtained from the 121°C data, was used in Eq. (1) for determination of the rate constants at other temperatures.

The rate constants for M- 1 and M-2 formation are summarized in Table 1. The activation energies were obtained by fitting the rate data with the Arrhenius equation.

Model Predictions

As pointed out above, the quadratic model of Ross (1993) enables one to predict bacterial destruction from the fractional marker yield (M/M_∞) as long as the peak temperature is given. To test the model, meatballs were heated from room temperature to a peak temperature ranging from 118 to 126°C, and a conservative estimate of bacterial destruction was calculated from the marker yields using 116°C as the peak temperature.

The D-value for the microorganism at 121°C was 0.57 min (Kim *et al.*, 1996a), which corresponds to k_n^r of 4.04 min^{-1}. k_m^r is 0.0302 and 0.285 for M-1 and M-2, respectively (Table 1). The z-value of 10°C was used for the microorganism to get the dimensionless activation energy, E_n, of 88.5 according to Eq. (7) (Ross, 1993):

$$E_n = 2.303[(T_r/Z) - 1] \tag{7}$$

The reported z-value of *Bacillus stearothermophilus* varies from 7.6 to 12°C (Lund, 1975; Feeherry *et al.*, 1987; David and Merson, 1990; Wescott *et al.*, 1995). The dimensionless activation energy values for M- 1 and M-2, calculated using Eq. (5) from the acti-

Table 1. Summary of rate constants and activation energies for M-1 and M-2 formation

	Rate constant (Abs Min^{-1})	
Temperature (°C)	M-1	M-2
110	0.0095	0.121
114	0.0160	0.176
118	0.0218	0.259
121	0.0302	0.285
126	0.0412	
131	0.0684	
Ea (kcal mol^{-1})	27.9	24.6

vation energies in Table 1, are 35.6 and 31.4, respectively. For peak temperature of 116°C, x_p becomes 0.0131 according to Eq. (6).

Substituting these values in Eq. (4), we get Eqs. (8) and (9) for M-1 and M-2, respectively:

$$\log(N_0/N) = -42.5 \log(1 - M/M_\infty) \tag{8}$$

$$\log(N_0/N) = -3.99 \log(1 - M/M_\infty) \tag{9}$$

The decadic bacterial destruction predicted by Eqs. (8) and (9) from the M- I and M-2 yields are compared with actual observations in Table 2.

Chemometric Scheme

The chemometric model connecting the marker yield to lethality is based on the assumption that the shape of the temperature pulse is quadratic in the dimensionless temperature descriptor, x. The temperature pulse, which is quadratic in x, where the temperature increases from room temperature to a peak temperature of 116°C over a 10 min interval, is shown in Fig. 2 as well as the actual meatball center temperature measured with a thermocouple. The actual temperature approaches the target temperature almost exponentially and is always higher than the temperature of the quadratic model. Therefore, the model prediction underestimates the actual lethality and can be used as a conservative estimate.

Let's suppose that the target F_0-value is 3 min, which corresponds to 12D reduction in *Clostridium botulinum*. The F_0-value of 3 min is equivalent to 5.3D reduction in *B. stearothermophilus* with a D-value of 0.57 min at the reference temperature. What we are looking for is the minimum chemical marker yield in M-1 or M-2 that will assure a minimum of 5.3D reduction in our test microorganism. Theoretically, there are infinite num-

Table 2. Comparison of bacterial destruction estimated from M-1 and M-2 yields assuming a teak temperature of 116°C with the observed destruction

			Log(N_0/N)				
			estimate from				
Case	M-1 yield (Abs)	M-2 yield (Abs)	M-1	M-2	observed	Temperature[1] (°C)	Time[2] (min)
1	0.02	0.21	0.4	0.9	0.6	121	8
2	0.01	0.02	0.2	0.1	1.4	126	6
3	0.03	0.37	0.6	2.3	1.1	118	10
4	0.03	0.38	0.6	2.5	5.5	121	10
5	0.12	0.41	2.3	2.9	>6[3]	126	8.5
6	0.25	(0.37)[4]	5.2	(2.3)[4]	>6	121	15
7	0.24	(0.44)	4.9	(3.6)	>6	126	11
8	0.31	(0.33)	6.7	(1.9)	>6	118	25
9	0.35	(0.28)	7.7	(1.4)	>6	121	19
10	0.39	(0.30)	8.9	(1.6)	>6	126	19

[1]Peak temperature reached.
[2]Time to reach the peak temperature from 23°C.
[3]Complete destruction of spores corresponding to greater than 6-log reduction.
[4]Numbers in parentheses are M-2 values in the decreasing region; therefore, bacterial destruction based on the M-2 yield is an underestimate.

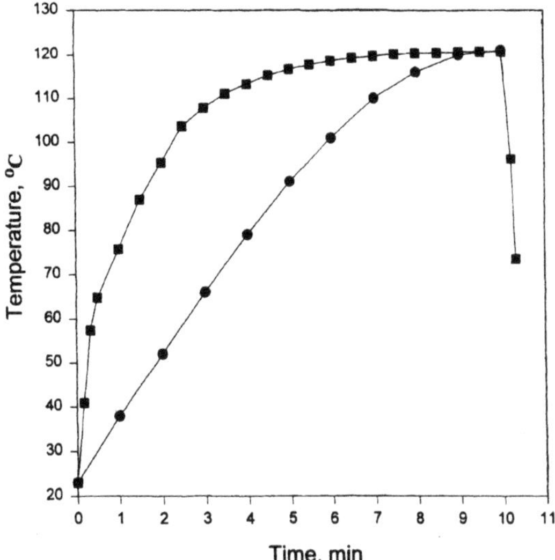

Figure 2. Temperature profile of the quadratic model (-●-) and the actual meatball center temperature (-■-).

bers of time-temperature histories that will yield the required 5.3 D reduction. In reality, the time the particulate spends in the heating and the holding section is limited by the bulk flow rate and the geometry of the aseptic processing line. The particulate center temperature cannot exceed the fluid temperature in conventional aseptic processing. (However, the particulate center temperature can be higher than the fluid temperature in ohmic heating and microwave sterilization.) Given sufficient time, the temperature profile at the particulate center will mimic that of the fluid with some time delay, and the particulate center will attain the maximum fluid temperature (Berry *et al.*, 1990). However, in aseptic processing where the heating and holding times may be too short to allow sufficient heat transfer, there may be a finite temperature difference between the fluid and the particulate center depending on the heat transfer properties and the holding time. Obviously, this aspect is of critical importance in any HTST process design.

In the present chemometric application, meatballs with the BIER according to 10 different temperature pulses such as shown in Fig. 2. The heating was done in such a way that the temperature increased from room temperature to the peak temperature, ranging from 118 to 126°C, over time ranging from 6 to 25 min. When the chamber was evacuated at the end of the pulse, the particulate center temperature dropped abruptly (Fig. 2). Choi also determined the M-1 and M-2 concentrations and the surviving spore population at the center of the meatball. The other author (Kim) applied the model to the marker yield data and estimated the bacterial destruction. The only information provided was the lower limit of the temperature range, 118°C. In using Eq. (4), 116°C was used as the peak temperature allowing a 2°C difference as a safety factor. The model predictions and observations of the decadic bacterial destruction are summarized in Table 2. Based on the results, the following chemometric scheme was devised to provide a "yes" or "no" answer regarding the attainment of F_0 of 3 at the particulate center if only the lower limit of the temperature range were given as 118°C.

Step 1

If both estimates of DR by M-1 and M-2 were smaller than 2 (less than half of the target DR of 5.3), the answer is "no". This applies to cases 1 and 2, where the M-1 yield is very low. Since the M-2 formation is faster than the M-1 formation, it is possible that the M-2 yield has reached 40% of the limiting value, and yet little lethality has been delivered as in case 1. Therefore, M-1 seems to be a more useful marker of lethality. Of course, if the peak temperature were much higher than 126°C in actual aseptic processing, sterility would be achieved at the low marker yields.

Step 2

If the DR predicted from M-1 were higher than 5 and the prediction from M-2 were lower than 5, clearly the M-2 yield exceeded the limiting value and is in the decreasing region (beyond 10 min in Fig. 1). Therefore, sterility can be assured, considering the safety factor, when the M-1 prediction is greater than 5 (cases 6–10 in Table 2) and the answer is "yes". The cut-off value of the M-1 yield is 0.25 Abs in the extract, which corresponds to 3.0 mM M-1 in the meatball sample.

Step 3

If the DR predicted from M-2 were higher than 2 and the prediction from M-1 were lower than 5 (cases 3–5), it is safe to say that the answer is "no" even though the actual lethality might exceed the requirement (cases 4, 5) due to the difference in the temperature profile (Fig. 2). This safety net will be greater at higher temperatures (cases 4, 5 vs 3). However, if one only knew the range of the peak temperature, one should use the lower limit of temperature which would yield a conservative estimate of lethality.

In summary, we conclude that a minimum M-1 yield of 3.0 mM from the initial glucose concentration of 0.56M (0.54% conversion) can be used to assure achievement of a minimum F_0-value of 3 in aseptic processing. The model suggests that the actual lethality will exceed the prediction based on the quadratic model due to the difference in the temperature profile. This chemometric approach needs to be tested using an industrial aseptic processing equipment. In ohmic heating, where the temperature increases much faster than in conventional aseptic processing, a different model needs to be used.

REFERENCES

Awuah, G.B.; Ramaswamy, H.S.; Simpson, B.K. Thermal inactivation kinetics of trypsin at aseptic processing temperatures. *J. Food Proc. Eng.* **1993**, *16*, 315–328.

Baiocchi, C.; Marengo, E.; Saini, G.; Roggero, M.A.; Giacosa, D. Reversed phase high-performance liquid chromatography and chemometrics, a combined investigation tool for complex phytochemical problems. *J. Chromatogr.* **1993,** *644*, 259-267.

Berry, M.F.; Singh, R.K.; Nelson, P.E. Kinetics of methytmethionine sulfonium salt destruction in a model particulate system. *J. Food Sci.* **1990**, *55*, 502–505.

David, J.R.D.; Merson, R.L. Kinetic parameters for inactivation of *Bacillus stearothermophilus* at high temperatures. *J. Food Sci.* **1990,** *55*, 488–493, 515.

Feeherry, F.E.; Munsey, D.T.; Rowley, D.B. Thermal inactivation and injury of *Bacillus stearothermophilus* spores. *Appl. Environ. Microbiol.* **1987,** *53*, 365-370.

Hicks, K.B.; Harris, D.W.; Feather, M.S.; Loeppky, R.N. Production of 4-hydroxy-5-methyl-3(2H)-furanone, a component of beef flavor, from a 1-amino-1-deoxy-D-fructuronic acid. *J. Agric. Food Chem.* **1974,** *22(4)*, 724–725.

Kim, H.-J.; Ball, D.; Giles, J.; White, F. Analysis of thermally produced compounds in foods by thermospray liquid chromatography-mass spectrometry and gas chromatography-mass spectrometry. *J. Agric. Food Chem.* **1994**, *42,* 2812–2816.

Kim, H.-J.; Taub, I.A. Intrinsic chemical markers for aseptic processing of particulate foods. *Food Technol.* **1993,** *47(1),* 91–99.

Kim, H.-J.; Choi,Y.-M.; Yang, A.P.P.; Yang, T.C.S.; Taub, I.A.; Giles, J.; Ditusa, C.; Chall, S.; Zoltai, P. Microbiological and chemical investigation of ohmic heating of particulate foods using a 5kW ohmic system. *J. Food Process. Preserv.* **1996a,** *20,* 41–58.

Kim, H.-J.; Taub, I.A.; Choi, Y.-M.; Prakash, A. Principles and applications of chemical markers of sterility in HTST processing of particulate foods. ACS Symposium Series 631, American Chemical Society, Washington, DC, 1996b.

Kim, M.-O.; Baltes, W. On the role of 2,3-dihydro-3,5-dihydroxy-6-methyl-4(H)-pyran-4-one in the Maillard reaction. *J. Agric. Food Chem.* **1996,** *44,* 282-289.

Lund, D. Heat Processing. In *Physical Principles of Food Preservation;*Fennema, O., Ed. Marcel Dekker, New York, 1975.

Maesmans,G.; Hendrickx, M.; De Cordt, S.; Van Loey, A.; Noronha, J.; Tobback, P. Evaluation of process value distribution with time temperature integrators. *Food Research Intl.* **1994,** *27,* 413–423.

Mulley, E.A.; Stumbo, C.R.; Hunting, W.A. Thiamine: A chemical index of sterilization efficacy of thermal processing. *J. Food Sci.* **1975,** *40,* 993–996.

Oka, K.; Oshima, K.; Inamoto, N.; Pishva, D. Chemometrics and spectroscopy. *Anal. Sci.* **1991,** *7, Suppl.,* 757–760.

Ross, E.W.. Relation of bacterial destruction to chemical marker formation during processing by thermal pulses. *J. Food Process Eng.* **1993,** *16,* 247–270.

Sastry, S.K.; Li, S.F.; Patel, M.; Konanayakam, M.; Bafna, P.; Doores, S.; Beelman, R.B. A bioindicator for validation of thermal processes for particulate foods. *J. Food Sci.* **1988,** *53,* 1528–1536.

Shaw, P.E.; Tatum, J.H.; Berry, R.E. Acid-catalyzed degradation of D-fructose. *Carbohyd. Res.* **1967,** *5,* 266–273.

Singh, B.; Dean, G.R.; Cantor, S.M. The role of 5-(hydroxymethyl)-furfural in the discoloration of sugar solutions. *J. Am. Chem. Soc.* **1948,** *70,* 517–522.

Taoukis, P.T.; Fu, B.; Labuza, T.P. Time-temperature indicators. *Food Technol.* **1991,** *45(10),* 70–82.

Van Loey, A.; Hendrickx, M.; De Cordt, S.; Haentjens, T.; Tobback, P. Quantitative evaluation of thermal processes using time-temperature integrators. *Trends in Food Sci. Technol.* **1996,** *7,* 16–26.

Weng, W.M.; Hendrickx, M.; Maesmans, G.; Tobback, P. Immobilized peroxidase: a potential bio-indicator for evaluation of thermal processes. *J. Food Sci.* **1991,** *56,* 567–570.

Wescott, G.G.; Fairchild, T.M.; Foegeding, P.M. *Bacillus cereus* and *Bacillus stearothermophilus* spore activation in batch and continuous flow systems. *J. Food Sci.* **1995,** *60,* 446–450.

10

HEATING RATE OF EGG ALBUMIN SOLUTION AND ITS CHANGE DURING OHMIC HEATING

T. Imai,[1] K. Uemura,[2] and A. Noguchi[2]

[1]University of Tsukuba
[2]National Food Research Institute
2-1-2, Kannondai
Tsukuba, Ibaraki, 305 Japan

Ohmic heating of egg albumin solution (10 w/v%) was examined at 50 -10 kHz under a constant 10 V/cm. The heating rate of the solution was almost constant and increased slightly as the frequency increased. The gel formation was observed at about 75°C and the heating rate increased above this temperature irrespective of the frequency used. The solution and gel showed almost the same impedance at the examined temperature (20–90°C) and frequency (10 Hz-100 kHz). When the concentration of egg albumin was reduced to 2 w/v%, no gel was formed and a constant heating rate at over 75°C was observed. The breaking strength of the gels showed little difference among the gels prepared by boiling water or Ohmic heating. These results suggest that the liquid components are not compartmentalized in the gel and that the sudden increase of heating rate above 75°C was caused by the reduction of heat transfer of the gel at its phase change to the gel. Ohmic heating was also applied to the fresh egg white at the same conditions as that of the egg albumin solution. The fresh egg white did not show any sudden increase of heating rate until it reached 90°C. However, the homogenized fresh egg white and its soluble part separated beforehand showed a slightly reduced heating rate and a sudden increase at about 60°C. These results suggest that the gelatinous component of fresh egg white such as ovomucin represses the transfer of generated heat during Ohmic heating.

INTRODUCTION

The thermal processing of food products, in general, involves heat transfer into the food from hot surroundings, the driving force being the temperature difference between

Process-Induced Chemical Changes in Food
edited by Shahidi *et al.* Plenum Press, New York, 1998

the food products and its surrounding. Heat transfer in liquid foods takes place mainly by convection, provided the liquid is not too viscous. However, in processing particulate foods, the solid phase will only heat by conduction, which is a much slower process than convection, because of the low thermal conductivity of most foods (Sweat, 1986). The quality deterioration of food products during thermal processing can be minimized by an HTST (high temperature short time) process. However, there may be an upper temperature limit to this, to avoid overcooking of the food's surface.

When food products contain sufficient water and electrolytes to pass electric current, Ohmic heating can be used to generate heat within the food products by the passage of an alternating current. This method enables the solid phase or viscous liquids to heat as fast as in thin liquids, thus making it possible to use HTST techniques on solid or viscous foods. Ohmic heating rates are critically dependent on the electrical conductivity of the food products (Halden et al., 1990). Solid foods such as plant tissue is composed of individual cell units and these cell units are isolated by their walls and cell membranes. The major components of cell membrane are phospholipids and can be regarded, electrically, as a condenser. Therefore, it is likely that the frequencies of alternating current will become one of the key parameters for quick heating by reducing the impedance of plant tissue. On the other hand, the fish protein gel prepared from Alaskan pollack is reported to have the heating rate dependent on the frequency of alternating current in spite of the fact that the gel does not have any membrane structure, and shows 7.5 times the heating rate at 10 kHz, compared with that in hot water (90°C). These results suggest that the existing state of liquid components in the matrix will be also one of key points for quick Ohmic heating.

The objectives of the present study were to examine the effect of various frequencies of alternating current on the heat generation in the egg albumin solution and its gel, and to also find the possible reasons for the sudden change in heating rate, if any, at the phase change from the liquid to the gel during Ohmic heating.

EXPERIMENTAL

Egg Albumin Solution

The egg albumin used in this study was chemical grade and was dissolved in the distilled water to make a 10% (w/v) solution. This solution had a pH of 6.7. When a fresh egg was used, the yolk was carefully removed and the egg white was used as is or, if necessary, gently agitated without any air bubbles. If required, the thin egg white was separated from the thick one by passing them through a 20 mesh sieve.

Equipment for Ohmic Heating

The selected sine wave was generated by the generator (model: 8904A, H.P. Co., USA) and amplified with the amplifier (model:4025, NF Circuit Block Co., Ltd., Japan) and then applied to the electrodes in the heating box.

Figure 1 shows a schematic representation of the ohmic heating setup. The applied voltage was kept to 10 V/cm, if not specified, within from 50 Hz to 10 kHz until the egg white solutions achieved the temperature of 90°C at the center. This temperature was monitored with an insulated T type thermocouple (1.6 mm diameter). The impedance of the solutions and the gels formed after heating was measured with the signal analyzer

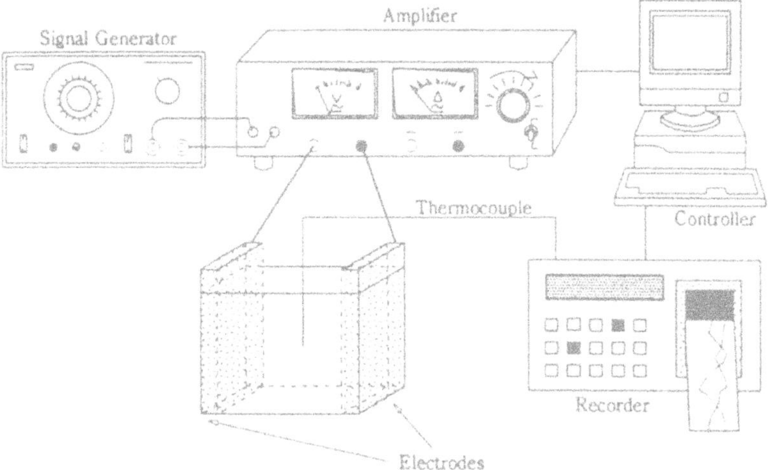

Figure 1. A schematic diagram of Ohmic heating set-up.

(model:3562A, H.P. Co., USA) from 10 Hz to 100 kHz. The absolute value |Z| of the complex impedance was calculated, according to the equation: $|Z| = (R^2 + X^2)^{1/2}$ where R was the real parts of complex impedance (Ω) and X is the imaginary parts of complex impedance (Ω). It should be noted that the impedance of the samples observed at less than 1 kHz was not so reliable because of the polarization of electrodes at a low frequency of alternating current.

Breaking Strength of Gels

After achieving 90°C at the center, the gels were quickly transferred into the plastic bag and cooled in ice water. Then, their breaking strength was measured using a rheometer (model: NRM 3002D, Fudoh Co., Ltd., Japan) with a 3 mm diameter plunger at 60 mm/cm. The peak force to penetrate the gel, was recorded and expressed as the breaking strength per unit area (g/cm^2).

RESULTS AND DISCUSSION

Effects of Frequency on the Heating Rate and the Breaking Strength of Gels

The heating rate at the center of the egg white solution for various frequencies and a constant voltage of 10 V/cm is shown in Figure 2. When the solution was transferred to the metal box with same dimension as the ohmic heating unit and dipped into the hot water at a temperature of 92°C, the center of the solution took about 1300 seconds until it reached 90°C, compared with about 380 seconds of ohmic heating at 10 kHz.

The solutions gave almost the same linear rise in temperature until about 75°C, when the gel formed. The heating rate was suddenly increased above this temperature, and it

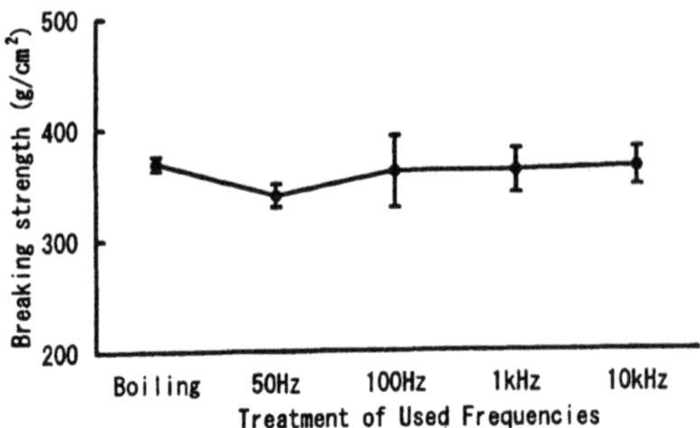

Figure 2. Heating rates of egg albumin solution during Ohmic heating at various frequencies. voltage applied : 10V/cm egg albumin : 10 w/v% (—— 50Hz; — — 100Hz; - - - - 1kHz; ——— 10kHz).

seemed to be more constant than before and almost independent of the used frequencies. Matsuda et al. (1981) reported that the egg albumin had its transition point at 74–82°C from the liquid to the solid phase. This phase changing to the gel (solid) may be the result of the change of heating rate at around 75°C.

Figure 3 shows the breaking strength of gels prepared by ohmic heating at various frequencies or hot water bath and the gels gave no particular difference. However, there could be a different arrangement of protein molecules in the gels due to their heat history and the small angle X ray scattering method will reveal them.

Effects of the Concentration of Egg Albumin Solution on the Heating Rate

The egg albumin used in this study was found to contain trace amount of salts and therefore, the applied voltage was adjusted at a fixed frequency, 10 kHz to have a similar heating rate among the examined solutions. The solutions, except for the 2%, showed the change of heating rate at around 75°C and the gel formation above this temperature. The 2% solution showed a constant, linear and similar heating rate to the other solutions below about 75°C and maintained the liquid phase and heating rate even above this temperature.

Doi (1993) reported that the gel formation of egg albumin required more than 2% of the material. These results also suggest that the change in heating rate was caused by the phase change of the solutions. The following speculation can be presented to explain the sudden change of heating rate, with gel formation:

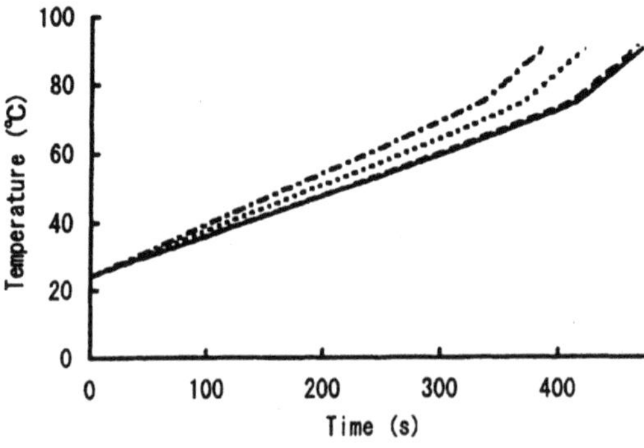

Figure 3. Effect of boiling and Ohmic heating on the breaking strength of egg albumin gel. Heat-induced gel was prepared by heating to 90°C in the hot water (92°C) or by Ohmic heating at various frequencies and 10V/cm until the mid-center of gel achieved 90°C.

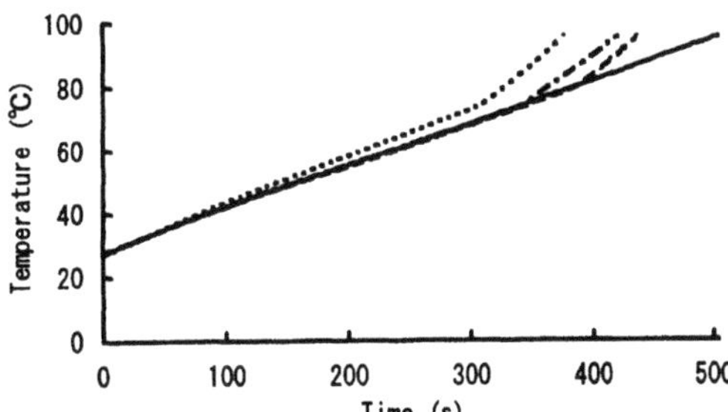

Figure 4. Effect of egg albumin concentration on the heating rate. Ohmic heating was carried out at 10kHz (—— 2% solution : 21V/cm; — — 5% solution : 14V/cm; - - - - 10% solution : 10V/cm; ——— 12% solution : 9V/cm).

1. Some salts are released from the protein upon denaturation as the gel forms at about 75°C. The released salts are likely to increase the electroconductivity of the sample and bring about the sudden increase in heating rate.
2. Gels have a lower impedance or larger dielectric loss than solutions and contributes to the sudden increase in heating rate.
3. The heat conduction at the wall of the vessel creates a temperature distribution which causes the heat convection of the sample as long as the sample maintains the liquid phase. When the gel is formed, heat conduction becomes predominant for heat transfer and the heating rate increases.

Impedance Changes of Egg Albumin Solution after Ohmic Heating

It is well known that the heat denatured egg albumin would expose its hydrphobic amino acids and shift its hydrophobic-hydrophilic balance (HLB) to be more hydrophobic, which results in coagulation and subsequent gel formation. It is likely that some of the salts are released from the protein molecules during their coagulation and gel formation and these released salts reduce the impedance of the solution which results in the sudden change of heating rate at around 75°C.

To confirm the first speculation, the following measurement was carried out; an egg albumin solution of 10% (w/v) was prepared at 25°C and heated to 70 and 80°C at 10 kHz. If formed during heating, the gel was crushed and removed by centrifugation and the resulting supernatant was examined. As shown in Figure 5, there was little change of impedance among these three samples and therefore, the other two possibilities should be checked to find the reason behind the change of heating rate.

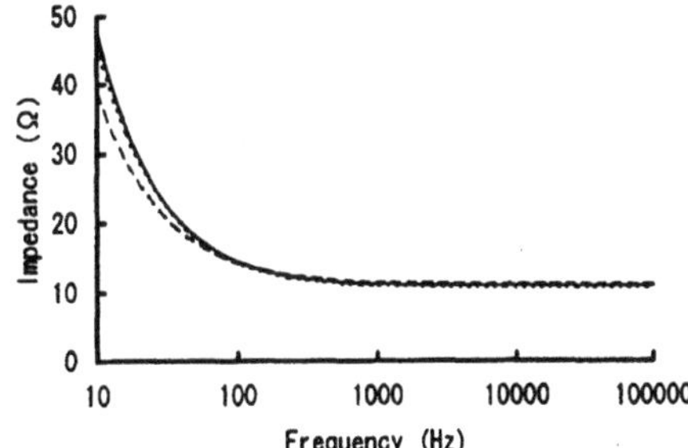

Figure 5. Impedance of solution after crashing and removing the aggregated protein gel caused by Ohmic heating. Ohmic heating carried out at 10kHz and 10V/cm (—— 25°C; — — — 70°C; - - - - - 80°C).

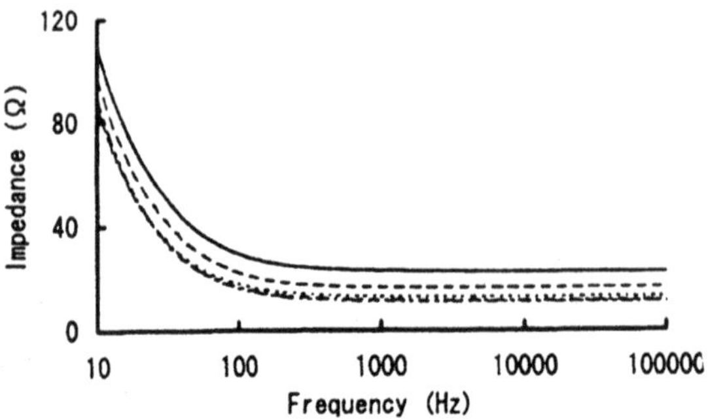

Figure 6. Impedance of egg albumin solution after achieving the selected temperature during Ohmic heating at 10kHz and 10V/cm (– • • – 90°C; – • – • 80°C; • • • • • 70°C; - - - - 50°C; —— 20°C).

Impedance of the Solutions and Gels

The egg albumin solution and the gel formed with ohmic heating to 90°C, were studied for their impedance at the selected temperature during ohmic heating (Figures 6 and 7). Both of them showed a slight decrease of impedance profile within the examined frequency as the temperature rose, which could likely contribute to the slight increase of heating rate at higher temperatures. It should be noted that little change was observed in the impedance between 70 and 80°C. These results suggest that the third speculation is most probable.

Ohmic Heating of Fresh Egg White

The white and yolk of a fresh egg were carefully separated and the egg white was studied for its ohmic heating. This fresh egg white gave a similar change in heating rate to that of the egg albumin solution at around 75°C when it was gently agitated before ohmic heating. However, no change was observed without agitation (Fig. 8). The fresh egg is composed of the thin and thick parts, both of which have similar impedance to each other. When the fresh egg white was separated into thin and thick parts, and these separate parts were examined for their ohmic heating, the thin part gave change of heating rate, but the thick part did not (Fig. 9). These results suggest that convection causes heat loss in the thin part and that heat conduction plays the main role in the rise of temperature in the thick part.

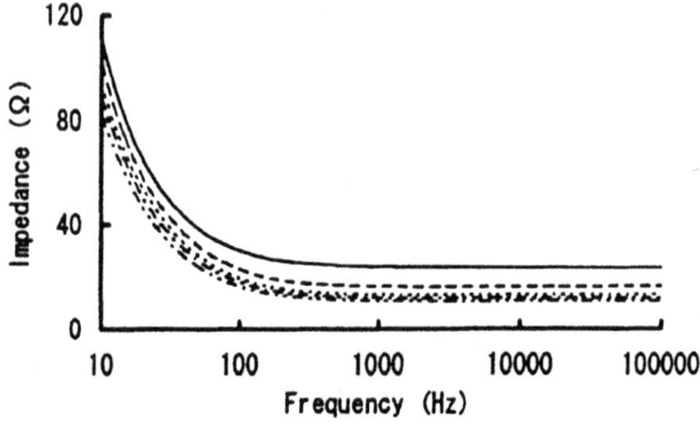

Figure 7. Impedance of egg albumin gel after achieving the selected temperature during Ohmic heating at 10kHz and 10V/cm. The gel was prepared by Ohmic heating at 10kHz and 10V/cm until the 90°C (– • • – 90°C; – • – • 80°C; • • • • • 70°C; - - - - 50°C; —— 20°C).

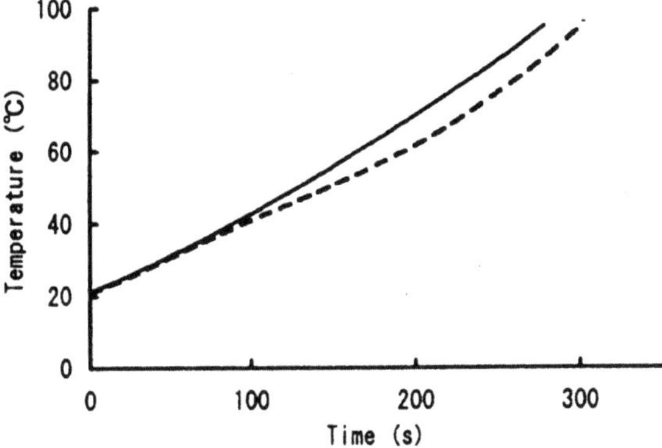

Figure 8. Heating rates of raw egg white during Ohmic heating at 10kHz and 10V/cm (—— native;— — mixing).

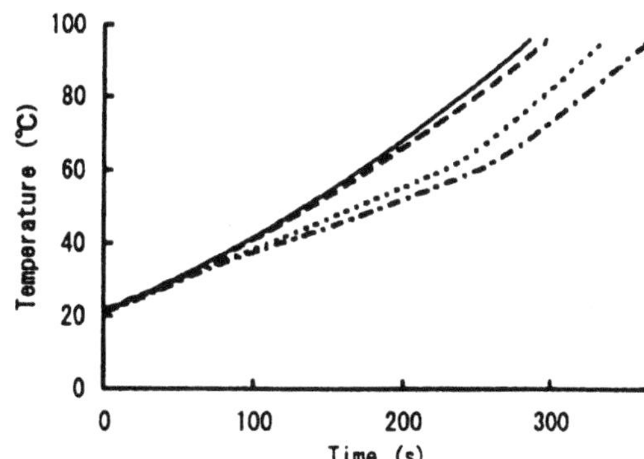

Figure 9. Heating rates of thin and thick egg whites during Ohmic heating at 10kHz and 50Hz and 10V/cm (—— thick : 50Hz; — — thick : 10kHZ; - - - thin : 50Hz; —•—•— thin : 10kHz).

CONCLUSION

It is clear that ohmic heating provides a quick and linear temperature rise to the protein solution. The phase change of the solution to a solid appeared to have the reasonable increment of heating rate. These results indicate that the viscosity or fluidity of materials will be one of the important parameters to control the temperature when a batch type system of ohmic heating is applied. The small change of impedance before and after gel formation of egg albumin solution suggests that the water and salts in the protein matrix of the gel are not compartmentalized but distributed continuously. Ohmic heating will provide the proteins homogeneous quick heating and therefore, they will be evenly heat-denatured to have a more aligned or oriented gel network and a better and smoother textured gel.

REFERENCES

Doi, E. J. *Food Engineering Soc. Japan* **1993**, *13*, 229.
Halden, K.; de Alwis, A.A.; Fryer, P.J. *Int. J. Food Sci. Technol.* **1990**, *25*, 9.

Matsuda, T.; Watanabe, K.; Sato, Y. Heat-induced aggregation of egg white proteins as studied by vertical flat-sheet polyacrylamide gel electrophoresis. *J. Food Sci.* **1981**, *46*, 1829–1834.
Sweat, V.E. *Engineering Properties of Foods*, Marcel Dekker: New York, 1986; p.49.

CHEMICAL CHANGES DURING EXTRUSION COOKING

Recent Advances

Mary Ellen Camire

Department of Food Science and Human Nutrition
University of Maine
5736 Holmes Hall
Orono, Maine 04469-5736

Cooking extruders process a variety of foods, feeds, and industrial materials. Greater flexibility in product development with extruders depends upon understanding chemical reactions that occur within the extruder barrel and at the die. Starch gelatinization and protein denautration are the most important reactions during extrusion. Proteins, starches, and non-starch polysaccharides can fragment, creating reactive molecules that may form new linkages not found in nature. Vitamin stability varies with vitamin structure, extrusion conditions, and food matrix composition. Little is known about the effects of extrusion parameters on phytochemical bioavailability and stability. Reactive extrusion to create new flavor, antioxidant and color compounds will be an area of interest in the future.

INTRODUCTION

Extrusion cooking is used today to process a variety of foods and feeds. Extrusion differs from other food processing methods in that several unit operations are performed simultaneously. Ingredients in a powder or granular form are added to the extruder at the feed end, then transported along the length of the barrel by one or two screws. Some extrusion systems require mixing with water to the desired moisture level before extrusion; newer models meter water with dry materials at the feed port. As the material moves along the barrel, it is subjected to heat (applied externally or by friction), mixing and shear. These processes may be modified by selection of various screw elements. At the die end

Process-Induced Chemical Changes in Food
edited by Shahidi *et al.* Plenum Press, New York, 1998

Major Changes Occurring During Extrusion Cooking

Chemical	Food Compounds Affected
Thermal degradation	Sugars, amino acids
Depolymerization	Starch, dietary fiber, protein
Recombination of fragments	Starch, dietary fiber
Physico-chemical	
Binding	Flavor
Volatilization	Flavor
Change in native structure	Starch (gelatinization) Protein (denaturation)

Figure 1. Classes of chemical changes occurring during extrusion cooking.

of the barrel, the heated food is compressed under pressure. As it exits the die, moisture flashes off due to a pressure drop. The flash-off results in expansion of the extrudate since steam stretches the still-molten food material.

Several monographs have been dedicated to extrusion cooking (Harper, 1981; Hayakawa, 1992; Kokini *et al.*, 1992; Mercier *et al.*, 1989; O'Connor, 1987), and several reviews of chemical and nutritional changes during extrusion cooking of foods have been published (Björck and Asp, 1983; Camire *et al.*, 1990; Cheftel, 1986; de la Gueriviere *et al.*, 1985). This review will focus on research published since 1990. Extrusion and flavor compounds are addressed in another chapter in this book monograph.

The major chemical changes during extrusion cooking fall into three categories: thermal degradation, depolymerization, and recombination of fragments (Figure 1). Binding of flavors and other smaller molecules by macromolecules is also common. Most reactions occur in the portion of the barrel just before the die. Thermally-labile compounds such as flavors may be injected at this site to minimize exposure to heat and shear. However, volatile compounds are distilled off with steam as material exits the die.

INTERPRETATION OF EXTRUSION STUDIES

Although many studies have focused on 2 or 3 extrusion processing variables, many factors are important. Screw speed, screw configuration, feed rate, die geometry, barrel temperature, and feed moisture influence shear, product viscosity in the barrel, and residence time. Meuser and van Lengerich (1984) developed the concept of relating product characteristics back to specific mechanical energy (SME), product or mass temperature (PT) and pressure. These secondary extrusion parameters result from the unique combinations of extrusion operating conditions that are selected by the extruder operator. These relationships are illustrated in Figure 2. Feed composition and prior processing history are frequently overlooked as sources of experimental variation. Too many papers have focused on production of unusual blends of materials, rather than examining chemical changes with subsequent physical, sensory and nutritional effects.

Another factor to consider when considering extrusion research is the type of extruder used. Extruders are costly. Many universities that conduct extrusion research have small laboratory-scale extruders; single screw machines are most common. The food industry, on the other hand, uses primarily large twin screw extruders. Can we really apply

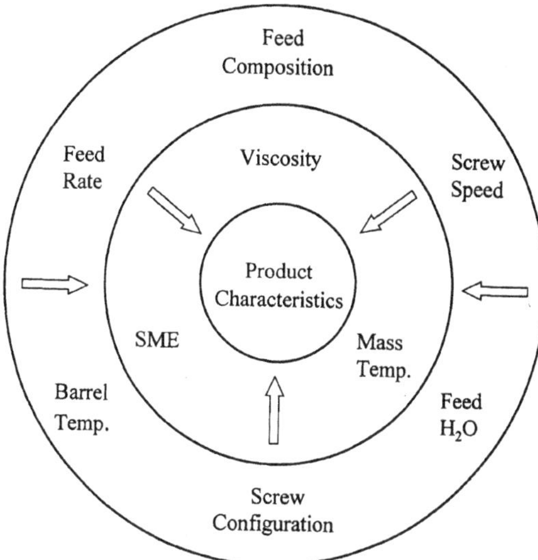

Figure 2. Inter-relationships among extrusion variables.

results from a meter-long single screw extruder to a production-sized model with a production output of several hundred kilograms per hour? Similar trends can be found, but better predictive models would allow for direct comparison, saving time and money in exploratory experiments by industry scientists and engineers.

Another potential pitfall is the application of information collected on one food system to a similar food material. For example, our research group conducted several experiments on the feasibility of extruding waste potato peels as a dietary fiber-rich food additive. No papers specifically dealing with extruding this material existed, but papers were published on extrusion of potato flakes and cereal brans. Despite differences among food systems, general trends often apply.

Researchers in academia often cannot afford purchasing common food ingredients such as wheat flour and corn meal on the scale required for experimentation. However, food processing by-products such as potato peels, soybean hulls, and cattle lungs are usually available in ample supply at little or no cost to researchers. These materials contain components found in "normal" foods, although some interpretation of results may be needed before they could be applied to common food items.

LIPIDS

Most extruded materials are fairly low in lipid content due to several problems inherent to extrusion. Lipid content greater than 5–6% affects extruder performance by reducing torque and subsequently, product expansion. Lipids literally grease the internal parts of the extruder, increasing slippage. Extrusion can enhance yields of oil from oilseeds by cooking and disrupting cell walls.

Lipid content of extrudates may appear lower than that found in the original material prior to processing. This apparent loss is most likely due to complex formation with amylose molecules or protein. Lower lipid content has primarily been reported for starchy materials. Digestion of extrudates with acid or amylase, followed by solvent extraction, can recover this "lost" lipid. Roughly half the ether-extractable lipid in whole wheat was recovered after extrusion, but total fat obtained by acid hydrolysis was not significantly affected (Wang *et al.*, 1993). Wheat bran, with a slightly higher initial lipid content, had higher free lipid levels after extrusion in that study. These apparently contradictory findings may have been caused by the lower starch level in bran, which was not available to complex with the lipid.

Hydrolysis of triacylglycerides is possible with sufficient moisture, but extrusion can denature hydrolytic enzymes sufficiently to prevent formation of free fatty acids after extrusion. Free fatty acids are more susceptible to oxidation and can contribute to off-flavors in extrudates.

Another concern with lipids is increased oxidation due to increased levels of iron and other metals from the extruder itself as metals can act as prooxidants. This "screw wear" phenomenon will be discussed later. Semwal *et al.* (1994) found greater lipid oxidation in extruded rice and dahl compared to oven-dried and drum-dried sample. Both iron content and peroxide values were higher in extruded products.

Expansion is desirable for snacks and breakfast cereals, but the increased surface area formed by the numerous air cells favors oxidation. However, antioxidants derived from Maillard reaction products may help reduce oxidation rates. Oatmeal cookies containing potato peels showed lower peroxide values than control cookies, and extruded peels had greater antioxidant activity than did non-extruded peels (Arora and Camire, 1994).

Nutritional value does not appear to be compromised by extrusion. Complexed lipids are digestible by monogastric and ruminant animals. Essential and omega fatty acids in extruded foods are more susceptible to lipid oxidation, and thus, reduced nutritional value, but few studies have examined this potential problem.

STARCH

Extrusion effects on starch are significant since starchy cereals and tubers provide the bulk of calories consumed by most people, particularly those living in less-developed nations. Gelatinization occurs at lower moisture levels (12–22%) than is possible with other food processing methods. Qu and Wang (1994) hypothesized that waxy corn starch extrusion is dominated by melting, which follows a zero order kinetics, rather than gelatinization, a first order reaction.

Third generation snack pellets are formed and partially cooked by extrusion, then puffed by frying or baking. Product moisture levels are often somewhat higher than for extrusion-puffed snacks. A critical temperature of 70–80 °C was required to gelatinize cassava-based shrimp pellets (Seibel and Hu, 1994).

Extrusion greatly increased wheat bran and whole flour starch susceptibility to enzymes, but no samples were fully gelatinized under the extrusion conditions employed in the study (Wang *et al.*, 1993). Addition of sucrose, salt or fiber to starchy foods such as cornmeal may affect gelatinization, and thus, expansion (Jin *et al.*, 1994).

The branched structure of amylopectin branches is susceptible to shear. Both amylose and amylopectin molecules may decrease in molecular weight. Politz and colleagues (1994a) studied relative molecular weight distributions in corn flour containing 60% amylopectin. Large amylopectin molecules were more likely to degrade. Low die temperature

(160° versus 185°C) and feed moisture 16 vs. 20%) significantly reduced average starch molecular weight in wheat flour, but protein content of flour was not an important factor (Politz *et al.*, 1994b). The researchers observed that the magnitude of starch degradation was greater for wheat than for corn. After extrusion, no molecules with molecular weight $> 10^7$ were found, and most samples exhibited an increase in the fraction with molecular wights of 10^5–10^7.

This molecular degradation may be exploited to produce dextrins and/or free glucose for syrups or subsequent fermentation processes. Successful conversion of starch to glucose by extrusion can be achieved by maximizing conditions to favor shear. Use of thermostable amylase considerably accelerates the process. Although enzymes are typically inactivated during extrusion, Termamyl® may be added to starch prior to extrusion; enzyme activity appears to be enhanced in the barrel. Glucose production from starch has been studied in barley (Linko *et al.*,1983), cassava (Grossman *et al.*, 1988), corn (van Zuilichem *et al.*, 1990; Roussel *et al.*, 1991), and potato waste (Camire and Camire, 1994).

Starch degradation may reduce radial expansion, which is critically related to product texture. Chinnaswamy (1993) has summarized many factors that affect expansion of starchy materials. Extrusion pressure is a better predictor of expansion than is die nozzle length/diameter ratio. Starches with 50% amylose showed optimal expansion, and barrel temperatures close to 150°C and low feed moistures favor expansion. Irradiation and certain additives were thought to improve expansion, but only sodium chloride proved to be useful. Over a range of amylose contents, 1% NaCl consistently produced more expansion than did the starch alone.

Reducing sugars on detached branches may react with other starch molecules, forming indigestible anhydro linkages. Transglycosidation has been reported by Theander and Westerlund (1987) , but structural changes are difficult to document. Politz *et al.* (1994b), however, found no changes in 2,3-glucose linkages in extruded wheat flour after methylation analysis; differences in extrusion conditions may be responsible for this observed discrepancy. Although lower digestibility is undesirable for infant and weaning as well as or other specialized dietary foods, formation of resistant starch may have applications in reduced calorie products. Chiu and coworkers (1994) were awarded a patent for a process that subjected high amylose starch to digestion with pullulanase, followed by extrusion to remove moisture. Extrusion raised the resistant starch, measured as dietary fiber, to 30%. Adding fiber to starch could also affect digestibility. Corn starch solubility decreased when extruded with added cellulose, and longer cellulose fibers reduced solubility further, possibly due to transglycosidation (Chinnaswamy and Hanna, 1991).

Amylose-lipid complex formation is dependent upon both starch and lipid type. Monoacylglycerols and free fatty acids added to corn starch at a 4% level were complexed with those starches containing a high proportion of amylose (Bhatnagar and Hanna, 1994a), but no complexes were found for any mixture containing tristearin. The operating parameters for the laboratory-scale single screw extruder used for this study were evaluated as well. Low feed moisture (19%) and barrel temperature (110–140°C) induced the greatest amount of complex formation between stearic acid and normal corn starch with 25% amylose (Bhatnagar and Hanna, 1994b). Although dough viscosity and specific mechanical energy were not reported, these findings suggest that extruder conditions favoring low viscosity and reduced residence time are not compatible with complex formation. These findings need to be confirmed with larger twin screw extruders before any definite recommendations can be made.

Extrusion may be used to modify starch for other purposes. Reactive extrusion, the intentional addition of chemicals to modify the base feed material, could be used to produce charged and other types of modified starches. Enzymatic hydrolysis of amioca, a high amylopectin corn starch, results in materials that gel readily. Extrusion of this starch broke few α(1→4) glycosidic bonds (Orford *et al.*, 1993). However, the starch was readily dispersible (over 90%) in water as a result of extrusion.

DIETARY FIBER

The case of contradictory findings for the analysis of dietary fiber are due to differences among analytical methods employed. The AOAC total dietary fiber method measures all compounds not digested by amylase and protease and insoluble in 80% aqueous ethanol. While cellulose, pectin, hemicelluloses, gums and lignin do meet these criteria, extrusion-modified starches and proteins could also be measured as fiber. Sites formerly accessible to digestion by enzymes may be involved in new bonds or physically-hindered. Many materials used to add dietary fiber to foods contain far less than 100% fiber. Artz and co-workers (1990) found no difference in X-ray diffraction patterns of corn fiber-corn starch blends after extrusion, which was expected since very little crystalline cellulose was present. The corn bran isolate used as a fiber source actually contained only 16.6% cellulose. Amorphous hemicelluloses comprised the remainder of the dietary fiber fraction.

Total dietary fiber measurement also does not discriminate changes in fiber solubility. AOAC methods that measure water-soluble and insoluble fiber, or enzymatic-chemical methods are more sensitive to extrusion-related changes in dietary fiber constituents. An enzymatic-chemical method found differences among foods for lignin and nonstarch polysaccharides (NSP), but uronic acids were unaffected by extrusion (Camire and Flint, 1991). The ratio of soluble to insoluble NSP increased for oatmeal and potato peels, but not for corn meal. Extrusion most likely solubilizes large molecules in a manner similar to that reported for starch.

Extrusion tripled the water solubility of sugar beet pulp fiber, primarily by reducing the molecular weight of pectin and hemicelluloses molecules (Ralet *et al.*, 1991). Ferulic acid, a phenolic acid normally associated with plant cell walls, was also recovered from the soluble sugar beet fraction. Smaller fragments may be soluble in aqueous ethanol, and thus discarded during the extraction steps common to enzymatic-gravimetric and enzymatic-chemical methods of fiber analysis.

Many factors influence fiber solubility. Acid and alkaline treatment prior to extrusion increased soluble fiber slightly in corn bran (Ning *et al.*, 1991). Grinding doubled the soluble fiber of pea hulls to 8% (dry basis), but extruded hulls all contained over 10% soluble fiber (Ralet *et al.*, 1993). The sum of insoluble plus soluble fiber decreased due to extrusion.

Soluble fiber has certain health benefits, including binding of bile acids in the small intestine, leading to reduced serum cholesterol levels. However, extrusion-induced solubilization may not offer the same health benefits. This "new" soluble fiber is chemically distinct from naturally soluble fiber compounds such as pectin and gums. Extrusion nearly doubled soluble fiber in potato peels, and significantly more cholic acid and deoxycholic acid were bound by peels extruded at the lower temperature studied (Camire *et al.*, 1994).

Young rats fed extruded oats, barley or wheat had lower total serum and liver cholesterol levels than did rats fed a control diet or feeds containing unextruded grains (Wang and Klopfenstein, 1993). Soluble fiber increased due to extrusion in all feed samples, and soluble β-glucans increased slightly in extruded oats and barley. Aqueous suspensions of extruded grains also exhibited higher viscosities, lending support to the theory that bile acids are trapped in viscous fiber mixtures in the small intestines.

Another health benefit of soluble fiber is its contribution to increased intestinal contents viscosity. Glucose is more slowly absorbed from viscous solutions, allowing for even serum glucose levels. This action is now widely recognized as beneficial for diabetics. Extruded lemon and orange peels exhibited higher levels of soluble fiber after processing, which increased viscosity *in vitro* (Gourgue *et al.*, 1994). However, these changes did not affect starch digestion and glucose diffusion.

Insoluble fiber may protect against colorectal cancer by binding carcinogens and preventing their interaction with tissues. Extrusion conditions of 110° C and 30% feed moisture significantly decreased the ability of potato peels to *in vitro* bind benzo[a]pyrene, a polycyclic aromatic hydrocarbon (Camire *et al.*, 1995b). Other extrusion conditions did not affect binding ability. Many breakfast cereals are extruded and are high in fiber. Of sixteen commercial cereals studied, all products bound at least 40% of the benzo[a]pyrene added (Camire *et al.*, 1995c). Carcinogen binding was not correlated with a specific dietary fiber fraction.

PROTEIN

Heat and shear within the extruder barrel denature proteins; thus many enzymes are inactivated. As a function of denaturation, protein solubility in water or dilute salt solutions is reduced, thereby changing protein functionality. SME may influence solubility more than does barrel temperature (Della Valle *et al.*, 1994). Wheat protein solubility decreases even at the relatively low temperatures (< 100°C) used for pasta extrusion (Ummadi *et al.*, 1995a). Large proteins may become dissociated into smaller subunits. Denaturation also exposes enzyme-susceptible sites, thus improving digestibility.

Two reviews of extrusion effects on food proteins have recently been published (Arêas, 1992; Camire, 1991). Arêas (1992) provided a thorough discussion of the mechanisms occurring during extrusion. Disulfide bonds break and reform, while new electrostatic and hydrophobic interactions promote aggregate formation. Although new peptide bonds may form during extrusion, their contribution to protein insolubilization and texturization is not clear. Many reports published on this topic used extrusion temperatures under 150°C; different mechanisms may be more important at higher temperatures. Soy protein extruded at different barrel temperatures (140, 160 and 180°C) and feed moistures (30 or 40%) exhibited very different solubilities in pH 7.6 phosphate buffer containing compounds known to disrupt protein-protein interactions (Table 1). Infrared spectroscopy revealed that all extruded samples contained β-pleated sheet structures (Prudêncio-Ferreira and Arêas, 1993).

Lysine is lost during Maillard and other thermal reactions. Maillard reactions may occur even if no reducing sugars are added to the feed, since starch and fiber fragments can react. Sucrose may be hydrolyzed during extrusion. In a model system of wheat starch, glucose and lysine, pH had the greatest effect on Maillard reactions, as measured by changes in color (Bates *et al.*, 1994). High feed moistures protect lysine in grains and foods intended for special nutritional needs. Since many extruded foods are not intended to be used as a primary source of protein, loss of lysine and other essential amino acids may be of little consequence. Extrusion texturization of soy isolate did not affect rat serum cholesterol, fecal steroid excretion, protein digestibility or biological value (Fukui *et al.*, 1993).

Table 1. Soy protein solubility (%) changes as a result of extrusion cooking[a]

Solubility category	Not extruded	Extruded					
		30 % Moisture			40 % Moisture		
		140°C	160°C	180°C	140°C	160°C	180°C
Soluble	15.01	3.93	6.78	8.41	4.50	3.46	6.96
Insoluble due to disulfide bonds	4.94	56.34	38.29	28.46	65.22	56.41	35.84
Insoluble due to noncovalent forces	17.68	32.82	32.16	31.84	30.42	32.88	28.29
Insoluble due to disulfide and noncovalent forces	58.31	2.45	13.51	25.67	Trace	Trace	23.64
Insoluble by unknown forces	4.06	4.46	9.26	5.67	Trace	7.15	5.32

[a]Adapted from Prudencio-Ferreira and Arêas (1993).

Traditionally, extrusion texturization of soy protein has been used to create meat analogues. Addition of sodium hydroxide during extrusion does not aid in texturization, with worse product quality at high pH (Dahl and Villota, 1991). Precipitation within the extruder at the soy isoelectric point may be necessary for adequate texturization. Huang and colleagues (1995) at Iowa Sate University devised a process in which soy protein isolate could be extruded into textile fibers. Brittleness was offset by addition of glycerol during extrusion and by various chemical treatments post-extrusion. A similar process could be developed to produce superior meat-like fibers.

Protein extrusion at high moisture contents can produce novel food products and ingredients (Cheftel *et al.*, 1992). Gels and emulsions may be created by extrusion. Relatively little is known about the effects of high moisture levels on chemical changes, particularly with respect to proteins. Extrusion was used successfully to acid coagulate skim milk powder, followed by a second extrusion to neutralize the acid casein to sodium caseinate (Barraquio and van de Voort, 1991). Whey protein isolate was extruded (pH 3.5–3.9, barrel temperature 90–100°C, and screw speed 100–200 rpm) to form coagulated semi-solid spreads that could function as fat substitutes (Queguiner *et al.*, 1992).

VITAMINS

This diverse group of chemicals has been widely studied since many extruded foods are intended for consumption by children and other groups with high nutritional needs. In addition to the reviews on nutritional changes during extrusion referred to earlier, a thorough review has focused on vitamin retention (Killeit, 1994). This review summarizes extrusion experiments since 1969. Killeit (1994) determined that certain extrusion parameters may increase vitamin loss: higher temperatures, screw speed, and specific energy input; and decreased feed moisture, die diameter, and throughput. Increasing moisture and throughput can protect vitamins by reducing mass temperature and thus limiting thermal degradation. Larger dies reduce pressure within the extruder, preventing molecular shear.

Vitamin D is the most stable of the oil-soluble vitamins, but few extrusion studies have examined this nutrient. Carotenoids and tocopherols are susceptible to oxidation and thermal degradation during extrusion and storage. Although β-carotene is not a vitamin, it has vitamin activity and is an important antioxidant and coloring agent. The content of added all trans β-carotene decreased by more than half when the temperature of wheat flour extrusion was 200°C compared with 125°C; increased levels of the 9-cis and 13-cis isomers were recovered (Guzman-Tello and Cheftel, 1990). The formation of these isomers suggests that thermal degradation is the primary cause of carotenoid loss. In the same study, extrusion under a nitrogen atmosphere or addition of BHT provided some protection against color loss, which was used as an indicator of oxidation.

Thiamine is particularly sensitive to thermal processing; the published retention values range from 0 to 95% (Killeit, 1994). Wheat flour extruded with no added water suffered large thiamine losses, and increased barrel temperature decreased the vitamin only at the slowest feed rate used (250 g/min versus 500 and 750 g/min) (Andersson and Hedlund, 1990). In the same study, extrusion conditions did not affect the content of riboflavin (B_2) or niacin; ascorbic acid (C) decreased with higher temperatures at 10% moisture. Since enriched wheat flour is an important source of B vitamins in the United States, care must be taken to protect these vitamins. Relatively little is known about the stability of synthetic vitamins compared with natural sources, or the effects of added vitamins as opposed to endogenous nutrients.

MINERALS

Despite the importance of minerals for health, relatively few studies have examined extrusion cooking effects on these nutrients. Minerals are heat-stable and unlikely to become lost in the steam distillate at the die. Binding or entrapment of minerals is possible, but animal feeding studies have not demonstrated impaired growth from mineral deficiencies.

On the other hand, food materials can gain minerals during extrusion. Iron and other metals used in fabrication of the barrel and screws literally wear off into the food. Materials with higher dietary fiber content appear to produce the greatest screw wear. Potato peel iron increased 38–83%, with significantly higher levels recovered from samples extruded at 143 °C versus 104 °C (Camire *et al.*, 1994). Rats fed extruded corn or potato utilized the screw wear iron as well as endogenous iron (Fairweather-Tait *et al.*, 1987) and extrusion did not appear to reduce iron and zinc absorption from wheat bran and flour fed to human volunteers (Faiweather-Tait *et al.*, 1989).

Extrusion may improve absorption of minerals by reducing other factors that inhibit absorption. Phytate, which may form insoluble complexes with minerals, was reduced by extrusion of wheat products (Fairweather-Tait *et al.*, 1989). Extrusion did not significantly reduce phytate levels in five types of Italian legumes (Lombardi-Boccia, *et al.*, 1991). Although total iron increased in all but white beans, dialyzable (available) iron decreased, possibly due to complex formation with phytate. Ummadi and coworkers (1995b) found that low shear extrusion conditions significantly increased dialyzable iron in navy beans, chickpeas, cowpeas, and lentils, as compared with samples that were boiled or extruded under high-shear conditions. Phytic acid was degraded under all processing conditions, but total phytate showed no major change. Tannins were significantly reduced by extrusion, leading the researchers to conclude that other factors may be responsible for changes in iron dialyzability.

NATURAL TOXINS

Extrusion cooking offers the opportunity to reduce or even completely eliminate natural toxins and antinutrients. Many otherwise nutritious foods, particularly legume seeds, contain compounds that are poisonous or impair nutrient utilization. The combination of chemical treatment with extrusion cooking is particularly promising. Previously published reviews have discussed extrusion studies to reduce aflatoxins, pathogens, and other undesirable food contaminants.

Trypsin inhibitors (TI) are present in many plant foods, but soybeans are one of the most widely consumed crops containing these antinutrients. TI prevent protein digestion and eventually consumption can lead to pancreatic hypertrophy or poor growth. Higher extrusion temperatures (138–154°C) were required for conventional soybeans to achieve the same nutritional quality as extruded Kunitz trypsin inhibitor-free beans processed at 121–138°C (Zhang *et al.*, 1993). Soybeans extruded at these temperatures were comparable to commercial solvent-extracted soy meal. Urease index decreased and chick growth performance increased as extrusion temperature was raised.

Extrusion alone cannot reduce the antigenicity of soybeans. Soy allergies are becoming more common as more people consume soy products. Twin screw extrusion with forward feed screw elements reduced soy meal antigenicity as barrel temperature was increased from 70 to 134°C (Ohishi *et al.*, 1994). The addition of kneading disc elements toward the die end of the screws produced comparable reductions at only 66°C. Feed rate and screw speed did not affect antigenicity reduction.

Tepal *et al.* (1994) extruded jack bean powder to remove several undesirable compounds; screw speed had no effect. Extrusion significantly decreased hemagglutinins and urease activity, but not canavanine levels. Since nitrogen solubility was also reduced by extrusion, this index could be used as a marker for protein-based unwanted materials. Other workers have reported that all antinutrients could not be destroyed simultaneously or to the same extent. Gujska and Khan (1991) found that nearly all hemagglutinin activity was eradicated in navy, pinto and garbanzo bean high starch fraction, but trypsin inhibitors were reduced only 70–85% under the extrusion conditions used.

Rapeseed glucosinolates impair animal growth, but canola varieties are generally low in these compounds and thus more suitable as feed for monogastric species. Darroch *et al.* (1990) extruded canola screenings, a mixture of canola and weed seeds and chaff, with added ammonia. Ammoniation significantly reduced total glucosinolates, but residual ammonia levels may restrict this treatment for feeds only.

Potatoes are highly nutritious, yet several natural toxicants are present in these tubers. Maga (1980) reported that extrusion reduced glycoalkaloids in potato flakes, a popular base material for extruded snacks. Twin screw extrusion of potato peels, which contain much higher levels of glycoalkaloids, did not change concentrations of either α-chaconine or α-solanine (Zhao *et al.*, 1994). Under *in vitro* digestion conditions, only 3–5% of glycoalkaloids were soluble, a condition required for absorption. Potato trypsin inhibitors are greatly reduced during steam peeling procedures, but the abrasion peeling process used by potato chip manufacturers does not affect these chemicals. Extrusion did not further reduce TI in steam peels, but did significantly reduce levels in abrasion peels (Zhao and Camire, 1995). Potatoes are treated in storage with a variety of chemicals. Chlorpropham, a sprouting inhibitor, and thiabendazole, a fungicide, were not reduced by extrusion of potato peels (Camire *et al.*, 1995a). The use of extrusion to reduce pesticide levels has not yet been reported by others.

Alkylresorcinols inhibit animal growth, yet toxicity to humans is not documented. Baking and fermentation degrade these compounds, which are concentrated in cereal brans. Extrusion decreased alkylresorcinols by over 50%, but varying feed moisture, barrel temperature and screw speed did not significantly influence the extent of reduction (Al-Ruqaie and Lorenz, 1992). Rye and triticale brans experienced a greater loss than did wheat bran, presumably due to existing differences in alkylresorcinol homologues in each species.

PHYTOCHEMICALS

Extrusion research is just now providing clues as to the fate of nutrients during extrusion. As nutrition science begins to unravel the importance of non-nutrient chemicals in foods, it is clear that extrusion effects on these compounds must be studied. For example, genistein and phytoestrogens in soy may help prevent cancer, yet extrusion texturization of soy might significantly reduce these compounds.

Phenolic compounds in grains, fruits and vegetables act as antioxidants and may have health benefits. Total free phenolics, primarily chlorogenic acid, decreased significantly due to extrusion in potato peels produced by steam or abrasion peeling (unpublished data). We suspect that the lost phenolics react with themselves or with other compounds to form larger insoluble materials.

FUTURE DIRECTIONS

Improved predictive models are needed to better predict chemical changes in extruded foods, but any model should include food composition and prior processing history of raw material. Food scientists and engineers should focus on the relationships between composition changes and product quality, both nutritional and sensory. Very different mechanisms may occur during high-moisture extrusion, creating a new line of research objectives.

REFERENCES

Al-Ruqaie, I.; Lorenz, K. Alkylresorcinols in extruded cereal brans. *Cereal Chem.* **1992,** *69,* 472–475.

Andersson, Y.; Hedlund, B. Extruded wheat flour: correlation between processing and product quality parameters. *Food Qual. Prefer.* **1990,** *2,* 201–216.

Arêas, J.A.G. Extrusion of food proteins. *Crit. Rev. Food Sci. Nutr.* **1992,** *32,* 365–392.

Arora, A.; Camire, M.E. Performance of potato peels in muffins and cookies. *Food Res. Intl.* **1994,** *27,* 14–22.

Artz, W.E.; Warren, C.C.; Villota, R. Twin screw extrusion modification of corn fiber. *J. Food Sci.* **1990,** *55,* 746–750, 754.

Barraquio, V.L; van de Voort, F.R. Sodium caseinate from skim milk powder by extrusion processing: physicochemical and functional properties. *J. Food Sci.* **1991,** *56,* 1552–1556, 1561.

Bates, L.; Ames, J.M.; MacDougall, D.B. The use of a reaction cell to model the development and control of colour in extrusion cooked foods. *Lebensm.-Wiss. u. Technol.* **1994,** *27,* 375–379.

Bhatnagar, S; Hanna, M.A. Amylose-lipid complex formation during single-screw extrusion of various corn starches. *Cereal Chem.* **1994a,** *71,* 582–587.

Bhatnagar, S; Hanna, M.A. Extrusion processing conditions for amylose-lipid complexing. *Cereal Chem.* **1994b,** *71,* 587–593.

Björck, I; Asp, N.-G. The effects of extrusion cooking on nutritional value - a literature review. *J. Food Eng.* **1983,** *2,* 281–308.

Camire, M.E. Protein functionality modification by extrusion cooking. *J. Am. Oil Chem. Soc.* **1991,** *68,* 200–205.

Camire, M.E; Camire, A.L. Enzymatic starch hydrolysis of extruded potato peels. *Starch/Stärke* **1994**, *46,* 308–311.

Camire, M.E; Flint, S.I.. Thermal processing effects on dietary fiber composition and hydration capacity in corn meal, oat meal, and potato peels. *Cereal Chem.* **1991**, *68,* 645–647.

Camire, M.E.; Camire, A.L.; Krumhar, K. Chemical and nutritional changes. *Crit. Rev. Food Sci. Nutr.* **1990,** *29,* 35–57.

Camire, M.E.; Zhao, J.; Violette, D.A. *In vitro* binding of bile acids by extruded potato peels. *J. Agric. Food Chem.* **1994,** *41,* 2391–2394.

Camire, M.E.; Bushway, R.J.; Zhao, J.; Perkins, B; Paradis, L.R. Fate of thiabendazole and chlorpropham residues in extruded potato peels. *J. Agric. Food Chem.* **1995a,** *43,* 495–497.

Camire, M.E.; Zhao, J.; Dougherty, M.P.; Bushway, R.J. *In vitro* binding of benzo[a]pyrene by extruded potato peels. *J. Agric. Food Chem.* **1995b,** *43,* 970–973.

Camire, M.E.; Zhao, J.; Dougherty, M.P.; Bushway, R.J. *In vitro* binding of benzo[a]pyrene by ready-to-eat breakfast cereals. *Cereal Foods World* **1995c,** *40,* 447–450.

Cheftel, J.C. Nutritional effects of extrusion cooking. *Food Chem.* **1986,** *20,* 263–283.

Cheftel, J.C.; Kitagawa, M.; Queguiner, C. New protein texturization processes by extrusion cooking at high moisture levels. *Food Rev. Intl.* **1992,** *8,* 235–275.

Chinnaswamy, R. Basis of cereal starch expansion. *Carbohydrate Polymers* **1993,** *21*, 157–167.

Chinnaswamy, R; Hanna, M.A.Physicochemical and macromolecular properties of starch-cellulose fiber extrudates. *Food Structure* **1991,** *10,* 229–239.

Chiu, C.-W.; Henley, M.; Altieri, P. Process for making amylase resistant starch from high amylose starch. U.S. Patent 5,281,276, Jan. 25, 1994.

Dahl, S.R; Villota, R. Twin-screw extrusion texturization of acid and alkali denatured soy proteins. *J. Food Sci.* **1991,** *56,* 1002–1007.

Darroch, C.S.; Bell, J.M.; Keith, M.O. The effects of moist heat and ammonia on the chemical composition and feeding value of extruded canola screenings for mice. *Can J. Anim. Sci.* **1990,** *70,* 267–277.

de la Gueriviere, J.F.; Mercier, C.; Baudet, L. Incidences de la cuisson-extrusion sur certains parametres nutritionnels de produits alimentaires notamment céréaliers. *Cah. Nutr. Diet.* **1985,** *20,* 201–210.

Della Valle, G.; Quillien, L.; Gueguen, J. Relationships between processing conditions and starch and protein modifications during extrusion-cooking of pea flour. *J. Sci. Food Agric.* **1994,** *64,* 509–517.

Fairweather-Tait, S.J.; Symss, L.S.; Smith, A.C.; Johnson, I.T. The effect of extrusion cooking on iron absorption from maize and potato. *J. Sci. Food Agric.* **1987,** *39,* 341–348.

Fairweather-Tait, S.J.; Portwood, D.E.; Symss, L.L.; Eagles, J.; Minski, M.J. Iron and zinc absorption in human subjects from a mixed meal of extruded and nonextruded wheat bran and flour. *Am. J. Clin. Nutr.* **1989,** *49,* 151–155.

Fukui, K.; Aoyama, T.; Hashimoto, Y.; Yamamoto, T. Effect of extrusion of soy protein isolate on plasma cholesterol level and nutritive value of protein in growing male rats. *J. Jap. Soc. Nutr. Food Sci.* **1993,** *46,* 211–216.

Gourgue, C.; Champ, M.; Guillon, F.; Delort-Laval, J. Effect of extrusion-cooking on the hypoglycaemic properties of citrus fibre: an *in vitro* study. *J. Sci. Food Agric.* **1994,** *64,* 493–499.

Grossman, M.V.E.; El-Dash, A.A.; Carvalho, J.F. Extrusion cooking of cassava starch for ethanol production. *Starch/Stärke* **1988,** *40,* 300–307.

Gujska, E; Khan, K. Feed moisture effects on functional properties, trypsin inhibitor, and hemagglutinating activities of extruded bean high starch fractions. *J. Food Sci.* **1991,** *54,* 443–447.

Guzman-Tello, R; Cheftel, J.C. Colour loss during extrusion cooking of beta-carotene-wheat flour mixes as an indicator of the intensity of thermal and oxidative processing. *Intl. J. Food Sci. Technol.* **1990,** *25,* 420–434.

Harper, J.M.*Extrusion of Foods.* CRC Press, Inc., Boca Raton, FL, 1981.

Hayakawa, I. *Food Processing by Ultra High Pressure Twin Screw Extrusion.* Technomic Publ. Co., Lancaster, PA, 1992.

Huang, H.C.; Hammond, E.G.; Reitmeier, C.A.; Myers, D.J. Properties of fibers produced from soy protein isolate by extrusion and wet spinning. *J. Am. Oil Chem. Soc.* **1995,** *72,* 1453–1460.

Jin, Z.; Hsieh, F.; Huff, H.E. Extrusion cooking of corn meal with soy fiber, salt, and sugar. *Cereal Chem.* **1994,** *71,* 227–234.

Killeit, U. Vitamin retention in extrusion cooking. *Food Chem.* **1994,** *49,* 149–155.

Kokini, J.L.; Ho, C.-T.; Karwe, M.V., Eds.; *Food Extrusion Science and Technology;* Marcel Dekker, New York, 1992.

Linko, P.; Hakulin, S.; Linko, Y.-Y. Extrusion cooking of barley starch for the production of glucose syrup and ethanol. *J. Cereal Sci.* **1983,** *1,* 275–284.

Lombardi-Boccia, G.; Di Lullo, G.; Carnovale, E. *In vitro* iron dialysability from legumes: influence of phytate and extrusion cooking. *J. Sci. Food Agric.* **1991,** *55,* 599–605.

Maga, J.A. Glycoalkaloid stability during the extrusion of potato flakes. *J. Food Process. Preserv.* **1980,** *4,* 291–296.

Mercier, C.; Linko, P.; Harper, J.M., Eds.; *Extrusion Cooking.* Am. Assoc. Cereal Chem., St. Paul, MN, 1989.

Meuser, F; van Lengerich, B. Systems analytical model for the extrusion of starches. In *Thermal Processing and Quality of Foods*; Zeuthen, P.; Cheftel, J.C.; Eriksson, C.; Jul, M.; Leniger, H.; Linko, P.; Varela, G.; Vos, G., Eds.; Elsevier Applied Sci. Publ., London, 1984, pp. 175–179

Ning, L.; Villota, R.; Artz, W.E. Modification of corn fiber through chemical treatments in combination with twin-screw extrusion. *Cereal Chem.* **1991**, *68*, 632–636.

O'Connor, C., Ed., *Extrusion Technology for the Food Industry.* Elsevier Applied Sci. Publ., London, 1987.

Ohishi, A.; Watanabe, K.; Urushibata, M.; Utsuno, K.; Ikuta, K.; Sugimoto, K.; Harada, H. Detection of soybean antigenicity and reduction by twin-screw extrusion. *J. Am. Oil Chem. Soc.* **1994,** *71,* 1391–1396.

Orford, P.D.; Parker, R.; Ring, S.G.The functional properties of extrusion-cooked waxy-maize starch. *J. Cereal Sci.* **1993,** *18,* 277–286.

Politz, M.L.; Timpa, J.D.; Wasserman, B.P. Quantitative measurement of extrusion-induced starch fragmentation products in maize flour using nonaqueous automated gel-permeation chromatography. *Cereal Chem.* **1994a,** *71,* 532–536.

Politz, M.L.; Timpa, J.D.; White, A.R.; Wasserman, B.P. Non-aqueous gel permeation chromatography of wheat starch in dimethylacetamide (DMAC) and LiCl: extrusion-induced fragmentation. *Carbohydrate Polymers* **1994,** *24,* 91–99.

Qu, D; Wang, S.S.Kinetics of the formations of gelatinized and melted starch at extrusion cooking conditions. *Starch/Stärke* **1994,** *46,* 225–229.

Queguiner, C.; Dumay, E.; Salou-Cavalier, C.; Cheftel, J.C. Microcoagulation of a whey protein isolate by extrusion cooking at acid pH. *J. Food Sci.* **1992,** *57,* 610–616.

Ralet, M.-C.; Della Valle, G.; Thibault, J.-F. Solubilization of sugar-beet pulp cell wall polysaccharides by extrusion cooking. *Lebensm.-Wiss. u.-Technol.* **1991,** *24,* 107–112.

Ralet, M.-C.; Della Valle, G.; Thibault, J.-F. Raw and extruded fibre from pea hulls. Part I: composition and physico-chemical properties. *Carbohydrate Polymers* **1993,** *20,* 17–23.

Roussel, L.; Vielle, A.; Billet, I.; Cheftel, J.C. Sequential heat gelatinization and enzymatic hydrolysis of corn starch in an extrusion reactor. Optimization for a maximum dextrose equivalent. *Lebensm.-Wiss. u.-Technol.* **1991,** *24,* 449–458.

Seibel, W; Hu, R.Gelatinization characteristics of a cassava/corn starch basèd blend during extrusion cooking employing response surface methodology. *Starch/Stärke* **1994,** *46,* 217–224.

Semwal, A.D.; Sharma, G.K.; Arya, S.S.Factors influencing lipid autoxidation in dehydrated precooked rice and Bengalgram dhal. *J. Food Sci. Technol.* **1994,** *31,* 293–297.

Tepal, J.A.; Castellanos, R.; Larios, A.; Tejada, I. Detoxification of jack beans (*Canavalia ensiformis*): I- Extrusion and canavaline elimination. *J. Sci. Food Agric.* **1994,** *66,* 373–379.

Theander, O; Westerlund, E. Studies on chemical modifications in heat-processed starch and wheat flour. *Starch/Stärke* **1987,** *39,* 88–93.

Ummadi, P.; Chenoweth, W.L.; Ng, P.K.W. Changes in solubility and distribution of semolina proteins due to extrusion processing. *Cereal Chem.* **1995a,** *72,* 564–567.

Ummadi, P.; Chenoweth, W.L.; Uebersax, M.A.The influence of extrusion processing on iron dialyzability, phytates and tannins in legumes. *J. Food Process. Preserv.* **1995b,** *19,* 119–131.

van Zuilichem, D.J.; van Roekel, G.J.; Stolp, W.; van't Riet, K. Modelling of the enzymatic conversion of cracked corn by twin-screw extrusion cooking. *J. Food Engin.* **1990,** 12, 13–28.

Wang, W.-M; Klopfenstein, C.F. Effect of twin-screw extrusion on the nutritional quality of wheat, barley, oats. *Cereal Chem.* **1993,** *70,* 712–715.

Wang, W.-M.; Klopfenstein, C.F.; Ponte, J.G. Effects of twin-screw extrusion on the physical properties of dietary fiber and other components of whole wheat and wheat bran and on the baking quality of the wheat bran. *Cereal Chem.* **1993,** *70,* 707–711.

Zhang, Y.; Parsons, C.M.; Weingartner, K.E.; Wijeratne, W.B.Effects of extrusion and expelling on the nutritional quality of conventional and Kunitz trypsin inhibitor-free soybeans. *Poultry Sci.* **1993,** *72,* 2299–2308.

Zhao, J; Camire, M.E.Glycoalkaloid content and *in vitro* solubility of extruded potato peels. *J. Agric. Food Chem.* **1994,** *42,* 2570–2573.

Zhao, J; Camire, M.E. Destruction of potato peel trypsin inhibitor by peeling and extrusion cooking. *J. Food Qual.* **1995,** *18,* 61–67.

12

SUCROSE LOSS AND COLOR FORMATION IN SUGAR MANUFACTURE

Les A. Edye and Margaret A. Clarke

Sugar Processing Research Institute, Inc.
1100 Robert E. Lee Blvd.
New Orleans, Louisiana 70124

The chemical reactions contributing to sucrose loss and color formation in evaporators in sugar manufacture (*i.e.*, hydrolysis of sucrose and degradation of monosaccharides at acid pH) are reviewed. A case study of a sugar factory's evaporator system demonstrates that the measurement of small but real losses of sucrose across the process is not possible by conventional sugar factory analyses. Alternative, more accurate techniques (*e.g.*, capillary gas chromatography or high performance ion chromatography [HPIC] with pulsed amperometric detection [PAD]) are considered. In the case study, glucose:sucrose ratios are determined by HPIC, and sucrose loss across the evaporator is estimated to be 1.39% of total sucrose. Loss measurements are thought to be underestimates; reasons for underestimation and the sources of errors are discussed. An approach to a more definitive loss measurement is proposed.

INTRODUCTION

Both the manufacture and refining of raw cane sugar use steam and reduced pressures to evaporate water from sugar containing process streams to yield crystalline sucrose. This water removal is achieved in continuous multiple effect evaporators followed by batch or continuous crystallization in vacuum pans. The physical principles of evaporators are well understood and have been the subject of standard texts. Control of a typical multiple effect evaporator is based on the measurement of liquid (*i.e.*, sucrose containing process stream) and steam feed rates to the first effect, evaporator body liquid level, product liquor Brix, and last effect steam (or vapor) pressure. These measurements can be used to calculate evaporator performance, and to optimize energy consumption. The same depth of understanding can not be claimed for the chemical changes occurring

Process-Induced Chemical Changes in Food
edited by Shahidi *et al.* Plenum Press, New York, 1998

in the process streams during evaporation. Measurements of change in pH and sucrose concentration (even by pol, let alone a more specific technique) across the evaporation process are not commonly performed in either the manufacturing or the refining of raw sugar. Even if these measurements were routine, the interpretation of the results can be difficult. Literature on thermal decomposition and color formation in aqueous sucrose solutions was summarized in *Sugar Technology Reviews* by Kelly and Brown (1978/79). They covered topics such as acid catalyzed decomposition of sucrose and hexoses, and included work contained in 189 references dating from 1932 to 1974. The earlier work of Mauch (1971) on the chemical properties of sucrose, an exhaustive review containing 272 references, is also worthy of note. We refer the reader to these two reviews for a thorough historical perspective on this subject. This paper reviews the current understanding of the acid hydrolysis of sucrose, the acid degradation of monosaccharides and the concurrent formation of colored compounds, and will concentrate on work reported subsequent to the review of Kelly and Brown. The reactions of sucrose and monosaccharides in aqueous alkaline solutions are not discussed since they have no relevance to sucrose loss and color formation in evaporators. The effect of evaporator conditions on the course of the acid catalyzed reactions is discussed and an example of sucrose loss and color formation measurements in a typical triple effect calandria evaporator with a single effect pre-evaporator is presented.

REACTIONS OF SUCROSE IN ACIDIC AQUEOUS SOLUTION

The Acid Hydrolysis of Sucrose

The investigation of the reaction of sucrose in aqueous acid solution has a long history; it was the subject of several kinetic studies in the early 19th century and Arrhenius later developed the equation describing the effect of temperature on reaction rate using data from sucrose hydrolysis experiments. It is generally accepted that the acid-catalyzed sucrose hydrolysis mechanism involves protonation of the glycosidic oxygen followed by heterolysis of the glycosidic bond to form the two monosaccharides, with one monosaccharide in the form of a cyclic carboxonium ion. Mega and Van Etten (1988) reported the use of ^{18}O shift in ^{13}C nuclear magnetic resonance to elucidate the point of bond cleavage in the acid-catalyzed hydrolysis of sucrose. Sucrose was hydrolyzed in the presence of $H_2{}^{18}O$, and the incorporation of ^{18}O into the products was determined. The results clearly indicated fructosyl-oxygen bond cleavage. Therefore, acid-catalyzed hydrolysis of sucrose initially yields D-glucose and a fructose carboxonium ion (Fru^+) which can react (see Figure 1) with either water to form D-fructose, or another Fru to form difructose dianhydrides, or another saccharide (*e.g.*, sucrose to form kestoses), or possibly even polyhydroxy phenolic colored compounds (PPCC). All these reactions regenerate the H^+ catalyst, and the contribution of each of these alternatives is dependent on the rate of reaction and concentrations of the hydroxyl oxygen donors.

In the first effect water is present at a significantly higher concentration than the solutes (*viz.*, sucrose, invert and PPCC), and therefore, the reaction of Fru with water to produce D-fructose predominates. In fact, in dilute solution, prior to evaporation, the above mentioned alternative reactions are unlikely. In later effects, as the concentration of solutes increases, the alternative reactions become more likely, *i.e.*, small amounts of kestoses and difructose dianhydrides may form in the later stages of evaporation if there is any further sucrose hydrolysis.

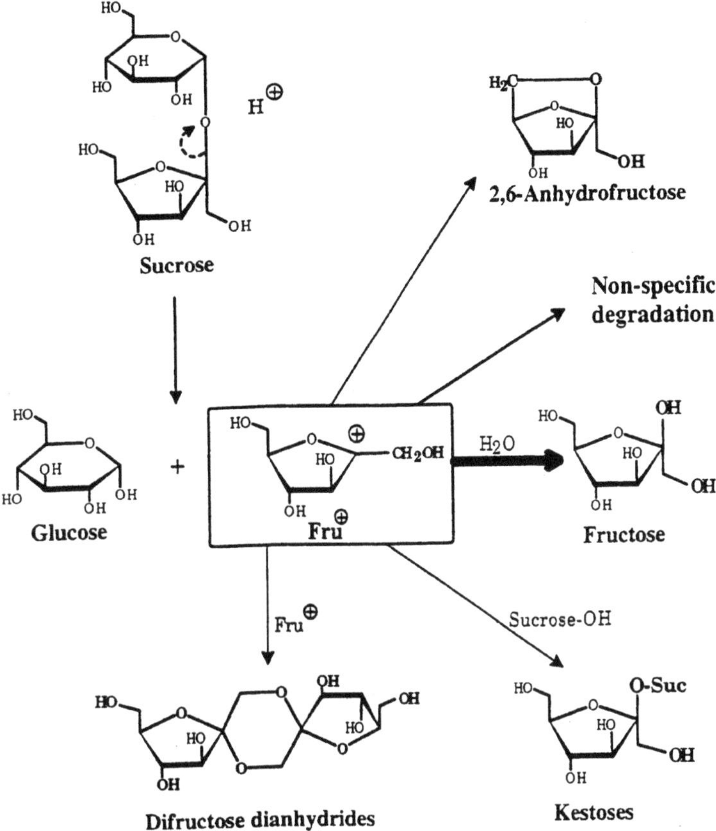

Figure 1. The hydrolysis of sucrose in acidic, aqueous solution.

Acid Degradation of Monosaccharides

Under mild acid conditions (*viz.*, pH 5–6 at 0–60°C) reducing sugars ionize and mutarotate, at lower pH (*viz.*, down to pH 3 or 4) and at higher temperatures (*viz.*, up to *ca.* 100°C) enolization and isomerization occurs. In acid solution enolization is initiated by direct protonation of the carbonyl group (see Figure 2). In fact, acids are far less effective enolization catalysts than alkalies and as a consequence D-glucose and D-fructose in aqueous solution show maximum stability between pH 3 and 4 (*e.g.*, McDonald, 1950).

Kelly and Brown's review (1978/79) indicates that, under conditions described above, the further acid-catalyzed reactions of reducing sugars (*e.g.*, dehydration to 5-(hydroxymethyl)-2-furaldehyde [HMF]) are extremely slow. For example (Kharin and Sapronov, 1969), while sucrose (2.0 M) hydrolysis at pH 5.6 and 100 C was measurable in a few hours, the further decomposition of the invert was demonstrated in a time scale of over 200 hours. Similarly (Wolfrom and Shilling, 1951) an 80% wt/wt solution of D-fructose after refluxing for 16 hours at pH 6.9 yielded only 0.1% D-glucose and 0.6% HMF, calculated on the basis of the original d-fructose; however, the fact that the authors did not

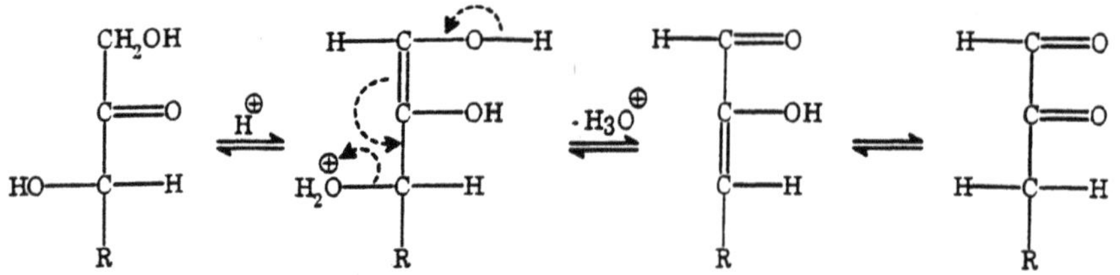

Figure 2. Enolization of reducing sugars in acidic, aqueous solution.

specify the amount of d-fructose that survived questions the relevance of the data. A typical product profile for D-fructose decomposition is shown in Table 1. Here it is sufficient to note that acid-catalyzed degradation of reducing sugars proceeds in a complex reaction network to products of isomerization, dehydration, fragmentation, and condensation.

It may be concluded that with the possible exception of isomerization, none of the above acid degradation reactions would be likely to occur to any large extent under conditions found in evaporators in the sucrose manufacturing industry. However, these studies were conducted in quite pure sugar solutions, whereas in factory operations, the sugar solutions are relatively complex mixtures of sugars, and other organic and inorganic compounds. In fact, in a series of experiments on the effect of salts on sucrose hydrolysis (Eggleston *et al.*, 1996), we observed rapid color formation in solutions containing sucrose (*ca.* 65 Brix) and $AlCl_3$ (0.05 M); after 30 min at 100°C the sucrose was completely hydrolyzed, at 60 min the solution was pale yellow, and after 180 min the solution was dark brown and the viscosity had increased significantly. Obviously, the degradation (dehydration, fragmentation and conden-sation to PPCC or HMWCC) of invert had progressed far more rapidly than would have been predicted from the studies on pure solutions (containing no salts). In evaporators, under mild acid conditions, some reducing sugar degradation may also proceed *via* Maillard reactions with amino acids; these reactions further complicate the product profile, and contribute to sucrose loss and undesirable color formation.

Table 1. Reported products of fructose decomposition in water at elevated temperatures

Dehydration	Fragmentation	Condensation
5-(hydroxymethyl)-2-furaldehyde[1]	formic acid[1]	"humin" or high
5-methyl-2-furaldehyde[2]	lvulinic acid[1]	molecular weight
-angelica lactone[2]	dihydroxyacetone[1]	colored compounds
-angelica lactone[2]	glyceraldehyde[1]	(HMWCC)
2-(2-hydroxyacetyl)furan[2]	2-furaldehyde[1]	
2-(2-hydroxyacetyl)furan formate[2]	pyruvaldehyde[1]	*Isomerization*
isomaltol[2]	lactic acid[1]	
4-hydroxy-2,3,5-hexanetrione[2]	acetol[1]	
4-hydroxy-2-(hydroxymethyl)-5-methyl-3(2H)-furanone[2]	glycoaldehyde2	D-glucose[1]
	acetic acid[2]	
	2,3-butanedione[2]	

[1]Major products (generally >1% absolute yield).
[2]Minor products.
Antal *et al.* (1990)

Inversion Rates and Sucrose Loss

The literature over many years has tabulated rates for sucrose hydrolysis and invert formation, at various concentrations, pH's and temperatures (Parker, 1970; Spencer and Meade, 1963; Silin, 1964). These studies, in general, were conducted in solutions of sucrose only, with pH adjusted by addition of acid or base, yielding solutions of very low ionic strength, inorganic content or ash content. The much-quoted tables of King and Jison result from measurement of hydrolysis rates in 0.5% to 2.5% w/v solutions of sucrose; these tables are reproduced in both the *Cane Sugar Handbook* (Spencer and Meade, 1963) and the *Handbook of Sugars* (Pancoast and Junk, 1980), in the former case with the unfortunate accompanying indication that the same rates could be applied in the 65 to 70 Brix (or % w/w) range.

Part of the foundation for the above-mentioned extrapolation of hydrolysis rate data from low sucrose concentration to refinery liquors at high sucrose concentrations is the assumption that impurities have no effect on reaction rate. Clarke (1977), in early studies on the use of HPLC to measure sucrose hydrolysis rates, observed hydrolysis rates greater than those previously reported in the literature (Parker, 1970; Spencer and Meade, 1963; Silin, 1964; Pancoast and Junk, 1980), when reactions were run at 0.1M KCl (*ca.* 0.7% KCl). After 6 hours, at pH 7, 90° C, and 60 Brix sucrose in water, 98.62% of the initial sucrose remained, whereas in 0.1M KCl, 95.15% sucrose remained (a *ca.* 3.5 fold increase in hydrolysis rate at 0.1M KCl). The increased rate of hydrolysis in solutions having an ash content in the 0.0–0.5% range reflects cane sugar refinery and raw sugar factory (sugar end) processing conditions more accurately than the rates in pure sucrose solutions of low concentration. Earlier studies did not simulate ash levels in processing conditions. The kinetics of sucrose hydrolysis, using simulated refinery liquors and factory syrups, is currently under investigation by the authors in collaboration with researchers at the United States Department of Agriculture (Eggleston *et al.*, 1995a,b,; 1996).

Measurement of pH and Sucrose Loss

In the case of laboratory research with dilute sucrose solutions the measurement of pH at high temperatures is affected not only by the change in water activity ($K_w = 1 \times 10^{-14}$ at 24° and 5.13×10^{-13} at 100 C or $pK_{water} = 12.29$ at 100° C) (Marshall and Frank [1981]), but also by a change in the characteristics of the electrode used in the measurement. In investigations where a pH electrode is calibrated at 25° and pH measurements are made at much higher temperatures the differences between calibration and operation temperatures certainly results in some error in the pH value. Furthermore while many pH electrodes may give stable pH readings even in boiling alkaline solutions, the relationship between these observed pH values and the actual hydrogen ion activities are often undetermined.

The measurement of pH is further complicated by the effect of high concentrations of sucrose (*e.g.*, 60 Brix or 60%w/w) on hydrogen ion activity. Clarke (1970) has discussed the effect of sucrose solution structure on pH and calcium ion electrode processes and shown a decreased response of these electrodes to changes in ionic activity in sucrose solutions at 60 Brix and 24 °C. This reduced electrode response can in part be explained by the structural order of the sucrose-water mixture (molecular association in sucrose-water systems has been reviewed by Allen *et al.* (1974). In a 60 Brix sucrose solution the ratio of water molecules to sucrose molecules is 12.7:1, with water molecules hydrogen-bonded to sucrose (*i.e.*, in the solvation shell) in dynamic equilibrium with free water. Therefore, the concentration of free water molecules and dissociated ions is much less than in dilute sucrose solutions. The number of water molecules in the sucrose solvation shell is dependent on the extent of intramolecular hydrogen bonding in the sucrose molecule. The effect of sucrose concentration and temperature on intramolecular hydrogen bonding in aqueous sucrose solutions is still not completely resolved, although recent advances in molecular modelling (Poppe and van Halbeek, 1992; Perez *et al.*, 1992) suggest that sucrose has far more freedom of rotation about the glycosidic linkage (*i.e.*, less intramolecular hydrogen bonding) than previously thought (Bock and Lemieux, 1982).

Therefore in most cases pH values measured at high temperatures in dilute solution should be considered approximate values only. In cases where the investigators address this problem and are careful to select a suitable electrode (*viz.*, one that manufacturers claim to have almost hysteresis-free pH measurement and a stable isopotential point over the temperature range) the error associated with electrode performance will be small, and differences in reported pH values will correspond to differences in actual pH. In cases where pH is measured in concentrated sucrose solutions the reported pH value should be considered as a nominal value only, and the differences in nominal pH values may or may not correspond to actual differences in hydrogen ion activity.

In sugar manufacture control operations, pH electrodes should not, as unfortunately they sometimes are, be calibrated with standard buffer solutions and then placed on stream in sugar liquors and assumed to read equivalent pH. Implicit in this operation is the equivalence of pH electrode response in dilute aqueous buffers at *ca.* 24 C and in high Brix sugar solutions at elevated temperatures. Such equivalence does not exist. Therefore, it would be unwise to conclude that a nominal pH value (from a pH electrode in a sugar refinery) greater than 8.5 indicates hydrogen ion activities that are too low to catalyze sucrose hydrolysis at a significant rate. The use of pH electrodes in the sugar industry should be based on experiential evidence that maintenance of nominal pH above a minimum value will prevent sucrose loss to invert.

Vukov (1965) has developed equations based on experimental data that predict the effect of temperature, pH and ionic strength on rate constants of sucrose decomposition in acid and alkaline medium. Other workers (de Bruijn *et al.*, 1991) report that Vukov's equation generally agrees with their experimental rate data, but only if the data is generated by the same analytical techniques. Clarke (1977) in early studies on the use of HPLC with refractive index detection to measure sucrose hydrolysis rates, observed hydrolysis rates greater than those determined by polarimetry (Parker, 1970; Spencer and Meade, 1963; Silin, 1964). It is our observation that sucrose hydrolysis rates determined by polarimetry are consistently lower than those determined by ion-exchange chromatography with pulsed amperometric detection. Considerable differences between sucrose determination by the phenol/H_2SO_4 test, polarimetry, HPLC and gas chromatography with flame ionization detection in mixed juice and molasses have also been reported (Schäffler and Smith, 1978; Vercellotti and Clarke, 1994).

Based on the above inadequacy of pH measurements and the variability in sucrose determinations by different analytical methods we stress a cautious approach in the interpretation of fundamental research data in the industrial environment. It seems likely that empirical equations for estimation of sucrose loss have little application, and obviously, the results of experimental determination of sucrose loss during evaporation will be dependent on the choice of analytical methods.

A CASE STUDY OF SUCROSE LOSS AND COLOR FORMATION

The case study of sucrose loss and color formation was performed at a north American sugar cane factory. The factory has a single effect calandria type pre-evaporator, and a triple effect calandria type evaporator. The mixed juice target pH was 7.0 ± 0.2. The temperature in the evaporator first effect was 102° ± 3° C, in the second effect 83° ± 3° C, and in the third effect 57° ± 3° C. The average residence time across the triple effect was 13 ± 2 min.

METHODS

Samples of mixed juice (MJ), syrup leaving the second effect (2ndS) and syrup leaving the third effect (3rdS) were obtained hourly for 14 hours, with a 10 min delay between sampling the MJ and the 2ndS and another 10 min delay before sampling the 3rdS. The sample pol (at 589 nm and 880 nm, both after clarification with $AlCl_3/Ca(OH)_2$), Brix and ambient pH values were obtained by conventional sugar laboratory methods, and color was determined by the ICUMSA method GS1–7 (ICUMSA Methods Book, 1994). The samples were also analyzed by HPLC (for sucrose - Biorad Aminex HPX-87C, 85° C, 0.8 mL min^{-1}, 0.25 mM $Ca(C_2H_3O_2)_2$, refractive index detection), by HPIC (for sucrose, glucose and fructose - Dionex CarboPac PA1, 1.0 mL min^{-1}, 100 mM NaOH, pulsed amperometric detection), and after drying (40° C, 0.1 mm Hg) and silylation by gas chromatography (GC) with mass selective detection (MS) (30 m x 0.25 mm HP-5 M.S. capillary column, 100° for 3 min + 6 min^{-1} to 250° + 250° for 10 min, or 250° for 10 min + 5° min^{-1} to 310° + 310° for 10 min, HP 5972 Mass Selective Detector at 70 eV).

RESULTS AND DISCUSSION

Figures 3 and 4 show the changes in refractometric dissolved solids and sucrose by pol at 589 nm in MJ, 2ndS and 3rdS over the 14 hour sampling period. In general, small increases or decreases in MJ Brix or pol values effect similar changes in 2ndS and 3rdS. It would appear that the delays in sampling between MJ and 2ndS or 3rdS make reasonable account for the average residence time across the evaporator process. The purity change across evaporator effects is shown in Figure 5, purities were calculated using polarimetric

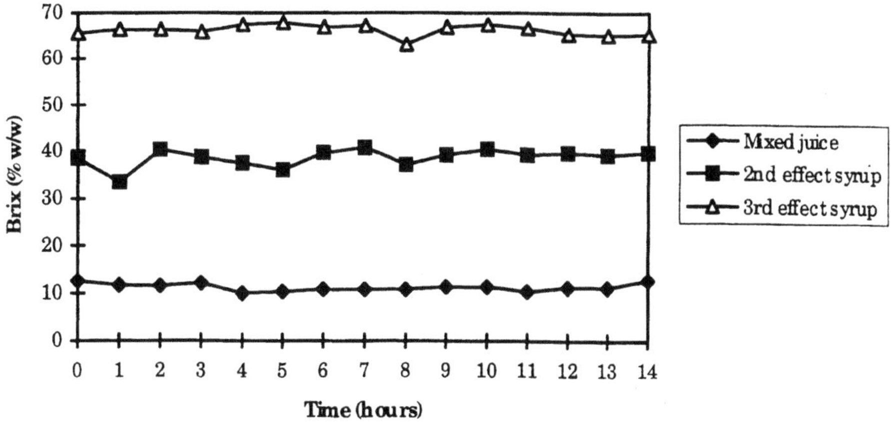

Figure 3. Mixed juice and evaporator Brix values during a 14 hour sampling period.

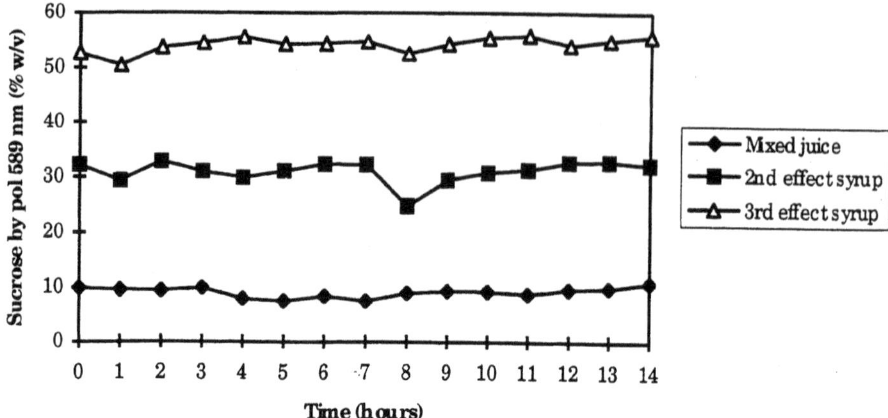

Figure 4. Mixed juice and evaporator pol (589 nm) values during a 14 hour sampling period.

sucrose measurements at 589 nm. No useful information can be obtained from these purity measurements, in fact a single measurement of purity change across the evaporator may lead to an erroneous conclusion on sucrose loss (*cf.*, 2 hour and 13 hour samples).

The variability in sucrose determinations by different analytical methods is shown in Figure 6. Obviously, purity values are dependent on the choice of analytical methods. As a digression, it is interesting to note that the 880 nm pol measurement is consistently lower than the 589 nm pol, HPLC and HPIC sucrose determinations.

The random variation in purity values across evaporators is caused, not only by errors in measurement, but also by the compounded effects of several physical and chemical changes. Brix values will decrease with scale formation. The chemical composition of scale has frequently been studied (*e.g.*, Sugar Milling Research Institute (South Africa) Annual Report, 1993–94); with some regional variation, scale from 1st and 2nd effects is calcium phosphates and sulphates, and some organic matter; the harder scale from later ef-

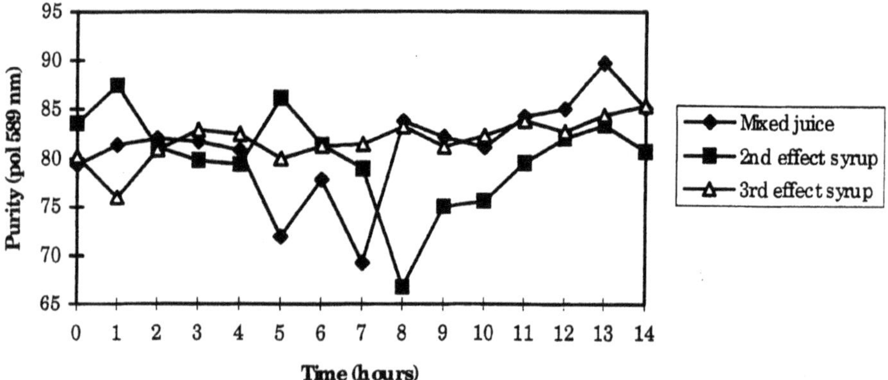

Figure 5. Purity change across evaporator effects.

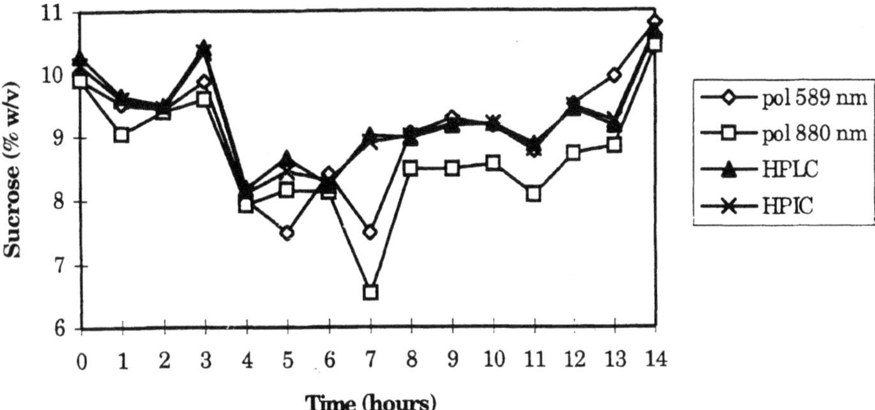

Figure 6. Mixed juice sucrose by pol at 589 nm and 880 nm, by HPLC and by HPIC.

fects contains calcium oxalates in a silicate matrix. Since very little sucrose is present in the scale, the formation of scale will decrease Brix and increase purity. Entrainment of juice droplets in vapor will cause loss of all soluble solids equally to the vapor phase and, therefore, should have no effect on purity. Sucrose hydrolysis will decrease the pol value and the purity, but subsequent loss of d-fructose (discussed in section 2.2) will result in an increase in pol value and an underestimate of sucrose hydrolysis. These physical and chemical changes, along with measurement error make it extremely difficult to accurately measure sucrose loss in evaporators.

In the above discussion it is manifest that sucrose degradation in evaporators (operating under normal conditions) can not be measured by the conventional sugar factory analytical methods. However, evidence of sucrose decomposition can be obtained by measuring pH drop and color formation across the evaporation process (see Table 2). The pH drop is due to formation of organic acids (*e.g.*, from monosaccharide degradation) and the scaling out of basic calcium salts. The color formation is due to condensation reactions of color precursors (*e.g.*, 5-(hydroxymethyl)-2-furaldehyde) and colored compounds, and occurs under conditions where sucrose is likely to decompose. Table 2 shows that both p H drop and color formation are occurring across the evaporation process in this study.

To the best of our knowledge, the only published meaningful attempt at measurement of sucrose degradation in evaporators is the comparative study of Purchase *et al.* (1987). In their study they measured the sucrose inversion by increases in glucose to sucrose ratios (as glucose % sucrose [Glc%Suc] determined by GC with flame ionization detection [FID]) across 6 evaporator systems. They concluded that sucrose decomposition

Table 2. Color formation and pH drop across the evaporation process

Sample (hour)	Mixed juice pH	pH drop	Mixed juice (ICU)	2nd effect syrup (ICU)	3rd effect syrup (ICU)	Color increase
3	7.07	0.95	21217	22625	23095	1878
8	7.11	1.00	39861	40240	41920	2059
12	6.36	0.24	36531	39829	39856	3325

Table 3. Change in glucose:sucrose ratios across the evaporation process

Sample time (hour)	Mixed juice (Glc%Suc)	2nd effect syrup (Glc%Suc)	3rd effect syrup (Glc%Suc)	Glc%Suc increase
3	1.31	1.65	2.05	0.74
8	1.47	1.85	2.14	0.67
12	1.16	1.78	1.94	0.78

during evaporation occurred mostly in the first two effects. The increases in Glc%Suc ranged from 0.00 to 0.36%, with typical pH drops of *ca.* 1.0. The Glc%Suc of 0.36% corresponds to 0.68% of the total sucrose decomposing during evaporation. This figure is an underestimate, since it is based on the assumptions that there is no glucose decomposition across the evaporator and that compounds that coelute with sucrose by GC analysis are not formed across the evaporator.

A similar analysis of Glc%Suc for the evaporator in this study is shown in Table 3. Sucrose and glucose concentrations in MJ, 2ndS and 3rdS for the 3, 8 and 12 hour samples were determined by HPIC analysis. The mean increase in Glc%Suc was 0.73%, which corresponds to 1.39% of the total sucrose decomposing during evaporation (*ca.* twice the maximum loss reported by Purchase *et al.*, 1987). Although the conclusion of Purchase *et al.* (1987) that most loss occurred in the first two effects is reasonable, we can not confirm it in this study.

Certainly, the most accurate determination of sucrose loss across any unit process in sugar manufacture would consist of analysis for a marker compound that is an end product of sucrose hydrolysis and monosaccharide degradation, and that could be directly related to sucrose loss. Unfortunately, no likely marker compound has been identified. Preliminary GC-MS analysis of MJ and 3rdS samples from this study indicate that most identifiable compounds (*e.g.*, kestoses) are present in both MJ and 3rdS.

Difructose dianhydrides (DFDA) are one class of compounds that are present in 3rdS but absent in MJ. These DFDAs which probably form late in the evaporation process (when water concentration is lowest, see section 2.1) coelute with sucrose under most capillary GC conditions. Formation of DFDAs may significantly reduce the sucrose loss estimated by GC-FID (Glc%Suc) in the 3rd effect.

CONCLUSION

In addition to reviewing the chemistry of sucrose hydrolysis, and monosaccharide degradation as it pertains to sugar loss and color formation in factory processes, this paper has attempted to point out the difficulties in accurately determining chemical loss in evaporators. Purchase *et al.* (1987) report an attempt at measurement of sucrose degradation in evaporators; in their comparative study of glucose:sucrose ratios (by GC-FID) in five evaporators they measured a maximum loss of 0.68% total sucrose. Using the same concept, but with a different analytical method (*viz.*, HPIC), we observe 1.39% of the total sucrose decomposing across the evaporator in the case study in section 3 (*ca.* twice the maximum loss reported by Purchase *et al.*, 1987). It is conceded that all measurements of sucrose loss based on glucose:sucrose ratios are underestimates. A search for a marker compound to unequivocally determine sucrose loss is briefly mentioned in section 3. To date a marker compound has not been discovered, but with continuing elucidation of the complex degradation reactions of monosaccharides a marker may eventually be realized.

ACKNOWLEDGMENTS

The authors wish to thank Mr. Xavier Miranda and Ms. Glori-Lyn Cargel for technical assistance, and Ms. Mary-An Godshall for GC-MS analysis.

REFERENCES

Allen, A.T.; Wood, R.M.; McDonald, M.P. Molecular association in the sucrose-water system. *Sugar Tech. Rev.* **1974**, *2*, 165–180.

Antal, Jr., M.J.; Mok, W.S.L.; Richards, G.N. Kinetic studies of the reactions of ketoses and aldoses in water at high temperatures. 1. Mechanism of formation of 5-(Hydroxymethyl)-2-Furaldehyde from D-fructose and sucrose. *Carbohydr. Res.* **1990**, *199*, 91–109.

Bock, K.; Lemieux, R.A. The conformational properties of sucrose in aqueous-solution-intromolecular hydrogen bonding. *Carbohyd. Res.* **1982**, *100*, 63–74.

Clarke, M.A. The effect of solution structure on electrode process in sugar solutions. *Proc. 1970 Technical Session on Cane Sugar Refining Research*, Boston, MA., 1970; 179–188.

Clarke, M.A.; Brannan M.A.; Carpenter F.G. A study of sugar inversion losses by high pressure liquid chromatography. *Proc. 1976 Technical Session on Cane Sugar Refining Research*, New Orleans, LA, 1977; pp. 46–56.

de Bruijn, J.M.; van der Poel, P.W.; Heringa, R.; van den Bliek, M. Sugar degradation in sugar beet extraction, a model study. *C.I.T.S.*, Cambridge, U.K., 1991; 379–390.

Eggleston, G.; Vercellotti, J. R.; Edye L. A.; Clarke, M. A. Behavior of water structure-breaking and structure-enhancing solutes on the thermal degradation of concentrated solutions of sucrose. *J. Carbohyd. Chem.* **1995a,** *14*, 1035–1042.

Eggleston, G.; Vercellotti, J. R.; Edye L. A.; Clarke, M. A. The chemistry of salt-catalyzed degradation of sucrose in concentrated aqueous solutions of sucrose. *International Society of Sugar Cane Technologists*, XXII Congress, Cartagena de Indias, Colombia (Sept, 1995), 1995b.

Eggleston, G.; Vercellotti, J. R.; Edye L. A.; Clarke, M. A. Effects of salts on the initial thermal-degradation of concentrated aqueous-solutions of sucrose. *J. Carbohyd. Chem.* **1996**, *15*, 81–94.

International Commission for Uniform Methods of Sugar Analysis (ICUMSA) Methods Book, ICUMSA, England, 1994.

Kelly, F.H.C.; Brown D.W. Thermal decomposition and colour formation in aqueous sucrose solutions. *Sugar Technol. Rev.* **1978/79**, *6*, 1–48.

Kharin, S.E.; Sapronov, A.R. Effect of pH and temperature on the stability of sucrose solutions. *Int. Sugar J.* **1969**, *71*, 122.

Marshall, W.L.; Frank, E.U. Equation for the ion product of water. *J Chem. Phys. Ref. Data*, **1981**, *10*, 295–304.

Mauch, W. The chemical properties of sucrose. *Sugar Technol. Rev.* **1971**, *1*, 239–290.

McDonald, E.J. Invert formation in sucrose solutions. *J. Res. Nat. Bur. Stand.* **1950**, *45*, 200.

Mega, T.L.; Van Etten, R.L. The O-18 isotope shift in C-13 nuclear magnetic-resonance spectroscopy. 12. Position of bond cleavage in the acid-catalyzed hydrolysis of sucrose. *J. Amer. Chem. Soc.* **1988**, *110*, 6372–6376.

Pancoast, H.M.; Junk, W.R. in *Handbook of Sugars*, 2nd edn, AVI Publishing Co., Westport, Connecticut, 1980.

Parker, K.J. Chemical problems in the sucrose industry. *La Sucrerie Belge*, **1970**, *89*, 119–126.

Perez, S.; Meyer, C.; Imberty, A.; French, A. in M. Mathlouthi, J.A. Kanters and G. Birch (Eds.), *Sweet Taste Chemoreception*, Elsevier, Amsterdam, 1992; pp. 55–73.

Poppe, L.; van Halbeek, H. The rigidity of sucrose: just an illusion? *J. Am. Chem. Soc.* **1992**, *114*, 1092–94.

Purchase, B.S.; Day-Lewis, C.M.J.; Schäffler, K.J. A comparative study of sucrose degradation in different evaporators. *Proc. South African Sugar Technol. Assoc.* 1987; pp. 8–13.

Schäffler, K.; Smith, J. True sucrose versus pol - The effect on cane quality and factory balance data. *Proc. South African Sugar Technol. Assoc.* 1978; pp. 59–63.

Silin, P.M. *Technology of Beet-sugar Production and Refining*, Published for the U.S. Dept. Agric. & the Nat. Sci. Fdn., 1964, Washington D.C. by the Israel Program for Scientific Translations.

Spencer, G.L.; Meade, G.P. in *Cane Sugar Handbook*, 9th Edn., John Wiley & Sons, New York, 1963; pp. 26–27.

Sugar Milling Research Institute (South Africa) Annual Report (1993–94), pp. 9–10.

Vercellotti, S.V.; Clarke, M.A. Methods of sugar analysis: comparison of modern and traditional methods. *Internat. Sugar J.*, **1994**, *96*, 437–445.

Vukov, K. Kinetics aspects of sucrose hydrolysis. *International Sugar J.* **1965**, *67*, 172–175.

Wolfrom, M.L.; Shilling W.L. Action of heat on D-fructose. III. Interconversion to D-glucose. *J. Am. Chem. Soc.* **1951**, *73*, 3557.

13

PROCESS-INDUCED CHANGES IN EDIBLE OILS

P. K. J. P. D. Wanasundara and F. Shahidi

Department of Biochemistry
Memorial University of Newfoundland
St. John's, NF, Canada, A1B 3X9

Lipids are one of the main dietary components that serve several functions in foods and nutrition. They could be endogenous or deliberately included in food. The basic molecules of lipids undergo different chemical reactions during refining, processing and storage. Some of these chemical reactions enhance the usage and functionality of food lipids. This chapter discusses the chemical changes of lipids during various processing operations. Specific changes in the minor constituents of lipids are also included.

INTRODUCTION

Lipids comprise a substantial fraction of our daily nutrient intake. Food lipids may originate from plant (*e.g.*, vegetable oils and their derived products), algal (algal oils), or animal (*e.g.*, milk, meat and fish) sources. Dietary lipids provide energy and essential fatty acids and also serve as transfer medium for fat soluble vitamins. In addition, lipids provide flavor, texture and mouthfeel to foods. Food lipids could be endogenous or added intentionally to obtain specific flavor or texture. Chemically, edible oils of plant or animal origin are mainly composed of triacylglycerols (TAG) with phospho- and glycolipids (PL and GL, respectively) comprising a small fraction. Sterols, waxes, lipid soluble vitamins, phenolics and other minor chemical constituents comprise the unsaponifiable matter of oil. During processing and storage, food lipids undergo chemical and physical changes. Often, process-derived chemical and physical changes of lipids are necessary to manifest specific characters of the food, however, these changes should not exceed the desirable limits. Isolation of lipids from sources is often required for their edible use. Processing steps involved to transform crude lipids to edible products may also change their chemical and physical characteristics. This contribution discusses process-induced physical and chemical changes in vegetable and marine oils.

Process-Induced Chemical Changes in Food
edited by Shahidi *et al.* Plenum Press, New York, 1998

CHANGES DURING EDIBLE OIL REFINING

Oils after extraction from the raw material require purification or refining and modification. The refining treatments are needed to remove or reduce the content of contaminants of the crude oil which may adversely affect the quality of the end product and the efficiency of lipid modification process. At the same time these processes should not adversely affect the nutritional value of the oil or lead to the formation of undesirable artifacts.

Impurities that negatively affect product quality or processing efficiency are moisture, dust, protein degradation products, free fatty acids, partial acylglycerols, phosphatides, oxidation products, pigments and compounds containing trace elements (*e.g.*, copper, iron, sulfur and halogens), polysaccharides and chlorinated pesticide residues (Young *et al.*, 1994). Table 1 summarises the major refining stages of edible oil processing and the type of impurities removed during each step.

Extraction of Oil

Oils are first separated from tissues by mechanical or chemical means or their combination. The raw material is subjected to cooking followed by mechanical separation of the oil, especially when dealing with animal tissues (rendering). Wet screw pressing or expelling (*e.g.*, olive, palm), high pressure screw pressing of seeds with high oil content (*e.g.*, copra, palm kernel, peanut, olive), pre-pressing followed by solvent extraction (*e.g.*, copra, palm kernel, peanut, sunflower, canola, cottonseed) and solvent extraction of seeds with low oil content (*e.g.*, soybean, rice bran) or some highly permeable seeds with high oil content (*e.g.*, palm kernel) are practised (Young *et al.*, 1994).

Heat treatment is necessary to denature the protein and to break cell walls so that oil and water can be easily removed from animal tissues (Bimbo, 1989). The raw material is first cooked with water in a continuous cooker at 50–70°C. The denatured protein in stickwater may then be separated from the oil by centrifugation or removal of liquors (oil and water) by applying pressure in a screw-type continuous press in which oil can be separated by centrifugation (Lee, 1963). The separated crude oil is then used for further processing/refining.

Degumming

Degumming is a pre-treatment designed to remove phosphatides, poly-saccharides, pigments and trace metals from crude oils. In this process, crude oils are treated with

Table 1. Refining stages of edible oils and the major impurities removed (adapted from Young *et al.*, 1994)

Refining stage	Major impurities removed or reduced
Degumming	Phospholipids, trace metals, pigments, carbohydrates, proteins
Neutralization	Fatty acids, phospholipids, pigments, trace metals, sulfur, insoluble matter
Washing	Soap
Drying	Water
Bleaching	Pigments, oxidation products, trace metals, traces of soap
Filtration	Spent bleaching earth
Deodorization	Fatty acids, mono- and diacylglycerols, oxidation products, pigment decomposition products, pesticides
Physical refining	Fatty acids, mono- and diacylglycerols, oxidation products, pigment decomposition products, pesticides
Polishing	Any residual traces of oil insolubles

water or dilute acids such as phosphoric acid to remove such impurities. According to List *et al.* (1978a) phosphoric acid degumming produces a better quality soybean oil in terms of flavor and oxidative stability than water-degummed oil (degummed without phosphoric acid). Removal of prooxidants such as iron has been partly responsible for the high oxidative stability and inhibition of off-flavor development in food lipids (List *et al.,* 1978b). Phosphoric acid may convert non-hydratable phosphatides into a form that can be removed. Gums left in the oil may cause high loss of oil during refining and also produce sediments in the storage tanks. In addition, phosphatides have a poisoning effect on catalysts used for hydrogenation. Marine oils are not usually degummed before being refined. The removal of phospholipids and mucilaginous materials in such oils, is concurrent with the removal of free fatty acids by alkali-refining.

Alkali Refining and Neutralization

Alkali refining may cause both physical and chemical changes in the oil. The alkali (normally dilute sodium hydroxide) added to the oil reacts with free fatty acids present to form soap. Gums absorb alkali and are coagulated by hydration, much of the pigments are degraded, adsorbed on the gums or made water-soluble by alkali and insoluble matters are entrained with other coagulated material (Bimbo and Crowther, 1991). After adding alkali to crude oils, the mixture is slightly heated to break the emulsion and then the soap stock is removed by centrifugation (Cowan, 1976; Carr, 1978). The refined oil is washed with warm water to remove the last traces of soap (Kwon *et al.*, 1984). Such impurities, if not removed, could affect the color, foam and smoke characteristics and/or cloudiness of the oil in later stages of processing that involve high temperatures.

The soap which is removed in the refining process is diluted with water and acidified to form fatty acids. The fatty acids are centrifuged to remove the aqueous phase, dried and used for production of fatty acids, soap or feed manufacturing. This product is called acid oil or acidulated soap stock by the oil industry (Bimbo and Crowther, 1991).

Wiedermann (1981) has summarized the reactions involved in the degumming and alkali refining as given below. The α-lipoids are referred to the phospholipids that react easily with water and readily precipitate as oil-insoluble hydrates and the β-lipoids are non-water precipitable without some pretreatment such as alkali or acid.

Phospholipids (α-lipoids)	$\xrightarrow{\text{Water}}$	Hydrated gums
Phospholipids (β-lipoids)	$\xrightarrow[\text{Alkali}]{\text{Water}}$	Hydrated gums
Metal/phospholipid complex	$\xrightarrow{\text{Acid}}$	Hydrated Fe-gums
Calcium/magnesium	$\xrightarrow{\text{Phosphoric acid}}$	Insoluble phosphates
Free fatty acids	$\xrightarrow{\text{Alkali}}$	Soap

Table 2. Typical Lovibond red color values of oils during processing steps (adapted from Wiedermann, 1981)

Processing step	Lovibond red color value
Refining	7–8 R
Bleaching	2-2.5 R
Deodorization	0.4-0.8 R
Heat bleaching	<1.0 R

Bleaching

After alkali-refining, the oil is usually bleached. Bleaching is performed to improve the color, flavor and oxidative stability of the oil. Although the main objective of bleaching is to reduce the content of colored compounds and natural pigments (*e.g.*, carotenoids, chlorophyll, xanthophyll and polyphenols), some suspended mucilaginous and colloid-like matter are also removed (Chang, 1967). Table 2 provides typical Lovibond color values that are expected following each refining step. Traces of soap, if still present, are also adsorbed by the bleaching material. Commonly used bleaching materials are natural clay, activated clay and carbon (Cowan, 1976). Activated carbon is normally used at 5–10% in combination with the clay. Natural clay is used if the oil is readily bleachable. However, acid-activated clay is usually used because its bleaching power is greater than that of the natural clay (Boki *et al.*, 1989; Morgan *et al.*, 1985; Richardson, 1978).

Winterization

Winterization removes the high melting fraction (waxes and non-triacylglycerol substances also known as "stearin") of the oil at low temperatures. This process is also known as racking or cold clearing. Winterization involves chilling of the oil at a prescribed rate, allowing the "stearin" or solid portions to crystallize and finally separating the two phases, usually by filtration, while cold (Bimbo, 1989). Winterization is an old practice that evolved from the observation that storage of oils in outdoor tanks during cold weather caused deposition of high-melting TAG at the bottom and clear liquid oil on the top; the clear oil was then decanted and used as a light oil. In other words, winterization removes the more saturated fraction of the oil from its unsaturated/polyunsaturated fraction. Solvent winterization of oil has recently been developed. In this method, the viscosity of the oil is reduced by dissolving it in a solvent such as hexane. Crystals of saturated fats formed over a short period of time are readily separated from the low viscosity liquid phase. Crystallized and non-crystallized fractions are then separated by filtration. Nowadays, marine oils are winterized to remove (i) waxes and other non-TAG constituents, (ii) naturally-occurring high-melting TAG and (iii) TAG formed during partial hydrogenation (List and Mounts, 1980).

Deodorization

Deodorization is the last major step in the refining of edible oils. This processing operation removes undesirable odors and flavors from the oil and ensures its shelf-life stability

(Gavin, 1978). Prior to deodorization, the oil may contain volatile odor and flavor-active components originally present in the crude oil, the "soapy" odor created by alkali-refining, the "earthy" odor generated by bleaching and the typical hydrogenation odor caused by hardening (Chang, 1967). It should be noted that some compounds may impart undesirable odor or flavor to oils at concentrations of 1 to 10 ppm or even lower (Chang, 1967). The deodorization process is essentially a steam distillation process where the volatile compounds are stripped from the non-volatile oil (Bimbo and Crowther, 1991) and also destroys peroxides in the oil and removes any aldehydes or other volatile products which might have resulted from atmospheric oxidation (Lin *et al.*, 1990). This process also serves to decrease the free fatty acid content and improves the color of the oil. The conventional procedures used for deodorization (*i.e.* steam stripping at 200–240 °C) are not suitable for marine oils since they may cause oxidation of polyunsaturated fatty acids (PUFA) (Bimbo and Crowther, 1991). Therefore, modifications of conventional deodorization process have been sought to deodorize marine oils. Dinamarca *et al.* (1990) have developed a pilot-scale process for deodorizing fish oil by high vacuum distillation at low temperatures (below 150 °C) and produced a bland oil without destroying its long-chain PUFA consequently.

Since environmental contaminants have been a concern in the marine oil industry, in certain parts of the world, short path distillation under high vacuum or supercritical fluid extraction may be suitable to prepare pesticide-free and high quality marine oils (Breivik, 1992). Table 3 provides changes that occur in lipid composition and minor constituents of cod liver oil at different stages of processing.

Specific Changes Due to Refining

There are specific changes that may occur in minor constituents of edible oils during processing. In the process of preparation of extra virgin and virgin olive oil, fruits are pressed and the expelled oil is bottled without any further processing, except washing, decanting, centrifugation and filtration. The free fatty acid content is a major criterion used for categorization of edible oils. Extra virgin oil contains <1% while virgin oil contains <3% free fatty acids. Olive oils with 3% acidity are subjected to the refining process including alkali neutralization, bleaching and deodorization. Since extra virgin and virgin olive oils undergo minimal processing, most of the phenolic compounds, pigments and tocopherols that give the unique flavor and high oxidative stability to the oil are retained (Padley *et al.*, 1994).

Sesame oil is usually obtained by expeller pressing of roasted or unroasted sesame seeds. The oil from roasted seeds does not require extensive refining, the suspended meal

Table 3. Changes of neutral and polar lipids, tocopherol, acid value and color of codliver oil during processing (adapted from Wanasundara, 1996)[1]

Sample	Neutral lipids %	Polar lipids %	α-Tocopherol mg/100g	Acid value mg KOH/g oil	Color[2]
Crude	98.17 ± 0.11	1.83 ± 0.02	10.8 ± 0.15	1.87 ± 0.15	1.10 ± 0.05
Alkali-refined	99.44 ± 0.40	0.56 ± 0.04	11.7 ± 0.21	0.07 ± 0.00	0.95 ± 0.02
Refined-bleached (RB)	99.33 ± 0.10	0.67 ± 0.09	10.0 ± 0.13	0.04 ± 0.00	0.70 ± 0.01
Refined-bleached and deodorized (RBD)	99.47 ± 0.15	0.53 ± 0.02	7.4 ± 0.11	0.05 ± 0.00	0.12 ± 0.00

[1]All values are mean of three replicates ± standard deviation.
[2]As Lovibond Yellow.

Table 4. Changes of sesamolin, sesamol and sesaminol in unroasted sesame oil (adapted from Fukuda *et al.*, 1994)

	Content, mg/g oil[1]		
Refining step	Sesamolin	Sesamol	Sesaminol
Crude oil	5.10	0.043	ND
Alkali treatment	4.58	0.002	ND
Warm water treatment	4.24	0.007	ND
Bleaching	ND	0.463	0.919
Deordorizing	ND	0.017	0.727

[1]ND, not detected.

particles are removed by settling, screening, filtering and bleached with a relatively low quantity of bleaching earth as compared to that required for other vegetable oils. Oil from unroasted sesame seeds is refined by alkali treatment and deodorization (Namiki, 1990). It has been documented that the high oxidative stability of sesame oil and the stability conferred to foods fried in it are due to the antioxidative lignans and tocopherols present. Sesamolin and sesamin are the major phenolic lignans that are found in sesame seed (Namiki, 1990). Minor amounts of sesaminol are found in the seed but relatively high proportions of it are present in the purified oil obtained from unroasted seeds. Chemical changes induced during the refining process of the oil are responsible for presence of different lignans in the oil. Table 4 shows quantitative changes in sesamolin, sesamol and sesaminol during refining process of unroasted sesame oil. Sesamolin is considered as the precursor of sesaminol and sesamolinol. At the bleaching step, a drastic change in the amounts of sesamin, episesamin, sesamolin, sesamol and sesaminol occurs (Osawa *et al.*, 1985; Fukuda *et al.*, 1986a). Fukuda *et al.* (1986b) have suggested that sesaminol was formed from sesamolin under anhydrous conditions in the presence of an acid (acid clay used for bleaching) that acts as a catalyst at high temperatures. Sesamolin is first decomposed to sesamol by protonolysis to form an oxonium ion and then the carbon-carbon bond was formed, thus the hypothesis is that sesaminol is formed from sesamolin by intermolecular group transformation (Fukuda *et al.*, 1986b). Figure 1 provides suggested chemical changes that occur in sesamolin during the refining process of unroasted sesame seed oil.

LIPID MODIFICATION

Edible vegetable and marine oils are refined and subjected to modification processes in order to enhance their usage and storage stability. Modification processes employed include hydrogenation, interesterification and fractionation of a single oil or blends of different oils. Modified oils find a wide variety of applications in different food formulations.

Hydrogenation

Hydrogenation is direct addition of hydrogen to the double bonds of unsaturated fatty acids of the oil. Consequently the fatty acids become saturated and thus, less prone to oxidation, and attain a high melting point. Hydrogenation is used to convert liquid oils to semi-solid plastic fats that are suitable for margarine, shortening and specialty products

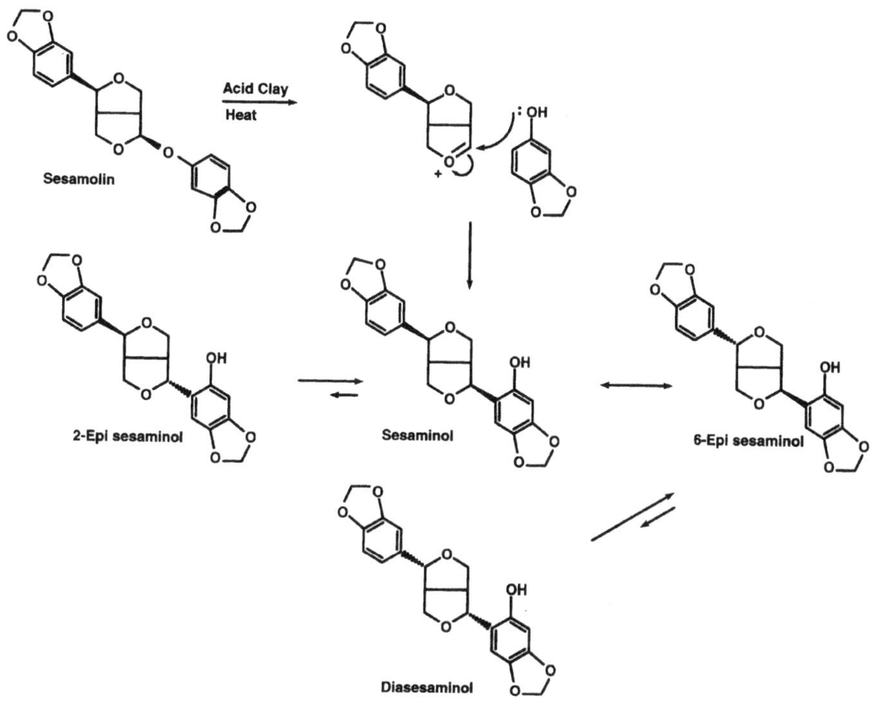

Figure 1. Reactions of converting sesamolin to sesaminol during bleaching process.

(deMan, 1990). The hydrogenation system essentially consists of an oil, a solid catalyst (*e.g.*, Ni) and hydrogen gas.

Generally hydrogenation of fat is not carried to completion and fats are hydrogenated only partially. Under these conditions hydrogenation may be selective or nonselective. Selectivity of hydrogenation is increased by increasing hydrogenation temperature and decreased by increasing pressure and agitation. In selective hydrogenation, hydrogen is first added to the most unsaturated fatty acids. Selectively hydrogenated oils are more resistant to oxidation due to preferential hydrogenation of polyunsaturated fatty acids such as linolenic acid (deMan, 1990).

Formation of positional and geometric isomers of unsaturated fatty acids occurs during hydrogenation as illustrated in Figure 2. The intermediates in the hydrogenation process may react with an atom of adsorbed hydrogen to yield the half-hydrogenated compound. Additional reaction of the half-hydrogenated compound with hydrogen results in the formation of a saturated compound. There is another possibility that the half-hydrogenated olefin may again attach itself to the catalyst surface at a carbon on either side of the existing bond with simultaneous loss of hydrogen. Upon desorption of this species, a positional or geometric isomer is formed. The proportion of *trans* acid is high because this is the more stable configuration. The double bond migration may also occur in both directions of the reaction center, but most extensively in the direction away from the ester group (deMan, 1990; Davidek *et al.,* 1990). Formation of *trans* isomers of fatty acids during hydrogenation is a health concern as hydrogenated plastic fats are widely used in food products. Therefore, several alternatives have been used to reduce *trans* fatty acid content in production of plastic fats.

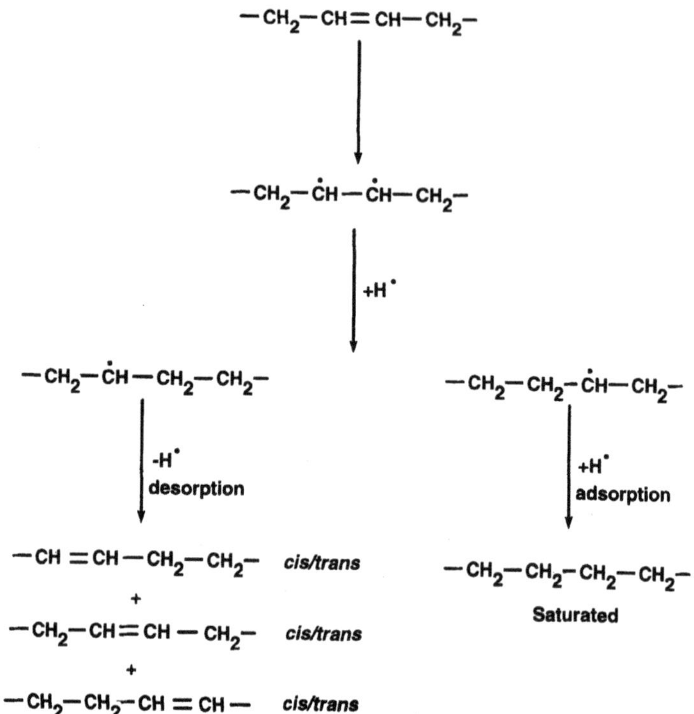

Figure 2. Possible products of hydrogenation of unsaturated fatty acid (adapted from deMan, 1990).

Figure 3 shows products of hydrogenation of oleic, linoleic and linolenic acids. The change of oleate to *iso*-oleate does not change unsaturation but affords a higher melting product. Hydrogenation of linoleate first produces some conjugated dienes, followed by formation of positional and geometrical isomers of oleic acid and finally stearate. Linolenic acid hydrogenation is more complex and greatly dependent on the reaction conditions. The solid isomers of oleic acids are formed by partial hydrogenation of polyunsaturated fatty acids or by isomerization of oleic acid which affects the consistency of partially hydrogenated oils (deMan, 1990).

Hydrogenation of highly unsaturated oils, especially marine oils, may be carried out in order to stabilize them against oxidative deterioration (Bimbo and Crowther, 1991). Hydrogenation also improves the color of the oil by decomposing objectionable color and taste-active components of the crude oil, particularly those in marine oils (Chang, 1967). Some constituents of the unsaponifiable matter are also affected by hydrogenation. Carotenoids, including vitamin A, are hydrogenated extensively. Some of the chlorine-containing pesticide contaminants are hydrogenated, however, sterols and tocopherols are not affected under the usual hydrogenation conditions (Belitz and Grosch, 1987).

Interesterification

Interesterification is the process by which the arrangement of fatty acid in a TAG molecule is changed and modification of physical and functional properties of the lipid is achieved. The term interesterification is often interchangeably used to describe reactions

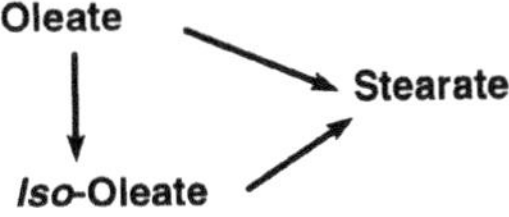

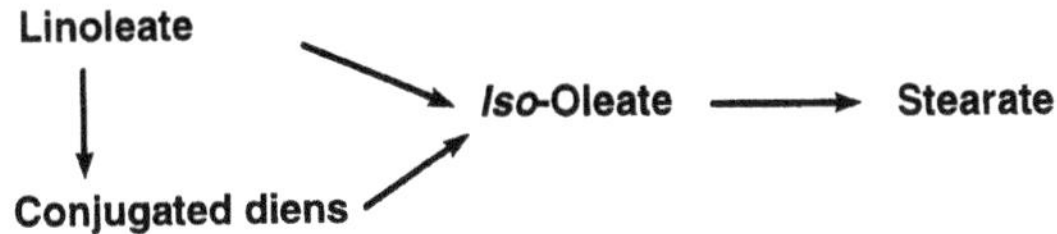

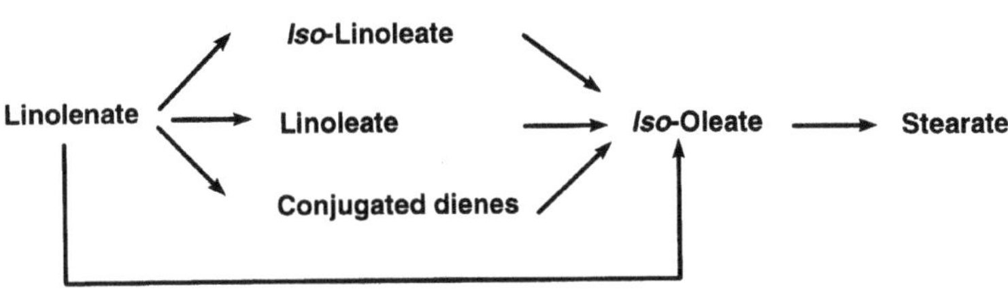

Figure 3. Products of hydrogenation of oleic, linoleic and linolenic acids (adapted from deMan, 1990).

that involve the exchange of acyl radical between ester and acid (acidolysis), an ester and an alcohol (alcoholysis) or an ester with another ester (transesterification). Figure 4 provides examples for these interesterification reactions. Interesterification can also be viewed as a break up of a specific triacylglycerol, removal of a fatty acid at random and shuffling it among the rest of the fatty acid pool and replacement at random by another fatty acid. This process involves the formation of an equilibrium mixture of the triacylglycerols formed by a random distribution of fatty acids on the glycerol molecule, thus referred as randomization or random interesterification (deMan, 1990; Gunstone, 1994).

Interesterification is performed either by chemical or enzyme catalysis and involves at least two oils that have different fatty acid compositions. In chemical interesterification, alkali metals (sodium, potassium) and alkali-metal alcoholates (*e.g.*, methylate, ethylate) are used as catalysts. Sodium methylate is the most widely used catalyst. The dried and deacidified oil is stirred at 80 to 100°C in the presence of alcoholate (0.1–0.3% of fat weight) and when the reaction is completed the catalyst is destroyed, by addition of water, and subsequently removed. The interesterified fat is recovered and then bleached and deodorized. Chemical interesterification progresses randomly with no regioselectivity (positional specificity) on the carbons of the glycerol moiety of the TAG (Gunstone, 1994).

Lipase-catalysed interesterification has found many applications in production of edible and specialty lipids due to mild reaction conditions, high catalytic efficiency, the inherent selectivity of natural catalysts and production of much purer products as compared to chemical methods (Sonnet, 1988). Lipases (hydrolases) are used for hydrolysis and ester synthesis. They are classified as non-specific or random, positional specific or 1,3-specific and acyl group- or structure-specific, depending on their activity towards fatty acids

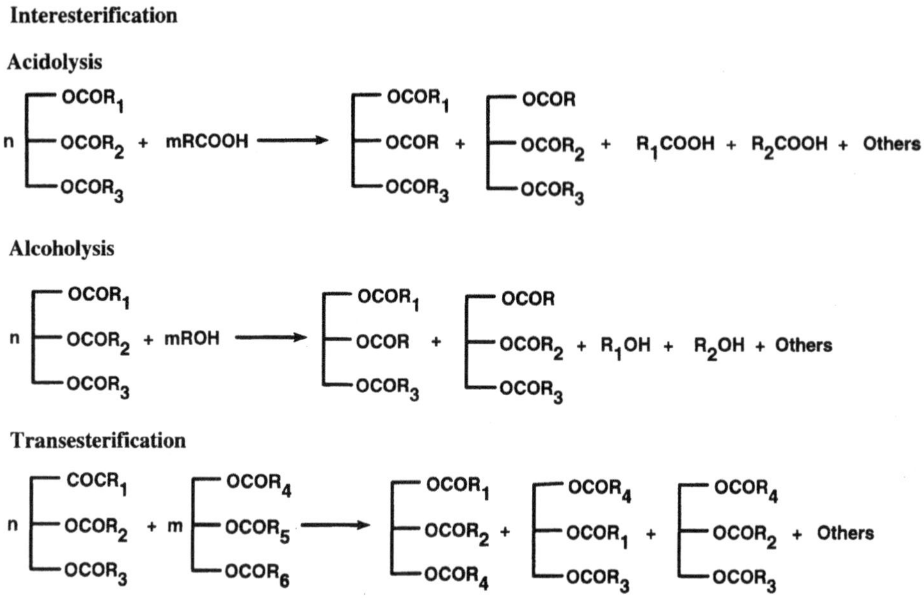

Interesterification

Acidolysis

$$n\left[\begin{array}{l}OCOR_1\\OCOR_2\\OCOR_3\end{array}\right. + mRCOOH \longrightarrow \left[\begin{array}{l}OCOR_1\\OCOR\\OCOR_3\end{array}\right. + \left[\begin{array}{l}OCOR\\OCOR_2\\OCOR_3\end{array}\right. + R_1COOH + R_2COOH + \text{Others}$$

Alcoholysis

$$n\left[\begin{array}{l}OCOR_1\\OCOR_2\\OCOR_3\end{array}\right. + mROH \longrightarrow \left[\begin{array}{l}OCOR_1\\OCOR\\OCOR_3\end{array}\right. + \left[\begin{array}{l}OCOR\\OCOR_2\\OCOR_3\end{array}\right. + R_1OH + R_2OH + \text{Others}$$

Transesterification

$$n\left[\begin{array}{l}COCR_1\\OCOR_2\\OCOR_3\end{array}\right. + m\left[\begin{array}{l}OCOR_4\\OCOR_5\\OCOR_6\end{array}\right. \longrightarrow \left[\begin{array}{l}OCOR_1\\OCOR_2\\OCOR_4\end{array}\right. + \left[\begin{array}{l}OCOR_4\\OCOR_1\\OCOR_3\end{array}\right. + \left[\begin{array}{l}OCOR_4\\OCOR_2\\OCOR_3\end{array}\right. + \text{Others}$$

Figure 4. Interesterification of triacylglycerol.

in TAG. Commonly studied lipases for interesterification are generally of microbial origin (*e.g.*, *Mucor miehei, Candida antarctica*).

Interesterification provides an alternative to hydrogenation for production of margarine and shortening fats with little or zero trans fatty acids and retaining high amounts of essential fatty acids (deMan, 1990). As the arrangement of fatty acids in the TAG is changed, the distinct physical properties of fat such as melting point, solid fat index, crystallization tendency and texture are also changed. Edible fats are polymorphic; the most important crystal forms in the production of plastic fats (*e.g.*, margarine and shortening) are the α, β' and β, of which the β' and β forms are stable. The desired crystal form for most plastic fats is the β' form. Individual fats and oils show tendencies towards crystal formation either in β or β' structure. The occurrence of polymorphism in fat crystals results in multiple melting points.

Lard crystallizes naturally in the β form (large crystals with a grainy texture) because of the high proportion of palmitic acid in the 2-position of its disaturated TAG. Upon random interesterification, the proportion of palmitic acid in position-2 is reduced from 64 to 24% and as a consequence the product crystallizes in the β' form which improves its shortening performance (Hustedt, 1976; Sreenivasan, 1978). Similarly, the crystal structure of margarine based on sunflower and canola oils could be stabilized in the β' form by random interesterification (Foglia *et al.*, 1993). Interesterification results in changes in the melting characteristics of fats and thus may be used to produce dietary margarines high in polyunsaturated fatty acids with zero to very low content of *trans* fatty acids (Gunstone, 1994) and an alternative way to produce plastic fats without hydrogenation. Table 5 illustrates the extent of changes in TAG brought about by interesterification and its influence on the melting point of products.

Table 5. Changes in the pattern of triacylglycerol in a partially hydrogenated palm oil by interesterification (Belitz and Grosch, 1987)

Component	Prior to interesterification	After interesterification	
		Single phase[1]	Direct[2]
Melting point	41	47	52
Triacylglycerol (mole %)[3]			
S_3	7	13	32
S_2U	49	38	13
SU_2	38	37	31
U_3	6	12	24

[1]Acyl residues are randomly distributed.
[2]Fat is separated into high and low melting fractions by reducing reaction temperature.
[3]S, saturated fatty acids; and U, unsaturated fatty acids. TAG, triacylglycerol.

Production of TAG enriched with polyunsaturated fatty acids (PUFA) via lipase catalysis in codliver oil has also been described (Haraldsson *et al.*, 1989). Preparation of γ-linolenic acid enriched TAG via rapeseed lipases (Syed Rahmatullah *et al.*, 1994; Jachmanian and Mukerjee, 1996) and microbial lipases (Akoh *et al.*, 1995, 1996) has also been reported. Incorporation of eicosapentaenoic and docosahexaenoic acids into peanut oil via transesterification catalysed by microbial lipases has been reported by Sridhar and Lakshminarayana (1992). According to Wada and Koizumi (1986) and Kimoto *et al.* (1994) interesterification of fish oils to form TAG containing PUFA in the 2-position rather than the 1,3-positions increases the oxidative stability of the modified oil as compared to its original counterpart.

Fractionation

Fractionation is carried out to remove undesirable lipid ingredients or to enrich desirable TAG. The melted fat is slowly cooled until the high melting TAG selectively crystallize without forming mixed crystals of high and low melting TAG. In the winterization process, fractional crystallisation via slow cooling of the oil is carried out in order to remove saturated fatty acids (Young *et al.*, 1994).

FRYING

Frying is one of the most popular methods for preparation of foods worldwide as it could be used for fast preparation and attaining specific fried flavor of products. The deep fat frying is a process where food is heated for a short time at a high temperature (about 180–200°C) and the fat or oil is used as the heat transfer medium. However, at high temperatures of frying, fats and oils are subjected to various chemical reactions which modify the flavor of the fried product and the quality of the remaining frying fat/oil.

During frying, hydrolytic, oxidative and pyrolytic (Pokorny, 1989) reactions occur and these lead to the formation of volatiles, nonpolymeric polar compounds of moderate volatility, dimeric and polymeric acids and partial acylglycerols as well as free fatty acids (Chang *et al.*, 1978). These reactions are responsible for a variety of physical and chemical changes that may be observed in the oil during frying. These physical and chemical changes lead to an increase in the viscosity, free fatty acid content, development of dark color and tendency to foam and decrease in iodine value, surface tension and changes in refractive index of the oil (Varela *et al.*, 1988).

Hydrolytic Reactions and Products

The extent of hydrolytic reactions occurring in an oil depends on the frying temperature used and the interface between the oil and the aqueous phase. The aqueous phase may be the frying food material as well as water droplets or bubbles of steam that originate from the frying food. The released moisture also agitates the oil and hastens the hydrolysis. Free fatty acids (FFA) arise from hydrolysis of triacylglycerols in the presence of water and heat (Nawar, 1996). During deep fat frying, the temperature may rise up to 150 to 200°C and water becomes miscible in oil. This may induce hydrolysis of triacylglycerols to free fatty acids, diacylglycerols (DAG), monoacylglycerols (MAG) and glycerol. However, partial acylglycerols do not accumulate in the frying medium and hydrolyze much faster than the TAG (Davidek *et al.*, 1990). Glycerol is volatile at 150–200°C and is partially lost from the frying oil by evaporation. The hydrolytic products reduce the thermal stability of frying oils. The blanket of steam formed above the surface of the oil tends to reduce the amount of oxygen available for oxidation.

Hydrolysis is an ionic reaction and thus, cations or anions catalyze the reaction. Free fatty acids and low-molecular weight acidic products that result from oil oxidation enhance the hydrolytic reactions in the presence of steam during frying (Yuki *et al.*, 1974). Acidic phospholipids which are oil-soluble are also active catalysts under frying conditions. Fried food material may contain ions of various organic acids, acidic amino acids or ester bound phosphoric acid, which may be at the interface and may induce acidic reactions (Pokorny, 1989). Hydrolytic process is also induced by partial acylglycerols and other surface active agents. Such surface active agents may emulsify the aqueous phase and increase the reactive interface of the frying medium. Oxidation products and salts of fatty acids also act as effective emulsifiers under deep fat frying conditions.

Thermal Oxidative Reactions and Products

Oxidation reactions that occur during frying have a great influence on chemical changes in frying oils. Oxidation of lipids is very rapid at frying temperatures. However, oxygen solubility is low at high temperatures of frying and becomes the rate-determining factor. The oxidation mechanism of frying oils is similar to their autoxidation at room temperature. Hydroperoxides, which are the primary products of lipid oxidation, are very unstable at frying temperatures and decompose into monomeric nonperoxidic products, polymers and volatile products (Pokorny, 1989). The decomposition products of hydroperoxides may further be oxidized, dehydrated, esterified, hydrolyzed or interact with the frying food. Due to the limitation of oxygen in the frying medium, di- or tri-substituted oxidation products are very low in frying oils compared to that in autoxidized samples at room temperature. According to Doi and Urakami (1976), soybean oil used continuously for frying contains dioxoacids, epoxides, oxolans and pyrans. Polymers produced under frying conditions are bound mainly by carbon-carbon or ether bonds (Paulose and Chang, 1978), but rarely by peroxide linkages (Pokorny, 1989).

Although saturated fatty acids, their esters and TAGs are relatively stable, when heated in the air at temperatures higher than 150°C, they produce a variety of oxidation products. Each TAG gives rise to a homologous series of hydrocarbons, aldehydes, ketones and lactones. The hydrocarbons formed are the same as those produced in the absence of oxygen. Therefore, a homologous series of *n*-alkanes and 1-alkenes, with the alkanes predominating, and decarboxylation products of the parent fatty acid are produced in substantial amounts. However, the amount of these compounds formed is much greater under oxidative conditions. The methyl ketones are generally produced in larger quantities than alkanals with the C_{n-1}-methyl ketones being the major carbonyl compounds formed. In each lactone series, the C_n-lactone is the most abundant. However, only δ-lactones are formed from those having a carbon number equal to that of the parent fatty acid (Nawar, 1985). The amounts of the C_n γ- and δ-lactones decrease with increased fatty acid chain length. Pokorny (1989) has reported a series of homologous compounds that can be detected in saturated esters oxidized at frying temperatures (Table 6). Relatively high molecular weight (1 or 2 C atoms less than original fatty acid) compounds were identified among alkanals, 2-alkanones and alkanoic acids (both free and esterified) (Endres *et al.*, 1962).

The accepted principal mechanism of thermal oxidation of saturated fatty acids involves the formation of monohydroperoxides and subsequent oxygen attack on all methylene groups of the fatty acid chain. The dominant oxidative products of saturated fatty acids are those with chain lengths near or equal to that of the parent fatty acids and oxidation occurs preferentially at the α, β and γ positions (Figure 5; Nawar, 1985). Figure 5 shows that subsequent cleavage in the region of the alkoxy radicals leads to a favorable formation of the long chain hydrocarbons, alkanals and methyl ketones.

Unsaturated lipids produce qualitatively similar products when thermally oxidized or autoxidized at low temperatures. These include a series of aldehydes, ketones, acids, esters, alcohols, hydrocarbons, lactones, cyclic compounds, dimers and polymers. However, quantitative pattern of the decomposition products formed at high temperatures is different from that of autoxidation, varying widely depending on the nature of the substrate and parameters of heat treatment (Nawar, 1985; Pokorny, 1989). Unsaturated fatty acids are much more susceptible to oxidation than their saturated analogs. According to Frankel (1980), at 25 to 80°C, relative proportions of isomeric hydroperoxides isolated from each substrate varies with the oxidation temperature, however, their qualitative pattern remains the same. At oxidation temperatures higher than 80°C, isolation and quantitation of hydroperoxide intermediates is difficult due to their extreme heat sensitivity. Furthermore, the primary decomposition products are unstable and rapidly undergo further oxidative decomposition. As the oxidative process continues, a variety of possible reaction mecha-

Table 6. Oxidation products of saturated fatty acids under frying conditions (adapted from Pokorny, 1989)

	Number of carbon atoms															
Compound	2	3	4	5	6	7	8	9	10	11	12	13	14	15	16	17
Alkanes	–	–	+	+	+	+	+	+	+	+	+	+	+	+	+	+
1-Alkenes	–	+	–	–	–	–	–	–	+	–	+	+	+	+	–	–
1-Alkanols	–	–	+	+	+	+	+	+	+	+	+	+	+	+	–	–
2-Alkanones	–	+	+	+	+	+	+	+	+	+	+	+	+	+	+	+
3-Alkanones	–	–	–	+	–	+	–	–	+	–	–	–	–	–	–	–
Alkanals	+	+	+	+	+	+	+	+	+	+	+	+	+	+	+	+
2-Alkenals	–	–	+	–	–	+	–	–	–	–	–	–	–	–	–	–
γ-Lactones	–	–	+	+	+	+	+	+	+	+	+	+	+	–	–	–
Alkanoates	+	+	+	+	+	+	+	+	+	+	+	–	–	–	–	–

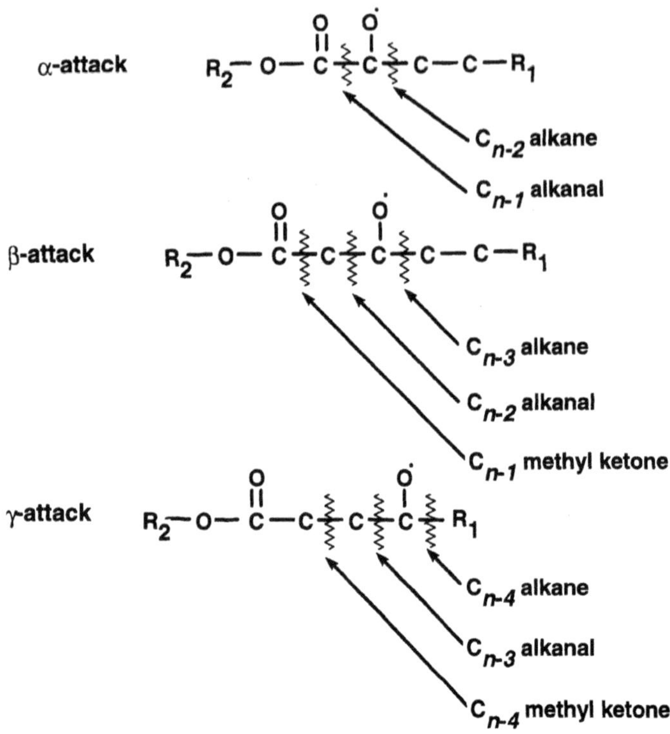

Figure 5. Thermal oxidation products of saturated fatty acid monohydroperoxide (n=number of carbon in fatty acid moiety).

nisms occur and lead to the formation of a relatively complex mixture of decomposition products. As the temperature of oxidation or the degree of unsaturation of fatty acid substrates increases, it is difficult to find a correlation between the pattern of end products and classical cleavage of C-C bonds of the expected hydroperoxide intermediates.

In a model thermal oxidation experiment heating of ethyl linoleate from 70 to 250°C, produced 11 predominant compounds in addition to numerous other products which were found in relatively small amounts. Typically, the expected oxidation products from ethyl linoleate are pentane and hexanal which arise from decomposition of 13-hydroperoxide and 2,4-decadienal, ethyl octanoate and ethyl-9-oxononanoate from decomposition of 9-hydroperoxide isomers. The C_8-aldehyde ester may also be produced by cleavage of 9-alkoxy radical followed by terminal hydroperoxidation. Further oxidation of the C_8-aldehyde ester may produce the corresponding acid ($HOOC(CH_2)_6$-$COOC_2H_5$) which upon decarboxylation gives rise to ethyl heptanoate. Both the C_8-dicarboxylic acid and the C_7-ethyl ester are the major products of linoleate thermal oxidation (Nawar, 1985). Several reports show that the decomposition products could not be predicted, especially at high temperatures (>80°C) (Noble and Nawar, 1975; Nawar, 1985; Pokorny, 1989). Hydrogen abstraction at allylic position outside pentadiene system has been speculated to be responsible for products such as aromatic compounds and cyclic monomers from methyl linoleate. Some of the hydroperoxides typical of linoleate, and even certain products of hydroperoxide cleavage, contain 1,4-pentadiene systems and would thus continue to undergo further hydroperoxide formation followed by scission at extremely rapid rates. Oxidation of linolenates is unique due to their tendency to form polyhydroperoxides, hydroperoxyepoxides and hydroperoxy cyclic peroxides. Radical combination of the alkoxy or peroxy radicals, formed in unsaturated lipids or their intra- and intermolecular addition to C-C double bonds, leads to the formation of oxydimers or polymers possessing hydroperoxide, hydroxide, epoxide, carbonyl or cyclic groups as well as ether and peroxide bridges (Nawar, 1985).

Thermolytic (Pyrolytic) Reactions and Products

When simple saturated substrates are exposed to heat in the absence of oxygen, predictable pattern of compounds such as hydrocarbons, free fatty acids or symmetric ketones are produced. The specific thermolytic products depend on the chain length of the parent fatty acid in the TAG molecule. A simple TAG produces (n=number of carbons in the fatty acid moiety) a series of normal alkanes and 1-alkenes with the C_{n-1} alkane predominating; a C_n fatty acid, a C_{2n-1} symmetric ketone; a C_n oxopropyl ester; C_n propene and propanediol diesters; and C_n diacylglycerols. In addition, acrolein, carbon monoxide and carbon dioxide are also formed (Nawar, 1985).

According to Lein and Nawar (1973), the pyrolysis of tricaprylate at 250 and 270°C produces acrolein, caproic acid and various diesters of glycols derived from diacylglycerols. Higman *et al.* (1973) have reported that pyrolysis of tripalmitate and tristearate at 400°C produced a mixture of saturated monocyclic and dicyclic acids containing 3 to 18 carbon atoms, and saturated hydrocarbons with 6 to 17 carbon atoms. In a "moisture-free system", TAG may undergo a chelate type of 6-atom ring closure by way of a hydrogen bridge. Rearrangement of the electrons would give rise to a fatty acid and an olefin as depicted in Reaction 1 (Figure 6). This also explains the formation of propenediol diesters. Expulsion of the acid anhydride from the TAG molecule produces 1- or 2-oxopropyl and the acid anhydride (Reaction 2; Figure 6). Decomposition of the 1-oxopropyl esters gives rise to acrolein and a C_n fatty acid (Reaction 3; Figure 6), while decarboxylation of the acid anhydride intermediate produces a symmetric ketone (Reaction 4; Figure 6). In the presence of moisture, free fatty acids are released from acylglycerols. Upon thermal hydrolysis, no evidence for positional specificity was apparent when the TAG in which the acid in the 2-position was labelled with ^{14}C (Buziassy and Nawar, 1968). However, a preferential release of the unsaturated and shorter chain acids was observed by Noble *et al.* (1967). If the esterified acids possess certain functional groups, such as hydroxyl, their hydrolysis by heat may result in the formation of compounds with a particular flavor significance such as lactones and methyl ketones. In the case of relatively high heating temperatures, the alkane and alkene series may be produced by homolytic cleavage of C-C bonds along the fatty acid chains.

When unsaturated fatty acids are subjected to thermolytic cleavage, dimers, polymers and cyclic compounds are produced. The dimers include alicyclic mono-, di-, and triene dehydrodimers, saturated dimers with cyclopentene structures, tetrasubstituted cyclohexanes and bicyclic and tricyclic dimers (Nawar, 1985). Cyclic mononmers exhibit toxicity in experimental animals (Van Gastel *et al.*, 1984), therefore, generation of cyclic monomers at high temperature frying with oils containing polyunsaturated fatty acids should be carefully monitored.

Combination of allyl radicals that result from hydrogen abstraction at methylene groups α to double bonds may give rise to dimeric compounds. Such radicals may also undergo disproportionation or inter- and intramolecular addition to C-C double bonds. Dimerization of unsaturated fatty acids can also occur via Diels-Alder reactions. In the case

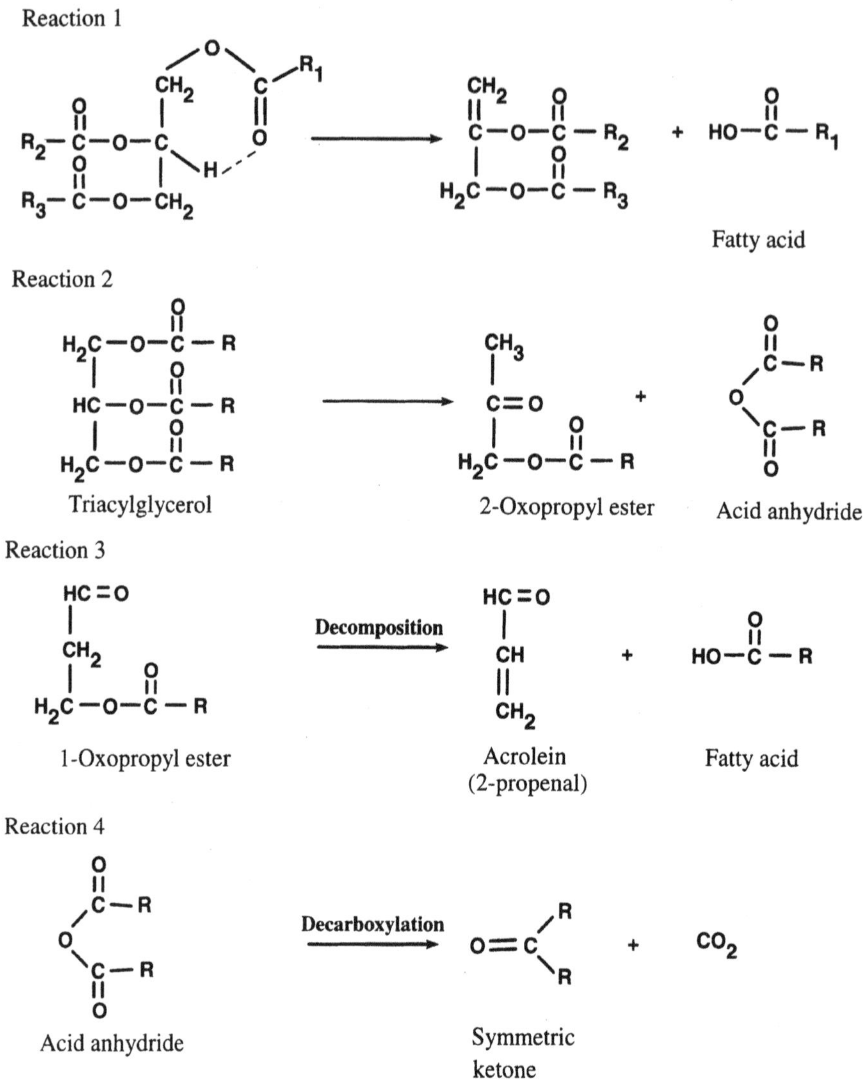

Reaction 1

Fatty acid

Reaction 2

Triacylglycerol

2-Oxopropyl ester

Acid anhydride

Reaction 3

Decomposition

1-Oxopropyl ester

Acrolein (2-propenal)

Fatty acid

Reaction 4

Decarboxylation

Acid anhydride

Symmetric ketone

CO_2

Figure 6. Thermal decomposition reactions of triacylglycerol.

of TAG dimerization can take place between acyl groups in two TAG molecules or between two acyl groups in the same molecule. If sufficient numbers of double bonds are available as in the dimers of polyunsaturated fatty acids, further reactions may take place in order to produce trimers and polymers. The larger the number of double bonds in the fatty acid chain and the higher the temperature of heating, the greater are the chances for thermal dimerization, polymerization and cyclization (Nawar, 1985).

Formation of polycyclic aromatic hydrocarbons via pyrolysis of lipids at temperatures of >400 °C has been observed. Although such temperatures are not common in food processing, the formation of such compounds is possible if certain portion of the food or its fat drippings come in contact with charcoal or very hot surfaces as may occur in broiling, barbecuing or roasting. Pyrolysis of oils starts at above 200°C to a measurable rate but overheating in contact with hot metal surfaces can also produce pyrolytic decomposition (Pokorny, 1989). Acrolein (2-propenal) is an easily perceived pyrolytic product, with pungent and irritating effect of overheated oils (Umano and Shibamoto, 1987). Acrolein is produced by pyrolysis or hydrolysis of monoacylglycerol (Lorant, 1977; Figure 7). When palm oil was pyrolyzed at 300°C, a mixture of alkanes, 2-alkenes, and cyclic hydrocarbons was produced (Alencar *et al.*, 1983). The main pyrolytic products of unsaturated oils

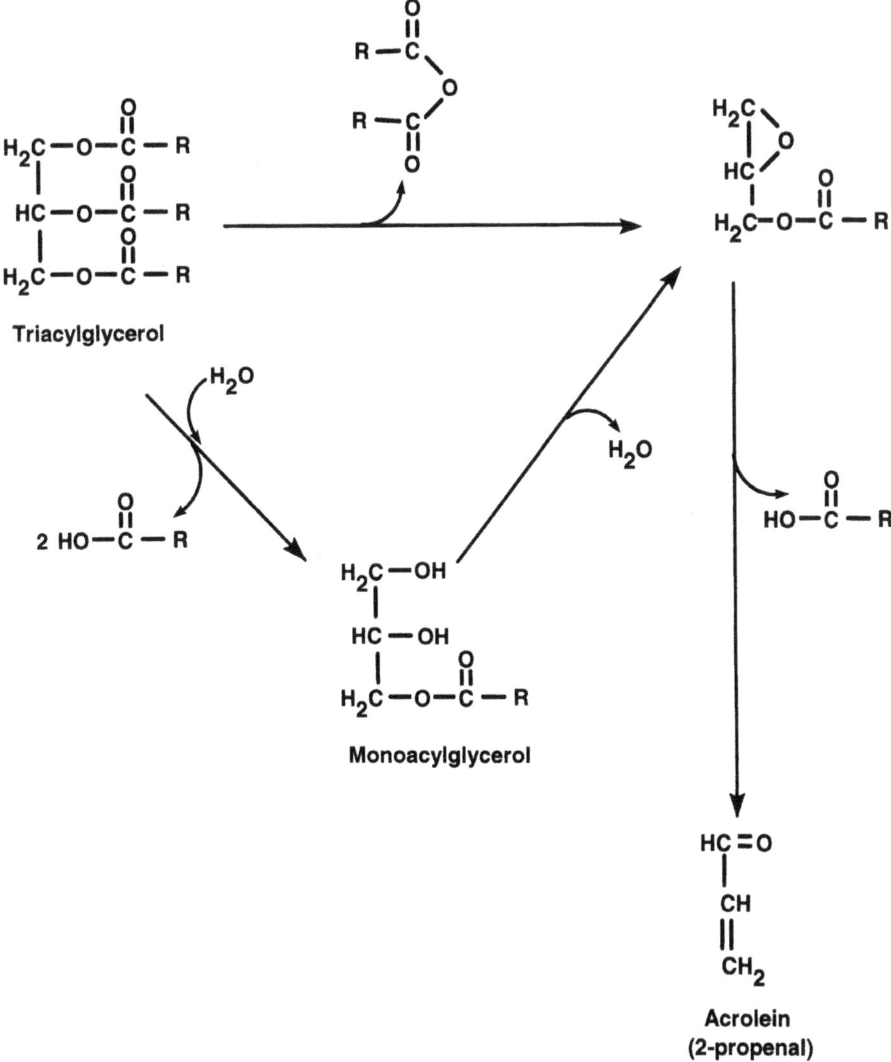

Figure 7. Formation of monoacylglycerol and acrolein by pyrolysis or hydrolysis (adapted from Pokorny, 1989).

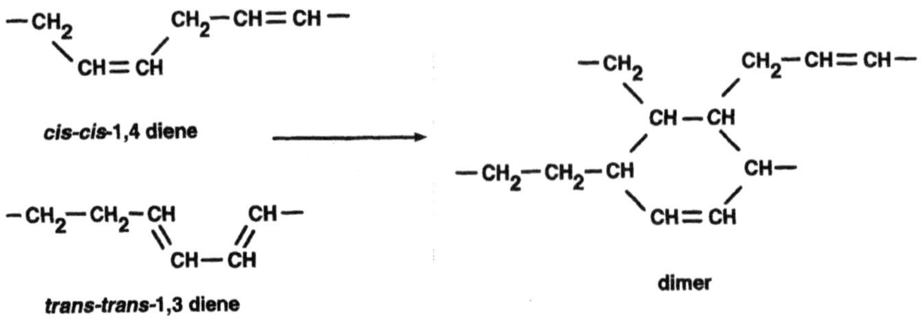

cis-cis-1,4 diene

trans-trans-1,3 diene

dimer

Figure 8. Polymerization of diene system to form dimers (adapted from deMan, 1990).

in nitrogen environment are free fatty acids, however, in air a mixture of oxidation products was obtained. The rate of pyrolysis substantially increases in the presence of oxidation products thus a rapid decrease in smoking point is observed (Pokorny, 1989). According to deMan (1990) polymerization reactions may take place by conversion of the *cis, cis*-1,4-diene system of linoleates to the *trans, trans*-conjugated 1,3-diene. The 1,4- and 1,3-dienes can combine via a Diels-Alder type addition reaction to produce a dimer (Figure 8). Polymerization results in a substantial increase in the viscosity of the frying oil which in turn may increase foaming (Nawar, 1995).

Evaporation of volatile compounds produced during frying also changes the composition of frying oil. Free fatty acids produced by hydrolytic reactions are partially evaporated during frying, mostly by steam produced from the frying food. Most volatiles that are expelled from oil are secondary oxidation products. Under usual frying conditions, 5 to 15% of the frying oil is converted to volatile compounds, but only less than 1% of volatiles remains in the oil and the rest are evaporated with the fumes during frying. The inclination to fuming is estimated on the basis of the smoking point of the oil. The smoking point of the fresh oil is usually above 200°C, but it decreases with increasing content of free fatty acids, monoacylglycerols, aldehydes and other oxidation products. When the smoking point is between 145 and 165°C, the oil is usually discarded (Pokorny, 1989).

These chemical and physical changes of oil and fried material which occur during deep fat frying are not always considered deleterious as some of these changes are necessary to provide the sensory qualities typical for fried food. However, extensive decomposition resulting from uncontrolled frying may be a potential source of adverse effects on both the sensory and nutritional quality of the oil and the fried material. The chemical and physical changes in the frying fat are influenced by a number of frying parameters and the compounds formed depend on the composition of both the oil and the fryied material. High temperature, long frying times and presence of metal contaminants favor extensive decomposition of the oil and also the design and the type of fryer and the heating pattern. Presence of antioxidants and related compounds may also affect decomposition of the oil (Pokorny, 1989).

ROASTING AND MICROWAVE HEATING

Heat treatment of oilseeds may be performed to enhance oil release and to improve the flavor of products (*e.g.*, sesame) or to inactivate antinutritional factors (*e.g.*, trypsin inhibitors in soybean). Traditionally, roasting of seeds (*e.g.* peanut, sesame) is carried out to obtain a favorite snack and/or an ingredient for different foods. Conventional roasting may use direct or convection heat while microwave heating may be an alternative to conventional modes of heating for institutional and domestic use. Changes of the intact and isolated lipids due to microwave treatment of oilseeds has been attributed to changes in oxidative stability, flavor and endogenous antioxidants present (Manley *et al.*, 1974; Yoshida and Kajimoto, 1988; 1989; Yen, 1990; Park *et al.*, 1996a).

Changes in Composition, Antioxidants, and Oxidative Stability

Microwave Heating. Microwaves (wavelength of 3 to 30 cm) are used to heat foods as a result of molecular excitation of food and their penetration power. Microwave heated foods do not have a heat gradient. Foods containing high moisture and fat readily absorb microwaves and are cooked or baked. Application of microwave for both home and institutional meal preparation has increased because of its convenience and rapid heating when compared to conventional means.

Chemical and oxidative stability changes of some commonly used vegetable oils subjected to microwave (0.5kW at 2450 MHz) heating have been studied by Fukuda *et al.* (1986c), Yoshida and Kajimoto (1988; 1989), Yen (1990), Yoshida *et al.* (1990; 1995), Shahidi *et al.* (1997a,b) and Abou-Gharbia *et al.* (1996; 1997). In general, the increase in the content of polyunsaturated fatty acids of the oil increases the oxidation of lipids upon microwave heating as evidenced by peroxide, 2-thiobarbituric acid reactive substances and carbonyl values (Yoshida *et al.*, 1991a, b). However, sesame oil showed an exceptionally high oxidative stability compared to soybean, corn and most other vegetable oils due to the presence of natural antioxidative compounds (Yoshida and Kajimoto, 1989;1994, Yen, 1990, Shahidi *et al.*, 1997a).

The fate of endogenous antioxidants in vegetable oils, especially tocopherols, due to microwave heating has also been studied. In corn and soybean oils, 90% of the original tocopherol level was retained after 8–10 min of microwave heating, however, in olive, palm and flaxseed oils the content of tocopherols was decreased substantially. When soybean oil was heated for 12 min, 40% of the total tocopherols was lost, but microwave treatment for 6 min retained 90% of their original amounts (Yoshida *et al.*, 1990; 1995).

According to Yoshida *et al.* (1991a) during microwave heating of fatty acid ethyl esters containing α, β, γ and δ tocopherols, stability of tocopherols decreased in the order of $\delta>\beta>\gamma>\alpha$, and this order did not depend on the type of ethyl esters of fatty acids examined. However, there was a greater reduction of individual tocopherols for the shorter chain length and less unsaturated fatty acid ethyl esters. Similarly, addition of individual tocopherols to tocopherol-stripped vegetable oils showed faster degradation of α-tocopherol followed by $\gamma>\beta>\delta$-tocopherols. Among the vegetable oils studied, coconut, palm and olive oils exhibited faster degradation of tocopherols than safflower, corn, rapeseed or soybean oils (Yoshida *et al.*, 1991a,b).

The endogenous tocopherols in sesame oil also showed a similar pattern of degradation upon microwave heating. The tocopherol content of sesame seed oil is in the order of $\gamma>\delta>\alpha$. During microwave heating of sesame seed, more than 80% of the original level of γ-tocopherol was retained after 30 min of heating. Other tocopherols (δ- and α-) were not detected after 16 min and 8 min of microwave heating, respectively (Yoshida and Kajimoto, 1994).

Effect of microwave heating on other major antioxidants (lignans) of sesame seed, sesamin and sesamolin (predominant lignans) and sesamol (minor lignan) has been studied

Table 7. Changes in the contents of sesamin, sesamolin and γ-tocopherol (mg/100g oil) of sesame oil due to different heat treatments of seeds (adapted from Shahidi *et al.*, 1997a)[1]

	Sesamin		Sesamolin		γ-Tocopherol	
Treatment	SC	EC	SC	EC	SC	EC
No treatment	580 ± 18	649 ± 20	349 ± 12	183 ± 7	358 ± 14	387 ± 7
Roasted at 200°C for 20 min	537 ± 20	576 ± 14	301 ± 10	146 ± 5	285 ± 9	261 ± 4
Steamed at 100°C for 20 min	541 ± 14	601 ± 18	331 ± 9	129 ± 5	309 ± 8	285 ± 5
Roasted at 200°C for 15 min and steamed for 100°C for 7 min	565 ± 21	583 ± 15	327 ± 8	146 ± 6	322 ± 10	366 ± 3
Microwaved for 15 min	544 ± 13	590 ± 17	267 ± 6	123 ± 3	320 ± 9	352 ± 4

[1]SC, Sudanese variety with seed coat; EC, Egyptian variety with seed coat.

by Yoshida and Kajimoto (1994) and Shahidi *et al.* (1997a). Sesamin and sesamolin contents decreased due to microwave heating by 5% after 12 min and 15% after 25 min. However, as much as 80% of both compounds remained unchanged after 30 min of heating. Conversely, the content of sesamol, which was the minor component in the original seed, increased gradually from two-fold after 16 min to ten-fold after 30 min heating. According to Shahidi *et al.* (1997a) decrease in the content of sesamin and sesamolin of sesame oil was more pronounced after 15 min of microwave heating than roasting (200°C for 20 min), steaming (100°C for 20 min) or roasting at 200°C for 20 min plus steaming for 7 min (Table 7). The pattern of changing of lignans in microwave heating was different from that of conventional conductive heating of the seed (Fukuda *et al.*, 1986c; Yen, 1990; Shahidi *et al.*, 1997a,b).

Studies on changes in lipid fractions of sesame oil due to microwave heating of seeds revealed that 30 min of heating (longer time) resulted in better extraction of total lipids. The yield of neutral lipid fraction increased from 96 to 98% after 20 min of heating and the amount of glycolipid fraction also increased. The glycolipid fraction contained a brown color due to the presence of polymerized products. The proportion of phospholipid fraction gradually decreased and at 30 min the level was 15% lower than its original amount in the oil (Yoshida and Kajimoto, 1989).

The fatty acid composition of major lipid classes showed a small, but significant ($P<0.05$), difference between total lipids, triacylglycerols and phospholipids of sesame oil after microwave heating (Yoshida *et al.*, 1995). Yoshida *et al.* (1995) have determined the optimum exposure time as 16 min for heat processing of sesame seeds by microwaving.

Roasting. Yen and Shyu (1989) and Abou Gharbia *et al.* (1996) have studied the effect of seed roasting temperature on the oxidative quality of sesame oil. Sesame oil did not show any apparent difference in hydroperoxide formation (peroxide value) due to different seed roasting temperatures, when stored in the dark. Seeds roasted at 200°C had higher stability than those roasted at lower temperatures, probably due to the presence of high levels of sesamol (converted from sesamolin at high temperatures). Tables 7 and 8 provide changes in the contents of sesamin, sesamolin and γ-tocopherol of sesame oil due to different heat treatments. There were no differences in the acid, iodine and saponification values and as well as refractive index of sesame oils prepared from seeds roasted at 180 to 220°C. The total polar content of oils increased in relation to increasing the roasting temperature and the color of the oil became dark. The phospholipid content of the oil was also reduced to undetectable when seeds were treated at 260°C. The content of unsaturated

Table 8. Changes of antioxidative lignans and total tocopherols in refined unroasted (RUSO) and roasted (RSO) sesame oils during heating at 180°C (adapted from Fukuda *et al.*, 1994)

	Fresh[1]		Heated at 180°C for 2 h[1]	
Compound	RUSO	RSO	RUSO	RSO
Sesamolin[2]	ND	3.5	ND	0.72
Sesaminol	0.95	ND	0.83	ND
Sesamol	trace	0.04	ND	0.76
Tocopherol (total)	0.34	0.42	0.28	0.38

[1]ND, not detected.
[2]Sesamolin is the precursor of sesaminol and sesamol.

fatty acids, especially oleic and linoleic acids, was reduced markedly when roasting temperature was greater than 220°C. The amounts of chlorophyll and sesamolin decreased with increasing roasting temperature. However, the highest level of sesamol and γ-tocopherol was found in oils prepared from seeds roasted at 200 to 220°C. Sesame oil prepared from seeds roasted at 200°C had the best flavor score when compared with other roasting temperatures studied (Takai *et al.* 1969; Yen and Shyu, 1989).

Sesame oil obtained from roasted seeds has a characteristic flavor and brown color and is not subjected to any further refining steps as for refined unroasted sesame oil. According to Fukuda *et al.* (1994), roasted sesame oil (RSO) is very resistant to oxidation. The degree of browning and oxidative stability of RSO increased with increasing roasting temperature. However, the content of sesamol, which is produced from degradation of sesamolin of seed during roasting of seed, did not increase considerably with roasting temperature (Fukuda *et al.*,1986c). During the frying process, large amounts of sesamol are produced from sesamolin (Figure 9) in RSO and these contribute to the high oxidative stability of the fried food and the frying oil (Table 8, Fukuda, 1992; Fukuda *et al.*, 1986c; 1988; 1994). The differences in the contents of sesamolin and sesaminol in refined unroasted sesame oil (RUSO) and RSO and proposed pathway of their interconversion have already been discussed elsewhere in this chapter.

Yoshida (1994) has also observed minor increases in oxidative stability indices, as reflected in acid, peroxide, anisidine and thiobarbituric acid reactive substances values, of sesame oil in relation to increasing temperature, from 120 to 180°C. The oil prepared from seeds roasted at 250°C, had a higher glycolipid content than unroasted seeds; consequently the color of the oil became darker. The total fatty acid content of the oil was also reduced as the content of triacylglycerols decreased due to their polymerization at high roasting temperatures (250°C).

Park *et al.* (1996a) have studied chemical changes of canola oil obtained from hexane extraction of seeds roasted at 180, 200, 220, 240 and 260°C for 8 or 10 min. The color of oil become dark as the seed roasting temperature increased similar to sesame seed oil. However, oxidative stability of canola oil, as determined by 2-thiobarbituric acid (TBA) numbers, decreased with increasing roasting temperature of the seeds, but the oil obtained from unroasted seeds had the highest oxidative stability.

However, contrasting results have been reported for the pressed canola oil obtained from heat treated seeds (Prior *et al.*, 1991a). As the temperature of heat treatment of seed prior to pressing increased the content of non-acylglycerol components (from browning reactions and polymerization), phosphorus and chlorophyll increased. The oxidative stability of the oil so obtained increased with increasing temperature of heating and also

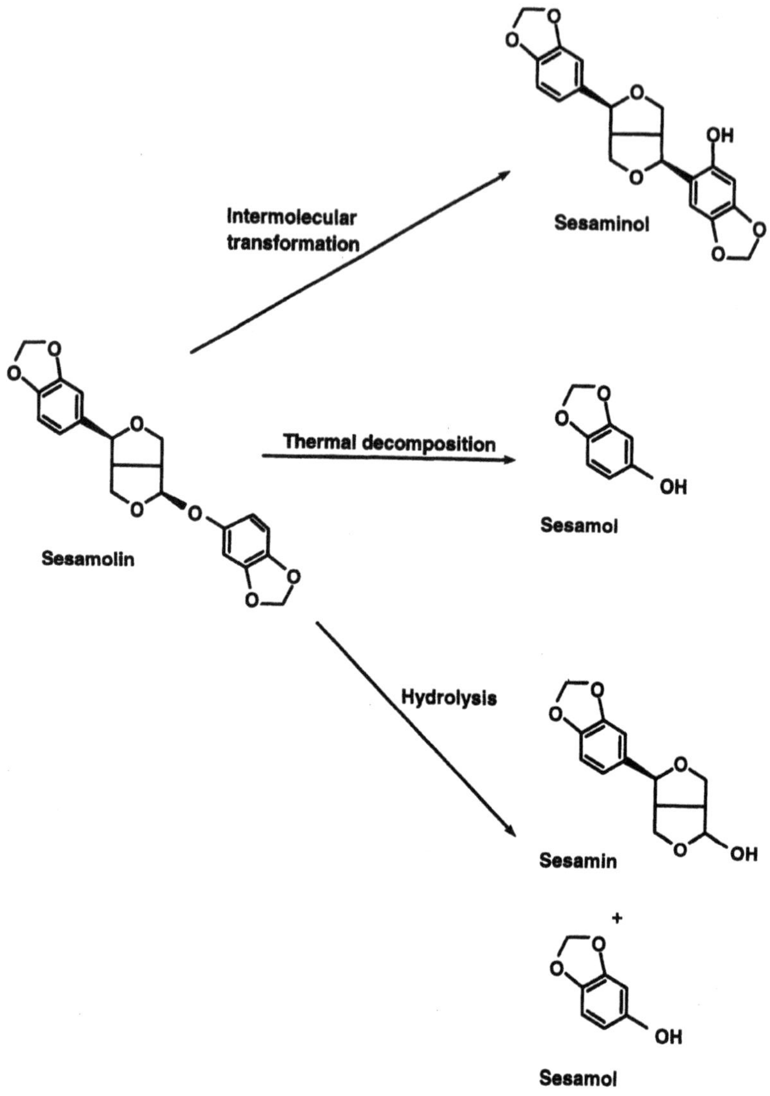

Intermoleculartransformation

Sesaminol

Thermal decomposition

Sesamol

Sesamolin

Hydrolysis

Sesamin

Sesamol

Figure 9. Possible changes of sesamolin due to processing of sesame seed (adapted from Fukuda *et al.*, 1994).

showed a direct relationship with increased content of non-acylglycerol fraction (Prior *et al.*, 1991b).

Lee *et al.* (1993) have examined the quality of sesame seed oil roasted for 30, 60, 90 and 120 min at 100, 200 and 300°C. Seeds roasted at 200°C for 90 min gave the highest yield of oil as well as overall aroma and taste quality. The oil contained low amounts of a brown-black precipitate and exhibited most favored sensory quality as evidenced by subjective analysis. Similarly, the oil obtained from seeds roasted at 200°C contained the highest number of volatile flavor compounds with relatively high concentrations of furfurals (sweet candy like) and pyrazines (roasted like) that are known to contribute to the typical flavor of sesame seed oil. Heterocyclic aroma compounds which are mainly formed due to Maillard reaction during roasting of sesame seeds are mainly alkylpyrazines (Kinoshita and Yamanishi, 1973; Manley *et al.,* 1974). Park *et al.* (1995b), Schieberle (1996) and Shahidi *et al.* (1997b) have reported detailed analysis of flavor compounds formed during roasting of sesame seed. Low concentrations of aldehydes (C5-C10) and ketones which are considered as contributing to oxidized fat-like and painty-like flavors were also present (Lee *et al.,* 1993). The major flavor volatiles of oils obtained from roasted canola seeds were identified as pyrazines and furans, in addition to two terpenols from cyclization and degradation of carotenoids of seeds (Park *et al.,* 1996a).

REFERENCES

Abou-Gharbia, H.A.; Shahidi, F.; Shehata, A.A.Y.; Youseff, M.M. Oxidative stability of extracted sesame oil from raw and processed seeds. *J. Food Lipids*, **1996,** *3*, 59–72.

Abou-Gharbia, H.A.; Shahidi, F.; Shehata, A.A.Y.; Youseff, M.M. Effects of processing on oxidative stability of sesame seed oil extracted from intact and dehulled seeds. *J. Am. Oil Chem. Soc.* **1997**, *74*, 215–221.

Akoh, C.C.; Jennings, B.H.; Lillard, D.A. Enzymatic modification of trilinolein; incorporation of n-3 polyunsaturated fatty acids. *J. Am. Oil Chem. Soc.* **1995,** *72*, 1317–1321.

Akoh, C.C.; Jennings, B.H.; Lillard, D.A. Enzymatic modification of evening primrose oil, incorporation of n-3 polyunsaturated fatty acids. *J. Am. Oil Chem. Soc.* **1996**, *73*, 1059–1062.

Alencar, J.W.; Alves, P.B.; Craveiro, A.A. Pyrolysis of tropical vegetable oils. *J. Agric. Food Chem.* **1983,** *31*, 1268–1270.

Belitz, H.D.; Grosch, W. *Food Chemistry*, Springer-Verlag, Berlin, 1987; pp. 128–200.

Bimbo, A.P. Technology of production and industrial utilization of marine oils. In *Marine Biogenic Lipids, Fats and Oils*; Ackman, R.G, Ed., CRC Press Inc. Boaca Raton, Florida, 1989; pp. 401–433.

Bimbo, A.P.; Crowther, J.B. Fish oils: Processing beyond crude oil. *Infofish Int.* **1991,** *6*, 20–25.

Boki, K.; Shinoda, S.; Ohno, S. Effects of filtering through bleaching media on decrease of peroxide value of autoxidized soybean oil. *J. Food Sci.* **1989**, *54*, 1601–1603.

Breivik, H. N-3 concentrates: A Scandinavian view-point. AOCS short courses. Modern Application of Marine Oils, 1992, May 7–8, Toronto, ON.

Buziassy, C.; Nawar, W.W. Specify in thermal hydrolysis of triglycerides. *J. Food Sci.* **1968,** *33*, 305–307.

Carr, R.A. Refining and degumming systems for edible fats and oils. *J. Am. Oil Chem. Soc.* **1978,** *55*, 765–771.

Chang, S.S.R.; Peterson, R.J.; Ho, C-T. Chemistry of deep fat fried flavor. In Lipids as a Source of Flavor; Supton, M.K., Ed. American Chemical Society, Washington DC. 1978; pp. 18–41.

Chang S.S. Processing of fish oils. In *Fish Oils*; Stansby, M.E., Ed., AVI Publishing Co., West Port, Connecticut, 1967; pp. 206–221.

Cowan, J.C. Degumming, refining, bleaching and deodorization theory. *J. Am. Oil Chem. Soc.* **1976,** *53*, 344–346.

deMan, J.M. *Principles of Food Chemistry*, Second Edition, Van Nostrand Reinhold. New York, 1990; pp. 36–88.

Davidek, J.; Velisek, J.; Pokorny, J. *Chemical Changes During Food Processing.* Elsevier Science Pub. Co. Inc. New York, 1990; pp. 169–229.

Dinamarca, E.; Garrido, F.; Valenzuela, A. A simple high vacuum distillation equipments for deodorizing fish oil for human consumption. *Lipids* **1990,** *25*, 170–171.

Doi, H.; Urakami, C. Evaluation of heated frying oils. II. Components in subfractions of polar triglycerides from heat-treated soybean oil. *J. Japanese Oil Chem. Soc.* [*Yakagaku]* **1976,** *25*, 831–841.

Endres, J.G.; Bhalerao, V.R.; Kumerow, F.A. Thermal oxidation of synthetic triglycerides II. Analysis of the volatile condensable and non-condensable phases. *J. Am. Oil Chem. Soc.* **1962**, *39*, 159–162.

Foglia, T.A.; Petruso, K.; Feariheller, S.H. Enzymatic interesterification of tallow-sunflower oil mixtures. *J. Am. Oil Chem. Soc.* **1993,** *55*, 281–285.

Frankel, E.N. Lipid oxidation. *Prog. Lipid Res.* **1980,** *19*, 1–22.

Frankel., E.N. Volatile lipid oxidation products. *Prog. Lipid Res.* 1982, *22*, 1–33.

Fukuda, Y.; Isobe, M.; Nagata, M.; Osawa, T.; Namiki, M. Acidic transfprmation of sesamolin, the sesame oil constituent into an antioxidant bisepoxylignan sesaminol. *Heterocycles* **1986a**, *24*, 923–926.

Fukuda, Y.; Nagata, M.; Osawa, T.; Namiki, M. Contribution of lignan analogues to antioxidative activity of refined unroasted sesame seed oil. J. Am. Oil Chem. Soc., **1986b,** *63*, 1027–1031.

Fukuda, Y.; Nagata, M.; Osawa, T.; Namiki, M. Chemical aspects of the antioxidative activity of roasted sesame seed oil and the effects of using the oil for frying. *Agric. Biol. Chem.* **1986c,** *50*, 857–862.

Fukuda, Y.; Osawa, T.; Kawakishi, S.; Namiki, M. Oxidative stability of foods fried with sesame oil. *Nippon Shokuhin Kogyo Gakkaishi* **1988,** *35*, 28–32.

Fukuda, Y. Antioxidants in sesame seeds and their applications. *Shokuhin* **1992,** *35(6)*, 30–37.

Fukuda, Y.; Osawa, T.; Kawakishi, S.; Namiki, M. Chemistry of lignan antioxidants in sesame seed and oil. *Phytochemicals for Cancer Prevention II*, Ho, C-T.; Osawa, T.; Huang, M-T; Rosen, R.T., Eds., American Chemical Society; Washington D.C. 1994; pp. 264–274.

Gavin, A.M.; Edible oil deodorization. *J. Am. Oil Chem. Soc.* 1978, *55*, 783–791.

Gunstone, F.D. Chemical properties. In *The Lipid Handbook.* 2nd edition; Gunstone, F.D.; Harwood, J.L.; Padley, F.B., Eds., Chapman and Hall, New York, 1994; pp. 561–604.

Haraldsson, G.G.; Hoskuldsson, P.A.; Sigurdsson, S.T.; Thorsteinsson, F.; Gudbjarnason, S. The preparation of triglycerides highly enriched with Δ-3polyunsaturated fatty acids via lipase catalysed interesterification. *Tetrahedron Lett.* **1989**, *30*, 1671–1674.

Higman, E.B.; Schmeltz, I.; Higman, H.C.; Chortyk, O.T. Studies on the thermal degradation of naturally occurring materials II. Products from the pyrolysis of triglycerides at 400 °C. *J. Agric. Food Chem.* **1973,** *21*, 202–204.

Hustedt, H.H. Interesterification of edible oils. *J. Am. Oil Chem. Soc.* **1976,** *53*, 390–392.

Jachamanian, I.; Mukherjee, K.D. Esterification and interesterification reaction catalysed by acetone powder from germinating rapeseed. *J. Am. Oil Chem. Soc.* **1996**, *73*, 1527–1532.

Kimoto, H.; Endo, Y.; Fujimoto, K. Influence of interesterification on the oxidative stability of marine oil triacylglycerols. *J. Am. Oil Chem. Soc.* **1994**, *71*, 470–473.

Kinoshita, S.; Yamanishi, T. Identification of basic aroma components of roasted sesame seeds. *J. Agric. Chem. Soc. Japan* **1973,** *47*, 737–739.

Kwon, T.W.; Snyder, H.E.; Brown, H.G. Oxidative stability of soybean oil at different stages of refining. *J. Am. Oil Chem. Soc.* **1984,** *61*, 1843–1846.

Lee, C.F. Processing of fish meal and oil. In *Industrial Fishery Technology*; Stansby, M.E., Ed., Reinhold Publishing Co. New York, 1963; pp. 219–235.

Lee, Y.G.; Lim, S.U.; Kim, J.O. Influence of roasting conditions on the flavor quality of sesame seed oil. *J. Korean Agric. Chem. Soc.* **1993**, *36*, 407–415.

Lein, Y.C.; Nawar, W.W. Thermal decomposition of tricaproin. *J. Am. Oil Chem. Soc.* **1973,** *50*, 76–78.

Lin, C.F.; Hsieh, T.C.Y.; Crowther, J.B.; Bimbo, A.P. Efficiency of removing volatiles from menhaden oils by refining, bleaching and deodorization. *J. Food Sci.* **1990,** *55*, 1669–1672.

List, G.R.; Mounts, T.L.; Warner, K.; Heakin, A.J. Steam-refined soybean oil I. Effect of refining and degumming methods on oil quality. *J. Am. Oil. Chem. Soc.* **1978a,** *55*:277–279.

List, G.R.; Mounts, T.L.; Heakin, A.J. Steam-refined soybean oil II. Effect of degummimg methods on removal of prooxidants and phospholipids. *J. Am. Oil. Chem. Soc. 55,* :280–284.

List, G.R.; Mounts, T.L. Partially hydrogenated-winterzed soybean oil. In *Handbook of Soy Oil Processing and Utilization*; Erickson, R.R.; Pryde, E.H.; Brekke, O.L.; Mounts, T.L.; Falb, R.A., Eds., American Oil Chemists' Society, Champaign, IL, 1980; Chapter 12.

Lorant, B. Kinetic studies on the thermal decomposition of fats. *Nahrung* **1977,** *21*, 491–496.

Manley, C.H.; Vallon, P.P.; Erikson, R.E. Some aroma components of roasted sesame seed *J. Food Sci.* **1974,** *39*, 73–76.

Morgan, D.A.; Shaw, D.B.; Sidebottm, M.J.; Soon, T.C.; Taylor, R.S. The function of bleaching earths in the processing of palm, palm kernal and coconut oils. *J. Am. Oil Chem. Soc.* **1985,** *62*, 292–299.

Namiki, M. Antioxidants/antimutagens in foods. *CRC Crit. Rev. Food Sci. Nutr.* **1990,** *29(4)*, 273–300.

Nawar, W.W. Thermal and radiolytic decomposition of lipids. In *Chemical changes in Food During Processing*; Richardson, T.; Finely, J.W., Eds., AVI Pub. Co. Westport, CT, 1985; pp. 79–105.

Nawar, W.W. Lipids In *Food Chemistry* Third edition; Fennema, O.R., Ed., Marcel Dekker, Inc. New York, 1996, pp. 225–319.

Noble, A.C.; Nawar, W.W. Identification of decomposition products from autoxidation of methyl 4,7,10,13,16,19-docosahexaenoate. *J. Am. Oil Chem. Soc.* **1975,** *52*, 92–95.

Noble, A.C.; Buziassy, C.; Nawar, W.W. Thermal hydrolysis of some natural fats. *Lipids,* **1967**, *2*, 435–436.

Osawa, T.; Nagata, M.; Namiki, M.; Fukuda, Y. Sesamolinol: A novel antioxidant isolated from sesame seeds. *Agric. Biol. Chem.* **1985**, *49*, 3351–3352.

Padley, F.B.; Gunstone; F.D.; Harwood, J.L. Occurence and characteristics of oils and fats. In *The Lipid Handbook.* 2nd edition; Gunstone, F.D.; Harwood, J.L.; Padley, F.B., Eds., Chapman and Hall, New York, 1994; pp. 47–224.

Paulose, M.M.; Chang, S.S. Chemical reactions involved in the deep-fat frying of foods VIII. Characterization of non-volatile decomposition products of triolein. *J. Am. Oil Chem. Soc.* **1978,** *55*, 375–380.

Park, D.; Maga, J.A.; Johnson, D.L. Chemical changes in toasted crude canola and sesame seed oil. *J. Food Lipids* **1996a,** *3*, 1–12

Park, D.; Maga, J.A.; Johnson, D.L.; Morini, G. Major volatiles in toasted canola oil. *J. Food Lipids* **1996b**, *2*, 249–258.

Prior, E.M.; Vadke, V.S.; Sosulski, F.W. Effect of heat treatment on canola press oils I. Non-triglyceride components. *J. Am. Oil Chem. Soc.* **1991a,** *68*, 401–406.

Prior, E.M.; Vadke, V.S.; Sosulski, F.W. Effect of heat treatment on canola press oils II. Oxidative stability. *J. Am. Oil Chem. Soc.* **1991b**, *68*, 407–411.

Pokorny, J. Flavor chemistry of deep fat frying in oil. In *Flavor chemistry of Lipid Foods*; Min, D.B.; Smouse, T.H., Eds., American Oil Chemists Society, Champaign, IL, 1989; pp. 113–155.

Richardson, L.L. Use of bleaching clay in processing of edible oils. *J. Am. Oil Chem. Soc.* **1978,** *55*, 777–780.

Schieberle, P. Odor-active compounds in moderately roasted sesame. *Food Chem.* **1996,** *55*, 145–152.

Shahidi, F.; Amarowicz, R.; Abou-Gharbia, H.A.; Shehata, A.A.Y. Endogenous antioxidants and stability of sesame seed oil as affected by processing and storage. *J. Am. Oil Chem. Soc.* **1997a,** *74*, 143–148.

Shahidi, F.; Aishima, T.; Abou-Gharbia, H.A.; Youssef, M.; Shehata, A.A.Y. Effect of processing on flavour precursor amino acids and volatiles of sesame paste (*tehina*). *J. Am. Oil Chem. Soc.* **1997b,** *74*, 667–678.

Sonnet, P.E. Lipase selectivities. *J. Am. Oil Chem. Soc.* **1988,** *65*, 900–904.

Sreenivasan, B. Interesterification of fats *J. Am. Oil Chem. Soc.* 1978, *55*, 796–805.

Syed Rahmatullah, M.S.K.; Shukla, V.K.S.; Mukherjee, K.D. γ-Linolenic acid concentrates from borage and evening primrose oil via lipase catalysed esterification. *J. Am. Oil Chem. Soc.* **1994,** *71*, 563–567.

Sridhar, R.; Lakshminarayana, G. Incorporation of eicosapentaenoic acid docosahexaenoic acids into groundnut oil by lipase-catalysed ester interchange. *J. Am. OIl Chem. Soc.* **1992,** *69*, 1041–1042.

Takai, Y.; Nakatani, Y.; Kobayashi, A. ; Yamanishi, T. Studies on the aroma of sesame oil: Intermediate and high boiling compounds. *J. Agric. Chem. Soc. Japan.* **1969,** *43*, 667–674.

Umano, K.; Shibamoto, T. Analysis of acrolein from heated cooking oils and beef fat. *J. Agric. Food Chem.* **1987,** *35*, 909–912.

Van Gastel, A.; Mathur, R.; Roy, V.V.; Rukmini, C. Ames mutagenicity tests of repeatedly heated edible oils. *Food Chem. Toxicol.* **1984,** *22*, 403–405.

Varela, G.; Bender, A.E.; Morton, I.D. *Frying of Food: Principles, Changes and New Approaches.* Ellis Harwood Ltd., Chichester, England, 1988.

Wada, S.; Koizumi, C. Influence of random interesterification on the oxidation rate of soybean oil triglyceride. *J. Japanese Oil Chem. Soc.* [*Yakagaku*] **1986,** *35*, 549–553.

Wanasundara, U.N. *Marine Oils, Stabilization, Structural Characterization and Omega-3 Fatty Acid Concentration.* Ph.D. thesis, 1996, Memorial University of Newfoundland, St. John's, Canada.

Wiedermann, L.H. Degummimg, refining and bleaching of soybean oil. *J. Am. Oil Chem. Soc.* **1981**, *58*, 159–166.

Yen, G-C. Influence of seed roasting process on the changes in composition and quality of sesame (*Sesamum indicum*) oil. *J. Sci. Food Agric.* **1990,** *50*, 563–570.

Yen, G-C.; Shyu, S-L. Oxidative stability of sesame oil prepared from sesame seed with different roasating temperatures. *Food Chem.* **1989**, *31*, 215–224.

Yoshida, H. Composition and quality characteristics of sesame (*Sesmum indicum*) seed oil roasted at different temperatures in an electric oven. *J. Sci. Food Agric.* **1994,** *65*, 331–336.

Yoshida, H.; Kajimoto, G. Effects of microwave treatment on the trypsin inhibitor and molecular species of triglycerides in soybean. *J. Food Sci.* **1988,** *53*, 1756–1760.

Yoshida, H.; Kajimoto, G. Effects of microwave energy on the tocopherols of soybean seeds. *J. Food Sci.* **1989,** *54*, 1596–1600.

Yoshida, H; Kajimoto, G., Microwave heating affects composition and oxidative stability of sesame (*Sesmum indicum*) oil. *J. Food Sci.* **1994,** *59*, 613–616, 625.

Yoshida, H.; Hirooka, N.; Kajimoto, G. Microwave energy effects on quality of some seed oils. *J. Food Sci.* **1990,** *55*, 1412–1416.

Yoshida, H.; Hirooka, N.; Kajimoto, G. Relationship between oxidative stability of vitamin E and production of fatty acids in oils during microwave heating. *J. Am. Oil Chem. Soc.* **1991a,** *68*, 566–570.

Yoshida, H.; Hirooka, N.; Kajimoto, G. Microwave heating effects on relative stabilities of tocopherols in oils. *J. Food Sci.* **1991b,** *56*, 1042–1046.

Yoshida, H.; Shigezaki, J.; Takagi, S.; Kajimoto, G. Variations in the composition of various acyl lipids, tocopherols and lignans in sesame seed oil roasted in microwave oven. *J. Sci. Food Agric.* **1995,** *68*, 407–415.

Young, F.V.K.; Poot, C.; Biernoth, E.; Krog, N., Davidson, N.G.J.; Gunstone, F.D. Processing of oils and fats. In *The Lipid Handbook.* 2nd edition, Gunstone, F.D.; Harwood, J.L.; Padley, F.B., Eds., Chapman and Hall, New York, 1994; pp. 249–318.

Yuki, E.; Ishikawa, Y.; Okabe, T. The composition of fatty acids in frying oils and hydrolysis during deep fat frying. *J. Japanese Oil Chem. Soc.* [*Yakagaku]* **1974,** *23*, 489–493.

EFFECTS OF PROCESSING STEPS ON THE CONTENTS OF MINOR COMPOUNDS AND OXIDATION OF SOYBEAN OIL

David B. Min, Tsung-Lin Li, and Hyung-Ok Lee

Department of Food Science and Technology
The Ohio State University
Columbus, Ohio 43210

The effects of minor compounds on the oxidative stability of soybean oil were studied by measuring the contents of peroxides, headspace oxygen and volatile compounds. The effects of processing on minor component contents were also studied. Fatty acids, mono- and diacylglycerols, thermal or oxidized triacylglycerols, oxidized tocopherols and peroxides acted as prooxidant in soybean oil during storage at 55°C. The phospholipids acted as prooxidant or antioxidant depending on the presence or absence of metals in the oil. The tocopherols acted as prooxidant or antioxidant depending on their concentration in the oil. The chlorophyll acted as a sensitizer to generate singlet oxygen in the photooxidation of soybean oils.

INTRODUCTION

Soybean oil is of major economic importance because its high quality and low cost, and represents the main source of plant-derived fat in our daily diet. Soybean oil production in the United States in 1987 was 5.7 million tons which corresponds to 86% of the total production of vegetable and marine oils (USDA, 1987). Soybean oil now dominates and will continue to dominate the U.S. oil market in spite of competition from other oils.

Soybean oil has a high linoleic acid content of around 50.8% and 7% of linolenic acid and is thus susceptible to oxidation. Adverse effects caused by oxidation of vegetable oils include loss of essential fatty acids, production of off-flavors and toxic compounds (Sonntag, 1979). The oxidative stability becomes an important quality control parameter for the manufactures and users of vegetable oils. Although the oxidative stability of soybean oil has been greatly improved through the concerted efforts of universities, industries and government research laboratories, attempts are constantly being made to further improve the oxidative stability of the oil.

Quality of soybean oil is affected by a number of factors such as quality of soybeans, content of moisture, storage temperature and processing conditions, exposure to air, prooxidants, light, minor compounds such as free fatty acids, mono- and diacylglycerols, phospholipids, oxidized triacylglycerols, tocopherols, triterpene alcohols, phytosterols, isoflavonones, and possibly some unknown compounds contained in soybean oil. Some of these minor compounds affect the oxidative stability of vegetable oils.

This contribution summarizes the effects of free fatty acids, mono- and diacylglycerols, phospholipids, thermal or oxidized triacylglycerols, tocopherols, chlorophyll, and peroxides on the oxidative stability of soybean oil and the effects of processing on the minor component contents.

EFFECTS OF FATTY ACIDS ON THE OXIDATIVE STABILITY

The free fatty acid contents of soybean oil ranged from 0.02 to 0.07% (Sonntag, 1979; Pryde, 1980; Jung *et al.*, 1989). Holman *et al.* (1947, 1954) and Miyashita and Takagi (1986) reported that free fatty acids oxidize more rapidly than their corresponding methyl esters. Miyashita and Takagi (1986) also reported that stearic acid accelerated the autoxidation of methyl linoleate and the decomposition of methylinoleate hydroperoxides. They proposed that the higher oxidation rates of free fatty acids compared to their methyl ester counterparts could be due to the catalytic effect of the carboxyl groups on the formation of free radicals via the decomposition of hydroperoxides.

Mistry and Min (1987) further pointed out that stearic, oleic, and linoleic acids were prooxidants in soybean oil, and their prooxidant activities were not significantly different from one another. They also suggested that the prooxidant activities of free fatty acids were due to carboxyl groups of the fatty acids. Figure 1 shows that the purified soybean oil containing stearic, oleic, linoleic or linolenic acid produced more volatile compounds than the purified oil during storage at 55°C. The oil containing octadecane produced a similar amount of volatile compound as the control during a 5-day storage. The formation of volatile compounds in oils containing the same level of different fatty acids was not statistically different ($p \geq 0.05$).

Mistry and Min (1987) also reported that headspace oxygen reacted more rapidly with oils containing 0.5% fatty acids than the control. Figure 2 clearly shows that the peroxide value of the control was lower than that of the oil containing 1.0% fatty acid. The oil containing 1.0% fatty acid generated significantly more peroxides than the control, and the oil containing octadecane produced the same peroxides as the control ($p \geq 0.05$). Octadecane, which had the same number of carbons as these fatty acids but no carboxylic group, did not act as a prooxidant. The carboxylic acid groups of the fatty acids were involved in the prooxidant activity of free fatty acids.

EFFECTS OF MONO- AND DIACYLGLYCEROLS ON THE OXIDATIVE STABILITY

Mistry and Min (1987) isolated minor components from soybean oil by using a combination of silicic acid column chromatography and low temperature fractional crystallization. One of the minor components showing prooxidant activity was positively

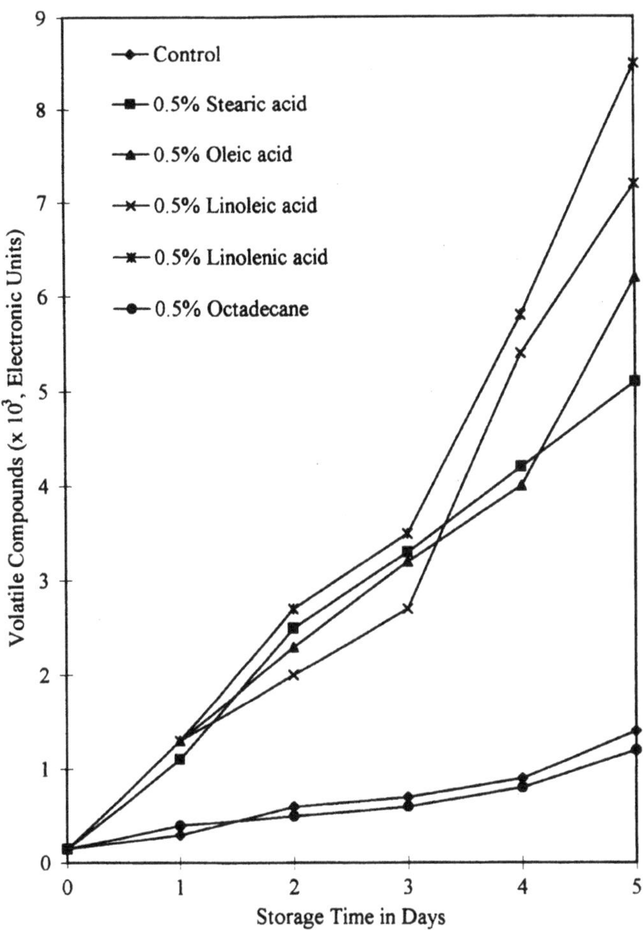

Figure 1. Effects of fatty acids and octadecane on volatile compounds formation in purified oil during storage at 55°C.

identified as a SN-α-monolinolein by a combination of infrared (IR) spectra, mass (MS) spectra, nuclear magnetic resonance (NMR) and two dimensional NMR spectra.

Mistry and Min (1987a) reported the effect of authentic monolinolein on the oxidative stability of soybean oil during storage at 55°C. As the level of SN-α-monolinolein in purified soybean oil increased from 0, 0.5 to 1.0%, volatile compounds and peroxide values of the oil increased (Figure 3). Duncan's Multiple Range Test for the effects of different levels of monolinolein on the volatile compound formation during storage (Table 1) showed that there was a significant difference in the volatile compound formation in purified oil containing 0, 0.5 to 1.0% monolinolein. The results suggested that SN-α-monolinolein was a prooxidant in purified soybean oil during storage.

Mistry and Min (1976; 1988) further studied the effects of mono- and diacylglycerols on the oxidative stability of purified soybean oil. The rates of disappearance of headspace oxygen in the oils containing monostearin, distearin, monolinolein or dilinolein

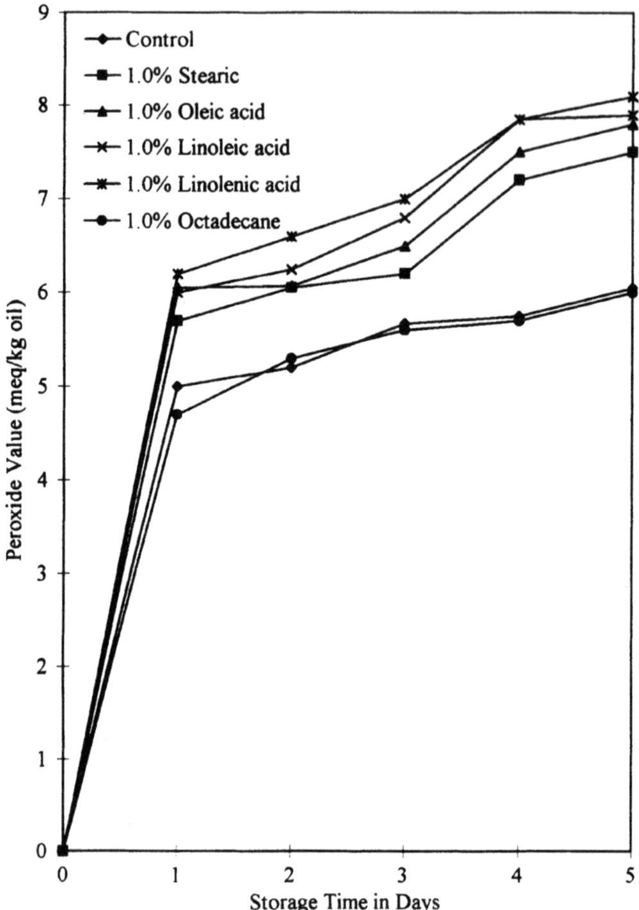

Figure 2. Effects of fatty acids and octadecane on peroxide values of purified oil during storage at 55°C.

were more rapid than the control sample. Monostearin, distearin, monolinolein or dilinolein acted as prooxidants in soybean oil (Figure 4). Table 2 demonstrates the rates of headspace oxygen consumption in the purified oil containing 0, 0.25, 0.5% monostearin, distearin, monolinolein or dilinolein during storage at 55°C. As the levels of the mono- and diacylglycerols in the oil increased from 0, 0.25, 0.5%, the rates of headspace oxygen of the sample decreased and were significantly different at $p = 0.05$.

EFFECTS OF PHOSPHOLIPIDS ON THE OXIDATIVE STABILITY

Crude soybean oil contained about 1.5% phospholipids (Jung *et al.*, 1989). The common phospholipids include phosphatidylcholine (PC), phosphatidylethanolamine (PE), and phosphatidylinositol (PI). Phospholipids constitute one of the important minor components which affect the oxidative stability of the oil. The effects of phospholipids on the

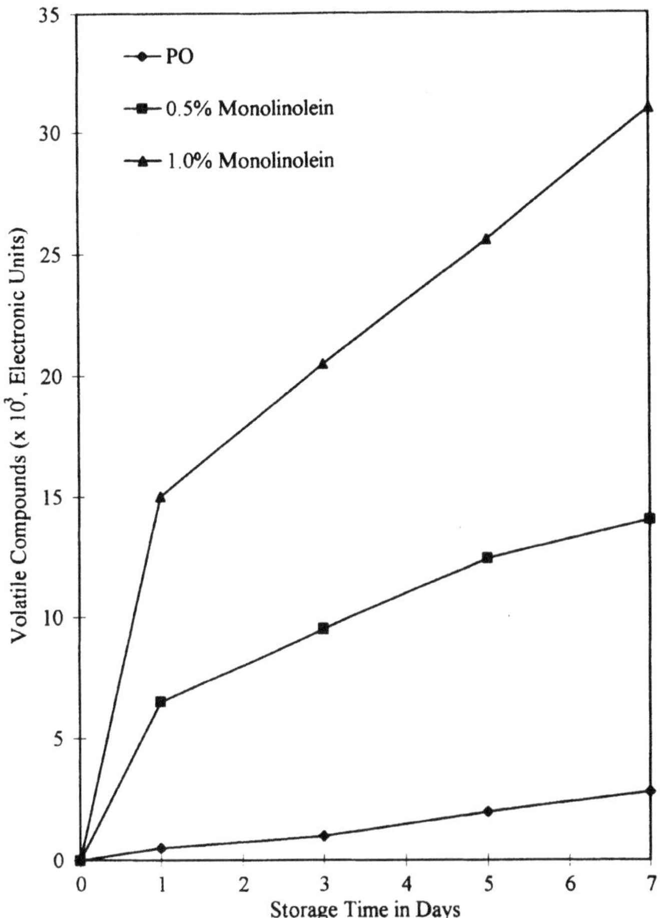

Figure 3. Effects of 0, 0.5 and 1.0% concentrations of SN-α-monolinolein on volatile compounds formation in purified soybean oil during storage at 55°C.

Table 1. Duncan's multiple range test for the effects of different concentrations of monolinolein on volatile compounds formation in purified oil during storage at 55°C

% Added Monolinolein	Means of GC peak areas of volatiles	Duncan grouping
0.0	1185	A
0.5	10921	B
1.0	23410	C

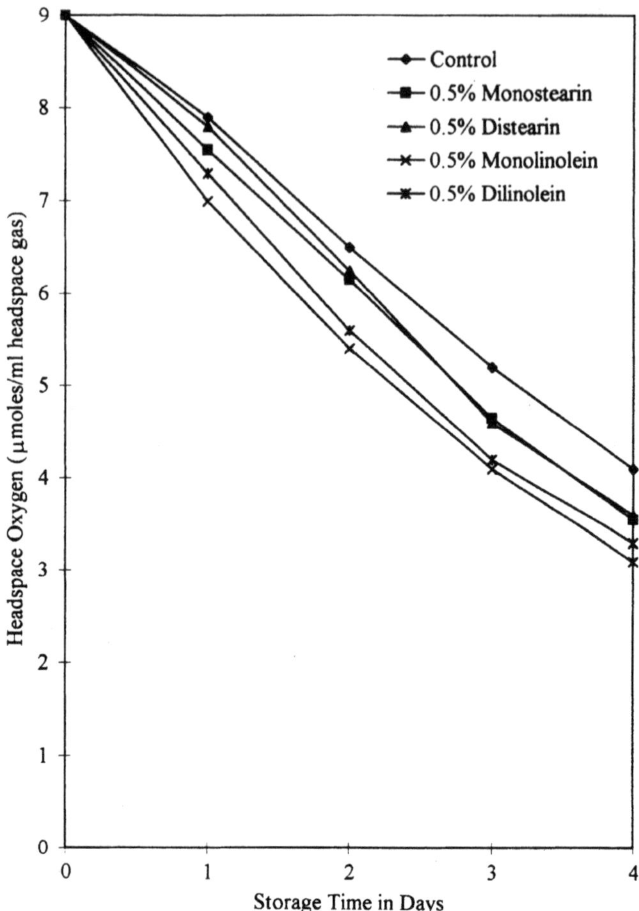

Figure 4. Effects of monostearin, distearin, monolinolein, and dilinolein on the headspace oxygen contents of purified soybean oil during storage at 55°C.

oxidative stability of oil often showed conflicting results. Those were attributed to the heterogeneity of the substrates and the phospholipids studied, as well as the variety of methods used. Phospholipids have been recognized as antioxidant synergists (Hildebrand, 1984; Dziedzic, 1984; Kwon, 1984; Hudson, 1981), chelating agents (Lunde *et al.*, 1977; Yoon *et al.*, 1987), prooxidants (Yoon *et al.*, 1987; Parke, 1981) or off-flavor precursors (Evans, 1954; Wilson, 1976; Love, 1974) in oil. Hudson and Mahgoub (1981) reported that PC and PE were not antioxidants, but they were powerful synergists in conjunction with tocopherols and flavonols, Kwon *et al.* (1984) also reported that phospholipids and tocopherols showed a definite synergism of antioxidant activity in soybean oil. They further reported that phospholipids significantly improved the oxidative stability of the oil containing tocopherol.

Lunde *et al.* (1977) reported that metal ions were transferred from water into vegetable oil containing phospholipids. They found that the sequestering action in the oil strongly depended on the content of phopholipids. Jung *et al.* (1989) observed that the re-

Table 2. Rates of disappearance of headspace oxygen in purified soybean oils containing 0, 0.25 and 0.5% monostearin, distearin, monolinolein or dilinolein during storage at 55°C

Samples	Rate of disappearance of headspace oxygen (μmoles O_2/mL headspace gas)
PO (control)	1.26
0.25% Monostearin	1.36
0.25% Distearin	1.40
0.25% Monolinolein	1.40
0.25% Dilinolein	1.41
0.5% Monostearin	1.40
0.5% Distearin	1.42
0.5% Monolinolein	1.47
0.5% Dilinolein	1.44

moval of iron in soybean oil during processing was closely related to the removal of phospholipids. Jung *et al.* (1989) reported that the correlation coefficient between the removal of phospholipids and iron in soybean oil during processing was 0.99. The removal of phospholipids by degumming will reduce metals like calcium, magnesium, and iron. Yoon and Min (1987) further reported the chelating activity of phospholipids in purified soybean oil by adding iron. Figure 5 shows the effects of 300 ppm PC and/or 1 ppm ferrous ion on the formation of volatile compounds of purified soybean oil (Yoon *et al.*, 1987). The formation of volatile compounds in purified oil was greatly accelerated by the addition of 1 ppm ferrous ion. The addition of 300 ppm PC reduced the formation of volatile compounds in the purified soybean oil containing 1 ppm ion. The 300 ppm PC reduced the volatile compounds formation in the purified soybean oil by 40% during storage. These authors also reported that 300 ppm PC minimized the oxygen consumption in purified soybean oil containing 1 ppm ion (Figure 6). Table 3 shows that the addition of 300 ppm phosphatidic acid (PA) or PE to purified soybean oil containing iron minimized the volatile compounds formation. The PC, phosphatidylglycerol (PG), cardiolipin (CL) and phosphatidylinositol (PI) seemed to be less effective than PA or PE in retarding the formation of volatile compounds in the purified soybean oil containing 1 ppm ion.

Yoon and Min (1987) observed that the 300 ppm PC accelerated the formation of volatile compounds in purified oil containing no iron (Figure 5). The addition of 300 ppm PC did not greatly affect the volatile compound formation in soybean oil. Figure 6 shows that 300 ppm PC accelerated the oxygen consumption in purified oil containing no iron. Table 4 shows that CL, PG, PA, PI or PC accelerated the formation of volatile compounds in the purified soybean oil containing no iron. CL had the highest prooxidant activity in terms of formation of volatile compounds followed by PG, PA, PI and PC, in a decreasing order.

The studies showed that 300 ppm phospholipids acted as antioxidant or prooxidant depending on the presence or absence of 1 ppm ferrous ion in the sample. Phospholipids had antioxidant activity in the oil containing 1 ppm ion.

Evans *et al.* (1954) proposed that the addition of phospholipids to soybean oil prior to decolorization caused significant darkening and introduced undesirable storage flavors

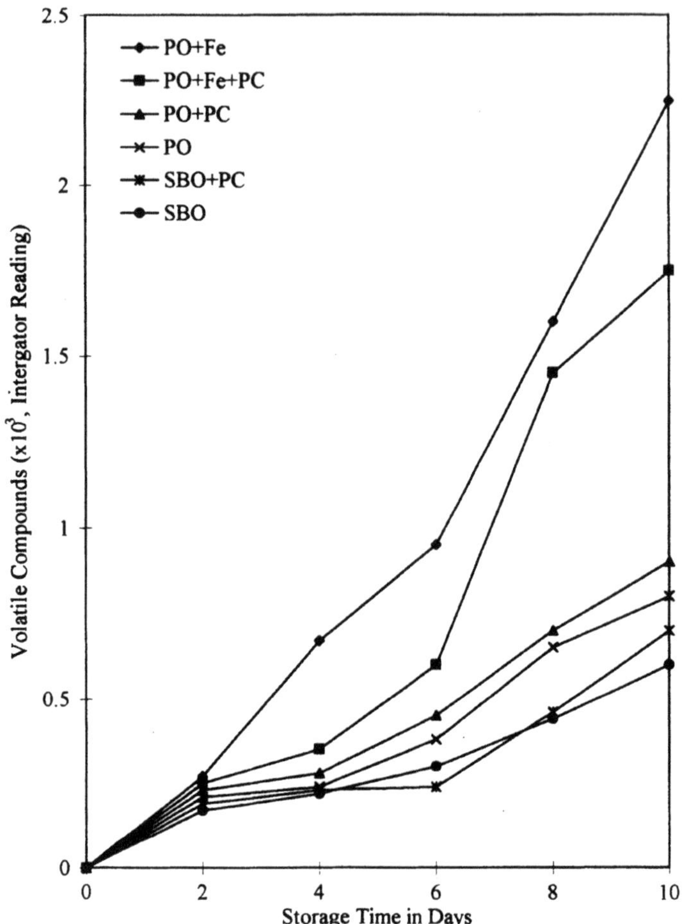

Figure 5. Effects of iron (Fe) added and phosphatidylcholine (PC) on volatile compounds formation in RBD soybean oil (SBO) and purified soybean oil (PO).

in the oil. Phospholipids play a major role in the development of warmed-over flavors in products such as poultry, mutton, beef and pork (Wilson *et al.*, 1976; Love *et al.*, 1974).

EFFECTS OF THERMALLY OXIDIZED TRIACYLGLYCEROLS ON THE OXIDATIVE STABILITY

The flavor of frying oils and fried products is mainly due to the volatile compounds which influence the aroma. Nonvolatile products may also change the taste and the mouthfeel. Billek *et al.* (Billek *et al.*, 1978) reported that RBD soybean oil contains about 1.2% oxidized triacylglycerols. The nonvolatile oxidized compounds include cyclic carbon-to-carbon linked dimers, noncyclic hydroxyl dimers, carbon-to-carbon or carbon-to-oxygen

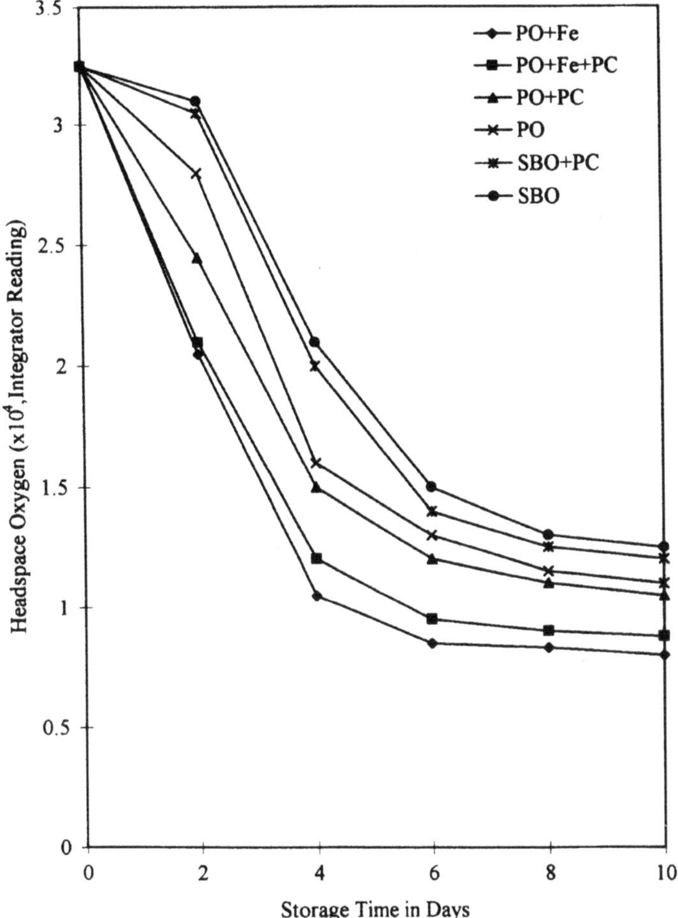

Figure 6. Effects of iron added and phosphatidylcholine on oxygen disappearance in headspace of RBD and purified soybean oil.

linkage, and trimers joined through carbon-to-carbon linkage (Billek, 1978; Paulose, 1973; Rojo, 1987; Perkins, 1967). Oxidized triacylglycerols could be formed by oxidation and polymerization of oils during processing, handling and/or storing (Paulose, 1973; Rojo, 1987; Perkins, 1967). Paulose and Chang (1973) have confirmed that 26.3% non-volatile oxidized compounds were derived from trilinolein after being heated for 72 hr at 180°C and these compounds contained hydroxyl groups, carbonyl groups and trans double bonds. Billek *et al.* (1978) also isolated 26.2% thermally oxidized compounds from soybean oil heated for 64 hr at 180°C.

Yoon *et al.* (1988) studied the effects of thermally oxidized triacylglycerols on the oxidative stability of purified soybean oil. Soybean oils were thermally oxidized in a 250 mL beaker at 180°C for 96 hr. The thermally oxidized compounds eluted from silicic acid chromatography were viscous and dark yellowish-red.

Figures 7 and 8 show that presence of thermally oxidized compounds accelerate the formation of volatiles and oxygen disappearance in the headspace of purified soybean oil,

Table 3. Duncan's multiple range test for the effects of phospholipids on volatile compounds formation (integrator reading) in purified soybean oil (PO) containing 1 ppm ferrous iron (Fe)[1]

	Storage time, days						
Sample	2	4	6	8	10	Mean[2]	Duncan's group[2]
PO	99	198	425	636	840	440	B
PO+Fe+PA	99	276	482	734	1101	538	B
PO+Fe+PE	100	247	502	800	1133	556	B
PO+Fe+PC	162	344	603	1433	1771	863	A
PO+Fe+PG	98	354	638	1479	1800	874	A
PO+Fe+CL	91	385	922	1290	1700	878	A
PO+Fe+PI	113	448	742	1608	1880	958	A
PO+Fe	151	588	930	1609	2085	1073	A

[1]Abbreviations are the same as described in the text.
[2]Means with the same letter are not significantly different.

respectively. The higher the amount of thermally oxidized compounds in the soybean oil, the greater the formation of volatile compounds and the faster the disappearance of oxygen in the oils. Duncan's multiple range tests (Table 5 and 6) show that the oxidative stability of oils containing 0, 0.5, 1.0, 1.5 and 2.0% oxidized triacylglycerols were significantly different from one another ($p = 0.05$). The results indicated that thermally oxidized compounds acted as prooxidants and decreased the oxidative stability of the purified soybean oil.

EFFECTS OF TOCOPHEROLS ON THE OXIDATIVE STABILITY

Soybean oil, cottonseed oil, and germ oil are rich in tocopherols. Commercial soybean oil contains about 1000–1500 ppm of tocopherols and the compositions of α-, β-, γ- and δ-tocopherols represent 4.0, 1.1, 67.1 and 28.8% in soybean oil, respectively (Jung *et al.*, 1989). The antioxidant activity of tocopherol has been well known at low concentrations (0.01–0.01%) (Cort, 1974; Labuza, 1969; Parkhurst *et al.*, 1968). Ke *et al.* (1977) reported that 0.1% α-tocopherols effectively inhibited the oxidation of mackerel skin oil.

Table 4. Duncan's multiple range test for the effects of phospholipids on volatile compounds formation (integrator reading) in purified soybean oil (PO)[1]

	Storage days					
Sample	2	4	6	8	Mean[2]	Duncan's group[2]
PO	99	198	425	636	340	C
PO+PE	97	200	425	638	340	C
PO+PC	108	215	482	698	376	BC
PO+PI	126	230	462	729	387	ABC
PO+PA	103	268	529	710	403	ABC
PO+PG	95	245	596	892	457	AB
PO+CL	102	280	585	907	469	A

[1]Abbreviations are the same as described in the text.
[2]Means with the same letter are not significantly different.

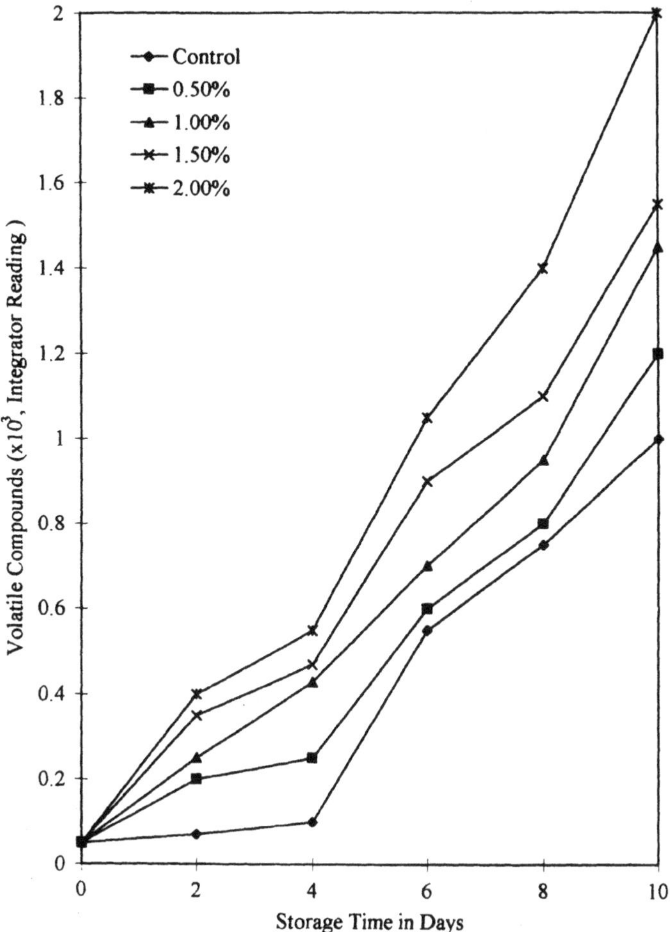

Figure 7. Effects of different levels of the thermally oxidized compounds on the volatile compounds formation in purified soybean oil during storage at 55°C.

These authors reported that antioxidant activity of 0.1% α-tocopherol was lower than the 0.02% TBHQ, but higher than 0.02% BHA or BHT. Parkhurst *et al.* (1968) further reported that the antioxidant efficiency decreased with increasing concentrations of tocopherols such that the addition of any single tocopherol above the concentration of 0.025% had little effect on oxidative stability of the oil.

High concentrations (>0.15%) of tocopherols in vegetable oils have been reported as having prooxidant activity (Lazuba, 1969; Cillard *et al.*, 1980; Frankel *et al.*, 1959). Frankel *et al.* (1959) reported that the optimum antioxidant activity of tocopherol was at a concentration of 0.04–0.06%, while higher concentrations showed antigonistic effects. Figure 9 shows that as the level of α-tocopherol in the oil increased from 0, 0.2 to 0.4%, disappearance of headspace oxygen of the oil increased.

Table 7 also shows that the mean content of volatile compounds and peroxide values of oils containing 0.2 or 0.4% tocopherol were significantly higher than those of the oil

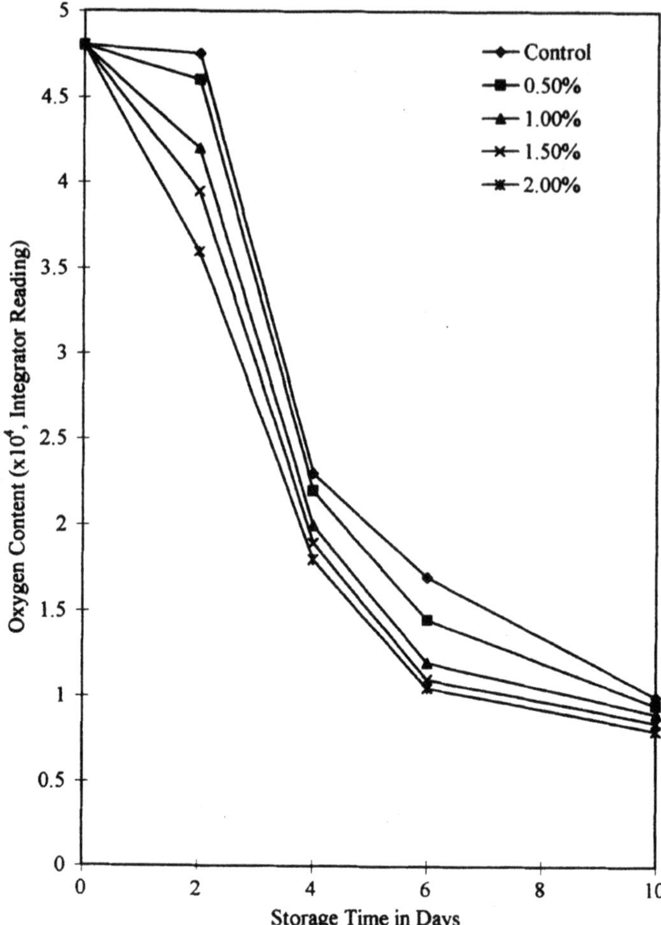

Figure 8. Effects of different levels of thermally oxidized compounds on oxygen disappearance in the headspace of purified soybean oil during storage at 55°C.

containing no tocopherol. The combined results of Figure 9 and Table 7 indicate that 0.2 and 0.4% α-tocopherol acted as a prooxidant in soybean oil.

RBD soybean oil contains about 0.11% tocopherols (Jung, 1989). Thus, further addition of tocopherol does not seem to be necessary to improve its oxidative stability. The addition of tocopherol to commercial RBD soybean oil could induce the lower oxidative stability of the oil.

Min and Jung (1992) further reported the effects of oxidized α-, γ- and δ-tocopherols on the oxidative stability of the oil. The peroxide values of purified soybean oil containing oxidized α-tocopherol were higher than those of the oil containing oxidized γ- or δ-tocopherols during storage. They also evaluated the effects of oxidized α-, γ- and δ-tocopherols on the headspace oxygen of purified soybean oils during storage at 55°C. The content of headspace oxygen of a fresh sample bottle was 20.73%. As the oxidized tocopherols increased from 0 to 100, 250, 500 and 1000 ppm, the headspace oxygen of soy-

Table 5. Duncan's multiple range test for the effects of different levels of thermally oxidized compounds on the volatile compounds formations in RBD soybean oil and purified soybean oil during 10 days storage at 55°C

Thermally oxidized compounds (%)	Volatile compounds in the headspace			
	RBD soybean oil		Purified soybean oil	
	Mean	Group[1]	Mean	Group[1]
0.0	355	A	501	A
0.5	501	B	596	AB
1.0	645	CD	740	CD
1.5	820	EF	873	E
2.0	926	F	1059	F

[1]Means with the same letter are not significantly different at p = 0.05. (Duncan's test were carried out for the RBD soybean oil and purified soybean oil, independently.)

bean oil decreased during storage. This suggests that oxidized tocopherols act as prooxidants. Furthermore, the headspace oxygen of soybean oil containing oxidized α-tocopherol was lower than that of oil containing oxidized γ- or δ-tocopherol. Therefore, it was concluded that the oxidized tocopherols act as prooxidants in soybean oil. The oxidized α-tocopherol had the highest prooxidant effect while their oxidized γ- and δ-tocopherols had lesser prooxidant effects. The prooxidant effects of tocopherols at high concentration might be due to their oxidation (Cillard *et al.*, 1980; Foot *et al.*, 1970). The prevention of oxidation of tocopherols and the removal of oxidized tocopherols during processing could improve the oxidative stability of soybean oil.

EFFECTS OF CHLOROPHYLL ON THE OXIDATIVE STABILITY

Chlorophyll has been reported to play an important role in the generation of singlet oxygen in model systems (Foot *et al.*, 1970; Endo *et al.*, 1984). Fakourelis and Min (1987) reported the effects of different levels of chlorophyll and β-carotene on the oxidative sta-

Table 6. Duncan's multiple range test for the effects of different levels of thermally oxidized compounds on the headspace oxygen disappearances in RBD soybean oil and purified soybean oil during 10 days storage at 55°C

Thermally oxidized compounds (%)	Oxygen contents in the headspace			
	RBD soybean oil		Purified soybean oil	
	Mean	Group[1]	Mean	Group[1]
0.0	29858	A	23862	A
0.5	24027	B	23164	B
1.0	22843	BC	20802	CD
1.5	20852	CD	19296	DE
2.0	19341	DE	18205	E

[1]Means with the same letter are not significantly different at p= 0.05 (Duncan's test were carried out for the RBD soybean oil and purified soybean oil, independently)

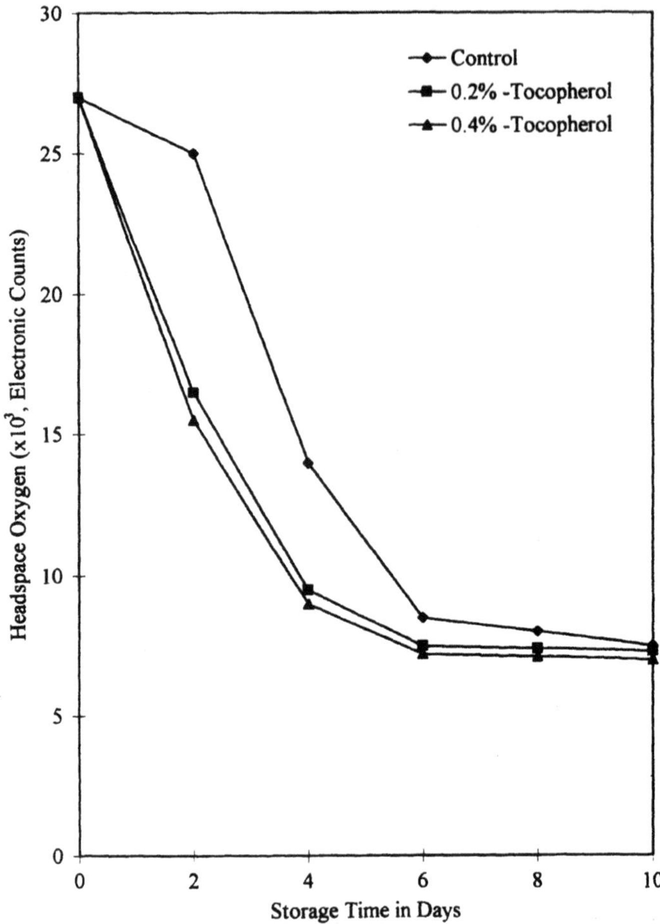

Figure 9. Effects of α-tocopherol on headspace oxygen content of soybean oil during storage at 60°C.

Table 7. Duncan's multiple range test for the effects of α-tocopherol on the headspace oxygen, volatile compounds and peroxide values or soybean oil and during storage at 60°C

	Oxygen[a]		Volatile compounds[1]		Peroxide values	
α-tocopherol (%)	Means	Duncan grouping	Means	Duncan grouping	Means	Duncan grouping
0.0	15120	A	508	A	8.6	A
0.2	12421	B	524	AB	11.9	B
0.4	12074	B	537	B	12.9	B

Means with the same letter are not significantly different ($p = 0.05$).
[1]Units in Electronic Counts.

Table 8. Effects of chlorophyll on the peroxide value and headspace oxygen of purified olive oil under light at 25 °C

Storage time (hr)	PV (meq/kg oil) Chlorophyll (ppm)			Headspace oxygen (μmoles O_2/mL headspace) Chlorophyll (ppm)		
	0	2	4	0	2	4
0	0.00	0.00	0.00	9.51	9.51	9.51
3	0.60	3.95	4.72	9.47	9.10	8.93
6	0.70	4.68	6.65	9.44	8.79	8.33
10	1.15	5.65	7.10	9.34	8.48	8.14
16	1.95	7.10	8.90	9.33	7.98	7.58
28	2.10	8.90	11.60	9.03	7.10	6.52
40	3.15	11.30	13.80	8.67	6.43	5.92
52	3.72	13.75	15.60	8.61	5.29	5.34
70	4.50	14.75	17.20	8.35	5.23	4.55

bility of purified olive oil during light storage at 25°C. Table 8 shows that the addition of chlorophyll resulted in higher peroxide values and greater oxygen disappearance in the headspace of the purified oil during storage. Thus, chlorophyll added to the purified oil acted as prooxidant when stored under light. It was further shown that different levels of chlorophyll added to the purified oil did not act as prooxidant when stored in the dark. These observations agree with the reports on chlorophyll which acted as a sensitizer to generate singlet oxygen in the photooxidation in model systems (Foot *et al.*, 1970; Endo *et al.*, 1984).

EFFECTS OF PEROXIDE ON THE OXIDATIVE STABILITY

Hydroperoxides are formed during processing, transportation and storage of fats and oils (Paulose and Chang, 1973; Billek *et al.*, 1978). The formation of hydroperoxide as primary oxidation products may affect oxidation of lipids via autoxidation, enzymatic and photosensitized oxidation (Labuza, 1971).

Mistry and Min (1988) studied the effects of peroxides on the content of oxygen and volatile compounds in the headspace of soybean oil during storage. As the peroxide value of soybean oil increased from 0, 2, 4, 6, and 8 to 10, the headspace oxygen content decreased and the volatile compounds of the soybean oil increased during storage (Figure 10 and 11). The headspace oxygen of the soybean oil containing an initial PV of 10 was 7.4% compared to 12.4% for the soybean oil with an initial PV of 0 after six days storage at 55 °C (Figure 10). The results clearly indicated that the higher the peroxide contents in soybean oils, the faster the reaction between the headspace oxygen and the soybean oil. The electronic counts of headspace volatile compounds of the fresh soybean oil having a PV of 0 was 250 compared to 950 for the soybean oil with a PV of 10 after six days storage at 55 °C (Figure 11). Thus, the higher the peroxide contents in soybean oils, the larger the volatile compounds formed by the lipid oxidation. Furthermore, Duncan's multiple range test showed that as the peroxide value increased from 0, 2, 4, 6, 8 to 10, headspace oxygen decreased and volatile compounds increased significantly at $P<0.05$ (Table 9). Obviously, the peroxides of oil acted as prooxidants and decreased the oxidative stability of soybean oil.

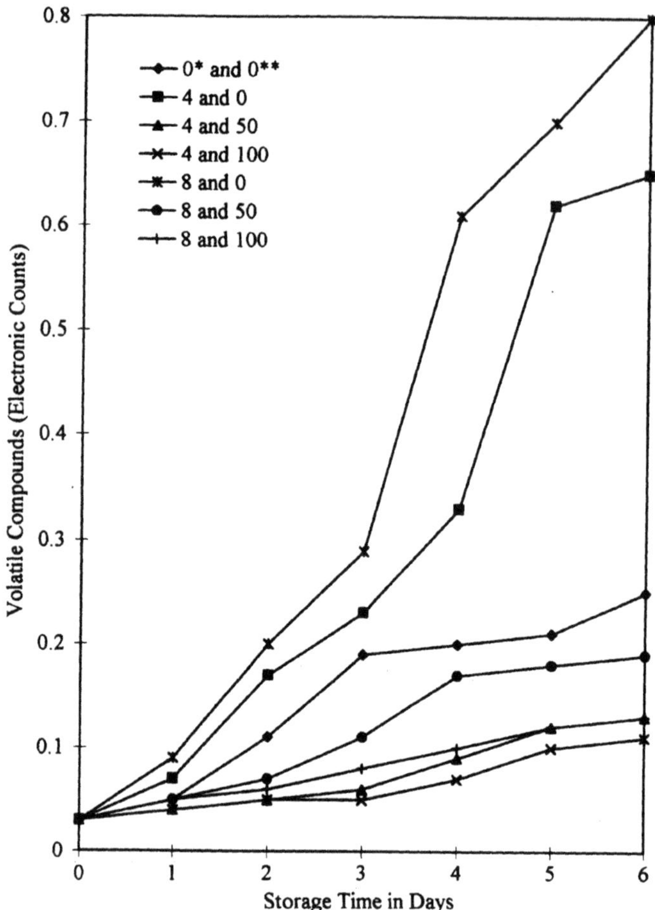

Figure 10. Effects of initial peroxide contents on the headspace oxygen contents of soybean oils during storage at 55°C.

EFFECTS OF PROCESSING ON MINOR COMPONENT CONTENTS

The effects of soybean oil processing steps on the content of unsaponifiable matter were studied by Gutfinger and Letan (1974). These authors reported that tocopherols were gradually removed during alkali refining, bleaching, and deodorizing. The sterol contents were reduced by the treatments of alkali refining and deodorizing (Table 10). The squalenes were removed primarily during the deodorizing step (Table 10).

The 90% of phospholipids in the oil removed during the degumming step. Caustic refining further removed more phospholipids but bleaching had little effect on phospholipid removal (Table 11) (List *et al.*, 1978).

List *et al.* (1978) reported the removal of phosphatide and iron by the caustic and steam refining. When they pretreated crude oil with phosphoric acid, the amount of iron

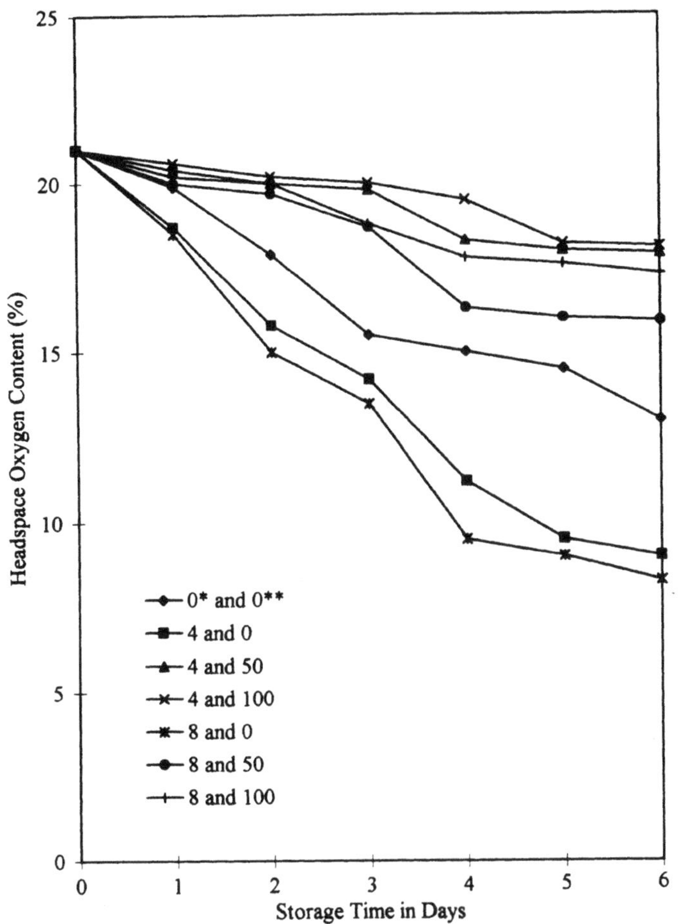

Figure 11. Effects of initial peroxide contents on the headspace volatile compounds of soybean oils during storage at 55°C.

Table 9. Duncan's multiple range test for the effects of peroxide contents on the headspace volatile compounds in soybean oils during storage at 55°C

	Headspace oxygen		Volatile compounds	
PV	Mean[a]	Group[b]	Mean[c]	Group[b]
0	15.7	A	176.7	A
2	14.5	B	243.8	B
4	13.2	C	354.2	C
6	12.8	C	397.0	D
8	12.3	CD	453.2	CD
10	11.8	D	508.7	D

[a]Means of headspace oxygen contents expressed in % after 1, 2, 3, 4, 5 and 6 days storage; [b]Means with the same letter are not significantly different at p=0.05; [c]Means of headspace volatile compounds expressed in the electronic counts of gaschromatographic peak after 1, 2, 3, 4, 5 and 6 days of storage.

Table 10. Changes in the contents of unsaponifiable components in soybean oil during processing

Soybean Oil	Contents of unsaponifiable component (μg/g)		
	Tocopherols	Sterols	Squalene
Crude	1132	3870	143
Degummed	1116	3730	142
Refined	997	3010	140
Bleached	863	3050	137
Deodorized	726	2620	97

Table 11. Changes in the contents of trace copper and iron during processing

Soybean oil	Copper (ppb)	Iron (ppm)	Phosphorus (ppm)
Crude	13.2	2.8	600.0
Degummed	3.7	2.1	58.0
Refining	2.5	0.2	13.1
Deodorized	1.1	0.2	0.2

was decreased from 1.4 ppm in the crude oil to 0.2 ppm in the degummed oil and further reduced to less than 0.1 ppm in the bleached oil. If crude soybean oils were not pretreated with phosphoric acid, their iron contents was 0.5 ppm in the degummed oil. However, the iron still can be further removed in the succeeding processing.

Sleeter (1981) reported that simple degumming was fairly effective in removing fatty acid, while most free fatty acids were removed at caustic refining step by neutralizing, alkalizing and washing. (Table 12). Likewise, the subsequent deodorizing step can further reduce the free fatty acid content of the oil.

Pritchett *et al.* (1947) reported that chlorophyll was reduced to 75% of the original amount by caustic refining. However, the effect of bleaching on the removal of chlorophyll depended mainly on the amount of earth used and the bleaching time. The addition of 0.25% bleaching earth removed 75% of chlorophyll, while a 91% reduction in chlorophyll content was noted when 0.75% bleaching earth was used. The deodorization had little effect on the removal of chlorophyll in the hydrogenated soybean oil. Hinners *et al.* (1946) also reported that the acidity of bleaching earth was critical to the removal of chlorophylls. The low pH (pH 3.04, 3.10) bleaching earth was more effective in removing

Table 12. Changes in the contents of free fatty acids during processing

Processing steps	Free fatty acid from two different runs (%)	
Crude	0.61	0.53
Degummed	0.31	0.44
Refined	0.05	0.05
Deodorized	0.02	0.03
Hydrogenated, nondeodorized	0.15	—
Hydrogenated, deodorized	0.02	0.03

Table 13. Contents of chlorophyll in soybean oil

Processing steps	Chlorophyll content (ppm)
Crude oil	0.4–3.0
Refined oil	0.1–2.7
Bleach oil (0.5% activated earth)	0.1 or less

chlorophyll than high pH (pH 7.40, 7.74) bleaching earth. The effects of processing on the content of chlorophyll in soybean oil are shown in Table 13.

REFERENCES

Billek, G.; Guhr, G.; Waibel, J. Quality assessment of used frying fats: A comparison of four methods. *J. Am. Oil Chem. Soc.* **1978**, *55*, 728–733.

Cillard, J.; Cillard, P.; Cormier, M. Effect of experimental factors on the prooxidant behavior of α-tocopherol. *J. Am. Oil Chem. Soc.* **1980**, *57*, 255–261.

Cillard, J.; Cillard, P.; Cormier, M.; Girre, L. α-Tocopherol prooxidant effect in aqueous media: Increased autoxidation rate of linoleic acid. *J. Am. Oil Chem. Soc.* **1980**, *57*, 252–254.

Cort, W.M. Antioxidant activity of tocopherols, ascorbyl palmitate, and ascorbic acid and their mode of action. *J. Am. Oil Chem. Soc.* **1974**, *51*,321–325.

Dziedzic, S.Z.; Hudson, B.J.F. Phosphatidyl ethanolamine as a synergist for primary antioxidants in edible oils. *J. Am. Oil Chem. Soc.* **1984**, *61*,1042–1045.

Endo, Y.; Usuki, R.; Kaneda, T. Prooxidant activities of chlorophylls and their decomposition products on the photooxidation of methyl linoleate. *J. Am. Oil Chem. Soc.* **1984**, *61*, 781–784.

Evans, C.D.; Gooney, P.M.; Scholfield, C.R.; Hutton, H.J. Soybean "Lecithin" and its fractions as metal-inactivating agents *J. Am. Oil Chem. Soc.* **1954**, *31*, 295–297.

Fakourelis, N.; Lee, E.C.; Min, D.B. Effects of chlorophyll and β-carotene on the oxidation stability of olive oil. *J. Food Sci.* **1987**, *52*, 234–235.

Foot, C.S.; Chang, Y.C.; Denny, R.W. Chemistry of singlet oxygen . X. Carotenoid quenching parallels biological protection. *J. Am. Chem. Soc.* **1970**, *92*, 5216.

Frankel, E.N.; Gooney, P.M.; Moser, H.A.; Cowan, J.C.; Evens, C.J.D. *Fette-Seifen-Anstrichmettel* **1959**, *61*, 1036.

Gutfinger, T.; Letan, A. Quantitative changes in some unsaponifiable compounds of soybean due to refining. *J. Sci. Fd. Agric.* **1974**, *25*, 1143–1147.

Hildebrand, D.H.; Terao, J.; Kiton, M. Phospholipids plus tocopherols increase soybean oil stability. *J. Am. Oil Chem. Soc.* **1984**, *61*, 552–555.

Hinners, H.F.; McCarthy, J.J.; Bass, R.E. The evaluation of bleaching earths. The absorptive capacity of some bleaching earths of various pH for chlorophyll in soybean oil. *J. Am. Oil Chem. Soc.* **1946**, *23*, 22–25.

Holaman, R.T.; Elmer, O.C. The rates of oxidation of unsaturated fatty acid and eaters. *J. Am. Oil Chem. Soc.* **1947**, *24*,127–129.

Holaman, R.T. In: *Progress in the Chemistry of Fats and other Lipids*; Holaman, R.T.; Lundberg, W.O.; Malkin, T., Eds.; Pergamon Press Ltd.: London, UK, 1954; Vol. II, p. 51.

Hudson, B.J.F.; Mahgoub, S.E.O. Synergism between phospholipids and naturally-occurring antioxidants in leaf lipids. *J. Sci. Food Agric.* **1981**, *32*, 208–210.

Jung, M.Y.; Min, D.B. Effects of oxidized α-, γ-, and δ- tocopherols on the oxidative stability of purified soybean oil. *Food Chem. Soc.* **1992**, *45*, 183–187.

Jung, M.Y.; Yoon, S.H.; Min, D.B. Effects of processing steps on the contents of minor compounds and oxidation of soybean oil. *J. Am. Oil Chem. Soc.* **1989**, *66*, 118–120.

Ke, P.J.; Nash, D.M.; Ackman, R.G. Mackerel skin lipids as an unsaturated fat model system for the determination of antioxidative potency of TBHQ and other antioxidant compounds. *J. Am. Oil Chem. Soc.* **1977**, *54*, 417–420.

Kwon, T.W.; Snyder, H.E.; Brown, H.G. Oxidative stability of soybean oil at different stages of refining. *J. Am. Oil Chem. Soc.* **1984**, *61*, 1843–1846.

Labuza, T.P. Kinetics of lipid oxidation in foods. *CRC Crit. Rev. Food Sci. Nutr.* **1971**, *2*, 355–405.

Lazuba, T.P.; Tsuyuki, H.; Karel, M. Kinetics of linoleate oxidation in model system. *J. Am. Oil Chem. Soc.* **1969**, *46,* 409–416.

List, G.R.; Mount, T.L.; Heakin, A.J. Steam-refined soybean oil: II. Effect of degumming methods on removal of prooxidants and phospholipids. J. Am. Oil Chem. Soc. **1978**, *55,* 280–284.

Love, J.D.; Pearson, A.M. Preparation of solutions for atomic absorption analyses of Fe, Mn, Zn, Cu in plant tissue. J. Agric. Food Chem. **1974**, *22,* 103–107.

Lunde, G.; Landmark, L.H.; Gether, J. *trans* Fatty acid content of commercial magarine samples determined by G. C. on OV-275. *J. Am. Oil Chem. Soc.* **1977,** *54,* 207–209.

Min, D.B.; Mistry, B.S. Isolation and identification of minor components and their effects on the flavor stability of soybean oil. In: *Frontiers of Flavor*; Charalambous, G., Ed.; Elsevier Scientific Publishing Co.: New York, NY, 1988.

Mistry, B.S.; Min, D.B. Effects of fatty acids on the oxidative stability of soybean oil. *J. Food Sci.* **1987**, *52,* 831–832.

Mistry, B.S.; Min, D.B. Isolation of SN-α-monolinolein from soybean oil and its effect on oil oxidative stability. *J. Food Sci.* **1987**, *52,* 786–790.

Mistry, B.S.; Min, D.B. Prooxidant effects of monoglycerols and diglycerols in soybean oil. *J. Food Sci.* **1988**, *53,* 1896–1897.

Miyashita, K.; Takagi, T. Study on the oxidative rate and prooxidant activity of free fatty acids. *J. Am. Oil Chem. Soc.* **1986,** *63,* 1380–1384.

Parke, D.W. Effects of phospholipid on the oxidative stability of soybean oil with emphasis on the use of infrared spectrophotometry as an indicator of oxidative. M.S. Thesis, The Ohio State University, 1981.

Parkhurst, R.M.; Skinner, W.A.; Sturm, P.A. The effect of various concentrations of tocopherol mixtures on the oxidative stability of a sample of lard. *J. Am. Oil Chem. Soc.* **1968,** *45,* 641–642.

Paulose, M.M.; Chang, S.S. Chemical reactions involved in deep fat frying of food: VI. Characterization of non-volatile decomposition products of trilinolein. *J. Am. Oil Chem. Soc.* **1973**, *50,* 147–154.

Perkins, E.G. Formation of non-volatile decomposition products in heated fats and oils. *Food Tech.* **1967**, *21,* 125–130.

Pritchett, W.C.; Taylor, W.G.; Carroll, D.M. Chlorophyll removal during earth bleaching of soybean oil. *J. Am. Oil Chem. Soc.* **1947**, *24,* 225–227.

Pryde, E.H. Composition of soybean oil. Handbook of Soy Oil Processing and Utilization; Erickson, D.R., Pryde, E.H., Brekke, O.L., Mounts, T.L., Falb, R.A. Eds., American Soybean Association, St. Louis, MO; American Oil Chemists Society: Champaign, IL, 1980, pp.13–31.

Rojo, J.A.; Perkins, E.G. Cyclic fatty acid monomer formation in frying fats. Determination and structural study. *J. Am. Oil Chem. Soc.* **1987**, *64,* 414–421.

Sleeter, R.T. Effects of processing on quality of soybean oil. *J. Am. Oil Chem. Soc.* **1981**, *58,* 239–247.

Sonntag, N.O.V. Reactions of fats and fatty acids. Bailey's Industrial Oil and Fat Products; Swern, D. Ed., Vol. I, 4th edn.; Wiley- Interscience Publication: New York, NY, 1979; pp.99–175.

USDA 1987 Foreign Agriculture Circular, Oilseed and Products; U.S. Dept. of Agriculture-Foreign Agricultural Service FOP, 1987; 12–87, p. 9.

Wilson, B.R.; Pearson, A.M.; Shorland, D.H. Effect of total lipids and phospholipids on warmed-over flavor red and white muscle from several species as measured by tribarbiouric acid analysis. *J. Agric. Food Chem.* **1976,** *24,* 7–11.

Yoon, S.H.; Min, D.B. Effect of thermally oxidized compounds on the flavor stability of soybean oil. *J. Am. Oil Chem. Soc.* **1986**, *61,* 425.

Yoon, S.H.; Jung, M.Y.; Min, D.B. Effects of thermally oxidized triglycerols on the oxidative stability of soybean oil. *J. Am. Oil Chem. Soc.* **1988**, *65,* 1652–1656.

ANTIOXIDIZING POTENTIALS OF BHA, BHT, TBHQ, TOCOPHEROL, AND OXYGEN ABSORBER INCORPORATED IN A GHANAIAN FERMENTED FISH PRODUCT

Toshiaki Ohshima, Vivienne V. Yankah, Hideki Ushio, and Chiaki Kiozumi

Department of Food Science and Technology
Tokyo University of Fisheries
Konan 4, Minato-ku, Tokyo 108, Japan

Raw whole fishes (Horse mackerel, Trachurus japonicus) were degutted and separately treated with antioxidants BHA, BHT, TBHQ and tocopherol before fermentation, by completely immersing the samples in 0.1% antioxidant solutions. Fish samples were then salted, fermented and dried to mark the end of processing. A portion of the control samples were packed in an oxygen absorber during storage for 2 months at room temperature to study the effect of oxygen scavenging on lipid oxidation. The moisture content and total lipid decreased with processing and storage. An inductive effect of the fish oil was observed with 2-thiobarbituric acid values and the formation of lysophos-phatidylcholine (LPC) for TBHQ and BHA treated fish during the fermentation process. Free fatty acid formation was detected in all fish samples throughout processing and storage. Residual antioxidant concentrations decreased with processing and storage. For samples treated with tocopherol, only *-tocopherol was detected at the end of 2 months storage. Although TBHQ showed the best antioxidative effect during processing, it was the fastest synthetic antioxidant to be depleted. However, relatively high levels of BHA were present in the sample after 2 months storage. Red color patches, suspected to be antioxidant degradative products, were observed around the operculum of TBHQ treated samples after processing.

INTRODUCTION

Recognition of the role of ω-3 fatty acids of fish in human nutrition has caused an upsurge of interest among consumers and a considerable expansion of the seafood indus-

Process-Induced Chemical Changes in Food
edited by Shahidi *et al.* Plenum Press, New York, 1998

try (Vondruska *et al.*, 1988). However, these desirable lipid sources are most vulnerable to oxidative destruction when some processing and storage methods such as heat applications, fermentation, and drying are used to prevent rapid deterioration of fish products. Lipid oxidation is, therefore, a major problem facing the fish processing industries because it results in physical deterioration, a reducing of the organoleptic properties of the food and the generation of potentially toxic oxidation compounds (Gurr, 1984).

Food antioxidants have gained an increasingly important role in the food industry in the continuing trend towards new forms of food processing and for long-term food storage. The different classes and corresponding levels of antioxidants allowed in foods are, however, subject to compliance with the governing food laws of different countries to ensure quality assurance (Kochhar and Russell 1990).

Fermented fish products, though very essential in most developing countries as the main source of protein in the diet, are still produced with little application of the modern technologies needed for product optimization and standardization. Most fermented fishery products are made from fatty fish (Essuman 1992). In previous reports, it has been shown that extensive oxidation occurred when methods as salting, fermentation and drying were applied in fish processing (Yankah *et al.*, 1993, 1996). In this study, antioxidant treatments including butylated hydroxyanisole (BHA), butylated hydroxytoluene (BHT), *tert*-butylhydroquinone (TBHQ), and tocopherols were applied to fish before processing, and then an oxygen absorber was used during storage; their efficacy to prevent lipid oxidation was evaluated.

EXPERIMENTAL

Sample Preparation

The horse mackerel *(Trachurus japonicus)* purchased from a local market was degutted and cleaned, and then separated into 5 groups representing control, and the four different batches for testing. Fish samples were immersed in their respective antioxidant solutions for different time periods.

BHA, BHT and TBHQ solutions were prepared by mixing a homogenized preparation of 0.75 g emulsifier (Sangen Company Ltd, Japan) dissolved in water at 70°C and cooled to room temperature with 1.5 g antioxidant in 1.5 g triacylglycerol. The mixture was homogenized at 1000 rpm to uniformity to give a final composition of 0.1% antioxidant and 0.5% emulsifier. Fish samples were immersed in the antioxidant solution for 1 h.

A solution of α-, γ- and, δ-tocopherols mixtures at a ratio of 0.78 : 4.3 : 4.8 with trace quantities of β-tocopherol making up a 10% based solution was used (Eisai Company Ltd.). Ten grams of the mixture were weighed and dissolved in 1000 mL distilled water. Fish samples were immersed in the solution for 3.5 h.

The 5 groups of fish were separately processed by salting with 15% refined domestic salt (w/w) and fermented at room temperature for 3 days. Fish were then sun-dried for 3 days to mark the end of processing. After drying, the control sample was divided into two groups, one part was packed in a cobalt based oxygen absorber, Ageless (Mitsubishi Gas Chemical Co., Japan), to study the effects during storage, while the other part was used to continue the study of the control sample. All samples were stored at room temperature for 2 months. Three pieces of fish taken from each sample treatment at the different stages of processing and storage were minced and analysed in triplicates.

Chemical Analysis

Moisture content determination was done according to the AOAC method (1990). Total lipids (TL) were quantified by the Bligh and Dyer method (1959) and classified into polar lipid (PL) and nonpolar lipids (NL). NL analysis was carried out according to the method of Ohshima and Ackman (1991), PL were analysed using the method of Ratnayake and Ackman (1985). TBA values of minced fish meat were measured using the method of Sinnhuber and Yu (1972).

Determination of Antioxidant Concentrations in Fish Samples

The antioxidant concentration of the fish was determined on its TL. The antioxidant in TL was extracted by shaking with acetonitrile and allowing the phases to separate. The upper acetonitrile layer was separated and filtered using a Millipore filter, then injected to an HPLC equipped with a spectrofluorometric detection (Yankah *et al*; submitted for publication).

The tocopherol concentrations in fish samples were determined by the AOCS method (1992). Briefly, extracted TL were dissolved in *n*-hexane to a known concentration and injected to the HPLC system equipped with a Shimadzu RF 550 spectrofluorometric detector. The column was a Lichrosorb Si 60 (4 mm id x 250 mm, Merck, Darmstadt, Germany). A mixture of propanol and *n*-hexane (5:95, v/v) was used as the mobile phase at a flow rate of 1.2 mL/min. Tocopherol concentrations were intrapolated from standard curves.

Statistical Analysis

Data were analysed using the Statgraphics software (STCC Inc., Rockville, Maryland). Multifactor analysis of variance, and multiple range analysis (Least significant difference, LSD) were done to determine the sources of variation in the data and the effects of antioxidant types, processing treatments and storage on the oxidation of the fish products. Tests of significance were done at the 99% confidence level ($p < 0.01$).

RESULTS AND DISCUSSION

Moisture

Change in moisture content of salted fermented fish during processing and storage are shown in Figure 1. Moisture values varied significantly at each stage of processing and storage from initial values ranging between 76 and 79 g/100 determined in raw fish. Though the changes were similar in the control during processing, during storage the oxygen absorber sample had significantly different moisture values than all the other samples ($p < 0.01$). This can be explained because of the use of oxygen absorber which eliminated some of the moisture. Moisture changes were similar in the control and TBHQ-treated samples during storage ($p < 0.01$).

Total Lipid

The total lipid composition of fermented fish is shown in Figure 2. The total lipid composition of raw fish samples decreased significantly during processing ($p < 0.01$). The

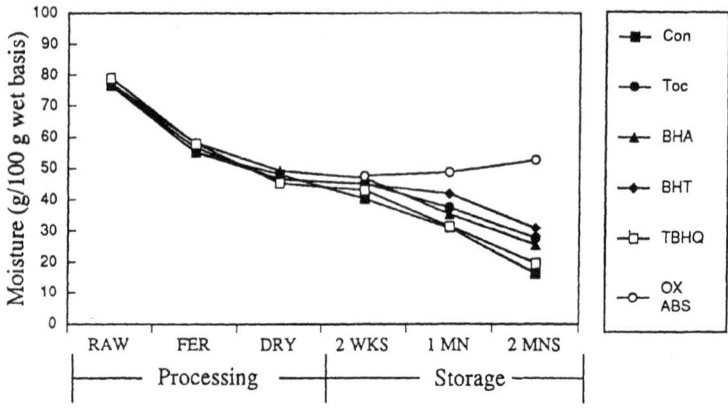

Figure 1. Changes in moisture content of salted fermented fish during processing and storage.

observed decrease could be due to losses from drip during fermentation, and the loss of subcutaneous fat during drying. At the end of processing, the lipid compositions of the various antioxidant treated samples were similar except for those containing tocopherol. After 2 months storage, however, significant differences were observed in the TL composition of the fish samples.

Residual Antioxidant Concentration and TBA Values

The amount of antioxidant in fish samples during processing and storage is shown in Figure 3. The control raw sample had an α-tocopherol content of 109 mg/kg TL. This value decreased to 2.32 mg/kg TL after fermentation and was not detected during storage. Tocopherol (α-, γ- and δ-) incorporated in the fish were at 256, 155 and 175 mg/kg, respectively. After fermentation, α- and γ-tocopherol were decreased to 3.98 and 13.9 mg/kg respectively, and were not detected during storage. δ-Tocopherol concentration de-

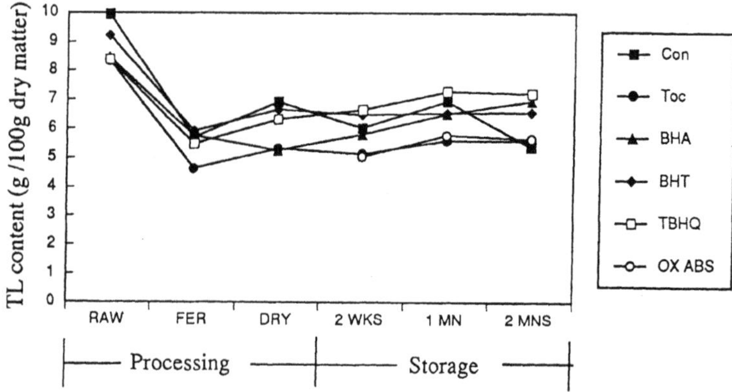

Figure 2. Changes in total lipid content of salted fermented fish during processing and storage.

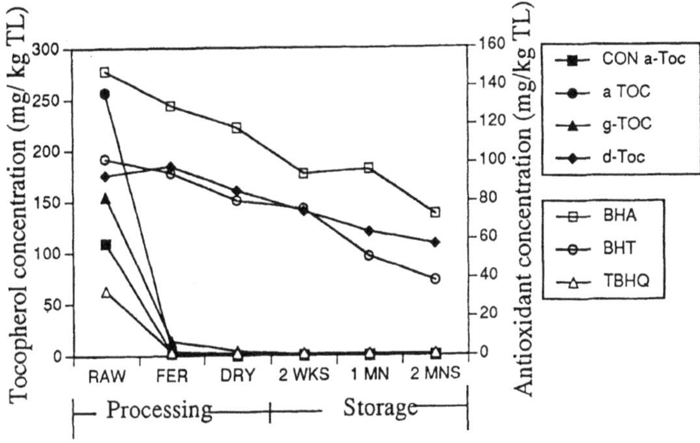

Figure 3. Residual antioxidant concentration in salted fermented fish during processing and storage.

creased with processing and storage, 108 mg/kgTL was present in the samples after 2 months storage.

Among the synthetic antioxidants, TBHQ concentration was the first to deplete in the samples from 132 mg/kg TL to 10.6 mg/kg TL after fermentation, and during storage only trace quantities of TBHQ were detected. Meanwhile, BHT concentrations decreased from 102 mg/kg TL in raw fish to 38.5 mg/kg TL after 2 months storage. The antioxidant least affected by processing and storage was BHA which decreased from 148 to 73 mg/kg TL after 2 months storage.

The TBA values showed that after fermentation there was an increase in TBA in all the samples studied except for those containing BHA and TBHQ (Figure 4). At the end of processing, the control sample had the highest TBA value at 152 mg MA/kg, about five times the value obtained for BHA or TBHQ samples after processing. Although TBHQ levels in the samples were decreased to trace quantities after processing, the TBA values

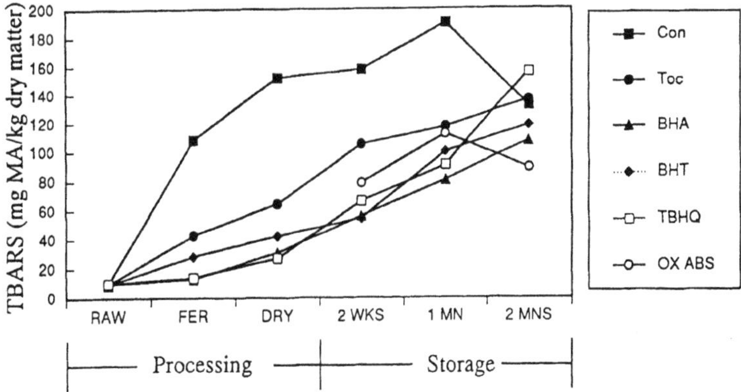

Figure 4. Changes in TBARS of salted fermented fish during processing and storage.

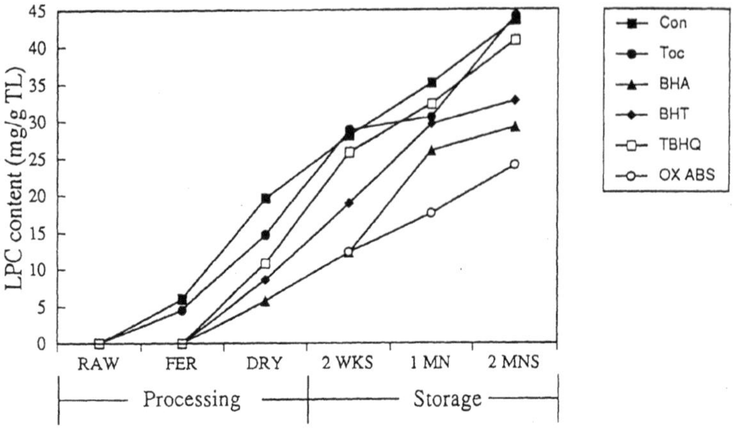

Figure 5. Changes in lyso-phosphatidyl choline (LPC) content of total lipid in salted fermented fish during processing and storage.

were comparable to those obtained for BHA- and BHT- incorporated samples at the end of one month storage. The degradation products of most antioxidants, including BHA, BHT, TBHQ and tocopherols, have been reported to show some antioxidant activity. The degradation products of TBHQ in particular have been observed to possess antioxidant activity higher than that of the parent TBHQ, depending on the substrate oil used in the study (Kikugawa *et al.*, 1990). At the end of storage, however, TBHQ samples had the highest TBA value. The oxygen absorber-packed samples had lower TBA values than the control samples during storage. This suggests that oxygen absorber has a good potential to reduce lipid oxidation in fish during processing and storage.

The formation of lysophosphatidylcholine (LPC) in salted fermented fish samples during processing and storage is shown in Figure 5. No LPC was detected in the raw fish and the synthetic antioxidant-treated samples after fermentation. However, increased LPC concentrations were observed in all samples until the end of storage. An induction period for the formation of LPC in the synthetic antioxidant treated fish was observed until after fermentation, then increases in LPC were found in all samples throughout processing and storage. High levels of LPC are usually found in salted and dried fish products since lysophospholipase, catalysing a further degradation of LPC to glycerophosphorylcholine and free fatty acid is inhibited at higher salt concentrations (Ohshima *et al.*, 1986).

The release of non-esterified fatty acids which can occur during processing and storage, normally by hydrolytic breakdown by lipases, is often a preliminary to their oxidative breakdown (Gurr 1984). The formation of free fatty acids in the fish samples during processing and storage is shown in Figure 6. The level of FFA in the extracted TL is normally used as a measure of lipid hydrolytic activity (Bligh *et al.*, 1988). Lipid hydrolysis by itself has no nutritional significance, but the accumulation of FFA in fish oils is undesirable due to the secondary reactions catalysis, such as increased susceptibility to oxidation and the consequent development of off-flavors (Lovern, 1962).

The lack of convincing information on biological effects of synthetic antioxidants and their degradation products has limited their wide application in food products. Evi-

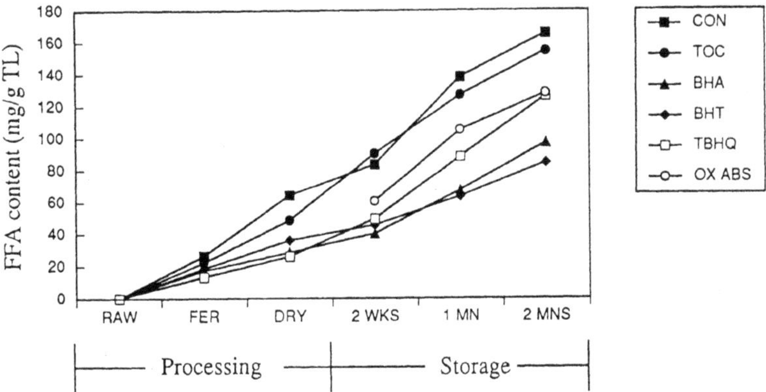

Figure 6. Changes in free-fatty acid (FFA) content of total lipid in salted fermented fish during processing and storage.

dence from reports on the effects of antioxidants on fish oils have however indicated that synthetic antioxidants including BHA, BHT and TBHQ, if administered at the recommended concentrations, have a relatively better antioxidative effect on food oils as compared to natural antioxidants such as tocopherols (Kaitaranta, 1992). In most cases, TBHQ has been considered the most effective antioxidant (Tatum and Chow 1992; Hawrysh *et al.*, 1990). However, in this work, the observation of red color patches on the TBHQ treated fish will limit its application on whole fish samples.

Antioxidants and Lipid Analysis

Synthetic antioxidants BHT and TBHQ did not affect the color development in TBA analysis. With the analysis of extracted TL to lipid classes of nonpolar lipid (NL) and polar lipid (PL) fractions, the antioxidants did not interfere with the PL analysis. With NL analysis, the peak for TBHQ overlapped with that of sterol and BHT overlapped with triacylglycerols. Peaks for BHA and free fatty acids (FFA) were almost inseparable. However, because the levels of antioxidants measured in the fish TL are very low, the effect of BHA in the determination of FFA in fish samples is expected to be negligible.

CONCLUSION

The application of antioxidants BHA, BHT, TBHQ, and tocopherol may help reducing lipid oxidation in fish during processing and storage. The lower TBA values, and LPC and corresponding FFA levels measured during processing, indicate that TBHQ is the most effective antioxidant. However, formation of red color pigments on fish due to degradation of the antioxidants is not desirable for its use on whole fish samples. BHA could be used during processing, together with oxygen absorber packing during storage in order to reduce lipid oxidation in fermented fish during processing and storage.

REFERENCES

AOAC. Official Methods of Analysis. 15th ed. Association of Official Analytical Chemists, Washington, DC, 1990.

AOCS Official Method. Determination of tocopherols and tocotrienols in vegetable oils and fats by HPLC. AOCS Official Methods and Recommended Practices. Champaign, IL, **1**, 1992.

Bligh, E.G.; Dyer W.J. A rapid method of total lipid extraction and purification. *Can. J Biochem. Physiol.* **1959**, *37*, 911–917.

Bligh, E.G.; Shaw, S.J.; Woyewoda A.D. Effects of drying and smoking on lipids of fish. In *Fish smoking and drying-The effect of smoking and drying on the nutritional properties of fish.* Burt, J.R. (Ed.), Elsevier Applied Science: New-York, 1988; pp. 41–52.

Essuman, K.M. Fermented fish in Africa. A study on processing, marketing and consumption. FAO Fisheries Technical Paper 329. Rome, 1992.

Gurr, M.I. Determination of the amounts and types of fats in foods. In *Role of fats in food and nutrition,* Gurr, M.I. (Ed.) p. 43–57, Elsevier Applied Science: London, 1984; pp. 43–57.

Kaitaranta, J.K. Control of lipid oxidation in fish oil with various antioxidative compounds. *J. Amer. Oil Chem. Soc.* **1992,** *69*, 810–813.

Kikugawa, K.; Kunugi, A.; Kurechi, T. Chemistry and implications of degrdation of phenolic antioxidants In *Food Antioxidants,* Hudson, B.J.F. (Ed.), Elsevier Applied Science, New-York, 1990; pp. 65–98.

Kochhar, S.P.; Russell, J.B. Detection, estimation, and evaluation of antioxidants in food systems. In *Food Antioxidants,* Hudson, B.J.F. (Ed.),Elsevier Applied Science: NewYork, 1990; pp. 19–64.

Lovern, J.A. The lipids of fish and changes occurring in them during processing and storage. In *Fish in Nutrition,* Heen, E.; Kreuzer R. (Eds.), Fishing News Books: London, 1962; pp. 86–111.

Ohshima, T.; Wada, S.; Koizumi, C. Lipid deterioration of salted gonads of sea urchin during storage at 5. *Bull. Japan Soc. Sci. Fish.* **1986,** *52*, 511–517.

Ohshima, T.; Ackman, R.G. New developments in chromarod/iatroscan TLC-FID: Analysis of lipid class composition. *J. Planar Chromatogr.* **1991**, *4*, 27–34.

Ratnayake, W.M.N.; Ackman R.G. Rapid analysis of canola gum lipid composition by Iatroscan thin layer chromatography-flame ionization detection. *Can. Inst. Food Sci. Technol. J.* **1985**, *18*, 284–289.

Sinnhuber R.O.; Yu T.C. The 2-thiobarbituric acid reaction, an objective measure of the oxidation deterioration occurring in fats and oils. *Yukagaku* **1977**, **26**, 259–267.

Tatum, V.; Chow C.K. Effects of processing and storage on fatty acids in edible oils. In *Fatty acids in foods and their health implications.* Chow, C.K. (Ed.), Marcel Deckker Inc.: NewYork, 1992; pp. 337–351.

Vondruska, J.; Otwell, W.S.; Martin R.E. Seafood consumption, availability and quality. *Food Technol.* **1988**, *45*, 168–172.

Yankah, V.V.; Ohshima, T.; Koizumi C. Effect of processing and storage on some chemical characteristics and lipid composition of a Ghanaian fermented fish product. *J. Sci. Food Agric.* **1993**, *63*, 227–235.

Yankah, V.V.; Ohshima, T.; Ushio, H.; Fujii, H.; Koizumi, C. 1996. Study of the differences between two salt qualities on microbioligy, lipd and water-extractable components of momoni, a Ghanaian fermented fish product. *J. Sci. Food Agric.* **1996**, in press.

16

MINIMIZING PROCESS INDUCED PROOXIDANT STRESSES

R. J. Evans and T. S. Jones

Kalsec, Inc.
1127 Buchelew Drive
Kingsport, Tennessee 37663

The adverse effects of trace metals, heat, steam and other conditions encountered in food processing relate to the acceleration of development of rancidity. Measures to retard oxidation of lipids, vitamins, pigments and proteins include elimination of prooxidants, removal of oxygen and use of blends of inhibitors formulated for specific substrates.

INTRODUCTION

Optimization of the oxidative stability of stressed foods requires examination of processing techniques and parameters to identify desirable changes. It is also necessary to review some of the often unrecognized negative changes in properties of processed foods that are the result of oxidation of lipids and proteins, pigments and flavor components induced by process conditions. In addition, measures should be taken to offset the adverse process influences.

An analogy may be helpful in following concepts presented in this contribution. Fire prevention techniques teach that the elements essential to supporting combustion are oxygen, combustibles and an ignition source. Removal of any one of the three would lead to the fire being extinguished. Rancidity, nutritional losses and color fading in processed foods and muscle foods can be the result of slow rate of oxidation and the same factors are important in managing oxidation in foods.

OXYGEN

Oxygen levels in foods can be reduced in several ways. Sparging of bulk vegetable oils or liquefied fats is helpful. Oxygen can be absorbed in sealed packages by inclusion

Process-Induced Chemical Changes in Food
edited by Shahidi *et al.* Plenum Press, New York, 1998

Table 1. Resistance of selected lipids to oxidation

Cocoa butter	most stable
Coconut oil	↓
Palm kernel oil	
Palm oil	
100 IV soy oil	
Peanut oil	
Cottonseed oil	
Corn oil	
Soybean oil	
Butterfat	
Marine oils	least stable

of ascorbic acid or pouches containing oxygen absorbers. Blanketing fryers with an inert gas has also shown benefit. Inert atmosphere packaging is popular. Packaging techniques that reduce oxygen levels below 1% in further processed foods are expensive, especially if dissolved oxygen is present. Near total oxygen free environments may promote anaerobic bacterial growth.

LIPID CONTENT

Elimination of lipids in grains, meats and many other foods is difficult. Oxidation of less than 1% of the molecules in the lipid can result in detectable rancidity. Table 1 lists susceptibility of various fats to oxidation (Berger, 1989).

For increased shelflife, one should select a tropical fat but the U.S. press has portrayed a negative image for this type of oil. Partially hydrogenated soybean oil has been a good compromise but *trans*-fatty acid concerns in such oils is a problem. Lowering fat levels is desired, but it is difficult to make popular foods palatable without fats. The new technology in nutritionally designed foods is highlighting certain polyunsaturated fatty acids as essential which in turn poses unique stability problems. From a marketing point of view presence of fat in products is essential and consideration of health benefits may lead to the selection of sources from the lower part of Table 1.

INITIATION

Prooxidants, in essence, act as igniters and affect initiation of oxidation (Table 2). Flavor, aroma, color, viscosity, texture and nutritional value can suffer as a result of oxidation, but these deteriorations often are not recognized as oxidation problems.

Specifications on ingredients can be tightened to minimize incoming prooxidants. Process modifications such as pH adjustment, water deionization and enzyme deactivation should be considered. The shorter wavelengths light can induce the formation of singlet oxygen, an aggressive molecule, that directly attacks even monounsaturated fats (Min *et al.*, 1989). Therefore, skylights or florescent lamps in processing areas might be of concern. Complete elimination of prooxidants is challenging. Gupta (1993) has detailed the problems in achieving this in a simple oil but it is much more difficult in a formulated food. Common ingredients

Table 2. Prooxidants and their food sources

Prooxidant	Source
Energy	heat, UV (sunlight fluorescent), irradiation
Iron	blood and muscle tissue, chlorophyll, pipe, tanks, water, salt
Copper	gaskets, valves, pumps, vessels, process water
Enzymes	grains, muscle tissue, dairy products
Chlorine	cleaners, sanitizers, tap water
Fluorescents	riboflavin in supplements & cheese, FD&C Yellow #5
Catalysts	residual in some synthetic lipids, esters
Alkaline pH	
Acids	excessive levels of ascorbic, etc.
Time	delays in refining, addition of inhibitors, etc.
Water	
Fatty Acids	
Ozone	

such as grains and dairy products contain enzymes. Heating destroys some enzymes but may activate others. Salt may introduce troublesome trace metals. The residual metal catalyst in hydrogenated oils or emulsifiers must be considered. Vessels, pumps and lines must be of the right grade of stainless steel. Metals do catalyze oxidation; Pokorny (1987) has listed the relative rates of activity of various metals (Table 3).

Iron and other transition or divalent metal ions react with polyunsaturated fats, abstracting an electron or a hydrogen atom (see Equation 1). The reaction of a polyunsaturated lipid, LH, with a prooxidant such as a transition metal ion, X, generates a lipid free radical, $L^{\bullet}$. The initiation of oxidation by metals correlates to the ignition source, the match, in the combustion triangle.

$$LH + X \rightarrow XH + L^{\bullet} \tag{1}$$

The lipid free radical then reacts preferentially with the available oxygen in subsequent reactions to form an alkylperoxy free radical, $LOO^{\bullet}$ (Equation 2).

$$L^{\bullet} + O_2 \rightarrow LOO^{\bullet} \tag{2}$$

The peroxy free radical is capable of abstracting a hydrogen atom from another lipid molecule, creating a new lipid free radical and a hydroperoxide, LOOH. The hydroperox-

Table 3. Relative prooxidant activity of metals

Metal	Relative Activity
Aluminum	1.0
Inconel	1.3
Hastelloy	1.5
Nickel	1.7
Stainless steel T316	1.8
Stainless steel T304	2.2
Mild steel	3.1
Cooper	7.5

ide will eventually decompose creating additional radicals thereby starting the chain reaction of oxidation (Equation 3).

$$L_1OO^{\bullet} + L_2H \rightarrow L_1OOH + L_2^{\bullet} \quad (3)$$

When lipid free radicals are generated, they not only promote the chain reaction of oxidation in fats and oil, but may also attack proteins, some pigments, and vitamins. The catalytic metals can be deactivated by the use of a chelating agent such as citric acid. However, lipid free radicals created before the addition of citric acid are capable of maintaining the chain reaction of oxidation. Removal of the prooxidants is challenging but can be rewarding.

QUENCHERS

Minimizing oxygen and prooxidants in a retail oil is not too difficult. In a formulated food, it is very difficult and often impractical. The important point is that by minimizing prooxidants, quality benefits. Oxidation will continue but is significantly retarded. Use of a synthetic antioxidant or natural extracts with antioxidant properties, such as tocopherol, rosemary, sage and tea, can be very helpful. Use of these free radical interceptors (FRIs) does not ordinarily prevent the oxidation reaction of Equation (2) since kinetics favor reaction of the lipid free radical with oxygen rather than with the FRI.

However, the chain reaction of Equation (3) can be quenched (extinguished) by FRIs. The reaction of an FRI, AH, with the alkylperoxy radical $ROO^{\bullet}$ forms the relatively stable hydroperoxide, ROOH, and a relatively stable interceptor radical $A^{\bullet}$ (Equation 4).

$$AH + ROO^{\bullet} \rightarrow ROOH + A^{\bullet} \text{ [or A]} \quad (4)$$

While the FRI functions by forming a hydroperoxide, it eliminates the chain reaction initiating effect of the peroxyl radical. Peroxide value (PV) determinations measure both. The singlet oxygen resulting from light energy can be quenched by carotenoids present in carrot, paprika and several other natural pigments.

FRI SUBSTRATE SPECIFICITY

The relative effectiveness of any food grade FRI varies with the composition of the substrate. Figure 1 shows induction times of two formulated FRI blends in Norwegian herring oil from different sources. The reverse relationship of the FRI effectiveness reinforces their substrate specific nature. Induction times and iodine values of control samples of the oils were similar. Formulating the most effective FRI blend requires knowledge of the lipid fatty acid composition and impurities present.

PROCESS STRESS

Heat encountered in food processing poses severe stress on its lipids. It accelerates the loss of volatile components in flavors, naturally present FRIs, vitamins and color. Fig-

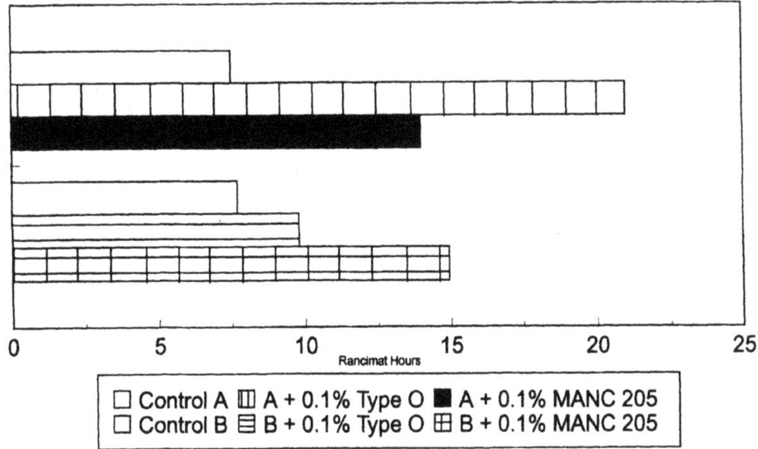

Figure 1. Herring oil from two different sources, A and B, had similar Iodine Values. The controls had similar Rancimat induction times. A rosemary extract, Herbalox Seasoning, Type O, was superior to a multi-component inhibitor blend, Duralox blend, MANC 205, in sample A. The reverse was observed in sample B.

ure 2 reflects the rapid loss of synthetic antioxidants BHA and TBHQ due to heat, and at a much slower rate the natural inhibitors, tocopherol and oleoresin rosemary.

Steam generated in food extrusion or use of injected steam to heat foodstuffs may compound the problem. Steam accelerates the stripping of lower molecular weight antioxidants out of foods. The use of superheated steam in the deodorization of vegetable oils reduces the content of naturally occurring tocopherol. Conditions can be controlled to optimize residual tocopherol levels in refined oil. Steam generated in extrusion, baking and deep fat frying may strip many of the added or naturally occurring FRIs from foods, thus reducing their shelflife.

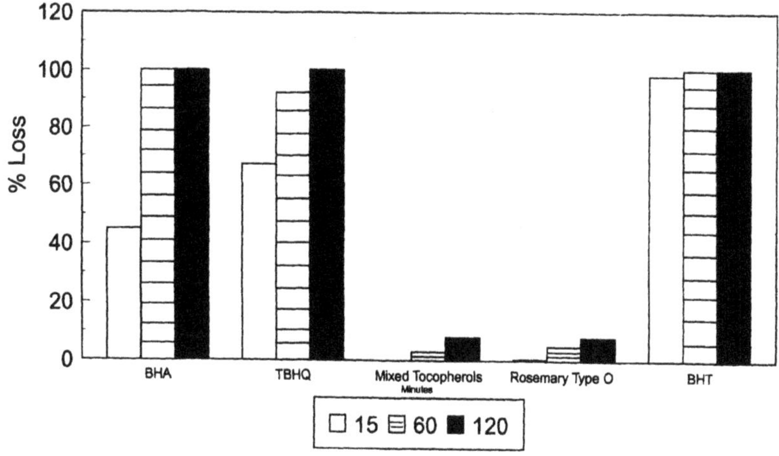

Figure 2. Various FRIs were held in open vessels at 200° C. Loss in weight versus exposure time is shown.

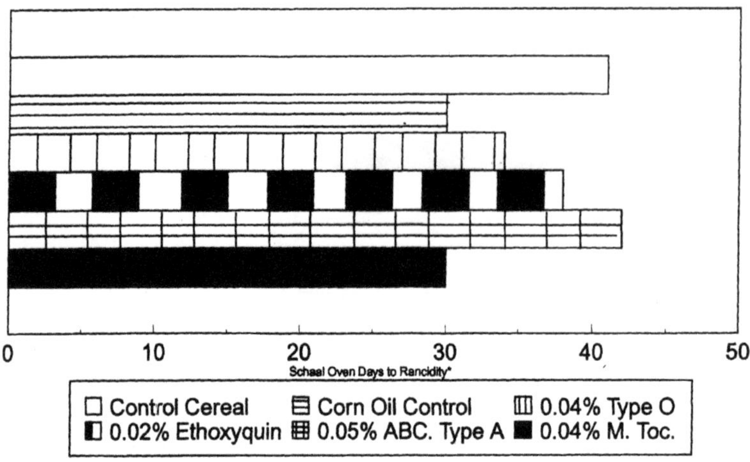

Figure 3. Low fat content is essential in extruded formulations. Five percent of supplemental corn oil was spray applied to a pet food after extrusion. The high surface area of extruded foods poses oxidative stress. The time at 65° C. required for samples to become rancid, as determined by sensory evaluation of odor, is shown. The corn oil was supplemented with rosemary Type O, ethoxyquin or a modified ascorbic acid.

Some foods subjected to heat or steam stress can be effectively protected by application of a FRI to their surfaces by spray or dip techniques. Application of the FRI to exposed surfaces immediately after processing can be very effective. Salt is another vehicle for post-processing application of FRIs. Oil roasted nuts, potato chips and crackers are routinely protected by FRIs, using salt as the carrier. Care must be taken to use salt with tightly-controlled specifications on its metal content.

The effectiveness of the appropriate FRI combination in stabilizing oils, normally sprayed onto expanded cereals, snacks and pet food is shown in Figure 3. The application of corn oil to an expanded pet food at a level of 5% posed an unusual stress. The oil is stressed by the high surface area of the expanded food. Inclusion of oleoresin rosemary or ethoxyquin, or a patented ascorbic acid (Todd, 1989, 1992) provides varying degrees of improvement in shelflife as determined by sensory evaluation of the aroma, a key factor in dog foods. Aroma development correlated well with the Active Oxygen Method (AOM) for peroxide value determination.

Gloss and dressing oils, applied to snacks and crackers after baking, require high stability. These high stability oils are also used as coating for dried fruits and release agents. High stability can be imparted to lower cost oils making them suitable for use in spray applications by selecting an appropriate FRI blend. FRI blends can eliminate the need for partial hydrogenation of oils. Figure 4 provides data showing that canola oil with a proprietary FRI blend can surpass the stability of the same oil, partially hydrogenated or stabilized with TBHQ.

MUSCLE FOODS

There is ample opportunity to make noticeable improvement in the acceptability of muscle foods where iron and enzymes pose problems. Current processing procedures in-

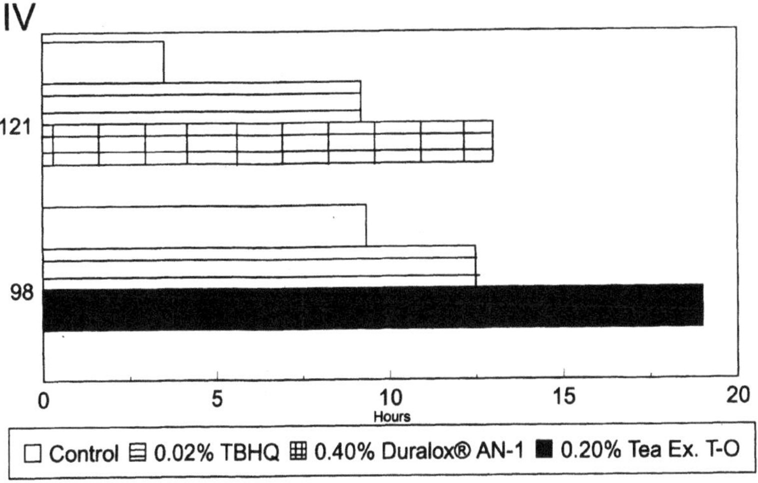

Figure 4. A control canola oil with an Iodine Value (IV) of 121 was stabilized with TBHQ or with AN-1 containing rosemary and a modified ascorbic acid. The control was then hydrogenated to a 98 IV and then stabilized with TBHQ. The rosemary/ascorbic blend provided stability superior to partial hydrogenation plus TBHQ.

duce oxidative deterioration. Cutting, grinding, comminution and other process steps rupture cells. This releases cellular iron and enzymes. The oxidation thus initiated can results in undesirable color changes, off flavor development and generation of offensive aromas in the cooking process. Iron induced oxidation is believed to cause polymerization of protein (Decker, 1993). Spray or dip application of a formulated blend to retard oxidation of muscle surfaces newly exposed to oxygen has impressive benefits.

Repeating the application of FRI blends after each step that exposes new surfaces is especially important in processing seafood. Shipboard operations allow for dipping fish, after gutting, into a water dispersed FRI blend. Addition of a chelating agent also has benefit. In processing plants one may utilize either dipping or spray application.

A solution of a water dispersible rosemary was applied to freshly fileted Alaskan salmon, by dipping. The typical color of salmon is the limiting factor in shelflife of refrigerated salmon filets. The shelflife of the treated salmon was increased 37% by stabilization of its pigments with rosemary.

In another study, frozen headed and gutted pink salmon were dipped in water to form a glaze with a proprietary blend of natural FRIs dispersed in the water. When 2-thiobarbituric acid (TBA) values reach 4–5 mg/kg sample, sensory approval of the product drops. Figure 5A shows the effectiveness of the FRI blend in retarding TBA increase during frozen storage.

At 6 months into this same study, some of the headed and gutted salmon samples were thawed and reprocessed to yield a mince which was then formed into large blocks used for production of restructured seafood items. TBAs for the blocks are shown in Figure 5B. The zero time in this block study contained mince salmon from the 6 month old glazed headed and gutted salmon in Figure 5A. In producing fish sticks and other restructured seafood, application of the appropriate FRI blend at each step to the new surfaces helps assure high quality of products.

In Figure 5C, pink salmon filets were prepared from fresh fish, dipped in the inhibitor blend and plate frozen in 16.5 lb. blocks. The controls were simply headed, gutted and

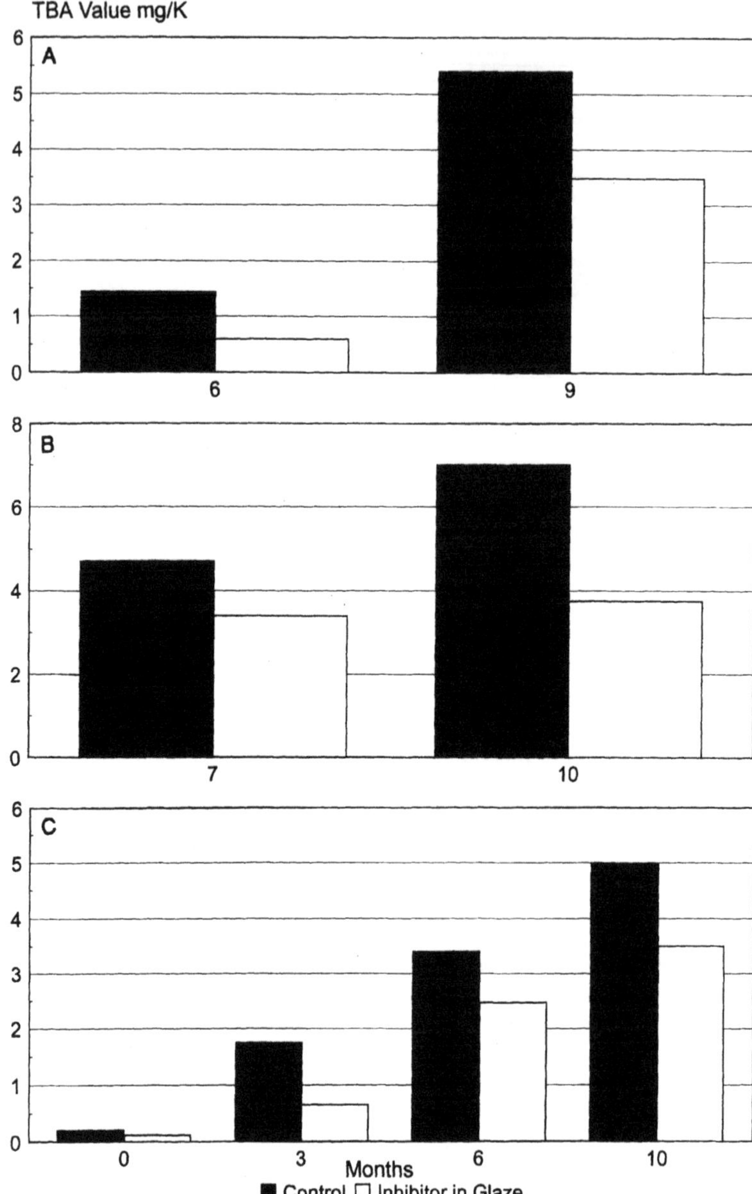

Figure 5A. Headed and gutted salmon were frozen in block and glazed. A formulated FRI blend, Duralox, was included in the experimental glaze. TBA values above 5 are associated with poor quality.

Figure 5B. Headed and gutted salmon was held for 6 month frozen in glazed block, then thawed, minced, refrozen and reglazed. TBA values were determined at 7 and 10 months after initial freezing. A formulated FRI blend was included in the glaze applied to the experimental blocks.

Figure 5C. Salmon filets were block frozen and fileted. An FRI blend, Duralox, was included in the experimental glaze. TBA values were maintained in the experimental samples at less than 4 for 10 months minimizing the seasonal economic impact.

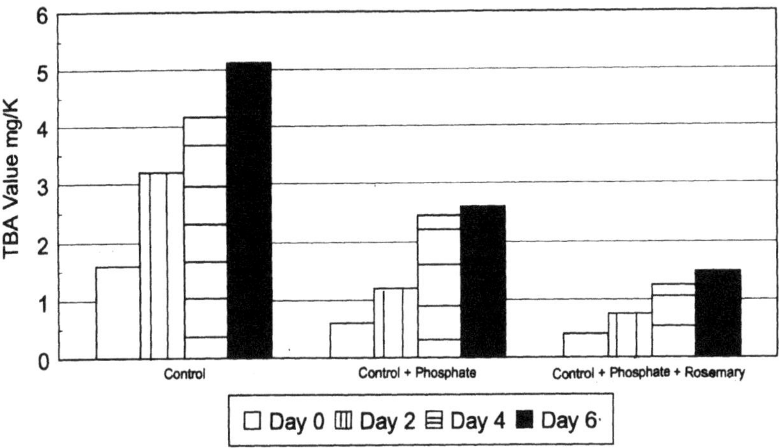

Figure 6. TBA values above 3 are indicative of off flavor in beef. The effect of rosemary and phosphate in extending refrigerated shelf life of ground beef patties is shown.

glazed. The inhibitor included in the dipping solution extended acceptable shelflife of product by over 60%.

Similar benefits in beef are shown in Figure 6. TBA values over 3 indicate off flavor in beef. High TBA values are sometimes seen in further processed poultry. Process modification to allow for spray application of suitable FRI blends can protect quality in a variety of lipid containing foods.

FRI DIVERSITY

As illustrated earlier (Figure 2), FRI performance is substrate specific. The impressive performance of the FRIs present in rosemary, sage and tea probably relates to the diversity of the structures in the many FRIs each contains. Thus far, more than 10 diverse FRI structures have been identified in tea, more than 20 are believed to be present in rosemary. Some of these structures are shown in Figure 7. The ability of these to react with the variety of radicals developing in lipids is far greater than that provided by the single configuration of each of the synthetic antioxidants. An illustration of the benefits of this molecular diversity is seen in Figure 8. Oxidation could generate a multitude of different free radicals and hence secondary oxidation products from different lipid fatty acids in foods. The variety of FRIs in the natural tea extract manages the radicals effectively.

Managing the many and diverse free radicals in a processed food requires complex formulations and appropriate application techniques. Thus formulation of custom-designed systems might be warranted.

CONCLUSION

Careful evaluation of process techniques can identify conditions that result in undesirable oxidative changes in processed foods. Remedial measures include minimizing

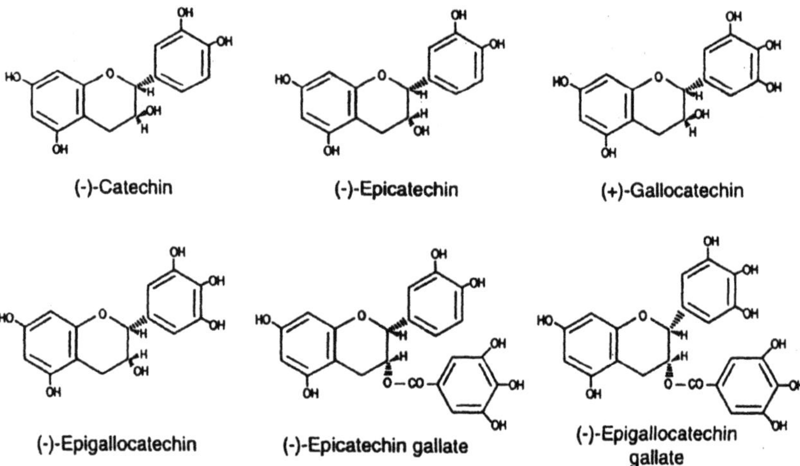

Figure 7. Structure of free radical interceptors present in green tea and in rosemary.

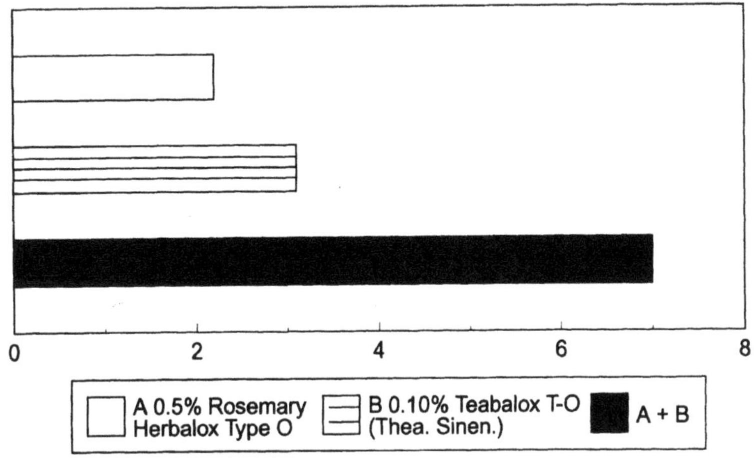

Figure 8. Synergistic effect of rosemary Herbalox and Teabalox.

prooxidant factors by process modification, tighter ingredient specification and proper selection and application of specific phenolic FRIs in combination with a modified ascorbic acid, acidulant, and quencher. Formulation of a blend offering superior results in minimizing prooxidant factors introduced in processing can be simplified by discussion with suppliers of such ingredients.

REFERENCES

Berger, K.B. Practical measures to minimize rancidity in processing and storage. In *Rancidity in Foods, Second Edition*, Allen, J.C.; Hamilton, R.J., Eds., Elsevier Applied Science: London, 1989; pp. 68–82.

Decker, E.A.; Xiong, Y.L.; Calvert, J.T.; Crum, A.D.; Blanchard, S.P. Chemical, physical and functional properties of oxidized turkey white muscle myofibrillar proteins. *J. Agric. Food Chem.* **1993**, *41*, 186–189.

Gupta, M.J. Processing to improve soybean oil quality. *Inform* **1993**, *4*, 1267–1272.

Min, D.B.; Lee, S.H.; Lee, E.C. Singlet oxygen oxidation of vegetable oils. In *Flavor Chemistry of Lipid Foods*, Min, D.B.; Smouse, T.H., Eds., American Oil Chemists' Society: Champaign, IL 1989; pp. 57–97.

Pokorny, J. Major factors affecting the autoxidation of lipids. In *Autoxidation of Unsaturated Lipids*, Chan, H.W.S., Ed., Academic Press: London, 1987; pp. 159–206.

Todd, P.T. Herb flavoring and/or antioxidant composition and process. U.S. Patents 4, 866, 635, 1989.

Todd, P.T. Activated ascorbic acid antioxidant compositions and carotenoids, fats, and foods stabilized therewith. U.S. Patents 5, 084, 293, 1992.

ANTIOXIDATIVE PROPERTIES OF PRODUCTS FROM AMINO ACIDS OR PEPTIDES IN THE REACTION WITH GLUCOSE

N. V. Chuyen, K. Ijichi, H. Umetsu, and K. Moteki

Department of Food and Nutrition
Japan Women's University
Tokyo 112, Japan

The livers of rats fed with a brownish "soybean paste" (Miso) or peptides-glucose reaction mixture showed lower TBA and chemiluminescence values than those of the control. From these results, it was clear that Miso and peptide-glucose reaction products also exhibited antioxidative effect *in vivo*. In order to explain this mechanism, the scavenging activity against reactive oxygen species (ROS) of various Maillard reaction products (MRPs) were studied. It was confirmed that peptide-glucose reaction products, Amadori rearrangement products, melanoidins, modified protein and its hydrolysate, brown pigments isolated from Miso and other foodstuffs showed strong scavenging activity against hydroxyl radical and superoxide anion. Consequently, it was estimated that scavenging activity of MRPs against ROS played an important role in the antioxidative effect of MRPs *in vivo*.

INTRODUCTION

Maillard reaction products (MRPs) generated by the reaction of amino acids or peptides with reducing sugars are known to have antioxidative effects in food systems (Kirigaya *et al.*, 1968; Lingnert *et al.*, 1980; Yamaguchi *et al.*, 1981). Previously, we reported that the reaction products from amino acids or proteins with glucose also exhibited their antioxidative effect *in vivo* (Chuyen *et al.*, 1990). However, the mechanism of their antioxidative effect is not yet clear. MRPs are also known for their scavenging of radicals and reactive oxygen species (Hayase *et al.*, 1990; Chuyen *et al.*, 1990; Okamoto *et al.*, 1992). Furthermore, Amadori rearrangement product (ARP), a key intermediate product in the Maillard reaction, is known to generate reactive oxygen species (ROS) in the presence of

Process-Induced Chemical Changes in Food
edited by Shahidi *et al.* Plenum Press, New York, 1998

transition metals (Kashimura and Morita, 1984; Sakurai *et al.*, 1990; Wolff *et al.*, 1990; Kawakishi *et al.*, 1990). However, the balance of generating and scavenging of ROS of ARP is not known. Thus, in this study, we investigated the antioxidative properties of a brownish foodstuff "soybean paste", peptide-glucose reaction products *in vivo* and the scavenging and generation of ROS of ARP, glycated protein, glycated protein hydrolysate, melanoidins and brown pigments isolated from foodstuffs.

MATERIALS AND METHODS

Preparation of Peptides from Casein

Casein from milk (Wako Pure Chemical Ind., Osaka) was purified by washing with 0.1N NaOH, acetic acid and ethanol as described in a previous paper (Utsunomiya *et al.*, 1990). The purified casein was hydrolyzed by pepsin-pancreatin as described by Kato *et al.* (1986) with pepsin for 4 h and pancreatin for 18 h at 37°C. The mixture prepared by this procedure consisted of 58.6% peptides with molecular weight of above 5,000 Da, 39.3% peptides with molecular weight of above 1,000 Da., and 2.1% peptides with molecular weight above 500 Da.

Preparation of Peptide-Glucose Reaction Mixture

The peptide mixture prepared above (casein hydrolysate, 100 g) was mixed with glucose (100 g) in 0.2 M $NaHCO_3$ (1000 ml) and heated at 100° C for 2 h. The whole mixture was then lyophilized to powder to be used as a test sample in this study.

Preparation of Modified Lysozyme and Its Hydrolysate

Lysozyme from egg white (Seikagaku Co., Ltd. Japan), crystallized 5 times, was mixed homogeneously with D-glucose at a ratio of 2:1 in a small quantity of water. The whole mixture was lyophilized and incubated at 50°C and 75%RH for 7 days. After the reaction, the browned lysozyme reaction mixture was dialyzed against water for 3 days at 4°C to remove the reaction products of low molecular weight, and then lyophilized again to obtain the nondialyzable modified lysozyme. Modified lysozyme hydrolysate was prepared by the reaction of pepsin (3 h) and pancreatin (20 h) at 37°C as described by Kato *et al.* (1986).

Synthesis of Amadori Rearrangement Product (ARP)

ARP from glycine, valine, diglycine, triglycine was synthesized by the method described by Sherr *et al.* (1980) by the reaction of amino acid or peptide with glucose in methanol. Then, the ARP was purified using Dowex 50W-X4 column chromatography. The chemical structures were confirmed by MS and NMR.

Synthesis of Melanoidins from Various Amino Acids

A mixture of 1 M of each amino acid (glycine, isoleucine, lysine, arginine, histidine, glutamic acid, serine, cysteine or phenylalanine) with 2 M glucose in 0.2 M $NaHCO_3$ solution was heated at 100°C for 7 h and then dialyzed for 3 days at 4°C, with renewing the outer water once a day. The resulting inner brown solution was lyophilized to obtain melanoidin powder.

Isolation of Brown Pigments from Foodstuffs

Brown pigments from coffee, soy sauce, sponge cake, black beer and roasted barley were extracted by distilled water. Then, the pigments were isolated from the extract by Sephadex G-50 with distilled water used as eluant. The brownish fraction was lyophilized to obtain the brown pigment powder.

Animals and Diets

Male Wistar growing rats were purchased from Nihon Clea Co., Ltd., Tokyo, housed in wire stainless steel cage at 23 ± 1 °C and 53% RH with a light and dark cycle of 12 h each. The animals had free access to water. After prefeeding for 3 days, the animals were divided into groups, each being fed with a different diet.

Experiment I. Growing rats weighing 40–45 g were divided into two groups of 7 rats each, and then fed for 6 weeks. One group was fed with soybean paste added at a level of 20% protein in the diet, the other group was fed with soybean paste control (soybean paste material). Other ingredients of the diet were: cellulose 5%, sucrose 25%, corn oil 5% (vitamin E free), vitamin mixture (Harper's mixture, vitamin E free), 1% mineral mixture (Harper's mixture) and α-corn starch was added to a total of 100%.

Experiment II. Growing rats weighing 40–45 g were divided into 2 groups of 5 rats each, and then fed for 6 weeks. One group was fed with peptide-glucose reaction mixture at a 10% of the diet. The unheated peptide-glucose mixture was used for the control group. Other ingredients of the diet were the same as those in Experiment I.

Measurement of Lipid Peroxidation Value in Rat Liver

Lipid peroxidation value in rat liver was measured as TBA value or chemiluminescence value. TBA values were measured by the method of Ohkawa *et al.* (1979). Chemiluminescence was measured by a synchronous single photon counting apparatus, a chemiluminescence detector CLD-100 (Tohoku Electronic Co., Ltd.,) in combination with a personal computer (NEC-9801VX) used for the detection of light emission from the liver homogenate at 40 °C for 500 sec .

Measurement of POV Value

Peptide (1 M)-glucose (1 M) reaction mixture (1 ml) or lyophilized soybean paste powder (10 mg) was mixed with ethanol (20 ml), 0.2 M phosphate buffer (25 ml, pH 7,0), then kept at 45 °C for 48 h. The titration was carried out with 0.01 M sodium thiosulfate.

ESR Measurement

ESR spectra were recorded on a JEOL JES-REIX spectrometer using an aqueous quartz flat cell (a JEOL LC-12 ESR cuvette, inner size 60 mm x 10 mm x 0.31 mm, effective volume 160 μl). Measurement conditions; modulation amplitude 0.063 mT (100 kHz), receiver gain 2000, recording range 5 mT, recording time 4 min, time constant 1.0 s, microwave power 8 mW (9.405).

Hydroxyl radical was generated from Fenton reaction with the mixing of $FeSO_4$ (2.27 μM) and H_2O_2 (227 μM). Superoxide radical was generated from a hypoxanthine-xanthine oxidase (HPX-XOD) reaction system. Both hydroxyl radical and superoxide were trapped by 5,5-dimethyl-1-pyrroline-1-oxide (DMPO).

RESULTS AND DISCUSSION

Antioxidative Effect of Soybean Paste in Vitro and in Vivo

Antioxidative activity of soybean paste against the oxidation of linoleic acid in vitro is shown in Table 1. Unaged Soybean paste also exhibited antioxidative activity, since soybean contain isoflavonoid such as genistin, daizin which show antioxidative activity. However, the antioxidative activity of aged soybean paste was much increased. Soybean paste contains a large quantity of amino acid, peptide, protein and reducing sugars. Maillard reaction might occur during the processing and aging process of soybean paste, since paste materials were heated at 100°C for several hours and the aging process was carried out for periods of 2 months to 2 years. Results on the antioxidative effect *in vivo* are shown in Table 2. The TBA value in the liver of rats fed with aging soybean paste was much lower than that of rats fed with unaged soybean paste. From this result, it can be considered that soybean paste also exhibited antioxidative effect *in vivo*.

ANTIOXIDATIVE EFFECT OF PEPTIDE-GLUCOSE REACTION PRODUCTS *IN VITRO* AND *IN VIVO*

The antioxidative effect of peptide-glucose reaction products prepared under various conditions against the oxidation of linoleic acid is shown in Table 1. The peptides pre-

Table 1. Antioxidant activity of peptide-glucose reaction mixture and soybean paste *in vitro*

Sample	POV value[1]
PG[2] 15-3[3,4]	177.7
15-4	108.3
18-3	111.8
18-4	97.6
21-3	194.8
21-4	153.4
Soybean paste	75.0
Soybean paste control	185.0

[1]POV, peroxide value was determined against the oxidation of linoleic acid.
[2]PG: peptide (casein hydrolyzate)-glucose reaction mixture.
[3]Casein was hydrolyzed by pepsin-pancreatin for 15 h.
[4]Peptide-glucose mixture was heated at 100°C for 3 h.

Table 2. TBA value and chemiluminescent intensity (CL) of liver homogenate of rats fed with peptide-glucose reaction product and soybean paste

Group	TBA value[1] (nmol/g wet liver)	CL value[2]
PC	145.8 ± 22.2[a]	24135.8 ± 2740.9[c]
PG	120.0 ± 11.4[b]	16987.0 ± 5618.6[d]
SPC	9.2 ± 4.5[a]	—
SP	23.0 ± 5.7[b]	—

PC: Control peptide-glucose group (n = 5).
PG: Peptide-glucose reaction product group (n=5).
SPC: Soybean paste control (n = 7).
SP: Soybean paste (n = 7).
[1]MDA = malonaldialdehyde.
[a,b]: Statistically significant ($p < 0.05$) between a and b.
[2]Measured by JASCO, CDL-100, 40°C for 500 sec.
[c,d]: Statistically significant ($p < 0.05$) between c and d.

pared by the hydrolysis of casein with pepsin for 3 h and pancreatin for 18 h, then reacted with glucose for 4 h, showed the strongest antioxidative effect. Lingnert *et al.* (1980) also reported that peptides and protein hydrolysates in the reaction with reducing sugars showed antioxidative effect *in vitro*. The casein hydrolysate peptides used thiş study consisted of 58% peptides of molecular weight above 5,000 Da and 40% peptides of molecular weight above 1,000 Da. These peptides were glycated in the reaction with glucose and exhibited their antioxidative effect *in vitro* and *in vivo*. The liver TBA and chemiluminescence values of rats fed with peptide-glucose reaction products showed much lower values than those of control rats fed with unheated peptide-glucose mixture. The same phenomenon was observed for rats fed with glycated casein and glycated soybean (Chuyen *et al.*, 1990).

There are various factors to be considered to account for the antioxidative mechanism of MRPs such as ARP, melanoidins and glycated peptides or protein. Hitherto, it was considered that the amino reductone structures, the electron donor, the chelating properties for metals and the reducing properties of melanoidins are the main causes for the antioxidative effect of MRPs. However, the scavenging activity of MRPs against ROS has recently been reported by Hayase *et al.* (1990, 1992) and Chuyen *et al.* (1990, 1996). Therefore, the scavenging activity against ROS of amino acid, peptide-glucose reaction mixture, ARPs, glucose modified protein, glucose modified protein hydrolysate, melanoidins and brown pigments isolated from foodstuffs was studied in this contribution.

SCAVENGING ACTIVITY OF AMINO ACID, PEPTIDE-GLUCOSE REACTION MIXTURES AGAINST ROS

Scavenging activity of amino acid (glycine), peptide (diglycine, triglycine)-glucose reaction products against hydroxyl radical is shown in Figure 1. The reaction mixture of glycine, diglycine or triglycine with glucose showed stronger scavenging activity than that of glycine, diglycine and triglycine alone. The glucose or its degraded moiety of MRPs of

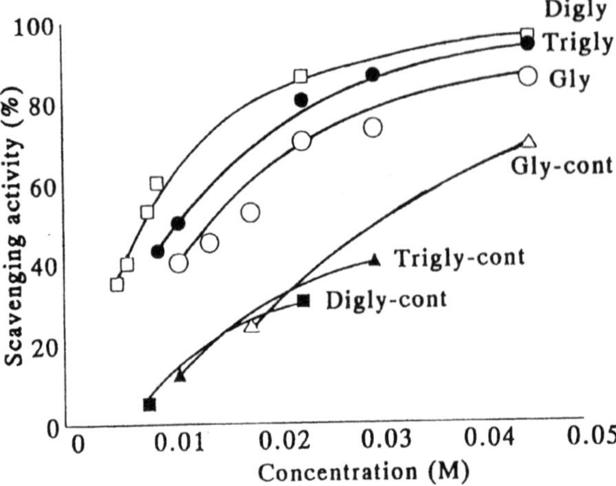

Figure 1. Scavenging activity of gly-glc, digly-glc, trigly-glc reaction mixture against OH radical. digly = diglycine + glucose; trigly = triglycine + glucose; gly = glycine + glucose; gly-cont, digly-cont, trigly-cont = control. Measurement conditions in details are the same as described in Table 3.

amino acid, peptide-glucose reaction mixture may be the cause. However, further investigations are necessary to clarify this situation.

SCAVENGING ACTIVITY OF ARP AGAINST ROS

Scavenging activity against hydroxyl radical by fructoseglycine is shown in Figure 2. Fructoseglycine strongly scavenged hydroxyl radical at a concentration of 227 ppm.

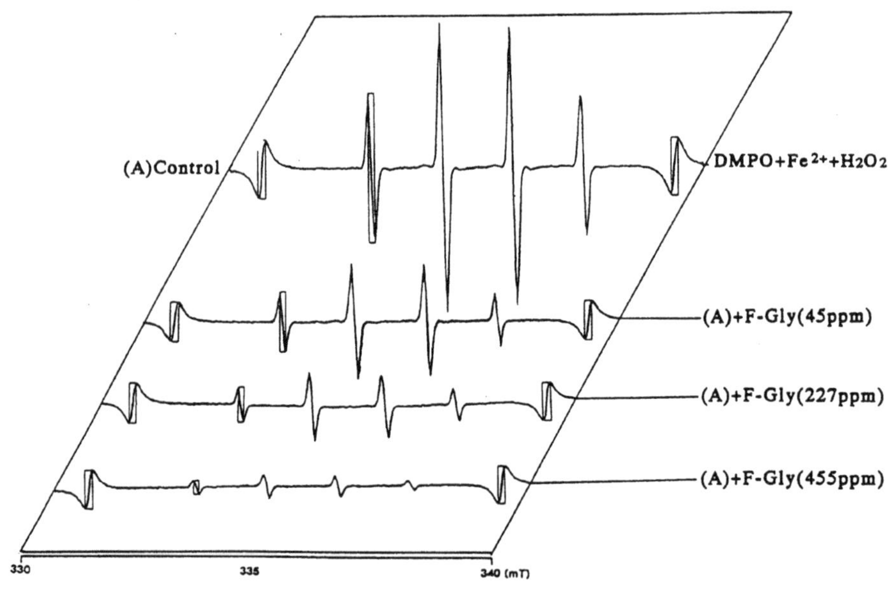

Figure 2. Radical scavenging activity of amadori compound (fructoseglycine, F-gly) against OH radical. Measurement conditions in details are the same as described in Table 3.

Table 3. Scavenging activity of Amadori compounds against hydroxyl and superoxide radicals

Sample	Concentration	Spin concentration	Inhibitory effect (%)
Hydroxyl radical			
1. Fructose-glycine	0.23	8.9	59
	0.9	4.0	81
2. Fuctose-valine	0.23	9.6	56
	0.9	2.7	87
3. Fructose-diglycine	0.045	14.1	35
	0.23	5.9	73
	0.9	5.2	76
4. Fructose-triglycine	0.045	10.4	52
	0.23	6.0	72
	0.9	2.5	88
Superoxide radical			
1. Fructose-glycine	0.13	22.1	0
	1.32	16.3	26
2. Fuctose-valine	0.13	17.6	10
	1.32	10.1	48
3. Fructose-diglycine	0.003	14.9	32
	0.13	7.0	68
	1.32	4.4	80
4. Fructose-triglycine	0.03	11.0	50
	0.13	6.1	72
	1.32	2.8	87

Hydroxyl radical was produced by Fenton reaction.
Fe^{2+} (22.7 μM) with H_2O_2 (227 μM) containing of DMPO (8.15 mM) was used as the trapping reagent.
Superoxide radical was produced by HPX-XOD reaction.
HPX (0.53 mM) with XOD (0.21 U/ml) containing of DMPO (0.7 M) was used as the trapping reagent.

At a concentration of 455 ppm, fructoseglycine scavenged almost the entire quantity of hydroxyl radical. Scavenging activity of various kinds of ARP against ROS is shown in Table 3. ARP from peptide showed stronger scavenging activity against hydroxyl radical and superoxide than that of the amino acids. ARP from valine also showed stronger scavenging activity against hydroxyl radical and superoxide than that of glycine. These is a difference of a methylene group in the structures of fructoseglycine and fructosevaline which may have some influence on their electron donor properties.

SCAVENGING ACTIVITY OF MELANOIDINS AGAINST ROS

Many kinds of melanoidins occur in foodstuffs. At present, the chemical structure of melanoidins are still unknown. However, it is known that many melanoidins in foodstuffs have molecular weights from 500 to 100,000 Da. Hayase *et al.* (1990) reported the scavenging activity of melanoidins from glycine-glucose system against ROS. In this study, we investigated the scavenging activity against ROS by various kinds of melanoidins prepared from amino acids such as glycine, isoleucine, arginine, lysine, histidine, cysteine, phenylalanine, serine and glutamic acid. The scavenging activity of various kinds of melanoidins against hydroxyl radical is shown in Figure 3. Melanoidins from glycine,

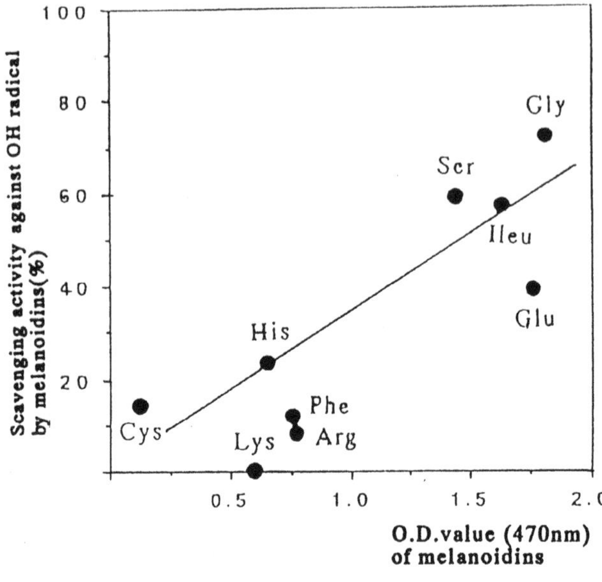

Figure 3. Relationship of scavenging activity of OH radical and browning of melanoidins. Measurement conditions in details are the same as described in Table 3.

isoleucine, serine showed a strong scavenging activity. There is also a correlation between the scavenging activity and browning of melanoidins. Melanoidins with darker color showed stronger scavenging activity. In general, melanoidins with higher molecular weight showed darker color, thus, it could be considered that melanoidins with higher molecular weight exhibited stronger ROS scavenging activity.

SCAVENGING ACTIVITY OF MODIFIED LYSOZYME AND ITS HYDROLYSATES AGAINST ROS

As a model system, we modified lysozyme with glucose. By the modification process, the decrease of lysine residue in lysozyme was about 50%. Then, the modified lysozyme was hydrolyzed by a pepsin-pancreatin system. The scavenging activity of intact lysozyme, modified lysozyme, intact lysozyme hydrolysate, modified lysozyme hydrolysate against ROS is shown in Figure 4. The scavenging activity of modified lysozyme and modified lysozyme hydrolysate was stronger than that of the intact lysozyme and its hydrolysate. The same tendency was observed for the scavenging activity against superoxide by these products (Fig. 5). The glucose moiety in the protein-glucose or peptide-glucose adduct may be the cause for the difference in the scavenging activity.

SCAVENGING ACTIVITY OF BROWN PIGMENTS ISOLATED FROM FOODSTUFFS

The scavenging activity of various kinds of Amadori compounds, model melanoidins, glycated protein and its hydrolysate was investigated. However, it is necessary to elucidate the scavenging activity of brown pigments from food-stuffs since the chemical

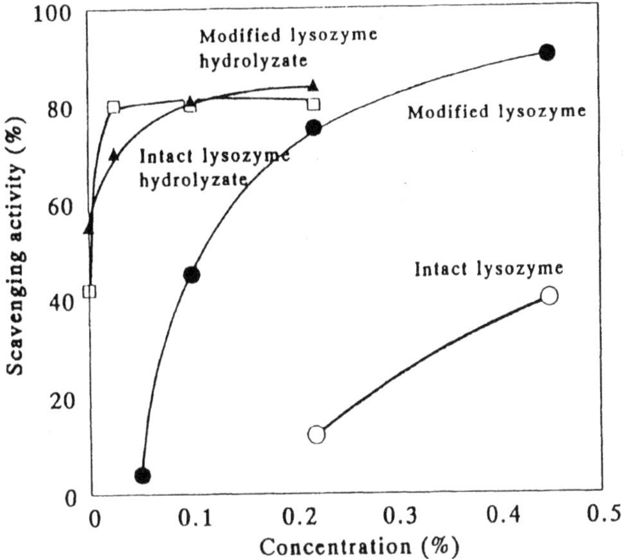

Figure 4. Scavenging activity of intact lysosyme, modified lysozyme and intact lysozyme hydrolysate, modified lysozyme hydrolyzate against OH radical. Measurement conditions in details are the same as described in Table 3.

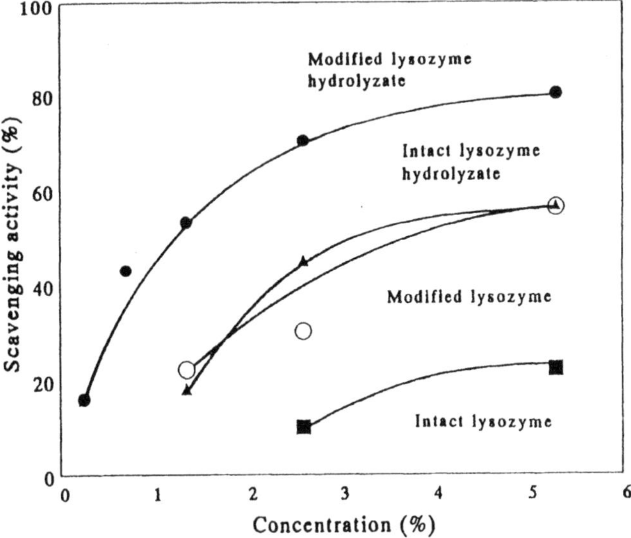

Figure 5. Scavenging activity of intact lysozyme, modified lysozyme and intact lysozyme hydrolyzate, modified lysosyme hydrolyzate against O_2^- radical. Measurement conditions in details are the same as described in Table 3.

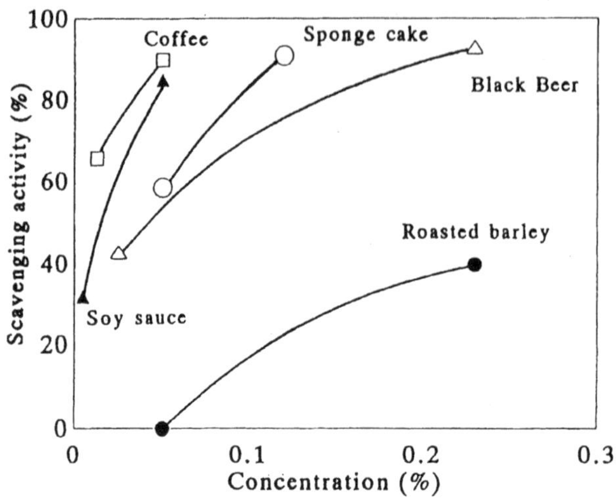

Figure 6. Scavenging activity of Brown Pigment against OH radical. Measurement conditions in details are the same as described in Table 3.

structure of melanoidins, glycated proteins, glycated peptides generated during the process of food processing and reservation may be more complicated and different from that of the model systems.

In this study, the scavenging activity against ROS of brown pigments isolated from coffee, soy sauce, sponge cake, black beer and roasted barley was investigated. The results are shown in Figures 6 and 7. These brown pigments showed strong scavenging activity against both hydroxyl radical and superoxide. Especially, in the case of hydroxyl radical, brown pigments exhibited strong radical scavenging activity at very low concentrations. The brown pigment from coffee exhibited strongest activity. Brown pigment (melanoidin) from coffee may be different with other brown pigment since in coffee bean, there are not only reducing sugar but a larger quantity of chlorogenic acid. This component is estimated to be a precursor of the coffee brown pigment. However, further studies in this area are necessary.

The generation of ROS by modified lysozyme in the presence of Fe^{2+} (50 ppm) was also studied by ESR. The results in Figure 8 show only a small quantity of generated radicals. Therefore, it could be concluded that the scavenging activity of modified protein against ROS was much stronger than the generation of ROS. Our study also showed that the scavenging activity against ROS of ARP, modified protein hydrolysate, melanoidins, brown pigments from foodstuffs was much stronger than their ROS generation capacity (data not shown).

As mentioned above, soybean paste, an example of a brownish foodstuff with abundant MRPs, showed antioxidative effect *in vitro* and *in vivo*. This antioxidative effect may be induced by the scavenging activity against ROS of peptide-glucose reaction products, ARPs, glycated proteins, glycated peptides, melanoidins and brown pigments. ROS are reported to be important factors which cause aging and cancer in human. Consequently, foodstuffs with brown color induced by the processing may play an important role for maintaining our health.

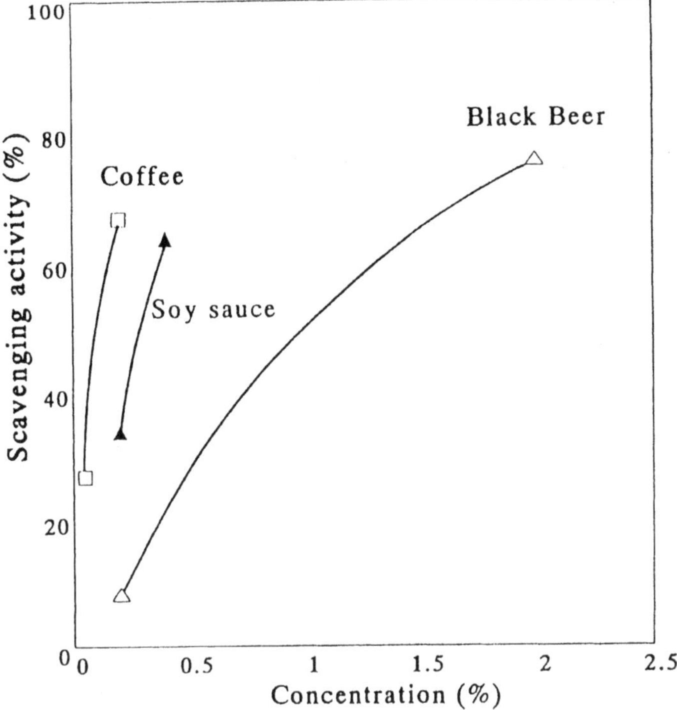

Figure 7. Scavenging activity of Brown Pigment against $O_2^{\cdot-}$ radical. Measurement conditions in details are the same as described in Table 3.

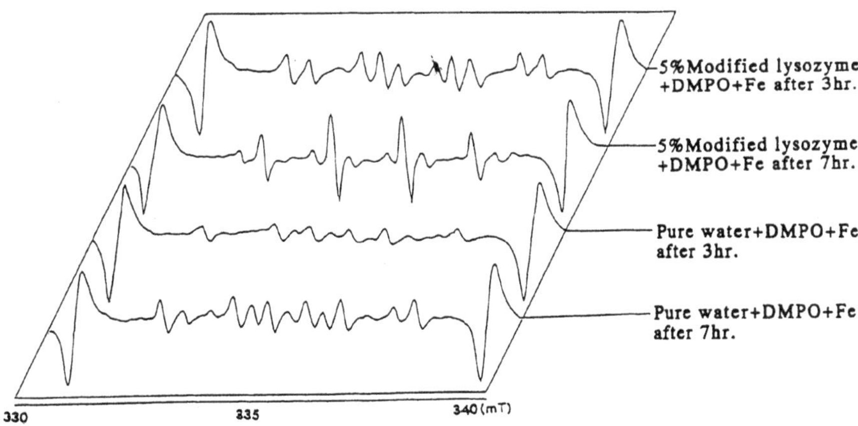

Figure 8. ESR spectra of modified lysozyme after incubation with iron for 3 h and 7 h. The radicals were measured after adding of Fe (50 pm) to the sample containing of DMPO (897 mM) used as trapping reagent after 3 h and 7 h. ESR measurement conditions in details are described in the text.

REFERENCES

Chuyen, N.V.; Utsunomiya, N.; Hidaka, A.; Kato, H. 1990. Antioxidative effect of Maillard reaction products *in vivo*. In *The Maillard Reaction in Food Processing, Human Nutrition and Physiology*; Finot, P.A.; Aeschbacher, H.U.; Hurrell, R.F.; Liardon, R., Eds.; Birkhauser Verlag Basel, 1990; pp 285–290.

Chuyen, N.V.; Ijichi, K.; Moteki, K. On the formation and scavenging of reactive oxygen species by Amino-carbonyl reaction products in model and food systems. *Nippon Kagaku Kaishi*. **1996**, *70*, 20.

Hayase, F.; Hirashima, S.; Okamoto, G.; Kato, H. Scavenging of active oxygens by melanoidin. In *The Maillard Reaction in Food Processing, Human Nutrition and Physiology*; Finot, P.A.; Aeschbacher, H.U.; Hurrell, R.F.; Liardon, R., Eds.; Birkhauser Verlag Basel, 1990; pp 361–366.

Kato, H.; Chuyen, N.V.; Utsunomiya, N.; Okitani, A. Changes of amino acids composition and relative digestibility of lysozyme in the reaction with α-dicarbonyl compounds in aqueous system. *J. Nutr. Sci. Vitaminol.* **1986**, *32*, 55–65.

Kashimura, N.; Morita. Nucleic acid cleavage activity of some carbohydrate derivatives. *J. Synth. Org. Chem., (Japan)* **1984**, *42*, 523–535.

Kawakishi, S.; Okawa, Y.; Uchida, K. Oxidative damage of protein induced by the Amadori compound-copper iron system. *J. Agric. Food Chem.* **1990**, *38*, 13–17.

Kirigaya, N.; Kato, H.; Fujimaki, M. Studies on antioxidant activity of nonenzymic browning reaction products. *Agric. Biol. Chem.* **1968**, *32*, 287–290.

Lingnert, H.; Eriksson. Antioxidative Maillard reaction products II. Products from sugars and peptides or protein hydrolysates. *J. Food Proc. Preserv.* **1980**, *4*, 173–181.

Okamoto, G.; Hayase, F.; Kato, H. Scavenging of active oxygen species by glycated proteins. *Biosci. Biotech. Biochem.* **1992**, *56*, 928–931.

Ohkawa, H,; Onishi.; Yagi, K. Assay for lipid peroxides in animal tissues by thiobarbituric acid. *Anal. Biochem.* **1979**, *95*, 351–358.

Sakurai, T.; Sugioka, K.; Nakano, M. Superoxide generation and lipid peroxidation during the oxidation of a glycated polypeptide, glycated polylysine in the presence of iron-ADP oxidative damage. *Biochem. Biophys. Acta.* **1990**, *1043*, 27–33.

Sherr, B.; Lee, C.M.; Jelesciewics, C. Absorption and metabolism of lysine Maillard products in relation to utilization of L-lysine. *J. Agric. Food Chem.* **1989**, *37*, 119–122.

Utsunomiya, N.; Chuyen, N,V.; Kato, H. Nutritional and physiological effects of casein modified by glucose, diacetyl, or hexanal. *J. Nutr. Sci. Vitaminol.* **1990**, *36*, 387.

Yamaguchi, N.; Koyama, Y.; Fujimaki, M. Fractionation and antioxidative activity of browning reaction products between D-xylose and glycine. *Prog. Food Nutr. Sci.* **1981**, *5*, 429–439.

Wolff, S.P. Diabetes mellitus and free radicals. *British Medical Bulletin*. **1993**, *49*, 642–652.

18

MAILLARD REACTION AND FOOD PROCESSING

Application Aspects

N. V. Chuyen

Department of Food and Nutrition
Japan Women's University
Tokyo 112, Japan

The Maillard reaction occurs widely in food and biological systems. This contribution reviews the relation between the Maillard reaction and food processing, particularly its contribution to flavor formation, antioxidative effects, desmutagenic activity and the improvement of protein functional properties. Proteins modified by glucose, and melanoidins are important components of foodstuffs while the reactions of amino acids or peptides with glucose or dicarbonyl compounds produce various kinds of flavor components. Melanoidins and Amadori rearrangement products play an important role in providing antioxidative effects, both *in vitro* and *in vivo*. Melanoidins also exhibit desmutagenic activity against carcinogenic compounds. Protein-polysaccharide conjugates, prepared by Maillard reaction at mild conditions, increase the emulsifying activity, as well as antioxidative and antimicrobial effects of the original proteins.

INTRODUCTION

Maillard is a chemical reaction involving the amino group and carbonyl groups which are common in foodstuffs. The reaction leads to browning and flavor production and is named after the French biochemist Louis Maillard, who first described the formation of brown pigment from the reaction of amino acids and reducing sugars (Maillard, 1912). The reaction is also called amino-carbonyl reaction, browning reaction, non-enzymatic browning reaction and sugar-amino reaction. The reaction occurs readily in food processing, storage and cooking process.

Process-Induced Chemical Changes in Food
edited by Shahidi *et al.* Plenum Press, New York, 1998

The amino compounds include amino acids, amines, peptides and proteins while carbonyl compounds include aldehydes, ketones, ascorbic acid, reducing sugars and polyphenols. In general, the Maillard reaction is divided into three distinct stages: early stage, advanced stage and final stage (Hodge, 1953).

The proposed mechanism of Maillard reaction is shown in Figure 1. It is generally agreed that in the early stage, a condensation occurs between an amino and a carbonyl group leading to the formation of a Schiff base and Amadori rearrangement product. Dur-

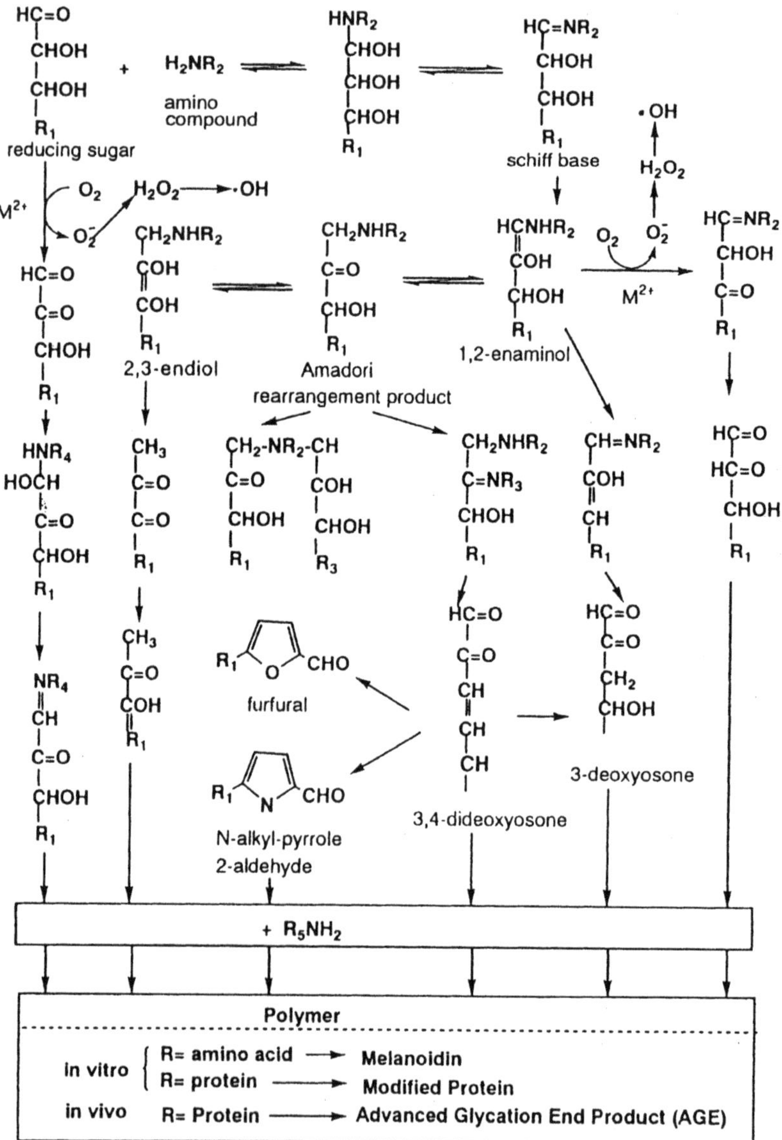

Figure 1. Proposed mechanism of Maillard reaction.

ing the early stage, there is no browning or generation of flavor compounds. In the advanced stage, starting from the Amadori compound, 2,3- and 1,2-enolization, Strecker degradation occurs with the generation of the keto-aldehydes, dicarbonyls, reductones, and N-heterocyclic flavor compounds such as 2-furaldehydes, pyrroles, pyrimidines, pyrazines and pyrrolidines.

Maillard reaction is known to occur in foods and biological systems (Eriksson, 1981; Waller and Feather, 1983; Fujimaki *et al.*, 1986; Baynes *et al.*, 1989; Finot , 1990; Finot *et al.*, 1990; Labuza *et al.*, 1994; Lee and Kim, 1996; Ikan, 1996). Recently, food browning and its prevention (Friedman, 1996) and the thermal generation of Maillard aromas (Ho, 1996) have been reviewed. Therefore, this overview, will concentrate on application aspects of the Maillard reaction relating to the generation of flavor, antioxidant properties, desmutagenic effects, protein-polysaccharide conjugates and other major reaction products.

THE GENERATION OF FLAVOR COMPOUNDS BY FOOD PROCESSING

Heating, either by cooking, roasting or other means, is very important in food processing. Temperatures ranging from 100 to 250°C are typically used and during this process, the Maillard reaction occurs with the generation of flavor compounds.

Generation of Flavor Compounds by the Reaction of Alkylamino Acid and Sugar

An example of flavor production from the reaction of alkylamino acids and sugars at 100° C in solution is shown in Table 1 (Mabrouk, 1979). The quality of flavor varies with the combination of amino acid and type of sugar. When a powdered amino acid and glucose was combined and heated at 180°C, chocolate and cacoa flavors were generated. The quality of the flavor depends largely on the ratio of amino acid and sugar, moisture content and pH. From the reaction of amino acids and reducing sugars in model systems, at least 4 types of compounds were generated *i.e.*, *O*-heterocyclics, *N*-heterocyclics, aldehydes and ketones. *O*-Heterocyclics are thought to be generated through the methyldicar-

Table 1. Aroma generated by the heating of amino acid with sugars[1]

Amino acid	Dihydroxyacetone	Glucose	Fructose	Maltose	Sucrose
Glycine	baked potato	caramerized sugar, faint beer	unpleasant caramel smell	weak	objectionable week NH_3
Methionine	baked potato	overcooked sweet potato	objectionable chopped cabbage	over cooked cabbage	unpleasant burned wood
Glutamic acid	chicken broth	old wood, pleasant	too week	too week	pleasant caramel
Lysine	strong dark corn syrup	backed sweet potato	objectionable fried butter	unpleasant wet wood	rotten wet potato
Phenylalanine	very strong, hyacinth	rancid caramel, unpleasant violets	stinging smell, very objectionable	pleasant sweet caramel	unpleasant sweet caramel

[1]Amino acid was heated with sugar at 100°C in solution.

bonyl intermediate. The degradation products of sugar with a caramel-like flavor such as maltol, isomaltol, furaneol, cyclotene were also reported. These compounds are generated from Amadori compounds through enolization, cyclization and dehydration. A component of brown sugar flavor, 4,5-dimethyl-3-hydroxy-2(5H)-furanone (sotolone) is generated by a similar mechanism.

Degradation Products from Amadori Compounds

In solution, most Maillard reaction products are derived from the Amadori compound. However, when amino acids and reducing sugars are heated at high temperatures and dry state, products from the degradation of Amadori compound, amino acid, reducing sugar and secondary reactions are produced. Shigematsu *et al.* (1977) and Kurata and Kato (1981) studied the thermal degradation products at 200°C of several Amadori such as 1-deoxy-1-L-alanino-D-fructose (DAF), 1-deoxy-1-L-valino-D-fructose (DVF), 1-de-

Table 2. Thermal degradation products from Amadori compounds

	1	2	3	4	5	6
DAF DVF	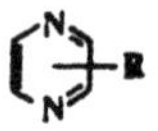	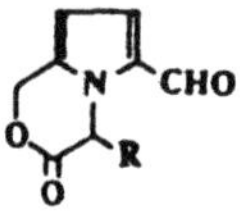	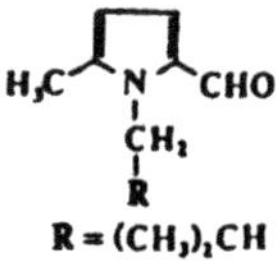 $R = (CH_3)_2CH$	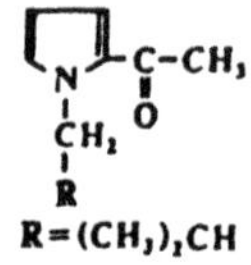 $R = (CH_3)_2CH$	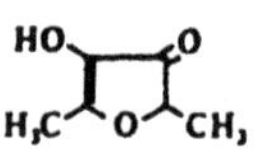	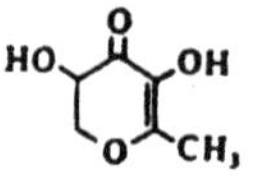
	Roasted Nut-like	Sweet Burnt note	Smoky Cinnamon-like	Smoky Medicine-like	Sweet Caramel-like	Odorless
		7	8	9	10	11
DPF	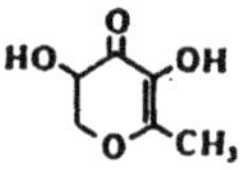	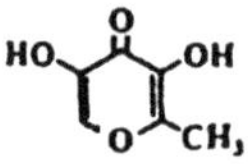	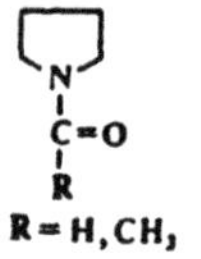 $R = H, CH_3$	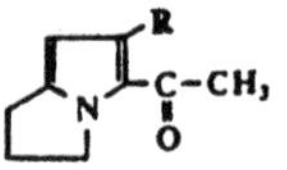	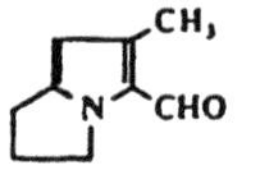	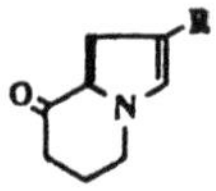
	Odorless	Slightly burnt note	Bread-like Weak pyrrolidine	Sweet smoky Medicine-like	Sweet smoky Cinnamon-like	Mild smoky
			12	13	14	
DBF	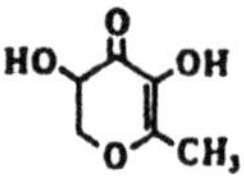	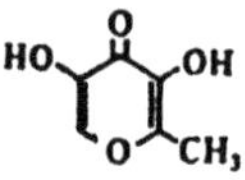	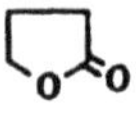	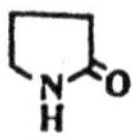	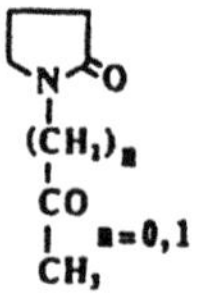 $n = 0, 1$	
	Odorless	Slightly burnt note	Burnt note	Sweet Weak pyrrolidine	Woody note	

DAF=1-deoxy-1-L-alanino-D-fructose.
DVF=1-deoxy-1-L-valino-D-fructose.
DPF=1-deoxy-1-L-prolino-D-fructose.
DBF=1-deoxy-1-(N-γ-aminobutyric acid)-D-fructose.

oxy-1-L-prolino-D-fructose (DPF) and 1-deoxy-1(N-γ-aminobutyric acid)-D-fructose (DBF), and a series of alkylpyrazines, pyrrole-lactones, pyrones, pyrrolidines, pyrrolidones, γ-butyrolactone, furans were identified (Table 2). These compounds play an important role in the formation of cooked and roasted flavor.

Generation of Flavor Compounds by the Reaction of Peptides and Carbonyls

Peptides are important components in food and biological systems and contribute to the taste of various foodstuffs. These peptides also contribute to the formation of aroma by reaction with carbonyls in food since the reaction rate of peptides is higher than that of amino acids (Chuyen *et al.*, 1973a). It was also reported that small peptides react with dicarbonyl compounds to generate pyrazinones and Strecker aldehydes (Chuyen *et al.*, 1972, 1973b; Figure 2). Rizzi (1989) reported that the reaction of dipeptides, and tripeptides with fructose generated Strecker aldehydes and alkylpyrazines.

Recently, Oh *et al.* (1991) also reported the formation of pyrazines from diglycine, triglycine, tetraglycine and suggested that tri- or tetrapeptides could be degraded through diketopiperazine. The formation of a series of alkyl-2(1H)-pyrazinones from pyruvaldehyde and diglycine was also reported (Oh *et al.*, 1992). This compound was suggested to derive from 2-(3'-alkyl-2'-oxopyrazin-1'-yl)alkanoic acid by decarboxylation at elevated temperatures (180°C). Shu and Lawrence (1994) also reported a series of alkyl-2(1H)-pyrazinones from the reaction of asparagine with monosaccharides and suggested that these compounds were formed by the condensation of α-dicarbonyl compounds with alanine amide (Figure 3). A reaction of tripeptide, glutathione with glucose at 180°C and pH 5.0 (Zhang and Ho, 1991) generating a variety of roasted sesame and cooked rice-like aromas was also studied. A total of 62 compounds were identified, including 7 furans, 8 carbonyls, 10 thiazoles, 19 thiophenes, 10 pyrazines and 6 cyclic polysulfides.

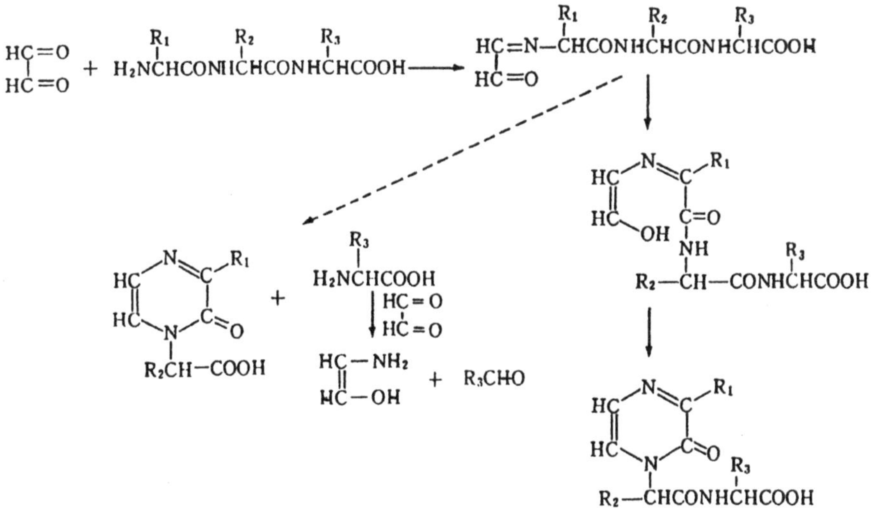

Figure 2. Possible reaction mechanism of tripeptides with glyoxal at 100°C, pH 5.0.

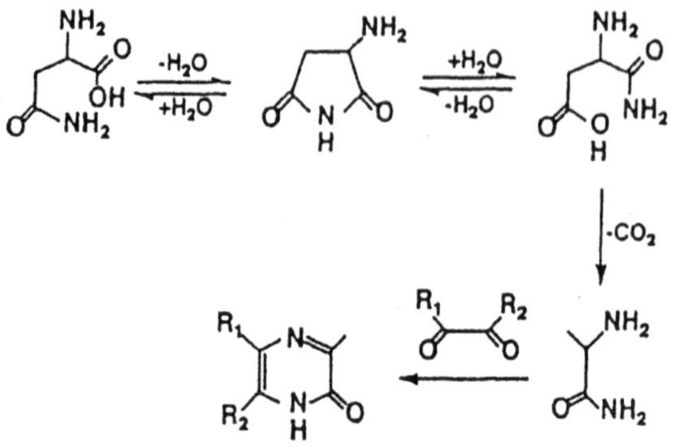

Figure 3. Mechanism for the formation of 3-methyl-2-(1H)-pyrazinone from the reaction of asparagine and glucose.

Generation of Strecker Aldehydes and Pyrazines

Volatile aldehydes and carbon dioxide from α-amino acids Strecker degradation, are considered to participate in the formation of cooked flavors of food and in flavor deterioration during food preservation. In the Strecker degradation, α-amino acids react with α-dicarbonyl compounds to form Schiff base which then enolizes and is decarboxylated and hydrolyzed to give carbon dioxide and an aldehyde with one fewer carbon than the original amino acid. The main α-dicarbonyls for this reaction are glyoxal, methylglyoxal, diacetyl and 3-deoxyosone. In reactions with these α-dicarbonyls, at high temperature, α-amino acids are easily decarboxylated leading to the generation of Strecker aldehydes. However, at low temperature (50°C), only a small quantity of the α-amino acid is decarboxylated (Fujimaki *et al.*, 1971), while a large quantity of the undecarboxylated form of the Schiff base leads to the formation of N-carboxymethyl amino acid (Chuyen *et al.*, 1973c) (Figure 4). This N-carboxymethyl amino acid is also reported to be generated *in vivo* (Wells-Knecht *et al.*, 1995).

Pyrazines are very important flavor components for cooked, roasted and toasted foods. The main formation mechanism of pyrazines is considered to be condensation of two molecules of α-aminoketone which are generated by the Strecker degradation. The reaction of α-hydroxyketone and ammonia also gives an α-aminoketone. Both α-dicarbonyls and α-hydroxyketones are degradation products of 1-deoxyosone or 3-deoxyosone (Figure 1). The contribution of α- and ε-amino nitrogen atoms to pyrazine formation in the reaction of labeled lysine with glucose was investigated by the use of lysine-α-amine-^{15}N (Hwang *et al.*, 1994). They found that both the α- and ε-amino group of lysine were involved in pyrazine formation. However, the nitrogen atom from the α-amino group of lysine reacts more readily with dicarbonyl to form pyrazine than the nitrogen atom from the ε-amino group. Hwang *et al.* (1995) also found that the reaction mixture containing lysine had the highest yield of pyrazine. They noted that lysine was able to increase the reactivity of glycine to produce pyrazines. Pyrazines give desirable aroma for beef products, toasted barley, peanut, popcorn, cocoa products, coffee and potato products, among others.

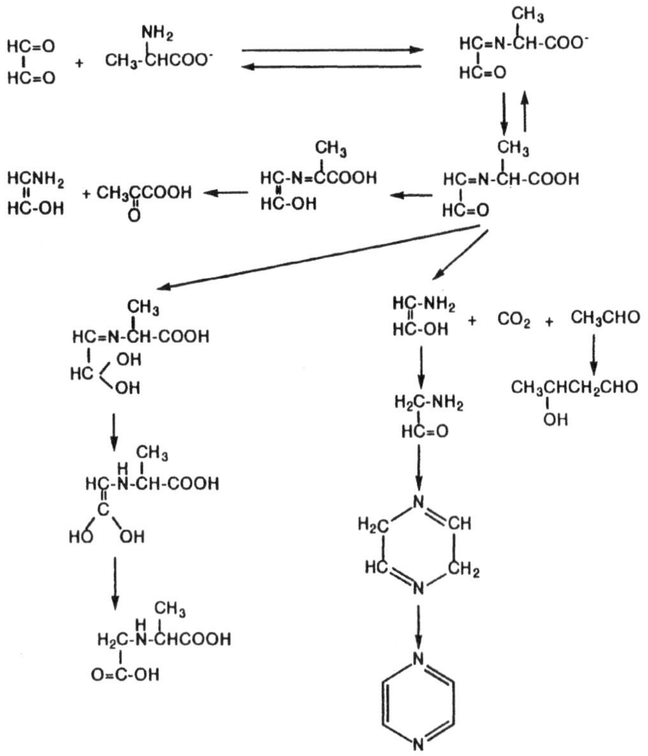

Figure 4. Strecker degradation and generation of N-carboxymethyl amino acid and pyrazine from the reaction of alanine with glyoxal.

Generation of *N*-Heterocyclic Compounds

N-heterocyclic compounds other than pyrazines such as pyrrolines, pyrrrolidines, piperidines and pyrroles are also very important flavor compounds. The formation of pyrrolines and pyrrolidines are reported to be generated from the reaction of proline with glucose (Shigematsu *et al.*, 1975; Tressl *et al.*, 1985a). The pyrrolidines possess smoky and roasty aromas while 2-acetyl-1-pyrroline was reported by Tressl *et al.* (1985b) to have a cracker-like odor. The pyrrole rings from proline and hydroxyproline are present in many of their reaction products. N-acetylpyrrole exhibits a cookie-like and mushroom-like odor (Tressl *et al.*, 1986).

Generation of Sulfur Containing-Compounds

Sulfur-containing flavor components are also very important in the food industry. Cysteine and cystine are considered indispensable components for the generation of meat-like aromas through their reaction with reducing sugars. When sulfur containing amino acid such as cysteine, cystine or methionine are used in the reaction with reducing sugar, sulfur-containing compounds such as thiophens, thiazoles are generated. When a mixture of cysteine-glucose was heated in dry state at 180°C, thiazolidine-4-carboxylic acid was

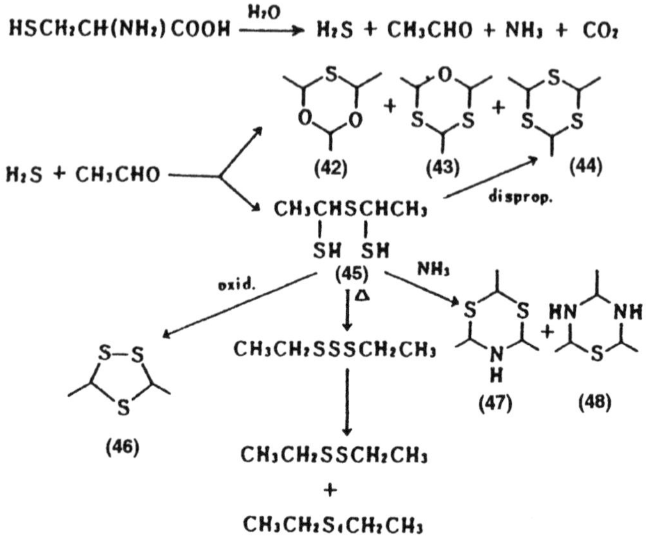
HSCH₂CH(NH₂)COOH
H₂O
H₂S + CH₃CHO + NH₃ + CO₂
H₂S + CH₃CHO
(42)
(43)
(44)
disprop.
CH₃CHSCHCH₃
SH SH
(45)
oxid.
NH₃
CH₃CH₂SSSCH₂CH₃
(46)
(47)
(48)
CH₃CH₂SSCH₂CH₃
+
CH₃CH₂S₄CH₂CH₃

Figure 5. Products from the reaction of ethanol, hydrogen sulfide and ammonia.

formed (Kato *et al.*, 1974). Cystamine, a degradation product of cysteine also reacts with glucose to form thiazolines and thiazolidines (Sakaguchi and Shibamoto, 1978). Several alkylthiazoles identified in fried chicken and french fried potato flavor have long-chain alkyl substituents on the thiazole ring (Ho and Carlin, 1989). Werkhoff *et al.* (1990) isolated various volatile sulfur-containing meat flavored components in model systems. They reported that a pH 5.0 aqueous solution of cystine, thiamin, glutamate and ascorbic acid heated at 120°C for 0.5 hr produced a complex mixture of compounds with an overall flavor resembling roasted meat. Various heterocyclic thioethers, disulfides and hemidithioacetals were identified for the first time in the volatiles of the heated meat flavor model mixture (Figure 6). The heterocyclic thioethers were estimated to be generated by the radical reaction between 2-methyl-4,5-dihydrothiophene and 2-methyl-3-furanthiol. Both these two compounds were generated from thiamin and hydrogen sulfide. Moreover, hydrogen sulfide, acetaldehyde, ammonia generated from cysteine react with one another to form thiolanes, thiadiazines, dioxathiane, trithiane and oxadithiane as shown in Figure 5 (Boelens *et al.*, 1974). These compounds can be formed in meats, allium species, rice and animal fats. Many patents concerning the production of synthetic meat flavor claim that cysteine is one of the most important precurssor (Okumura *et al.*, 1990). When methionine or methionine sulfoxide were mixed with glucose and heated at 180°C for 1 hr, various pyrazines, methional, dimethyl disulfide and dimethyl trisulfide were generated (Yu and Ho, 1995). Glucose has a catalytic effect on the formation of volatile compounds from thermal degradation of methionine or methionine sulfoxide (Yu and Ho, 1995).

The reaction medium is also very important for the production of flavor components. Okumura *et al.* (1990) and Okumura (1992) studied the volatile products formed from L-cysteine and dihydroxyacetone thermally treated in different solvents such as

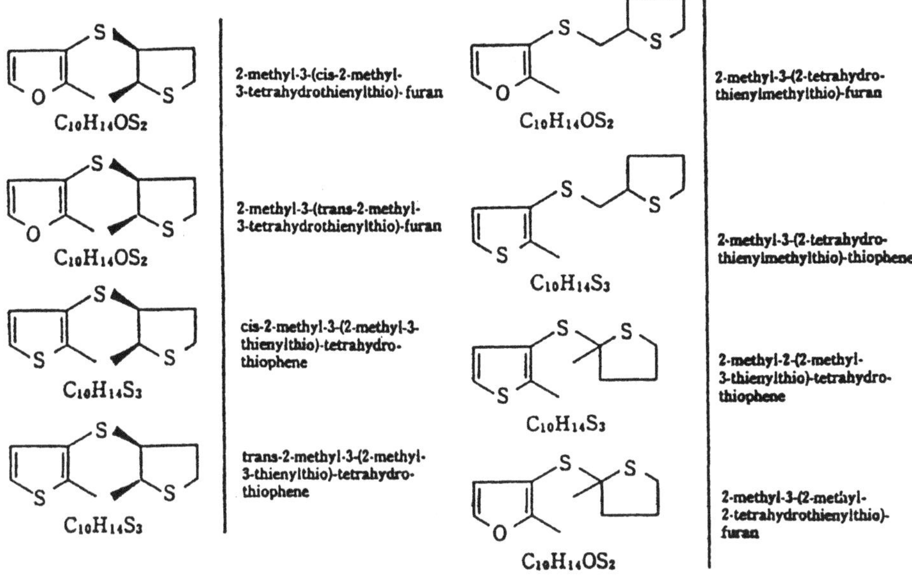

Figure 6. Heterocyclic thioethers in a model meat flavor system.

water, glycerin or triacylglycerols. They found the flavor from triacylglycerols was roasted rice and sulfury, while the flavor from glycerine was roasted, burnt rice and somewhat like roasted meat. The flavor generated in water was roasted rice, sweet sulfury roasted and somewhat like roasted chicken. The difference in generated products is shown in Figure 7.

ANTIOXIDATIVE EFFECT OF MAILLARD REACTION PRODUCTS

Maillard reaction products (MRPs) are known for their ability to retard lipid oxidation. The antioxidative effect of MRPs was first reported by Franzke and Iwainsky (1954), when they observed the antioxidative stability of margarine by the addition of the glycine-glucose reaction products. They estimated that the brown polymer pigment of the reaction mixture was the cause for the antioxidative stability. Later on, the antioxidative activity (AOA) of MRPs from model systems was extensively studied by various authors. Kirigaya *et al.* (1968, 1969) studied the relation of color intensity and reductones with AOA of MRPs from various amino acids and reducing sugars. They reported that the MRPs from arginine, histidine or cysteine with glucose had stronger AOA than that of other amino acid MRPs. Furthermore, the quantity of nitrogen and reductones had an important role in the AOA of the MRPs. MRPs with higher quantities of nitrogen and reductones exhibited stronger AOA. However, Rhee and Kim (1975) observed AOA from the products of a glucose caramelization-type browning reaction, and suggested that amino groups are not always necessary for the formation of antioxidative products. Thus indicating the need for further investigations in this area. Yamaguchi and Fujimaki (1974) reported that AOA

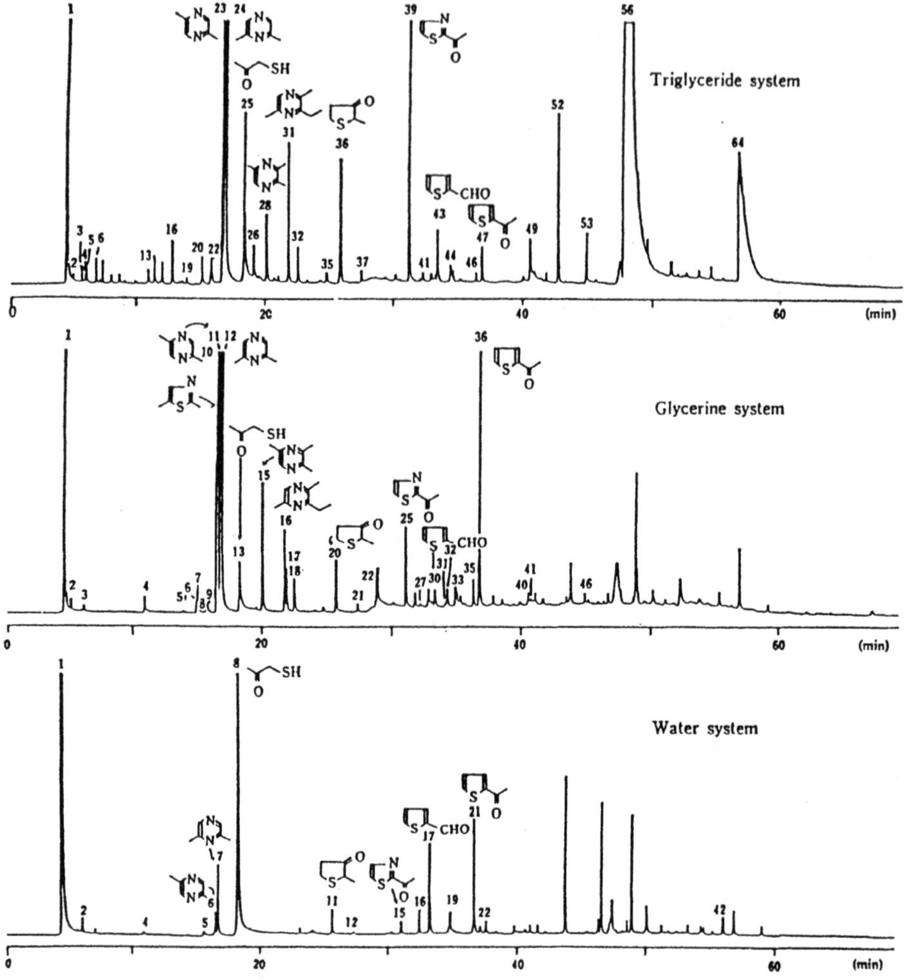

Figure 7. Gas chromatograms of volatiles formed by L-cysteine-dihydroxyacetone model system in various medium.

of MRPs from the reaction of glycine with xylose was comparable to the effect of butylated hydroxyanisole (BHA) but lower than that of butylated hydroxytoluene (BHT) (Figure 8). The synergy between melanoidins and tocopherol was observed (Yamaguchi *et al.*, 1981). Lingnert *et al.* (1980a) studied the AOA of MRPs from various amino acids with reducing sugars and concluded that MRPs from basic amino acids such as arginine, histidine, lysine showed strong AOA which increased as the concentration of amino acids and reducing sugars increased.

Besides, MRPs from amines (Yamaguchi, 1969), peptides (Lingnert *et al.*, 1980b) and protein hydrolysates (Utsunomiya *et al.*, 1983) also exhibit AOA. Lingnert *et al.* (1980b) observed that stronger AOA was obtained with MRPs from histidylglycine-xylose reaction than MRPs from histidine-glucose reaction, and that the AOA is dependent on the

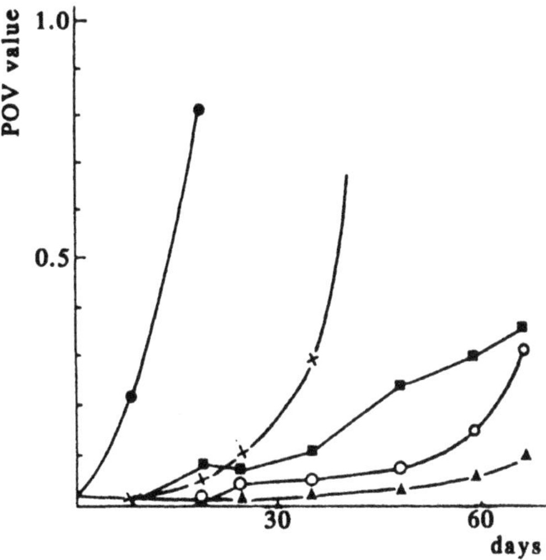

Figure 8. Comparison of antioxidative effect of various antioxidants with melanoidin. Symbols are: ×, control; ■, BHA; ●, erythorbic acid; ▲, BHT; and ○, melanoidin.

amino acid sequence in the peptides. Utsunomiya *et al.* (1983), examined the AOA of ovalbumin hydrolysate in the reaction with glucose (OVG) and found that OVG show potent AOA. The OVG also have strong synergistic AOA with tocopherols. Recently, Eiserich *et al.* (1992) observed some volatile antioxidants formed from an L-cysteine-D-glucose Maillard model system and found that 2,5-dimethyl-4-hydroxy-3(2H)-furanone, oxazole and thiazoles exhibit AOA. However, the antioxidative strength of these individual volatiles was weaker than those of α-tocopherol or BHA.

Antioxidative Effect of MRPs in Food Systems

Three methods are used in the food industry to add MRPs to foodstuffs: (a) amino acids and reducing sugars are added before processing to generate MRPs by the Maillard reaction, (b) amino acid-reducing sugar reaction mixture is added to foodstuff, and (3) melanoidins are added. Kato *et al.* (1976) prepared a browning oil by adding isoleucine and glucose to soybean oil and heated it at 175 °C for 5 min. They found that the antioxidative stability of potato chips was increased when this browning oil was added (Figure 9). Yamaguchi *et al.* (1964) also found that MRPs maintained the stability of fats contained in biscuit and cookies. Lingnert *et al.* (1980c) reported an increase in oxidative stability of cookies containing 0.1% of MRPs from histidine-glucose. The same tendency was also found for sausages (Lingnert *et al.*, 1980d). Bedinghaus and Ockerman (1995) prepared various kinds of MRPs from arginine, histidine, lysine, leucine and tryptophan by the reaction with glucose or xylose. The MRPs were added to fresh ground pork patties prior to cooking and after which the patties were cooked to 68 °C internal temperature and stored at 4 °C for 10 days. The results demonstrated that the MRPs were effective against lipid oxidation in ground pork patties. The most effective MRPs were from lysine-xylose and tryptophan-xylose when compared to control.

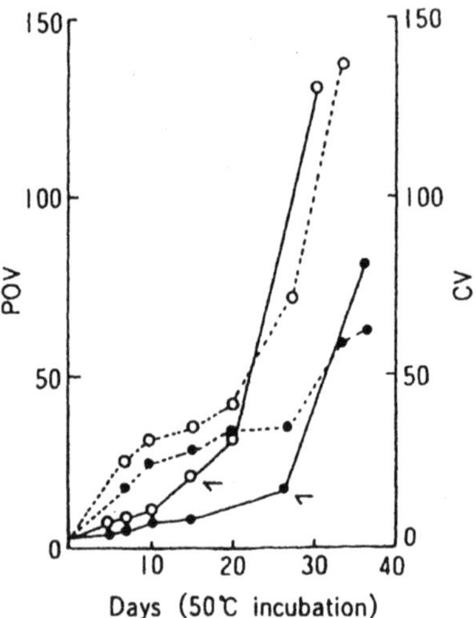

Figure 9. Effects of browning reaction products for the autoxidation of potato chips* at 50°C. Potatoes were sliced with deep-fat-fried in soybean oil with our without browning reaction products. Symbols are: ○ — ○, POV; and ○ --- ○, CV of soybean oil in potato chips (control), ● — ● , POV; and ● --- ●, CV of soybean oil ± browning reaction products (10%) in potato chips. Arrows indicate the development of oxidized flavor.

Antioxidative Effect of MRPs in Vivo

The MRPs not only exhibit antioxidative activity (AOA) *in vitro*, but also *in vivo* (Chuyen *et al.*, 1990). The MRPs of histidine, lysine, arginine and glycine were added to the diet of rats at a level of 6% and the rats were fed for 2 months. The 2-thiobarbituric acid (TBA) values of rats fed with MRPs showed significant lower values than that of the control (Table 3). The strong antioxidative strength order of these amino acids were as follows: His, Arg, Lys, Gly; similar to *in vitro* results. Consequently, it is estimated that the MRPs which are strong *in vitro*, are also strong *in vivo*. The MRPs also demonstrated synergistic effects with vitamin E (Table 4).

Moreover, MRPs from casein or soy protein also showed AOA *in vivo* after feeding to rats for 6 weeks. The liver TBA of rats fed with MRPs from casein or soy protein exhibited significant lower values than those of the control. From these data, it can be estimated that MRPs from proteins also exhibit AOA *in vivo*.

Antioxidative Mechanism of MRPs in Vitro and in Vivo

The antioxidative mechanism of MRPs *in vitro* and *in vivo* is currently unclear. However, it is estimated that the reductone structures, the electron donor property (Kato,

Table 3. TBA values[1] of livers of rats fed with Maillard reaction

Group	Control	His	Lys	Gly	Arg
TBA (nmol MDA/g wet liver)	69.4 ± 1.3	33.3 ± 1.7[2]	47.5 ± 1.6[2]	63.6 ± 1.6[3]	43.9 ± 1.5[2]
Inhibition (%)	—	52.0	31.6	8.4	36.7

MDA - Malondialdehyde.
[1]Mean ± SE (n=4).
[2]Statistically significant ($p<0.01$) against control.
[3]Statistically significant ($p<0.05$) against control.

Table 4. Hemolysis and TBA values[1] of livers of rats fed with His-Glc reaction products

Group	N	C	E_1	E_1C	E_2	E_2C
Hemolysis	81.2 ± 6.9	68.2 ± 11.7	28.2 ± 5.8[2]	11.6 ± 5.6[2]	9.2 ± 1.8_2	4.3 ± 1.5[2]
TBA (nmol MDA/g wet liver)	59.1_1 ± 6.1	49.6^b ± 3.6	49.7 ± 6.1	33.5^c ± 4.8	47.2^d ± 5.4	30.3^e ± 4.5
Inhibition (%)	—	16.0	15.9	43.3	20.1	48.7

Hemolysis was measured using dilauric acid (1 mg/mL) after incubation at 37° for 30 min. O.D. was read at 540 nm.
N = control; C - His-Glc, E_1 = Vitamin E (4 ppm), E_1C = (His-Glc + E_1,) E_2 = Vitamin E (8 ppm); E_2C = (His-Glc + E_2).
[1]Mean ± SE (n=4). [2]Statistically significant (p<0.01) against control. MDA = Malondialdehyde.
[a,c] = Statistically significant between a and c (p<0.05);
[a,b,d,e] = Statistically significant between a and e (p<0.01);
[b,e] (p<0.05); [d,e] (p<0.05).

1973) and the chelating properties of melanoidins against transition metals (Gomyo and Horikoshi, 1976; Terasawa *et al.*, 1991) are important factors contributing to the AOA of MRPs. Recently, the scavenging activity of MRPs such as Amadori compounds and melanoidins against reactive oxygen species was reported (Hayase *et al.*, 1989; Okamoto *et al.*, 1992, Chuyen *et al.*, 1990, 1996, Yen and Hsieh, 1995). This scavenging activity is also an important factor for the AOA of MRPs.

Since the melanoidins are brown in color, the application of melanoidins in foodstuffs is limited. In order to avoid this disadvantage, Yamaguchi (1991) decolorized melanoidins by hydrogenation, oxidation or microorganism (*Coriolus versicolor* IFO 30340) and found that the AOA of melanoidins decolorized by hydrogenation decreased their activity. However, melanoidins decolorized by ozone (Figure 10) or Coriolus versicolor maintained their original activity. This finding is very important for the food industry. Terasawa *et al.* (1996) also decolorized melanoidins from foodstuffs using *Coriolus versicolor* IFO 3034, *Paecilomyces canadensis* NC-1 and *Streptomyces werraensis* TT14, and found that the brown pigment in foods can be categorized by comparing the microbial decolorization pattern of the foods and model brown pigments.

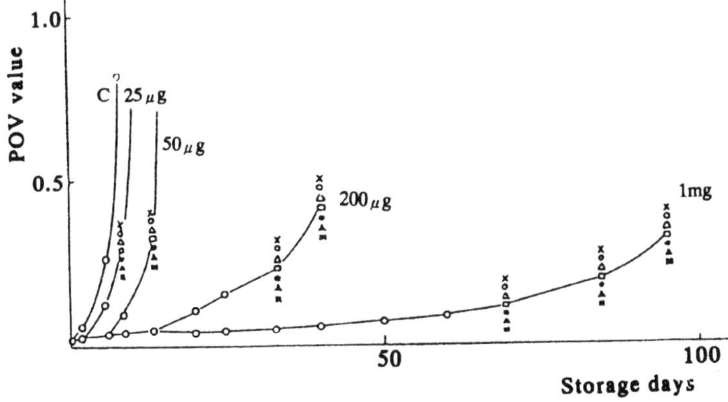

Figure 10. The comparison of antioxidative activity of melanoidin with that of oxidized melanoidins. The concentration of melanoidin was 20 mg/100 ml of distilled water.

DESMUTAGENICITY OF MAILLARD REACTION PRODUCTS (MRPs)

Both mutagenic and desmutagenic compounds are formed by the Maillard reaction. Ohmura *et al.* (1983) reported the generation of some weak mutagens such as 5-hydroxymethyl-2-furfural, 2-methylthiazolidine and products from triose reductones through the Maillard reaction at 100°C. Heterocyclic amines such as MeIQx and DiMeIQx are also generated by the reaction of amino acid, glucose and creatinine (Wakabayashi *et al.*, 1986; Jägerstad *et al.*, 1991; Jackson and Hargraves, 1995). Hiramoto *et al.* (1993) reported that some MRPs which generate reactive oxygen species lead to the cleavage of DNA strands.

However, the desmutagenic effect of MRPs are also reported. Chan *et al.* (1979) reported the desmutagenic effects of a heated lysine-fructose mixture and caramelized sucrose. Kato *et al.* (1985) reported the desmutagenic effect of melanoidins against the mutagens needing S9 mix such as heterocyclic amines *i.e.*, Trp-p-1, Trp-p-2, Glu-p-1, Glu-p-2, IQ, Aflatoxin B_1 and 2-aminoanthracene. Melanoidins also showed desmutagenic effect on mutagens not needing S9 mix such as 2-nitrofluorene, MNNG, 4-NQO, streptozotocin (Lee *et al.*, 1994). The desmutagenic mechanism against heterocyclic amines was also made clear. Melanoidins inhibit the mutagenic activity of activated heterocyclic amine but not the S9 enzymes (Lee *et al.*, 1994). Low molecular weight melanoidins were absorbed at a level of 12.8% in rats, and 80% of these melanoidins were present in a form bound to protein in the plasma. The protein-bound melanoidins also exhibited desmutagenic activity against Trp-p-1. Thus, it is considered that absorbed melanoidins may actually act as desmutagens *in vivo* (Lee *et al.*, 1992). Melanoidins prepared from peptides or egg albumin hydrolysate also showed desmutagenic activity against Trp-p-1 (Table 5). The brown pigment from foodstuffs or food materials such as soy sauce, black beer, cara-

Table 5. Desmutagenicity of nondialyzable melanoidins (MEL) prepared from glucose and various amino acids, peptides or egg albumin hydrolysate at 110°C, pH 7.0 for 1 h, 2 h or 95°C, pH 7.0 for 7 h against Trp-P-1

Nondialyzable melanoidins (2 mg/plate)		Desmutagenicity[1] Revertants/plate 110°C 1 h	110°C 2 h	95°C 7 h
Glu	MEL	198	225	115
Gly	MEL	121	145	151
Ala	MEL	612	79	84
Val	MEL	107	89	107
His	MEL	76	156	64
Lys	MEL	412	127	240
Trp	MEL	89	65	366
GC	MEL	—	—	128
GGG	MEL	—	—	278
EAH	MEL	—	—	141

[1]Desmutagenicity against Trp-P-1 (0.1 μg/plate) was assayed by the preincubation method, using *S. Typhimurium* TA98 in the presence of S9 mix.
The revertants for Trp-P-1 used as a control were 553±19.
GG: Glycylglycine; GGG: Glycylglycylglycine; EAH: Egg Albumin Hydrolyzate.

Table 6. Mutagenicity and desmutagenicity of browning foods and their high molecular weight fractions

Foods	Fraction No.	Mutagenicity[1] revertants per plate	Desmutagenicity[2] against Trp-P-1 Revertants per plate	(%)
Soy Sauce	I	127±20	314±7	(43.2)
	II	176±10	191±9	(65.3)
	III	102±5	129±8	(76.7)
Black Beer	I	107±15	365±10	(34.0)
	II	107±9	254±8	(54.1)
	III	105±5	149±3	(73.8)
Carmel	I	138±8	74±1	(86.7)
	II	132±10	62±3	(88.8)
	III	132±7	42±1	(92.4)
Molasses	I	—	103±2	(81.4)
	II	—	64±1	(88.5)
	III	—	62±1	(88.8)

[1]Mutagenicity was assayed by the preincubation method, using S. typhimurium TA100 in the absence of S9 mix. The revertants for distilled water used as a control were 112±8.
[2]Desmutagenicity against Trp-P-1 (0.1 μg/plate) was assayed by the princubation method, using S. typhimurium TA98 in the presence of S9 mix. The reverants for Trp-P-1 used as a control were 553±19.
Fraction numbers are: I, Lyophilized fraction; II, Nondialyzable fraction; and III, Melanoidins isolated from nondialyzable fraction by Sephadex G-25 and G-100 column chromatography.

mel and molasses also show strong desmutagenic activity (Table 6). Yen *et al.* (1992) and Yen and Lii (1992a, 1992b) reported the desmutagenic effect of MRPs derived from amino acids with sugars such as glycine-xylose, tryptophan-glucose. They also studied desmutagenic mechanism of MRPs from xylose-lysine, and concluded that desmutagenic effects of xylose-lysine MRPs to IQ might be due to the interaction of MRPs and IQ metabolites to form inactive adducts (Yen and Tsai, 1993; Yen and Hsieh, 1994). This result is in accordance with those reported by Lee *et al.* (1992).

Besides, melanoidins are also known to degrade nitrites and inhibit the formation of carcinogenic nitrosamines (Kato *et al.*, 1987). The inhibition was highest at pH 1.2 (Table 7). Nitrosamines are generally produced by an electrophilic reaction of nitrite with the

Table 7. Inhibition (%) of N-nitrosamine formation by melanoidins at various pH conditions

Melanoidins	pH 1.2	pH 3.0	pH 4.2	pH 6.1
Nondialyzable				
NDEA	99.1±0.4	99.0±0.2	30.9±7.6	66.7±4.6
NPYR	60.6±1.4	82.8±1.9	34.8±7.6	55.9±3.2
Reduced				
NDEA	38.0±1.6	9.9±2.2	—	—
NPYR	20.0±0.9	7.9±2.4	—	—

Melanoidins (10 mg) were treated with nitrite (34.5 mg) and diethylamine (9.1 mg) or pyrrolidine (8.9 mg) in 10 mL of each pH solution at 37°C for 2 h.
NDEA: N-nitrosodiethylamine.
NPYR: N-nitrosopyrrolidine.
Data are mean values of three determinations ± standard deviation.

corresponding secondary amine under acidic conditions in the human stomach. From these results, it can be estimated that melanoidin might be effective for the prevention of stomach cancer, and melanoidins can be used as additive for functional foods.

Melanoidins are also reported to exhibit antibacterial activity (Einarsson, 1990), however, this activity is not strong. In a Japanese patent, Yajima (1979) used melanoidin mixed with glycerol fatty acid ester such as glycerol monocaprylate to prolong the preservation of meat products such as ham, sausage, and agricultural food products.

IMPROVEMENT OF THE FUNCTIONAL PROPERTIES OF PROTEINS BY MAILLARD REACTION

Many investigations have attempted to convert food proteins into proteins with improved functional properties by chemical and enzymatic modifications. However, food protein modification by chemical reagents is limited because of safety concerns. Recently, experiments were carried out to prepare protein-polysaccharide conjugate by Maillard reaction (Kato *et al.*, 1990, 1991; Nakamura *et al.*, 1991, 1992a). A safe ovalbumin-dextran conjugate was prepared by covalent binding between amino groups in the protein and the reducing-end carbonyl group in the polysaccharide in the dry state without using any other chemical reagent (Kato *et al.*, 1990). The emulsifying properties of these conjugates were reported to be much better than those of commercial emulsifiers even under acidic pH or high-salt conditions (Table 5, Kato *et al.*, 1992). Since high salt conditions, acidic pH and heating processes are commonly used in industrial application, the lysozyme-galactomannan conjugate is suitable as a safe macromolecular emulsifier for food processing. This lysozyme-galactomannan also showed antimicrobial effect against Gram-negative bacteria such as E. Coli. In the presence of lysozyme-galactomannan, the living cells were drastically decreased by heating at 50°C, while the bactericidal effect was not observed in the presence of native lysozyme and in control medium. Under the reaction condition of 60°C and 79%RH in two weeks, the lysozyme was attached to two galactomannans per molecule (Kato *et al.*, 1993) (Figure 11). The emulsion stability of lysozyme-galactomannan conjugate was greatly decreased with the extent of the acetylation of lysyl residues of the conjugate. Thus, it is estimated that the electrostatic repulsion of positive charge in the lysozyme contributed to the emulsifying properties of lysozyme-galactomannan conjugate (Nakamura *et al.*, 1994).

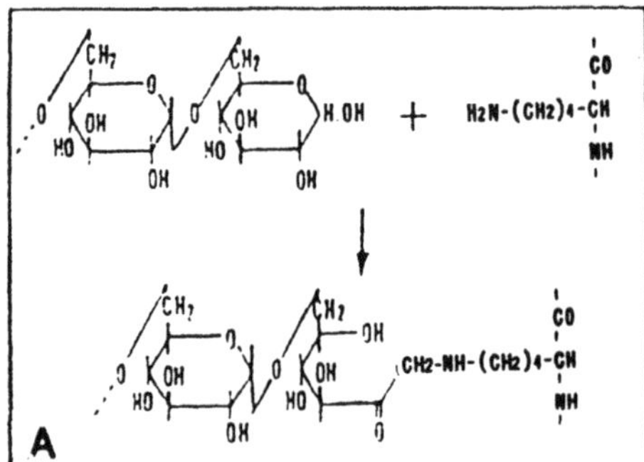

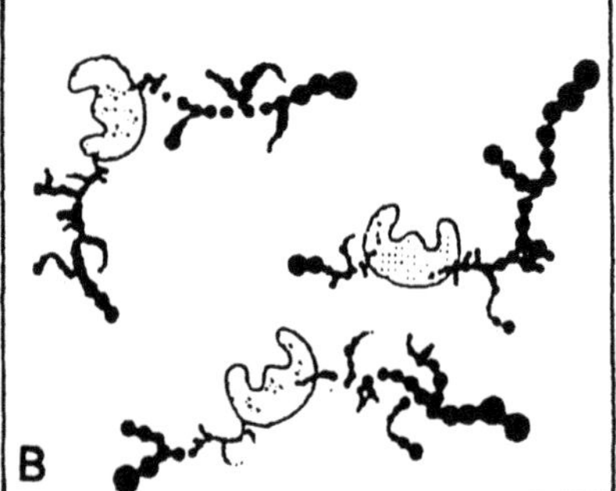

Figure 11. Scheme for binding of protein with polysaccharide through Maillard reaction (A) and the binding mode (B). Doted areas indicate protein molecules; branched solid circles represent polysaccharide molecules.

Table 8. Emulsifying properties of lysozyme-galactomannan conjugate and commercial emulsifiers

Conditions	LG(RH65)[a]	LG(RH79)[b]	Lz[c]	SE11[d]	Q18S[e]
pH 7.4[f]					
activity, OD_{500nm}	1.843	1.920	0.187	0.944	1.165
stability, min	>40	>40	3.5	4.1	2.5
pH 3.0[g]					
activity, OD_{500nm}	1.575	1.569	0.968	0.304	1.288
stability, min	38	>40	1.7	5.1	3.8
0.2 M NaCl[h]					
activity, OD_{500nm}	1.807	1.902	0.129	0.693	1.097
stability, min	39	>40	3.0	0.8	2.3
Heated[i]					
activity, OD_{500nm}	1.757	1.892	0.299	1.142	1.168
stability, min	>40	>40	0.9	2.4	1.8

[a]LG (RH65), lysozyme-galactomannan conjugate obtained from 2 weeks of incubation at 60°C under 65% relative humidity; [b]LG (RH79), lysozyme-galactomannan conjugate obtained from 2 weeks of incubation at 60°C under 79% relative humidity; [c]LZ, native lysozyme; [d]SE11, sucrose fatty acid ester (HLB11) from Taiyo Kagku Co., Ltd.; [e]Q18S, decaglyceryl monoesterate (HLB12) from Taiyo Kagaku; [f]1/15 M sodium phosphate buffer; [g]1/15 M sodium citrate buffer; [h]1/15 M sodium phosphate buffer (pH 7.4) containing 0.2 M NaCl; [i]Samples were heated to 90°C at a rate of 3°C/min in 1/15 M sodium phosphate buffer (pH 7.4) and then immediately cooled to 20°C.

Dried egg white was also covalently attached to galactomannan in dry state through the Maillard reaction at 60°C and 79% RH. The resulting protein-polysaccharide conjugate also had superior emulsifying properties compared to commercial emulsifiers (Nakamura *et al.*, 1992a) (Table 8). The sizes of the polysaccharides also had some effect on the emulsifying property of the conjugate. When various sizes of galactomannan polysaccharide, and xyloglucan oligosaccharide were used as materials for the conjugate, the emulsifying properties of the lysozyme-galactomannan conjugates increased with the length of the polysaccharide chain (Shu *et al.*, 1996) (Figure 12). As shown in Figure 12, the emulsion stability of the lysozyme-galactomannan (3500–6000) conjugate is very low. It appears that a protein galactomannan conjugate having a molecular size of more than 6000–12000 Da is essential for improvement of emulsifying properties. Since the emulsifying properties of lysozyme-xyloglucan was very low, polysaccharides are more effective than oligosaccharides in the formation of emulsifier conjugates.

Nakamura *et al.* (1992b) also reported that the antioxidative effect of ovalbumin was remarkably enhanced by covalent binding of dextran or galactomannan through a controlled Maillard reaction (Table 9). From the above results, it can be considered that conjugates formed by the interaction between protein and polysaccharide at mild conditions are very useful for the food industry in terms of their emulsifying, antioxidative and antimicrobial activities.

THE ROLE OF MODIFIED PROTEIN AND MELANOIDINS IN FOOD AND NUTRITION

When foods are heated during processing and storage, protein are often modified by carbonyl compounds and their nutritive value decreased (Mauron, 1981). Kimiagar *et al.* (1980) reported the effects of long term feeding of browned egg albumin to rats from 3 to

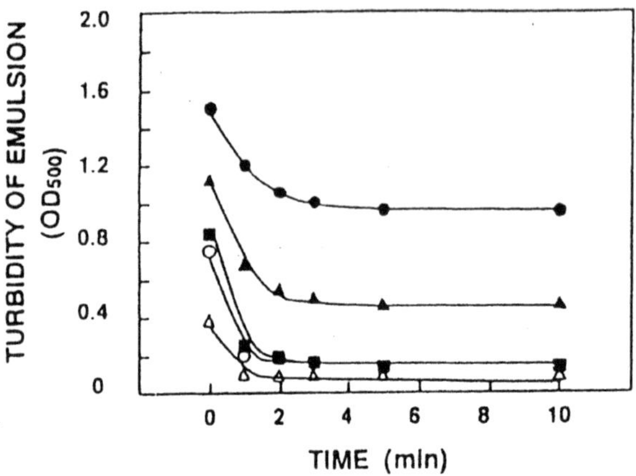

Figure 12. Emulsifying properties of lysozyme-polysaccharide conjugates prepared with various sizes of GM and XB: Δ, native lysozyme; ●, lysozyme-GM (24,000) conjugate; ▲, lysozyme-GM (6,000–12,000) conjugate; ■, lysozyme-GM (3,500–6,000) conjugate; ○, lysozyme-xyloglucane (1,400) conjugate. Data are from a representative experiment repeated three times with similar results. Symbols are: GM, galactomannan; and XG, xyloglucan.

12 months and confirmed that there were significant differences between the browned-protein-fed group and the control group in weight gain, cecum, kidney, liver, serum alkaline phosphatase and serum GOT. From these results, Kimiagar *et al.* (1980) suggested that supplementation of the diet with amino acid could not completely restore the biological value of the browned protein and there was a possibility that some inhibitory or antinutritional compounds might be formed during the Maillard reaction.

Chuyen *et al.* (1991) also studied the nutritional and physiological effect of casein modified by glucose under various conditions on growing of adult rats. They also had

Table 9. Relative oxidation of methyl linoleate in powder model system after days of incubation at 20°C[1]

Compound(s)	POV (%)	TBA (%)
Ovalbumin and dextran		
no addition (control)	100	100
mixture (1.25%)	63.3	41.6
mixture (2.5%)	59.1	34.0
conjugate (1.25%)	36.9	17.4
conjugate (2.5%)	31.6	15.7
Ovalbumin and galactomannan		
no addition (control)	100	100
mixture (1.25%)	67.1	42.3
mixture (2.5%)	63.6	37.8
conjugate (1.25%)	46.9	23.3
conjugate (2.5%)	40.2	19.5

[1]POV, peroxide value; TBA, thiobarbituric acid value.

similar conclusion as Kimiagar (1980). Finot *et al.* (1990) divided physiological effects of the MRPs into 3 types: (a) purely nutritional due to the loss of nutrients; this effect can be restored by supplying the destroyed nutrient, (b) MRPs dependent and modifying the metabolism of intact nutrients, and (c) MRPs dependent and modifying other metabolic functions. Much research has been conducted on these topics, however, the effects are very complicated. In order to explain the physiological effects, products released from the modified proteins must be examined.

Besides, Kaminogawa *et al.* (1984) reported that the allergen from lactose used for infant food, was a sugar-protein complex and the protein moiety of allergens contained high amounts of glutamic acid, glutamine, threonine, aspartic acid and proline. However, Öste *et al.* (1990) reported that the antigenicity of soybean trypsin inhibitor was suppressed by the reaction with reducing sugars such as glucose, lactose and maltose. Modified protein (BSA) was also reported to induce tumor necrosis factor (cachectin) and interleukin-1 *in vivo* (Vlassara *et al.*, 1988). Thus modified proteins had some effects on the immunological system *in vivo*. Since most foods contain modified proteins, it is necessary to make clear the degradation products of modified protein and their physiological effect *in vivo*.

The Maillard reaction has both merits and shortcomings in food processing. The challenge is to control the reaction to increase the positive while minimizing its negative aspects on a case by case basis. In this respect, many volatile compounds have been identified and quantified, however, when we mix all the identified components together, in most cases, we can not reproduce the real flavor of the food. There are various unknown factors to be understood, among which are the identification of nonvolatile Maillard reaction products. At present some products are actually used in food industry for their flavor, color and antioxidative effects. However, with the recent understanding about the functional properties of MRPs, more application studies are required. We consume various kinds of MRPs in the foods we eat everyday, however, the relation of MRPs to human health such as their antinutritional effects, safety and their beneficial aspects are still obscure. Fundamental research is required to elucidate fully the merits and drawbacks of the Maillard reaction.

CONCLUSIONS

Melanoidins are also found in most foodstuffs, and are especially prevalent in soy sauces, soy pastes, sauces, teas, coffees, and roasted products. Melanoidins exhibit antioxidative effect, desmutagenic effect, scavenging activity against reactive oxygen species and may lower serum cholesterol (Miura and Gomyo, 1990). Melanoidins do not affect the digestion or absorption of proteins and lipids (Homma and Fujimaki, 1981).

Since melanoidins have various physiological effects, it is necessary to measure quantitatively the amount of melanoidins in foods. Recently, Hirano *et al.* (1996) proposed a tentative method for measuring brown pigment in foods. The principle was based on any brown pigment with general absorption in the visible range is quantitatively equivalent to melanoidin prepared from a model system of the Maillard reaction. The spectral curve for general absorption was represented by the equation $dE/d\lambda = -kE$ (E, absorbance; λ = wavelength; k = constant > 0), of which the value k was found to correlate with the molecular weight and the molar extinction coefficient, and to serve as an emperical parameter for the measurement of brown substances in different foods.

REFERENCES

Baynes, J.W.; Monnier, V.C. (Eds.) *The Maillard Reaction in Aging, Diabetes, and Nutrition*. Alan R. Liss. Inc.: New York, 1989.

Bedinghaus, A.J.; Ockerman, H.W. Antioxidative Maillard reaction products from reducing sugars and free amino acids in cooked ground pork patties. *J. Food Sci.* **1995**, *60*, 992–995.

Boelens, M.; van der Linde, L.M.; de Valois, P.J.; van Dort, H.M.; Takken, H.J. Organic sulfur compounds from fatty aldehydes, hydrogen sulfide, thiol and ammonia as flavor constituents. *J. Agric. Food Chem.* **1974**, *22*, 1071–1076.

Chan, R.I.M.; Stich, H.F.; Rosin, M.P.; Powrie, W.D. Antimutagenic activity of browning reaction products. *Cancer Lett.* **1979**, *15*, 27–33.

Chuyen, N.V.; Kurata, T.; Fujimaki, M. Studies on the reaction of dipeptides with glyoxal. *Agric. Biol. Chem.* **1973a**, *37*, 327–334.

Chuyen, N.V.; Kurata, T.; Fujimaki, M. On the reaction of dipeptides with dicarbonyl compounds. *Agric. Biol. Chem.* **1972**, *36*, 1257–1258.

Chuyen, N.V.; Kurata, T.; Fujimaki, M. Formation of *N*-[2(3-alkylpyrazin-2-on-1-yl)acyl] amino acids or peptides on heating tri- or tetra peptides with glyoxal. *Agric. Biol. Chem.* **1973b**, *37*, 1613–1618.

Chuyen, N.V.; Kurata, T.; Fujimaki, M. Formation of *N*-carboxymethyl amino acid from the reaction of α-amino acid with glyoxal. *Agric. Biol. Chem.* **1973c**, *37*, 2209–2210.

Chuyen, N.V.; Utsunomiya, N.; Hidaka, A.; Kato, H. Antioxidative effect of Maillard reaction products in vivo. In *The Maillard Reaction in Food Processing, Human Nutrition and Physiology*; Finot, P.A.; Aeschbacher, H.U.; Hurrel, R.F.; Liardon, R., Eds.; Birkhäuser Verlag: Basel, 1990; pp 285–290.

Chuyen, N.V.; Utsunomiya, N.; Kato, H. Nutritional and physiological effects of casein modified by glucose under various conditions on growing and adults rats. *Agric. Biol. Chem.* **1991**, *55*, 657–664.

Chuyen, N.V.; Ijichi, K.; Moteki, K. On the formation and scavenging of reactive oxygen species by amion-carbonyl reaction products in model and food system. *Nippon Nogeikagaku Kaishi* **1996**, *70*, 20.

Einarson, H. The mode of action of antibacterial Maillard reaction products. In *The Maillard Reaction in Food Processing, Human Nutrition and Physiology*; Finot, P.A.; Aeschbacher, H.U.; Hurrel, R.F.; Liardon, R., Eds.; Birkhäuser Verlag: Basel, 1990; pp 215–220.

Eriksson, C. *Maillard Reaction in Foods*, Pregamon Press, Oxford, 1981.

Eiserich, J.P.; Macku, C.; Shibamoto, T. Volatile antioxidants formed from an L-cysteine/D-glucose Maillard model system. *J. Agric. Food Chem.* **1992**, *40*, 1982–1988.

Finot, P.A.; Aeschbacher, H.U.; Hurrel, R.F.; Liardon, R. (Eds.) 1990. *The Maillard Reaction in Food Processing, Human Nutrition and Physiology*, Birkhauser Verlag: Basel, 1990.

Finot, P.A. 1990. Metabolism and physiological effects of Maillard reaction products. In *The Maillard Reaction in Food Processing, Human Nutrition and Physiology*; Finot, P.A.; Aeschbacher, H.U.; Hurrel, R.F.; Liardon, R., Eds.; Birkhäuser Verlag: Basel, 1990; pp 259–272.

Franzke, C.; Iwainsky, H. Zur antioxydativen Wirksamkeit der Melanoidine. *Dtsch. Lebensm. Rundsch.* **1954**, *50*, 251–254

Friedman, M. Food browning and its prevention: an overview. *J. Agric. Food Chem.* **1996**, *44*, 631–653.

Fujimaki, M.; Chuyen, N.V.; Kurata, T. Studies on the decarboxylation of amino acids with glyoxal. *Agric. Biol. Chem.* **1971**, *35*, 2043–2049.

Fujimaki, M.; Namiki, M.; Kato, H. (Eds.) *Amino-carbonyl Reactions in Food and Biological Systems*. Kodansha, Ltd.: Tokyo, 1986.

Gomyo, T.; Horikoshi, M. On the interaction of melanoidin with metallic ions. *Agric. Biol. Chem.* **1976**, *40*, 33–40.

Hayase, F.; Hirashima, S.; Okamoto, G.; Kato, H. Scavenging of active oxygens by melanoidins. *Agric. Biol. Chem.* **1989**, *53*, 3383–5.

Hiramoto, K.; Kato, T.; Kikugawa, K. Generation of DNA - breaking activity in the Maillard reaction of glucose - amino acid mixtures in a solid system. *Mutat. Res.* **1993**, *285*, 191–198.

Hirano, M.; Miura, M.; Gomyo, T. A tentative measurement of brown pigments in various processed foods. *Biosci. Biotech. Biochem.* **1996**, *60*, 877–879.

Ho, C.-T. Thermal generation of Maillard aromas. In *The Maillard Reaction: Consequences for the Chemical and Life Sciences*; Ikan, R., Ed.; John Wiley & Sons Ltd.: Chichester, England, 1996; pp 27–53.

Ho, C.-T.; Carlin, J.T. Formation and aroma characteristic of heterocyclic compounds in foods, In *Flavor Chemistry: Trends and Developments;* Teranishi, R.; Buttery, R. G.; Shahidi, F., Eds.; ACS Symp. Ser. 388, American Chemical Society: Washington D.C. 1989; pp 92–104.

Hodge, J.E. Chemistry of browning reactions in model system. *J. Agric. Food Chem.* **1953**, *1*, 928–943.

Homma, S.; Fujimaki, M. Growth response of rats fed a diet containing nondialyzable melanoidin. *Prog. Food Nutr. Sci.* **1983**, *5*, 209–216.

Hwang, H.I.; Hartman, T.G.; Rosen, R.T.; Lech, J.; Ho, C.-T. Formations of pyrazines from the Maillard reaction of glucose and lysine-α-amine-^{15}N. *J. Agric. Food Chem.* **1994**, *42*, 1000–1004.

Hwang, H.I.; Hartman, T.G.; Ho, C-T. Relative reactivity of amino acids in pyrazine formation. *J. Agric. Food Chem.* **1995**, *43*, 179–184.

Ikan, R. (Ed.) *The Maillard Reaction - Consequences for the Chemical and Life Sciences*. John Wiley and Sons: Chichester, England, 1996.

Jagerstad, M.; Skog, K.; Grivas, S.; Olsson, K. Formation of heterocyclic amines using model systems. *Mutat. Res.* **1991**, *259*, 219–233.

Jackson, L.S.; Hargraves, W. A. Effects of time and temperature on the formation of MeIQx and DiMeIQx in a model system containing threonine, glucose and creatinine. *J. Agric. Food Chem.* **1995**, *43*, 1678–1694.

Kaminogawa, S.; Kumagai, Y.; Yamauchi, K.; Iwasaki, E.; Mukoyama, T.; Baba, M. Allergic skin reactivity and chemical properties of allergens in two grades of lactose. *J. Food Sci.* **1984**, *49*, 529–535.

Kato, H. Antioxidative activity of amino-carbonyl reaction products. *Shokuhin Eiseigaku Zasshi.* **1973**, *14*, 343–351.

Kato, S.; Yano, N.; Suzuki, I.; Ishi, T.; Kurata, T.; Fujimaki, M. Effect of L-cysteine on browning of egg albumin. *Agric. Biol. Chem.* **1974**, *38*, 2425–2430.

Kato, H.; Horie, T.; Fujimaki, M. Antioxidant activity of the browning oils prepared from D-glucose and L.-leucine. *Nippon Eiyo Shokuryo Gakkaishi* **1976**, *29*, 179–181.

Kato, H.; Kim, S.B.; Hayase, F.; Chuyen, N.V. Desmutagenicity of melanoidins against mutagenic pyrolysates. *Agric. Biol. Chem.* **1985**, *49*, 3093–3095.

Kato, H.; Lee, I.E.; Chuyen, N.V.; Kim, S.B.; Hayase, F. Inhibition of nitrosamine formation by nondialyzable melanoidins. *Agric. Biol. Chem.* **1987**, *51*, 1333–1338.

Kato, A.; Sasaki, Y.; Furuta, R.; Kobayashi, K. Functional protein - polysaccharide conjugate prepared by controlled dry heating of ovalbumin - dextran mixture. *Agric. Biol. Chem.* **1990**, *54*, 107–112.

Kato, A.; Shimokawa, K.; Kobayashi, K. Improvement of the functional properties of insoluble gluten by pronase digestion followed by dextran conjugation. *J. Agric. Food Chem.* **1991**, *39*, 1053–1056.

Kato, A.; Mifuru, R.; Matsudomi, N; Kobayashi, K. Functional casein - polysaccharide conjugate prepared by controlled dry heating. *Biosci. Biotechnol. Biochem.* **1992**, *56*, 567–571.

Kato, A.; Minaki, K.; Kobayashi, K. Improvement of emulsifying properties of egg white proteins by the attachment of polysaccharide through Maillard reaction in dry state. *J. Agric. Food Chem.* **1993**, *41*, 540–543.

Kimiagar, M.; Lee, T. C.; Chichester, C.O. Long term feeding effects of browned egg albumin to rats. *J. Agric. Food Chem.* **1980**, *28*, 150–155.

Kirigaya, N; Kato, H; Fujimaki, M. Studies on antioxidant activity of nonenzymic browning reaction products. Part I. Relations of color intensity and reductones with antioxidant activity of browning reaction products *Agric. Biol. Chem.* **1968**, *32*, 287–290.

Kirigaya, N.; Kato, H; Fujimaki, M. Studies on antioxidant activity of nonenzymic browning reaction products. Part II. Antioxidant activity of nondialyzable browning reaction products. *Nippon Nogei Kagaku Zasshi.* **1969**, *43*, 484–491.

Kurata, T; Kato, H. Formation of cooking flavor in foods. *The Koryo* **1981**, *132*, 11–26.

Labuza, T. P.; Reineccius, G. A.; Monnier, V. M.; O'Brien, J.; Baynes, J.W. (Eds.) *Maillard Reactions in Chemistry, Food and Health.* The Royal Society of Chemistry: Cambridge, U.K., 1994.

Lee, I.E.; Chuyen, N.V.; Hayase, F.; Kato, H. Absorption and distribution of ^{14}C-melanoidins in rats and the desmutagenicity of absorbed melanoidins against Trp-P-1. *Biosci. Biotechnol. Biochem.* **1992**, *56*, 21–23.

Lee, I.E.; Chuyen, N.V.; Hayase, F.; Kato, H. Desmutagenicity of mlanoidins against various kinds of mutagens and activated mutagens. *Biosci. Biotechnol. Biochem.* **1994**, *58*, 21–23.

Lee, T.C.; Kim, H.J. (Eds.) *Chemical Makers for Processed and Stored Foods*. ACS Symp. Ser. 631, American Chemical Society: Washington D.C., 1996.

Lingnert, H.; Ericksson, C.E. Antioxidative Maillard reaction products I. Products from sugars and free amino acids. *J. Food Process. Preserv.* **1980a**, *4*, 161–172.

Lingnert, H.; Ericksson, C. E. Antioxidative Maillard reaction products II. Products from sugars and peptides or protein hydrolysates. *J. Food Process. Preserv.* **1980b**, *4*, 173–1981.

Lingnert, H. Antioxidative Maillard reaction products III. Application in cookies. *J. Food Process. Preserv.* **1980c**, *4*, 219–233.

Lingnert, H.; Lundgren, B. Antioxidative Maillard reaction products IV. Application in sausage. *J. Food Process. Preserv.* **1980d**, *4*, 235–246.

Mabrouk, A.F. Flavor of browning reaction products. In *Food Taste Chemistry*; Boudreau, J.C., Eds.; ACS Symp. Ser. 115, American Chemical Society: Washington D.C., 1979; pp 205–245.

Maillard, L.C. Action des acides amines sur les sucres: formation des melanoidins par voie methodologique. *C. R. Acad. Sci.* **1912**, *154*, 66–68.

Mauron, J. 1981. The Maillard reaction in food, a critical review from the nutritional stand-point. *Prog. Food. Nutr. Sci.* ***1981***, *5*, 5–35.

Miura, M.; Gomyo, T. Effect of melanoidin on cholesterol in plasma, liver and feces in rats fed a high-cholesterol diet. In *The Maillard Reaction in Food Processing, Human Nutrition and Physiology*; Finot, P.A.; Aeschbacher, H.U.; Hurrel, R.F.; Liardon, R., Eds.; Birkhauser Verlag: Basel, 1990; pp 291–296.

Nakamura, S.; Kato, A.; Kobayashi, K. New antimicrobial characteristics of lysozyme - dextran conjugate. *J. Agric. Biol. Chem.* **1991**, *39*, 647–650.

Nakamura, S.; Kato, A.; Kobayashi, K. Bifunctional lysozyme-galactomannan conjugate having excellent emulsifying properties and bactericidal effect. *J. Agric. Food Chem.* **1992a**, *40*, 753–739.

Nakamura, S.; Kato, A.; Kobayashi, K. Enhanced antioxidative effect of ovalbumin due to covalent binding of polysaccharides. *J. Agric. Food Chem.* **1992b**, *40*, 2033–2037.

Nakamura, S.; Kato, A.; Kobayashi, K. Role of positive charge of lysozyme in the excellent emulsifying properties of Maillard-type lysozyme-polysaccharide conjugates. *J. Agric. Food Chem.* **1994**, *42*, 2688–2691.

Oh, Y.C.; Shu, C.K.; Ho, C.-T. Some volatile compounds from thermal interaction of glucose with glycine, diglycine, triglycine and tetraglycine. *J. Agric. Food Chem.* **1991**, *39*, 1553–1554.

Oh, Y.C.; Shu, C.K.; Ho, C.-T. Formation of novel 2(1H)-pyrazinones as peptide-specific Maillard reaction products. *J. Agric. Food Chem.* **1992**, *40*, 118–121.

Okamoto, G.; Hayase, F.; Kato, H. Scavenging of active oxygen species by glycated proteins. *Biosci. Biotechnol. Biochem.* **1992**, *56*, 928–931.

Okumura, J.; Yanai, T.; Yajima, I.; Hayashi, K. 1990. Volatile products formed from L-cysteine and dihydroxyacetone thermally treated in different solvents. *Agric. Biol. Chem.* **1990**, *54*, 1631–1638.

Okumura, J. Volatile flavor products from Maillard reaction. *Food Technol. (in Japanese)* **1992**, *30*, 41–52.

Ohmura, H.; Tadan, N.; Shinohara, K.; Murakami, H. Formation of mutagens by the Maillard reaction. In *The Maillard Reaction in Food and Nutrition*; Waller, G. R.; Feather, M. S., Eds.; ACS Symp. Ser. 215, American Chemical Society: Washington D.C., 1983; pp 537–544.

Öste, R.E.; Brandon, D.L.; Bates, A.; Friedman, M. Antibody-binding to a Maillard-reacted protein. In *The Maillard Reaction in Food Processing, Human Nutrition and Physiology*; Finot, P.A.; Aeschbacher, H.U.; Hurrel, R.F.; Liardon, R., Eds.; Birkhauser Verlag: Basel, 1990; pp 303–308.

Rhee, C.; Kim, D.H. Antioxidative activity of acetone extracts obtained from a camelization-type browning reaction. *J. Food Sci.* **1990**, *40*, 460–463.

Rizzi, G.P. Heat-induced flavor formation from peptides. In *Thermal Generation of Aromas*; Parliment, T.H.; McGorrin, R.J.; Ho, C.-T., Eds.; ACS Symp. Ser. 409; American Chemical Society: Washington D.C., 1989; pp 172–181.

Sakaguchi, M.; Shibamoto, T. Formation of heterocyclic compounds from the reaction of cysteamine and D-glucose, acetaldehyde, or glyoxal. *J. Agric. Food Chem.* **1978**, *26*, 1179–1183.

Shigematsu, H.; Shibata, S.; Kurata, T.; Kato, H.; Fujimaki, M. 5-Acetyl-2,3-dihydro-1H-pyrrolizines and 5,6,7,8-tetrahydroindilizin-8-ones, odor constituents formed on heating L-proline with D-glucose. *J. Agric. Food Chem.* **1975**, *23*, 233–237.

Shigematsu, H.; Shibata, S.; Kurata, T.; Kato, H.; Fujimaki, M. Thermal degradation products of several Amadori compounds. *Agric. Biol. Chem.* **1977**, *41*, 2377–2385.

Shu, C.K.; Lawrence, B.M. 1994. Presented at the 208th National Meeting of the American Chemical Society, Washington D.C.

Shu, Y.W.; Sahara, S.; Nakamura, S.; Kato, A. Effects of the length of polysaccharide chains on the functional properties of the Maillard-type lysozyme-polysaccharide conjugate. *J. Agric. Food Chem.* **1996**, *44*, 2544–2548.

Terasawa, N.; Murata, M.; Homma, S. Separation of model melanoidin into components with copper chelating sepharose 6B column chromatography and comparison of chelating activity. *Agric. Biol. Chem.* **1991**, *55*, 1507–1514.

Tressl, R.; Rewicki, D.; Helak, B.; Kampershröer, H.; Martin, N. Formation of 2,3-dihydro-1H-pyrrolizines as proline specific Maillard products. *J. Agric. Food Chem.* **1985a**, *33*, 919–923.

Tressl, R.; Rewicki, D.; Helak, B.; Kampershršer, H. Formation of pyrrolidines and piperidines on heating L-proline with reducing sugars. *J. Agric. Food Chem.* **1985b**, *33*, 924–928.

Tressl, R.; Helak, B.; Martin, N.; Rewicki, D. Formation of proline specific Maillard products. In *Amino-carbonyl Reaction in Food and Biolofical Systems;* Fujimaki, M.; Namiki, M.; Kato, H., Eds.; Elsevier-Kodansha: Tokyo, 1986; pp 235–244.

Utsunomiya, N.; Hayase, F.; Kato, H. Antioxidative activities of Maillard reaction products of D-glucose with ovalbumin hydrolyzed by proteases, and their synergistic effect with tocopherols. *Nippon Eiyo Shokuryo Gakkaishi.* **1983**, *36*, 461–465.

Vlassara, H.; Brownlee, M.; Manogue, K.R.; Dinarells, C.A.; Pasagian, A. Cachectin / TNF and IL-1 induced by glucose-modified proteins: role in normal tissue remodeling. *Sci.* **1988**, *240*, 1546–1548.

Wakabayashi, K.; Takahashi, M.; Nagao, M.; Sato, S.; Kinae, N.; Tomita, I.; Sugimura, T. Quantification of mutagenic and carcinogenic heterocyclic amines in cooked foods. In *Amino-carbonyl Reactions in Food and Biological Systems*; Fujimaki, M.; Namiki, M.; Kato, H., Eds.; Elsevier Kodansha, Tokyo, 1986.

Waller, G.W.; Feather, M.S. (Eds.) *The Maillard in Food and Nutrition*, ACS Symp. Ser. 215, American Chemical Society: Washington D.C., 1983.

Werkhoff, P.; Bruning, J.; Emberger, R.; Guntert, M.; Kopsel, M.; Kuhn, W.; Surburg, H. Isolation and characterization of volatile sulfur-containing meat flavor components in model systems. *J. Agric. Food Chem.* **1990**, *38*, 777–791.

Wells-Knecht, K.J.; Zyzak, D.V.; Litehfield, J.E.; Thorpe, S.R.; Baynes, J.W. Mechanisms of antioxidative glycosylation: Identification of glyoxal and arabinose as intermediates in the autooxidative modification of proteins by glucose. *Biochem.* **1995**, *34*, 3702–3709.

Yajima, M. Method of preservation and quality improvement of foodstuff. Japan Patent No. 1049789, 1979.

Yamaguchi, N.; Yokoo, Y.; Koyama, Y. Studies on the browning reaction products on the stability of fats contained in biscuit and cookies. *Nippon Shokuhin Kogyo Gakkaishi* **1964**, *11*, 184–189.

Yamaguchi, N. Effect of 3-deoxy-xylosone and its browning reaction product on the stabilities of fat and oil. *Nippon Shokuhin Kogyo Gakkaishi* **1969**, *16*, 94–96.

Yamaguchi, N.; Fujimaki, M. Studies on browning reaction products from reducing sugars and amino acids. Antioxidative activity of purified melanoidins and their comparison with those of legal antioxidants. *Nippon Shokuhin Kogyo Gakkaishi* **1974**, *21*, 6–12.

Yamaguchi, N.; Koyama, Y.; Fujimaki, M. Fractionation and antioxidative activity of browning reaction products between D-xylose and glycine. *Prog. Food Nutr. Sci.* **1981**, *5*, 429–439.

Yamaguchi, N. Antioxidative effect of decolorized melanoidin. *New Food Industry (in Japanese)* **1991**, *33*, 76–80.

Yen, G.C.; Tsai, L.C.; Lii, J.D. Antimutagenic effect of Maillard browning products obtained from amino acids and sugars. *Food Chem. Toxicol.* **1992**, *30*, 127–132.

Yen, G.C.; Lii, J.D. Influence of the reaction conditions on the antimutagenic effect of Maillard reaction products derived from xylose and lysine. *J. Agric. Food Chem.* **1992a**, *40*, 1034–1037.

Yen, G.C.; Lii, J.D. Antimutagenic effect of Maillard reaction products prepared from glucose and tryptophan. *J. Food Prot.* **1992b**, *55*, 615–619.

Yen, G.C.; Tsai, L.C. Antimutagenic of a partially fractionated Maillard reaction product. *Food Chem.* **1993**, *47*, 11–15.

Yen, G.C.; Hsieh, P.P. Possible mechanisms of antimutagenic effect of Maillard reaction products prepared from xylose and glycine. *J. Agric. Food Chem.* **1994**, *42*, 133–137.

Yen, G.C.; Hsieh, P.P. Antioxidative activity and scavenging effects on active oxygen of xylose - lysine Maillard reaction products. *J. Sci. Food Agric.* **1995**, *67*, 415–420.

Yu, T.H.; Ho, C.-T. 1995. Volatile compounds generated from thermal reaction of methionine and methionine sulfoxide with or without glucose. *J. Agric. Food Chem.* **1995**, *43*, 1641–1646.

Zhang, Y.; Ho, C.-T. Comparison of the volatile compounds formed from the thermal reaction of glucose with cysteine and glutathione. *J. Agric. Food Chem.* **1991**, *39*, 760–763.

GENERATION AND THE FATE OF C_2, C_3, AND C_4 REACTIVE FRAGMENTS FORMED IN MAILLARD MODEL SYSTEMS OF [^{13}C]GLUCOSE AND [^{13}C]GLYCINE OR PROLINE

Varoujan A. Yaylayan, Anahita Keyhani, and Alexis Huygues-Despointes

Department of Food Science and Agricultural Chemistry
McGill University
21,111 Lakeshore
Ste. Anne de Bellevue, Quebec, Canada, H9X 3V9

Model studies with pyrolysis/GC/MS using labeled [^{13}C] glucoses with labeled [^{15}N/^{13}C]glycines and proline have indicated that the Maillard model systems consisting of glucose and glycine or proline generate similar C_2, C_3, C_4 fragments such as acetic acid, and pyruvaldehyde. Furthermore, the labeling studies have enabled the identification of the origin of these reactive intermediates and their stable end-products such as N-acetylpyrrolidine, 1-(1′-pyrrolidinyl)-2-propanone amd 1-(1′-pyrrolidinyl)-2-butanone in proline model system and pyrazines and pyrazinones in glycine. In glycine model system, pyruvaldehyde and 2,3-butandione were found to be formed either from the degradation of the carbohydrate moiety (90 and 35%, respectively) or by an aldol-type interaction of glycine with α-ketoaldehydes. The same intermediates in proline system are formed exclusively from the carbohydrate degradation pathway.

INTRODUCTION

Chemical changes initiated by the Maillard reaction during thermal processing of food, are considered to be important contributors to the modification of color and sensory properties in the final product. The complexity of the Maillard reaction precludes its complete analysis through classical organic chemistry and requires the use of isotopically labeled starting materials such as sugars and amino acids to establish the origin and the fate

Process-Induced Chemical Changes in Food
edited by Shahidi *et al.* Plenum Press, New York, 1998

of a multitude of short chain reactive intermediates that form during the process. Identification of such common reactive intermediates can help to predict the formation of certain end-products and eventually can lead to the classification of Maillard reaction into its underlying elementary processes.

Pyrolysis coupled with gas chromatography/mass spectrometry (Py/GC/MS) is ideally suited to perform complex chemical reactions in the pyrolysis probe and to separate and identify the products formed in relatively short period of time. Py/GC/MS has been demonstrated to be a fast and convenient technique for the analysis of Maillard reaction products (Huyghues-Despointes et al., 1994; Yaylayan and Keyhani, 1996), especially arising from isotopically enriched compounds for mechanistic studies (Huyghues-Despointes and Yaylayan, 1996; Keyhani and Yaylayan, 1996a; Keyhani and Yaylayan, 1996b). Quartz tube Py/GC/MS is particularly suited to perform small scale reactions without the need to isolate or extract the reaction mixture that leads to the loss of valuable isotopically labeled products. The use of ^{13}C- and ^{15}N-enriched compounds for the elucidation of reaction mechanisms concerning the Maillard reaction has been well documented (Tressl et al., 1993; Amrani-Hemaimi, 1995). In this study Py/GC/MS was used to identify the origin of small carbon fragments using ^{13}C-enriched glucoses and $^{15}N/^{13}C$- enriched glycines or proline.

MATERIALS AND METHODS

All reagents, chemicals and d-[1–^{13}C]glucose, d-[2–^{13}C]glucose, d-[6–^{13}C]glucose were purchased from Aldrich Chemical Company (Milwaukee, WI). d-[3–^{13}C]glucose, d-[4–^{13}C]glucose, d-[5–^{13}C]glucose, [^{15}N]glycine, [1–^{13}C]glycine, [2–^{13}C]glycine and [1,2–^{13}C]glycine were purchased from ICON Services Inc. (Summit, New Jersey).

Pyrolysis-GC/MS Analysis

A Hewlett-Packard GC/mass selective detector (5890 GC/5971B MSD) interfaced to a CDS pyroprobe 2000 unit was used for the Py/GC/MS analysis. Solid samples (1–4 mg) of amino acid/glucose in different ratios, were introduced inside a quartz tube (0.3 mm thickness) and plugged with quartz wool and inserted inside the coil probe with a THT (total heating time) of 20 s. The GC column flow rate was 0.8 ml/min for a split ratio of 92:1 and a septum purge of 3 ml/min. The pyroprobe interface temperature was set at 250 °C at a heating rate of 50 °C/ms. Capillary direct MS interface temperature was 180 °C; ion source temperature was 280 °C. The ionization voltage was 70 eV and the electron multiplier was 1682 V. The mass range analyzed was 30–300 amu. The column was a fused silica DB-5 column (30m length × 0.25 mm i.d. × 25 um film thickness; Supelco, Inc.). Unless otherwise specified, the column initial temperature was -5 °C for 3 min and was increased to 50 °C at a rate of 30 °C/min; immediately the temperature was further increased to 270 °C at a rate of 8 °C/min and kept at 270 °C for 5 min. Products that were not found in the mass spectral libraries were identified by comparison with literature mass spectral data or by generating the products from their proposed precursors and comparing mass spectra and chromatographic retention times.

RESULTS AND DISCUSSION

Origin of the volatile and reactive short chain carbon fragments such as glyoxal, pyruvaldehyde and 2,3-butandione formed in Maillard reaction mixtures is relatively difficult to determine due to the multiple origin of these components. Some could arise directly from the sugar or Amadori product, others could be formed by chain elongation through aldol condensation with fragments arising from either sugar or the amino acid. In addition, detection of these components could also pose some difficulties due to their volatility and reactivity. However, tracing back the origin of each carbon atom in a product could be successfully achieved by GC/MS analysis of model reaction mixtures performed with separately ^{13}C-labeled sugars and amino acids at each of their carbon atoms. This approach entails, for example, the analysis of eight model reaction mixtures in the case of the simplest amino acid glycine (two carbon atoms) with glucose (six carbon atoms). The employment of Py/GC/MS technique developed in our laboratory, as an integrated reaction, separation and identification system, to perform such studies, not only accelerates the time required to accomplish the labeling studies but also reduces the high cost associated with expensive labeled starting materials since only few milligrams are required to perform each Py/GC/MS analysis.

To overcome the problem of direct detection of the short chain reactive species in the model systems, an alternate approach could be followed which entails the identification of stable end-products that incorporate these fragments into their molecular structure and subsequently could be detected by Py/GC/MS analysis. The structures of these stable end- products depend on the amino acid type. To determine the formation and the fate of these short chain carbonyl compounds, two model systems were selected, one containing a primary amino group (glycine) and the other a secondary amino group (proline). Py/GC/MS analysis of these two model systems have indicated that the volatile end-products formed that incorporate short chain fragments, were 1-(1-pyrrolidinyl)-2-propanone, 1,2-(1,1-dipyrrolidinyl)-1-propene, N-acetylpyrrolidine and 1-(1-pyrrolidinyl)-2-butanone in the case of proline model system and di- and trimethylpyrazines and pyrazinones (Oh et al., 1992) in the case of glycine model system. Plausible mechanisms of formation of these end-products have been proposed (Huyghues-Despointes and Yaylayan, 1996, Keyhani and Yaylayan, 1996b). Previous studies (Keyhani and Yaylayan, 1996b) have identified a novel aldol type reaction (Scheme 1) between the C-2 atom of glycine and α-ketoaldehydes resulting in the conversion of the aldehyde moiety into a methyl ketone, such as the conversion of pyruvaldehyde into 2,3-butandione. Scheme 2 summarizes all the fragments and their end product in both proline and glycine model systems identified by Py/GC/MS. Tables 1–3 indicate the percent label distribution in all the products using ^{13}C-labeled glucoses and the two amino acids. The distribution of the isotope labels was calculated from the intensities of the parent ions. The results were corrected to account for the natural ^{13}C content of the corresponding unlabeled reference compounds.

ORIGIN OF C_2 FRAGMENTS

Acetic acid (ACA) is a common C_2 fragment that could be detected in both model systems studied. The major source (92 %) of acetic acid in proline model system is the C1-C2 carbon fragment of glucose moiety and around 5 % arises from C5-C6 fragment (see Table 3). However, in glycine model system 17 % of acetic acid arises by deamination of glycine as indicated by the incorporation of C-2 atom from glycine into the acetic acid moiety (see Table 2). Similar to proline, glycine model system also generates acetic acid in large proportion (70 %) from glucose C1-C2 fragment. Furthermore, the formation of acetic acid could also be verified by its incorporation into stable end-products such as N-acetylpyrrolidine (NAP) in the case of proline model system and N-methylacetamide,

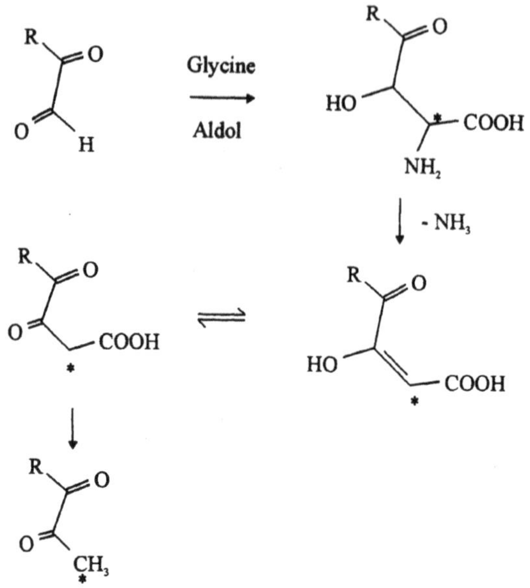

Scheme 1. Mechansim of transformation of α-ketoaldehydes into methyldiketones through aldol condensation.

dimethylpyrazinone (DMPN) in the case of glycine model system. In fact, 40% of DMPN could also arise through a three carbon unit fragment form glucose (pyruvaldehyde) as indicated in Scheme 2. The mechanism of formation of DMPN from two and three carbon sugar fragments is outlined in Scheme 3. Sixty percent of DMPN is formed from the reaction of glyoxal with three moles of glycine. One mole of glycine converts glyoxal into intermediate I as shown in Scheme 3 and the two subsequent moles of glycine converts intermediate I into DMPN.

ORIGIN OF C_3 FRAGMENTS

Pyruvaldehyde is an important and reactive three carbon fragment in Maillard model systems. It can be generated through retro aldol reaction of Amadori products (Huyghues-Despointes and Yaylayan, 1996) in both glycine and proline model systems (see Scheme 2). In the case of glycine, pyruvaldehyde could also be generated from the glyoxal interaction with glycine (Scheme 1), this path however, represents only 10% of the observed pyruvaldehyde. Similar to glyoxal, pyruvaldehyde can also undergo chain elongation and conversion into 2,3-butandione by the action of glycine as indicated by the incorporation of C-2 carbon atom of glycine into 2,3-butandione structure (see Table 2 and Scheme 2). According to labeling studies with [2–^{13}C]glycine, 65 % of 2,3-butandione produced in glycine/glucose mixture comes from such interaction and 35 % directly from carbohydrate moiety. Pyruvaldehyde, once generated in the glycine model system, can interact with two or three moles of glycine to produce di- and trimethylpyrazinones, respectively, or it can undergo Strecker degradation and eventually produce trimethylpyrazine (TMP) by interaction with 3-amino-2-butanone (Strecker degradation product of 2,3-butandione) as sum-

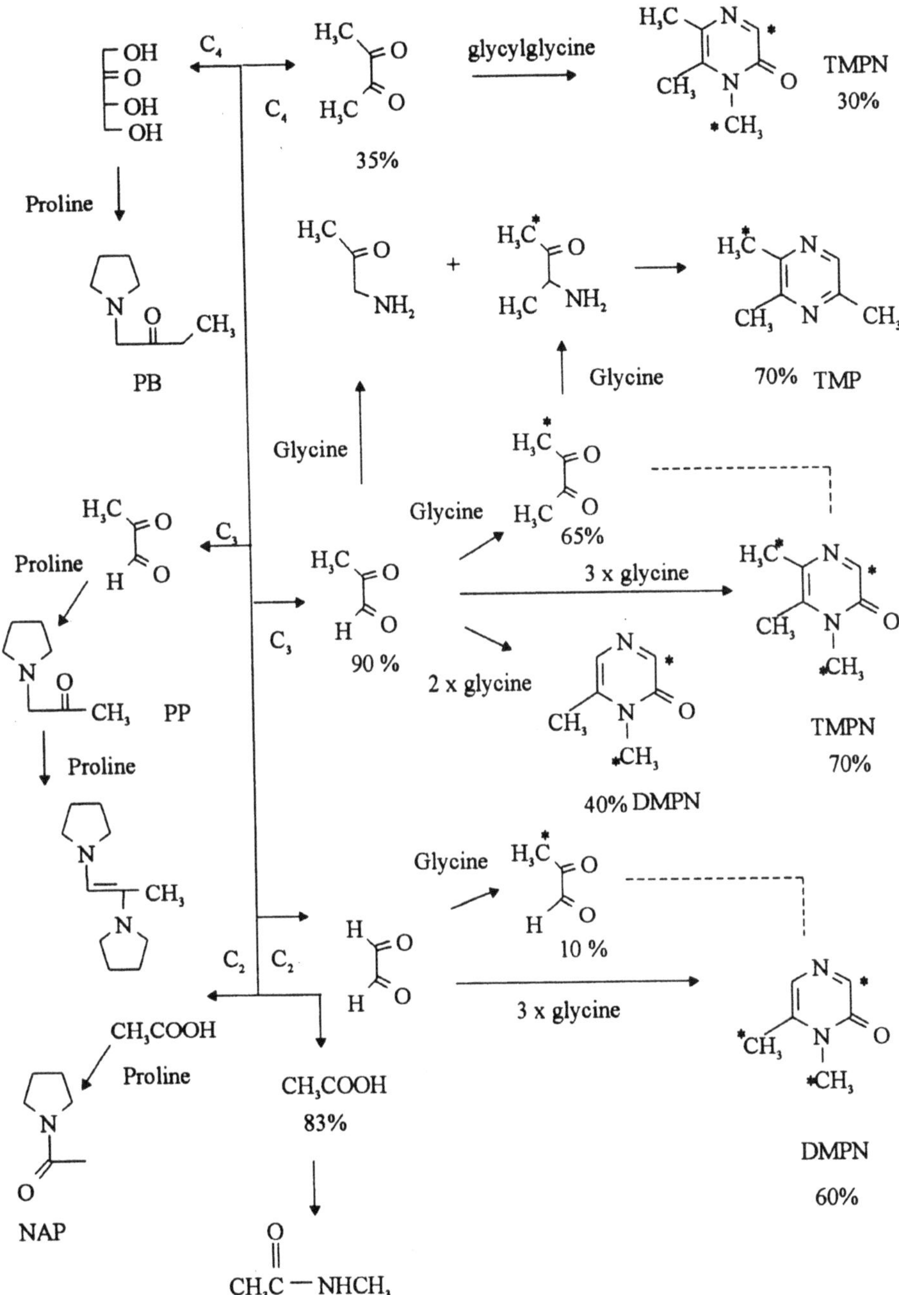

Scheme 2. Interaction of C_2, C_3 and C_4 sugar fragments with glycine and proline. Asterix indicates C-2 carbon atom of glycine.

Table 1. Percent label distribution in selected products from ^{13}C-labeled glucoses and glycine[1]

Model	DMPN C_2		TMPN C_3		TMP C_3			ACA C_2	
	M	M+1	M	M+1	M	M+1	M+2	M	M+1
Glycine/Glucose	100	0	100	0	99	1	0	100	0
Glycine/[1-^{13}C]Glucose	68	30	77	23	51	40	9	30	70
Glycine/[2-^{13}C]Glucose	68	31	77	23	50	40	10	30	70
Glycine/[3-^{13}C]Glucose	91	9	70	30	49	41	10	98	2
Glycine/[4-^{13}C]Glucose	88	12	27	73	14	42	44	98	2
Glycine/[5-^{13}C]Glucose	42	58	27	73	14	42	44	93	7
Glycine/[6-^{13}C]Glucose	38	57	26	74	14	42	44	93	7

DMPN = 1,6- and 1,5-dimethylpyarazinone mixture (4:1)
TMPN = 1,5,6-trimethylpyrazinone
TMP = trimethylpyrazine
ACA = acetic acid
[1]Ratio of amino acid /sugar (3:1).

marized in Scheme 2. Seventy percent of TMP is produced from 2,3-butandione generated through pyruvaldehyde pathway and 30 % from 2,3-butandione produced from the carbohydrate moiety.

In the case of proline system, pyruvaldehyde can only be produced through a retro aldol reaction of Amadori product or glucose and eventually reacts with one mole of proline to produce 1-(1′-pyrrolidinyl)-2-propanone (PP) which subsequently forms 1,2-(1′,1′-dipyrrolidinyl)-1-propene by reaction with another mole of glycine.

Table 2. Percent label distribution in selected products from [^{13}C] glycines and glucose[1]

	M	M+1	M+2	M+3	M+4
Dimethylpyarazinone					
[1-^{13}C]Glycine/Glucose	0	98	2	0	0
[2-^{13}C]Glycine/Glucose	0	0	40	60	0
[1,2-^{13}C]Glycine/Glucose	0	0	0	30	70
1,5,6-Trimethylpyrazinone					
[1-^{13}C]Glycine/Glucose	0	98	2	0	0
[2-^{13}C]Glycine/Glucose	0	0	30	70	0
[1,2-^{13}C]Glycine/Glucose	0	0	0	20	80
Trimethylpyrazine					
[2-^{13}C]Glycine/Glucose	20	70	10	0	0
[1,2-^{13}C]Glycine/Glucose	20	70	10	0	0
Acetic acid					
[2-^{13}C]Glycine/Glucose	83	17	0	0	0
N-Methylacetamide					
[2-^{13}C]Glycine/Glucose	34	57	9	0	0
2,3-Butandione					
[2-^{13}C]Glycine/Glucose	35	65	0	0	0

[1]Ratio of amino acid /sugar (3:1).

Table 3. Percent label distribution in selected products from ^{13}C-labeled glucoses and proline[1]

	ACA C_2		NAP C_2		PP C_3		PB C_4	
Model	M	M+1	M	M+1	M	M+1	M	M+1
Proline/Glucose	100	0	100	0	100	0	100	0
Proline/[1-^{13}C]Glucose	8	92	27	73	45	55	73	27
Proline/[2-^{13}C]Glucose	8	92	27	73	45	55	74	26
Proline/[3-^{13}C]Glucose	98	2	95	5	45	55	5	95
Proline/[4-^{13}C]Glucose	98	2	95	5	55	45	7	93
Proline/[5-^{13}C]Glucose	95	5	80	20	55	45	30	70
Proline/[6-^{13}C]Glucose	95	5	80	20	55	45	26	74

ACA = acetic acid
N-acetylpyrrolidine = NAP
1-(1'-pyrrolidinyl)-2-propanone = PP
1-(1'-pyrrolidinyl)-2-butanone = PB
[1]Ratio of amino acid /sugar (1:1).

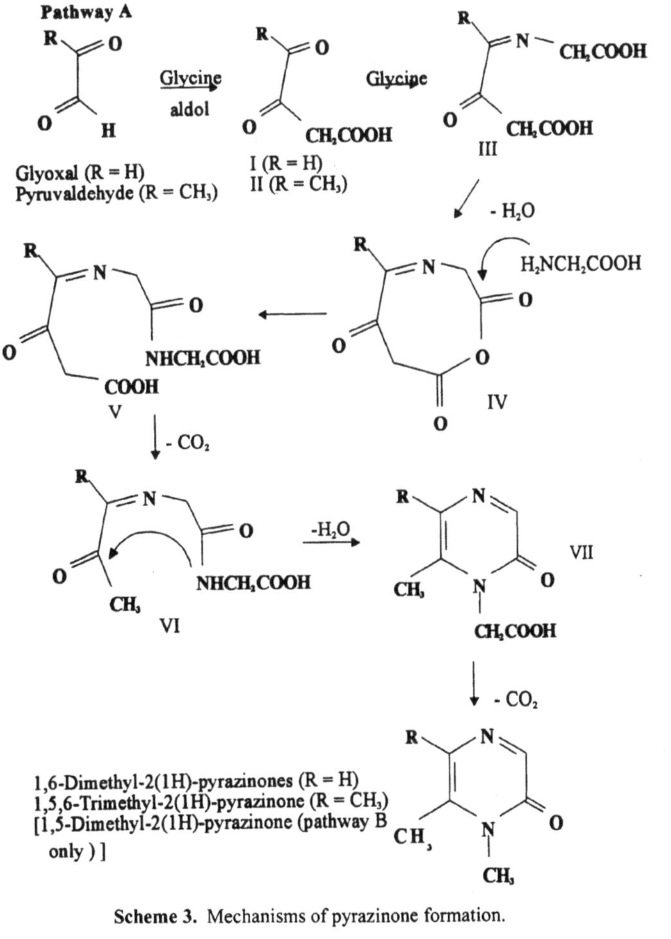
Pathway A
Glycine
aldol
Glycine
Glyoxal (R = H)
Pyruvaldehyde (R = CH₃)
I (R = H)
II (R = CH₃)
III
- H₂O
H₂NCH₂COOH
IV
V
- CO₂
VI
-H₂O
VII
- CO₂
1,6-Dimethyl-2(1H)-pyrazinones (R = H)
1,5,6-Trimethyl-2(1H)-pyrazinone (R = CH₃)
[1,5-Dimethyl-2(1H)-pyrazinone (pathway B only)]

Scheme 3. Mechanisms of pyrazinone formation.

ORIGIN OF C_4 FRAGMENTS

2,3-Butandione is the only C_4 fragment that could be identified in glycine model system. As indicated above, it could arise either from the sugar degradation in 35 % yield or from pyruvaldehyde reaction with glycine in 65 % yield (see Table 2 and Scheme 2). As mentioned above 2,3-butandione eventually produces TMP as a stable end-product or reacts with glycylglycine to produce TMPN (30 %) that incorporates only two C-2 carbons of glycine. The mechanisms of trimethylpyrazinone formation through 2,3-butandione (pathway B) and through pyruvaldehyde (pathway A) are illustrated in Scheme 3. The C_4 fragment formed in proline model system could only be identified by its stable end-product 1-(1′-pyrrolidinyl)-2-butanone (PB). The structure of the C_4 fragment generating PB was traced back to a tetrulose moiety (Huyghues-Despointes and Yaylayan, 1996) that undergoes a series of elimination and reductions to produce PB after interaction with proline.

CONCLUSION

The results of the labeling experiments with a primary and a secondary amino acids indicate that in both systems, same reactive fragments, interact similarly with the amino acids but produce structurally unrelated products (for example pyrazinone vs 1,2-(1′,1′-dipyrrolidinyl)-1-propene) due to the nature of the secondary amino group of proline that prevents further cyclizations.

ACKNOWLEDGMENT

V. Y. acknowledges funding for this research from the Natural Sciences and Engineering Research Council of Canada (NSERC).

REFERENCES

Amrani-Hemaimi, M.; Cerny, C.; Fay, B.F. Mechanisms of formation of alkylpyrazines in the Maillard Reaction. *J. Agric.Food Chem.* **1995**, *43*, 2818–2822.

Huyghues-Despointes, A.; Yaylayan, V. Retro-aldol and redox reactions of Amadori compounds: mechanistic studies with variously labeled d-[^{13}C]glucose. *J. Agric. Food Chem.* **1996** (in press).

Huygues-Despointes, A.; Yaylayan, V.; Keyhani, A. Pyrolysis/GC/MS analysis of 1-[(2′-carboxyl)pyrrolidinyl]-1-deoxy-d-fructose (Amadori Proline). *J. Agric. Food Chem.* **1994**, *42*, 2519–2524.

Keyhani, A.; Yaylayan, V. Pyrolysis/GC/MS analysis of N-(1-deoxy-d-fructos-1-yl)-l-phenylalanine: Identification of novel pyridine and naphthalene derivatives. *J. Agric. Food Chem.* **1996**, *44*, 223–229.

Keyhani, A.; Yaylayan, V. Elucidation of the mechanism of pyrazinone formation in glycine model systems using labeled sugars and amino acids. Submitted to *J. Agric. Food Chem.*

Oh, Y-C.; Shu, C-K.; Ho, C-T. Formation of novel 2(1H)-pyrazinones as peptide -specific Maillard reaction products. *J. Agric. Food Chem.* **1992**, *40*, 118–121.

Tressl, R.; Helak, B.; Kersten, E. Formation of proline and hydroxyproline-specific Maillard products from [1–^{13}C]glucose. *J. Agric. Food Chem.* **1993**, *41*, 547–553.

Yaylayan, V.; Keyhani, A. Py/GC/MS analysis of non-volatile flavor precursors: Amadori compounds. In *Contribution of Low and Non-Volatile Materials to the Flavor of Foods*. Pickenhagen; Spanier; C-T Ho, Eds.; Allured Publishing Company (in press), 1996.

20

METAL CHELATING AND ANTIOXIDANT ACTIVITY OF MODEL MAILLARD REACTION PRODUCTS

Arosha N. Wijewickreme and David D. Kitts

Department of Food Science
University of British Columbia
6650, N.W. Marine Drive
Vancouver, B.C., Canada, V6T 1Z4

Model glucose-lysine (Glu-Lys) and fructose-lysine (Fru-Lys) Maillard Reaction products (MRPs) were generated using 14 different reaction combinations, by varying four experimental parameters: time, temperature, initial water activity (aw), and initial pH. The synthesized MRPs were analysed for their potential copper chelating, antioxidant or prooxidant, and genotoxic activities. Yield of MRPs varied depending on the conditions used to generate them. Amount of copper bound to MRPs ranged from 0.031–1.574 and 0.016–2.267 :mol copper per mg MRP for different Glu-Lys and Fru-Lys MRPs, respectively, in different synthesis experiments. The assesment of antioxidant activity by the 2-thiobarbituric acid (TBA) method using a lipid system devoid of copper ions showed varying antioxidant activity for both Glu-Lys and Fru-Lys MRPs. Using an oxygen electrode for the measurement of oxygen depletion in a model lipid system containing copper, as an endpoint measure of lipid peroxidation, 7 antioxidant, 2 prooxidant, and 4 inactive MRPs and 5 antioxidant, 5 prooxidant, and 2 inactive MRPs were obtained for Glu-Lys and Fru-Lys MRP synthesis experiments, respectively. All derived MRPs showed DNA strand breaking activity above a concentration of 0.001% (w/v). The results indicate the importance of reaction conditions and the source of reducing sugar in the generation of MRPs with variable antioxidant, prooxidant, and genotoxic activities.

INTRODUCTION

The non-enzymatic interaction between reducing sugars and amino acids, peptides, or proteins, in foods typlifies the carbonylamine browning reaction, widely known as the

Process-Induced Chemical Changes in Food
edited by Shahidi *et al.* Plenum Press, New York, 1998

Maillard browning reaction. Maillard reaction products (MRPs) result from a complex network of chemical reactions which ultimately lead to the production of polymeric brown colored compounds known as melanoidins. The chemical reactions involved in the derivation of different MRPs have been the subject of considerable investigation for many years (Feather, 1981; Yaylayan *et al.*, 1995). Products of model melanoidins have been reported to exhibit mutagenic/non-mutagenic, clastogenic/non-clastogenic, and metal chelating properties (Rendleman, 1987; Terasawa, 1991; Kitts *et al.*, 1993). In many respects, an underlying cause or effect for the biological properties of MRPs reported by various investigators has been the antioxidant activity (Lingert and Eriksson, 1981; Chiou, 1992; Bedinghaus and Ockerman, 1995; Smith and Alfarvaz, 1995).

The purpose of the study presented herein, was to examine the copper chelating and antioxidant activity of model MRPs synthesized by reacting L-lysine with either an aldose sugar D-glucose or the ketose sugar D-fructose under fourteen different reaction combinations of initial pH, initial water activity (a_w), temperature and time durations.

EXPERIMENTALS

Materials

L-lysine, D-glucose, D-fructose, ethylenediaminetetraacetic acid (EDTA), trizma base, 1,1,3,3-tetraethoxypropane, 2-thiobarbituric acid (TBA), butylated hydroxyanisole (BHA), linoleic acid, and hexamine were obtained from Sigma Chemical Company (St. Louis, MO). Glycerol and sodium hydroxide (NaOH) were purchased from Fisher Scientific Company (Fair Lawn, NJ). Dialysis tubing (Molecular weight cut off = 6–8 kD and 3.5 kD) were purchased from Spectrum Scientific Company (Houston, TX). C_{18} Sep-pak cartridges were obtained from Waters Associates (Milford, MA). Tween 80 was purchased from Difco Laboratories (Detroit, MI). The distilled water used to prepare model MRPs was further purified by a Barnstead E-pure system.

Preparation of Model MRPs

MRPs were prepared by heating a solution of 0.8M D-glucose or D-fructose with 0.8M L-lysine in a hot air oven (Kitts *et al.*, 1993). The initial reaction conditions used in the preparation of model MRPs are given in Table 1. The experimental conditions were randomly selected using Random-centroid optimization program (Dou *et al.*, 1993). Four variable experimental parameters including, time, temperature, a_w, and pH were used in the centroid design. The initial a_w of the reactions were adjusted according to the method of Eichner and Karel (1972) and the initial pH values were adjusted with 5M NaOH. All experiments were conducted in 500 mL volumetric flasks containing 100 mL of reactants. The resulting brown solution was rapidly cooled on ice, dialysed against 20–30 changes of double distilled deionized water at 4 °C for seven days, and lyophilized. The derived glucose-lysine (*Glu-Lys*) and fructose-lysine (*Fru-Lys*) MRPs were weighed and stored in a dessicator at 0 °C until further use.

Metal Chelation Activity of MRPs

The metal chelating activity of model MRPs was determined by the combined use of atomic absorption spectroscopy and tetramethyl murexide (TMM) method (Terasawa *et*

Table 1. Initial and final experimental conditions [reaction time, oven temperature, initial water activity (a_w), and initial pH] used and obtained in the preparation of *Glu-Lys* and *Fru-Lys* non-dialysable model MRP mixtures and the yield gained at the end of the reaction

	Initial study conditions				Final study conditions				Yield (g)[2]	
					Glu-Lys		*Fru-Lys*			
Expt. no.	Time (min)	Oven Temp.[3]	pH	a_w	pH	a_w	pH	a_w	*Glu-Lys*	*Fru-Lys*
1	65	110	6.37	0.58	3.20	0.56	3.20	0.56	2.41 ± 4.2	0.75 ± 3.2
2	90	146	6.18	0.81	3.15	0.80	3.16	0.78	0.50 ± 6.0	0.05 ± 2.8
3	119	127	6.14	0.74	3.14	0.71	3.12	0.68	7.03 ± 2.1	0.55 ± 1.1
4	86	112	6.97	0.68	3.51	0.65	3.51	0.65	0.18 ± 1.1	1.48 ± 2.1
5	107	157	8.51	0.57	3.62	0.53	3.75	0.51	2.00 ± 3.2	5.11 ± 1.7
6	85	86	7.98	0.68	3.53	0.66	3.54	0.67	0.13 ± 2.0	0.01 ± 1.1
7	45	91	6.67	0.94	3.32	0.93	3.30	0.94	–	–
8	58	143	6.98	0.95	3.51	0.92	3.53	0.91	0.70 ± 1.5	0.51 ± 2.0
9	116	99	7.81	0.80	3.60	0.76	3.61	0.77	0.10 ± 1.3	0.45 ± 3.1
10	43	112	6.44	0.88	3.24	0.87	3.22	0.88	1.30 ± 4.1	–
11	43	159	8.57	0.62	3.54	0.59	3.51	0.61	4.10 ± 3.0	6.65 ± 2.8
12	30	144	7.14	0.81	3.53	0.80	3.53	0.79	0.23 ± 1.8	0.60 ± 1.6
13	71	129	8.41	0.78	3.86	0.74	3.84	0.73	7.50 ± 2.3	0.40 ± 3.2
14	108	80	7.66	0.88	3.58	0.84	3.56	0.85	1.21 ± 2.1	2.09 ± 3.7

[1]Equivalent reaction time calculated for T_{ref} = 393 K; z = 19.1 C°
[2]Yield in gram non-dialysable MRP per 100 mL of final reaction solution. Values represent mean ± SD (n= 3).
[3]Ambient temperature in the oven (C°).
(–)= Non-dialysable MRPs were not detected.

al., 1991). The atomic absorption spectroscopy measured total copper, while TMM method measured the amount of free copper present. The amount of copper bound to MRPs was determined from the difference between the two methods.

Copper sulphate ($CuSO_4$ 0.05- 0.4 mM), MRPs (100 µg/mL), and TMM (1 mM) were respectively dissolved in 10 mM hexamine. HCl (pH 5) buffer containing 10 mM potassium chloride (KCl). Assay consisted of 1 ml of MRPs, 1 ml of $CuSO_4$ (0.05–0.4 mM), and 0.1 ml of TMM. Total copper in the samples were measured by atomic absorption spectroscopy. Free copper in the samples were obtained by measuring the absorbance ratio, A_{460}/A_{530}, of the samples and reading the amount of free copper from a standard curve, where the absorbancy ratio (A_{460}/A_{530}) in a solution of 1 mL $CuSO_4$ (0.05–0.4 mM), 1 mL of hexamine. HCl buffer, and 0.1 mL of TMM was plotted against the amount of copper present. The amount of copper bound to MRPs was obtained by calculating the difference between total copper and free copper in the system.

Assessment of Antioxidant Activity of MRPs

A) Oxygen Consumption Measurements. Oxygen depletion in a linoleic acid emulsion system with added cupric [Cu(II)] ions and MRPs was measured according to the methods of McGookin and Augustin (1991) and Lingert *et al.* (1979), using a YSI Model 5300 Biological Oxygen monitor (Yellow Springs, OH). The reaction mixture consisted of 1.5 ml of linoleic acid emulsion (1.5 g linoleic acid and 0.4 g of Tween 80 mixed with 40 mL of potassium phosphate buffer; pH 6.8), and 0.6 ml of MRP solution (3 mg/ml buffer), 0.6 mL of $CuSO_4$ (2 mM). The reaction mixture was pumped into a jacketed reaction vessel containing an oxygen electrode at room temperature. Oxygen depletion was recorded immediately after the reaction mixture was introduced into the vessel. For the measurement of oxygen depletion in the absence of antioxidative compounds, the experiment was performed in an identical manner with the exception that 0.6 ml of MRP solution was replaced with 0.6 mL of buffer. Both antioxidant and prooxidant activity of MRPs were expressed in terms of a protective index (PI), which is defined as:

$$\Pi = \frac{[\textit{Time} \text{ for } 50\ \%\ O_2\ \textit{depletion with MRPs}]}{[\textit{Time} \text{ for } 50\ \%\ O_2\ \textit{depletion without MRPs}]} \tag{1}$$

where: PI < 1 - prooxidant activity; PI = 1 - no activity; PI > 1 - antioxidant activity (Lingert *et al.*, 1979).

B) Measurement of Thiobarbituric Acid Reactive Substances (TBARS). MRPs (2 mg) were incubated with 5 mL of linoleic acid emulsion (1.5g of linoleic acid and 0.4 g Tween 80 dissolved in 200 mL of potassium phosphate buffer; pH 6.8) and 5 mL of buffer at 45°C for 48 hours. Following incubation, the solution was diluted 10 times with 25 mM Tris buffer (pH 7.4), containing 0.02% sodium azide, and passed through a C_{18} Sep-pak cartridge to remove residual lipids. Lipid free homogenate was assayed by the 2-thiobarbituric acid (TBA) assay described by Buege and Aust (1978). One ml of TBA reagent containing 0.02 % freshly prepared BHA was added to 2 mL of clarified homogenate in test tubes with marble caps, and immersed in a boiling water bath for 15 min. After cooling, absorbance at 532 nm was recorded using a Shimadzu 160 UV-Vis spectrophotometer. Quantification of MDA content in samples was made from a standard curve prepared from 1,1,3,3-tetraethoxypropane in 1% sulfuric acid. All results were expressed as percent antioxidant activity (% AO) which is defined as:

$$\%\ AO = \frac{[\textit{TBA value of the control} - \textit{TBA value of the test sample}] \times 100}{[\textit{TBA value of the control}]} \tag{2}$$

C) DNA Nicking Studies. Pseudomonas Bal 31 bacteria was grown at 28 °C in an Erlenmeyer flask containing 1L of Bal-broth medium [20 mL 10 mM Tris HCl (pH 7.5); 24g $MgSO_4.7H_2O$; 52g NaCl; bacto-nutrient broth; 20 ml 1M $CaCl_2$; 7 ml 20% KCl per 1000 mL deionized water] stirring vigerously. At a density of 2x 10^7 bacteria / mL, the culture was infected with PM_2 phage with a known titer and the PM_2 bacteriophage DNA was purified and extracted as described by Espejo and Canelo (1968).

To detect the DNA strand breaking activity of *Glu-Lys* and *Fru-Lys* MRPs on PM_2 bacteriophage DNA, 2 μl of DNA (0.1 μg/mL) was incubated in a 500 μL Eppendorf tube with 2 μL of MRPs (10^{-4} % or10^{-1} %; w/v), 2 μL of potassium phosphate buffer (50 mM, pH 7.4), and 4 μL of double distilled deionized water, at 37°C for one hr. Following incubation, 2 μL of loading dye containing bromophenol blue (0.05%) was added into the incubated reaction mixture and 10 μL of that mixture was loaded into an agarose gel (0.7%; w/v) immersed in Tris Acetate EDTA (TAE) buffer (40 mM Tris acetate and 1 mM EDTA; pH 7.4). The electrophoresis was conducted for 90 min at 50 V, and the DNA bands were visualized under illumination of UV light following staining with ethidium bromide (0.5μg/mL) for 20 min. Gels were photographed using a Polaroid MP-4 camera loaded with Polaroid Type 55 professional films and the photographic negatives were scanned by a Bio-Rad model GS-670 imaging densitometer to quantitate the percentages of super coiled (S), nicked circular (NC), and linear (L) forms of DNA remaining following each incubation experiment.

RESULTS AND DISCUSSION

Yield of Glucose-Lysine (Glu-Lys) and Fructose-Lysine (Fru-Lys) MRPs

The initial reaction conditions (*e.g.*, pH, a_w, reaction time and ambient oven temperature), along with final pH and a_w values, and yields of non-dialysable MRPs are given in Table 1. The final pH of all experiments performed was acidic (*e.g.* range 3.12–3.84), regardless of the initial pH value recorded. The a_w values remained relatively unchanged from initial values. The yield of MRPs varied depending on the combination of reaction conditions and source of reducing sugar. The minimum and maximum yields of synthesized MRPs ranged from 0.1 to 7.5 g for *Glu-Lys* reactions and from 0.05 to 6.65 g for *Fru-Lys* reactions. Non-dialysable melanoidin could not be recovered in *Glu-Lys* experiment number 7 and experiment numbers 7 and 10 in the *Fru-Lys* reaction. The two experiments in the *Glu-Lys* reaction which produced the greatest MRPs yields were experiment numbers 3 and 13. In both experiments, although similar initial ambient temperature (127, 129°C) and water activity (0.74, 0.78) values respectively were used, both the initial pH and reaction times were substantially different. In the case of *Fru-Lys* reactions, the two experiments that produced highest MRP yields (*i.e.*, experiment numbers 5 and 11) involved both similar ambient temperature (157, 159°C), and initial pH (8.51, 8.57) values, respectively, but large differences in reaction times. These results are a further confirmation of earlier findings that the production of MRPs are influenced by a variety of factors including reaction conditions and the source of reactants involved (Eichner and Karel, 1972; Obretenov *et al.*, 1993; Cuzzoni *et al.*, 1988).

Copper Chelation Activity of Model MRPs

The amount of measured copper bound to different *Glu-Lys* and *Fru-Lys* MRPs is given in Table 2. Many of the MRPs derived from different synthesis experiments possessed detectable copper binding activity with the relative amount of copper binding varying with specific experimental conditions. The lowest and the highest amount of copper bound to MRPs ranged from 0.031 to 1.574 μmol copper per mg MRP for *Glu-Lys* MRPs and 0.016 to 2.267 μmol copper per mg MRP for *Fru-Lys* MRPs.

Copper chelating activity of glucose-glycine MRPs have been reported earlier by Terasawa *et al.* (1991). According to Hashiba (1985), hydroxypyranone and hydroxypyridone residues present in melanoidin have the affinity to combine ferric ions. Moreover, Rendleman (1987) showed that brown foods such as coffee and toasted bread also has the affinity to chelate certain metal ions. He also indicated that melanoidin can act as an anionic polymer in chelating metal ions.

The significance of metal chelating activity of MRPs to human health has two different aspects. On one hand, low bioavailability of trace metal ions could occur *in vivo* and on the otherhand, MRPs could be an effective antioxidant in retarding metal catalyzed lipid oxidations reactions.

Antioxidant/Pprooxidant Activity of Model MRPs

A) Oxygen Consumption Measurement. In the present study, PI values calculated using oxygen consumption measurements were used as a reliable chemical measure for

Table 2. Protective index (PI), percent antioxidant activity (% AO), and the amount of copper bound to different *Glu-Lys* and *Fru-Lys* MRPs

Synthesis experiment number	% AO[1]		PI[2]		Bound copper μmol bound copper/mg MRP	
	Glu-Lys	*Fru-Lys*	*Glu-Lys*	*Fru-Lys*	*Glu-Lys*	*Fru-Lys*
1	10.50 ± 5.1	5.6 ± 1.7	0.5	0.9	0.472 ± 0.21	0.467 ± 0.78
2	5.3 ± 2.2	3.2 ± 0.7	0.99	1.3	0.041 ± 0.17	0.0
3	20.8 ± 3.1	7.3 ± 1.5	2.7	2.5	1.102 ± 0.15	0.157 ± 0.90
4	9.9 ± 6.4	8.9 ± 1.2	1.1	1.1	0.188 ± 0.08	2.267 ± 0.14
5	12.1 ± 3.0	10.8 ± 1.0	1.4	1.9	0.314 ± 0.11	0.661 ± 0.11
6	3.3 ± 4.0	2.0 ± 1.1	1.0	0.3	0.063 ± 0.17	0.0
7	–	–	–	–	–	–
8	5.2 ± 0.1	1.6 ± 1.1	0.8	0.9	0.157 ± 0.21	0.047 ± 0.15
9	1.5 ± 1.0	3.1± 1.8	0.7	1.0	0.0	0.118 ± 0.35
10	2.0 ± 1.1	0.0	1.3	–	0.031 ± 0.25	–
11	4.0 ± 4.5	20.1 ± 1.5	1.0	1.2	0.470 ± 0.37	1.417 ± 0.07
12	8.5 ± 7.0	6.2 ± 1.1	1.0	0.3	0.630 ± 0.87	0.157 ± 0.01
13	25.3 ± 4.1	0.8 ± 2.3	1.9	0.4	1.574 ± 0.43	0.016 ± 0.05
14	10.2 ± 5.5	0.1 ± 0.7	1.2	0.2	0.157 ± 0.80	0.078 ± 0.03

[1]% AO = Percent antioxidant activity as measured by TBA method.
[2]PI = Protective Index.
(–) = non-dialysable melanoidin was not detected.

evaluating free radical reactions. Reactive oxygen species attack unsaturated fatty acids producing lipid peroxides which subsequently decompose to reactive aldehydes such as malonaldehyde. Autooxidation studies conducted with soybean oil have shown a linear relationship between oxygen consumption and peroxidation of unsaturated fatty acids (Kishida *et al.*, 1993).

The PI values calculated for each *Glu-Lys* and *Fru-Lys* MRP synthesis experiment are given in Table 2. *Glu-Lys* synthesis experiments produced 7 antioxidant, 2 prooxidant and 4 inactive MRPs. In comparison, *Fru-Lys* synthesis experiments produced 5 experiments of each having either antioxidant or prooxidant MRPs and 2 experiments with inactive MRPs. Former studies conducted with EDTA in the presence of Fe(II) also have demonstrated both prooxidant and antioxidant activities of EDTA on membrane phospholipids depending on the EDTA to Fe(II) concentration ratio (Gutteridge *et al.*, 1979). Among the different model MRP derivation experiments performed, antioxidative activity was found to occur in more instances with *Glu-Lys* MRPs compared to *Fru-Lys* MRPs. However, no relationship existed between the yield of non-dialysable MRP derived in different model *Glu-Lys* and *Fru-Lys* experiments and the characteristic antioxidant or prooxidant activity. Rather, identical experimental conditions used to generate MRPs in both *Glu-Lys* and *Fru-Lys* model systems were individually unique in producing a relationship between antioxidant or prooxidant activity and overall yield of non-dialysable material produced. For example, identical conditions used in experiment numbers 3 and 11, although producing different yields for *Glu-Lys* and *Fru-Lys* MRPs, respectively, manifested similar antioxidant activities. In contrast, *Fru-Lys* reaction experiment number 5 which produced the highest MRP yield exhibited less antioxidant activity, compared to MRPs generated from *Glu-Lys* experiment 5 using identical conditions. In addition, both *Glu-Lys* and *Fru-Lys* experiment numbers 6, 12 and 14, produced similar MRP yields under duplicated conditions of synthesis, but resulted in antioxidant activity for the *Glu-Lys* MRPs and prooxidant activity for the *Fru-Lys* MRPs.

Factors including pH, a_w, reaction time, temperature, and concentration of reactants have been regarded important in influencing the composition of derived MRPs (Pomeranz *et al.*, 1962; Rendleman and Inglett, 1990). Our study further demonstrates that varying the initial reaction conditions used to synthesize MRPs is also important in producing functional MRPs with characteristic antioxidant or prooxidant activity. Since copper was used as a promotor of lipid oxidation in the oxygen consumption study, one possible explanation for these findings could be the characteristic affinities of different MRPs to sequester transition metals. This was suspected in our study, since specific *Fru-Lys* experiment numbers 6, 12, and 14 which exhibited greater prooxidant activity (PI < 1) in the oxygen electrode study were also observed to possess no copper chelating activity.

B) TBARS Measurement in the Model Lipid Emulsion System. The TBARS generated in a model linoleic acid emulsion system containing *Glu-Lys* and *Fru-Lys* MRPs, in the absence of metal ions, is presented in Table 2 (*i.e.*, %AO). Unlike the oxygen depletion measurements which identified both the antioxidant or prooxidant activity of individual *Glu-Lys* and *Fru-Lys* model experiments, no potential prooxidant effect of MRPs was identified using the TBARs measurement. As such, all *Glu-Lys* and *Fru-Lys* MRPs derived from different experimental conditions, reduced lipid peroxidation in the lipid emulsion system. However, despite the finding that all experiments show antioxidant activity as assessed using TBARs endpoint measurements, many of the products derived from different *Fru-Lys* experiments exhibited relatively low (*e.g.* below 5%) antioxidant activity.

These results agree with the findings of Yamaguchi and Fujimaki (1974), who showed the antioxidant activity of xylose-lysine and xylose-arginine model MRPs in a linoleic acid model system. A decreased rate of lipid oxidation in cooked ground pork patties with added *Glu-Lys* model MRPs has been reported by Bedinghaus and Ockerman (1995).

In the absence of added metal ions, the effectiveness of MRPs as antioxidants in the lipid emulsion system could also signify an affinity to scavenge free radicals, and thus to reduce propagation of autooxidation reaction. Hyase *et al.* (1989) used electron spin resonance to demonstrate the effectiveness of glucose-glycine conjugate derived melanoidins to scavenge superoxide radicals. Taken together, these results indicate that the antioxidant activity of MRPs can be characterized as a combined effect of inhibiting autooxidation propagation reactions through chelation of metal ions as well as by possibly scavenging of free radicals.

C) In Vitro DNA Nicking Assay. In the present study, an *in vitro* genotoxic bioassay utilizing DNA was used as an endpoint measure of assessing the potential genotoxic activity of MRPs. The relative efficacy of different model MRPs to break supercoiled DNA into nicked circular or linear forms was shown to be specific to individual experimental conditions used to produce both *Glu-Lys* and *Fru-Lys* model MRPs, and was dependent on the concentration of individual MRPs incubated with DNA. Increasing the exposure concentration of both *Glu-Lys* and *Fru-Lys* model MRPs to DNA resulted in enhanced DNA strand breakage. In addition, a greater breakage of DNA strands generally occurred when *Fru-Lys* MRPs were used. DNA exposed to low concentrations of MRPs (*e.g.* 10^{-3}%; Figure 1 A) produced similar DNA strand breakage (approx. 1–10%) for both *Glu-Lys* and *Fru-Lys* MRPs across all experiments. As the concentration of MRPs incubated with DNA was increased to 10^{-1}% (w/v), not only did the magnitude of DNA strand breakage increase for both model MRPs (Figure 1B), but also the spread between common experiments within different *Glu-Lys* and *Fru-Lys* reactants also increased. In addition, the DNA strand cleavage caused by *Glu-Lys* MRPs remained relatively constant for all the experiments compared to *Fru-Lys* MRPs derived under similar experimental conditions. For ex-

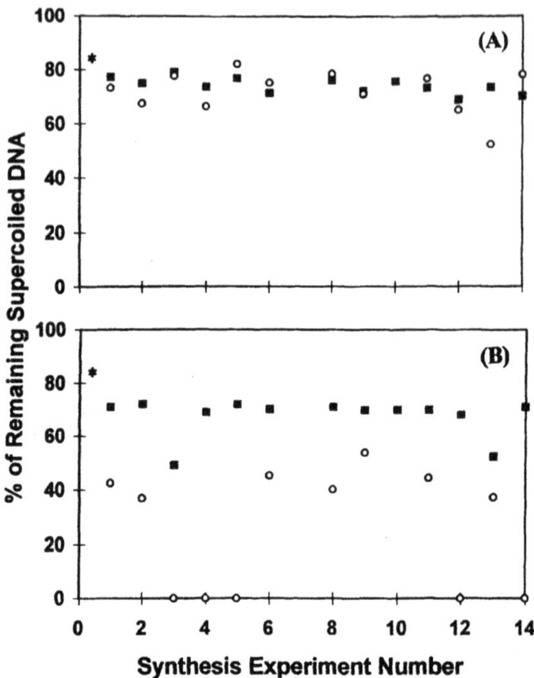

Figure 1. Incubation study conducted with model MRPs and PM_2 bacteriophage DNA at 37° C for one hour. **A** = DNA + 0.001% (w/v) *Glu-Lys Fru-Lys* MRPs, **B** = DNA + 0.1% (w/v) *Glu-Lys* and *Fru-Lys* MRPs. * = Percentage of circular DNA in the control sample; n = *Glu-Lys* MRPs; ∘ = *Fru-Lys* MRPs.

ample, at this concentration (10^{-1}%), no remaining supercoiled DNA was observed with certain specific *Fru-Lys* MRPs (*i.e.*, experiment numbers 3, 4, 5, 12, and 14) following the incubation treatment. In those experiments, all supercoiled DNA were completely broken to nicked circular and linear forms.

The dose-response curves depicting the effect of both model *Glu-Lys* and *Fru-Lys* MRPs to break DNA, support previous findings that have reported mutagenecity of browning mixtures (Shinohara *et al.* 1980; Kitts *et al.* 1993). Although the chemical nature of the complex mixture of compounds produced in these model MRPs remains unknown, the present findings show that varying the conditions to derive different MRPs can result not only in generation of specific antioxidant properties, but also potential prooxidant and genotoxic activities as well. Comparing these results with the apparent yield of MRP generated in each experiment revealed little indication of a relationship between DNA strand breakage and the amount of material generated in individual MRP experiments. These results do not totally agree with the findings of others that have implied that the more intensively browned solutions usually produce greater mutagenic activity (Shinohara *et al.* 1980). The results, however, agree with the findings of Aruoma (1993) who showed the DNA breaking activity of known antioxidants.

CONCLUSIONS

MRPs derived from *Glu-Lys* and *Fru-Lys* model reactions under varying initial reaction conditions (reaction temperature, time, a_w and pH) were examined for metal chelating affinity, antioxidant or prooxidant activity, and genotoxicity. Varying the initial reaction conditions had a marked effect on the yield of MRPs recovered from both sugar-amino acid reactants. Copper chelation affinity of MRPs also varied depending on the initial synthesis experimental conditions employed. Antioxidant activity was found to be greater for *Glu-Lys* MRPs compared to those MRPs formed from *Fru-Lys* precursors when assessed by oxygen consumption and TBAR formation in a model lipid system. MRPs synthesized by both *Glu-Lys* and *Fru-Lys* reactions nicked DNA strands in a concentration dependent manner, thus depicting genotoxicity of MRPs in a non-lipid system. The DNA strand breaking activity was remarkably greater in connection with specific *Fru-Lys* MRPs. The above results are important in evaluating the potential effect of different MRPs on development of racidity due to lipid oxidation reactions in thermally processed foods.

ACKNOWLEDGMENTS

Financial support provided by the Natural Sciences and Engineering Research Council of Canada to D.D.K. is greatefully acknowledged.

REFERENCES

Aruoma, O.I. Free radicals and food. *Chemistry in Britain.* **1993**, 29, 210–214.

Bedinghaus, A.J.; Ockerman, H.W. Antioxidative Maillard reaction products from reducing sugars and free amino acids in cooked ground pork patties. *J. Food Sci.* **1995**, 60, 992–995.

Buege, I.A.; Aust, S.D. Microsomal lipid peroxidation. *Methods in Enzymalogy* **1978**, 52, 303–311.

Cuzzoni, M.T.; Stoppini, G.; Gazzani, G. Influence of water activity and reaction temperature of ribose-lysine and glucose-lysine Maillard systems on mutagenicity, absorbancy, and content of furfurals. *Food Chem. Toxicol.* **1988**, 26, 815–822.

Chiou, R.Y-Y. Antioxidative activity of oils prepared from peanut kernels subjected to various treatments and roasting. *J. Agric. Food Chem.* **1992**, 40, 1958–1962.

Dou, J.; Toma, S.; Nakai, S. Random-centroid optimization for food formulations. *Food Res. Int.* **1993**, 26:27–37.

Eichner, K.; Karel, M. The influence of water content and water activity on the sugar amino browning reaction in model systems under various conditions. *J. Agric. Food Chem.* **1972**, 20, 218–222.

Espejo, R.T.; Canelo, E.S. Properties of bacteriophage PM_2, a lipid containing bacterial virus. *Virology* **1968**, 34, 738–747.

Feather, M.S. Amino-assisted sugar dehydration reaction. *Prog. Food Nutri. Sci.* **1981**, 5, 37–45.

Gutteridge, J.M.C.; Richmond, R.; Halliwell, B. Inhibition of the iron catalysed formation of hydroxyl radicals from superoxide and of lipid peroxidation by desferrioxamine. *Biochem. J.* **1979**, 184, 469–472.

Hashiba, H. Oxidative browning of Amadori compounds - colour formation by iron with Maillard reaction products. In *Amino-carbonyl Reactions in Food and Biological Systems*; Fujimaki, M., Namiki, M., Kato, H., Eds.;. Kodansha Ltd. Tokyo. 1985, pp. 155–164.

Hayase, F.; Hirashima, S.; Okamoto, G.; Kato, H. Scavenging of active oxygen by melanoidins. In *The Maillard Reaction in Food Processing, Human Nutrition, and Physiology*; Finot, P.A., Aeschbacher, H.U., Hurrell, R.F., Liardon, R., Eds.; Advances in Life Sciences. Birkhauser Verlag, Switzerland 1989.

Kishida, E.; Kamura, A.; Tokumaru, S.; Oribe, M.; Iguchi, H.; Kojo, S. Re-evaluation of malondialdehyde and thiobarbituric acid reactive substances as indices of autooxidation based on oxygen consumption. *J. Agric., Food Chem.* **1993**, 41, 1–4.

Kitts, D.D.; Wu, C.H.; Stich, H.F.; Powrie, D. Effect of glucose - lysine Maillard reaction products on bacterial and mammalian cell mutagenesis. *J. Agric Food Chem.* **1993**, 41, 2353–2358.

Lingert, H.; Vallentine, K.; Eriksson, C.E. Measurement of antioxidative effect in model systems. *J. Food Proc. Pres.* **1979**, 3, 87–103.

Lingert, H.; Eriksson, C.E. Antioxidative effect of Maillard reaction products. *Prog. Food Nutri. Sci.* **1981**, 5, 453–466.

McGookin, B.J.; Augustin, M. Antioxidant activity of casein and Maillard reaction products from casein - sugar mixtures. *J. Dairy Res.* **1991**, 58, 313.-320.

Obretenov, T.D.; Ivanov, S.; Peeva, D. Antioxidative activity of Maillard reaction products obtained from hydrolysates. In *Amino-carbonyl Reactions in Food and Biological Systems.* Fujimaki, M., Namiki, M., Kato, H., Eds.; Kodansha Ltd. Tokyo. 1993, pp. 281–290.

Pomeranz, Y.; Joghson, J.A.; Shellenberger, J.A. Effects of various sugars on browning. *J. Food Sci.* **1962**, 29, 350–354.

Rendleman, Jr.J.A. Complexation of calcium by melanoidin and its role in determining bioavailabilty. *J. Food Sci.* **1987**, 52, 1699–1705.

Rendleman, J.A. Jr.; Inglett, G.E. The influence of Cu^{2+} in the Maillard reaction. *Carbohyd. Res.* **1990**, 201, 311–326.

Shinohara, K.; Wu, R-T.; Jahan, N.; Tanaka, M.; Morinaga, N.; Murakami, H.; Omura, H. Mutagenicity of the browning mixtures by amino-carbonyl reactions on *Salmonella Typhimurium. TA 100. Agric. Biol. Chem.* **1980**, 44, 671–672.

Smith, I.S.; Alfarvaz, M. Antioxidant activity of Maillard reaction products in cooked ground beef, sensory and TBA values. *J. Food Sci.* **1995**, 60, 234–236, 240.

Terasawa, N.; Murata, M.; Homma, S. Separation of model melanoidin into components with copper chelating sepharose 6B column chromatography and comparison of chelating activity. *Agric. Biol. Chem.* **1991**, 55, 1507–1514.

Yamaguchi, N.; Fujimaki, M. Studies on the reaction products from reducing sugars and amino acids. Part XIV. Antioxidant activities of purified melanoidins and their comparison with those of legal antioxidants. *J. Food Sci. Technol. (Japan).* **1974**, 21, 6–12.

Yaylayan, V.A.; Heyhami, A.; Huyghues-Despointes, A. Thermal decomposition reactions of sugar - amino acids and Amadori compounds. 1997, this volume.

VOLATILE COMPONENTS FORMED FROM REACTION OF SUGAR AND β-ALANINE AS A MODEL SYSTEM OF COOKIE PROCESSING

S. Nishibori,[1] R. A. Berhnard,[2] T. Osawa,[3] and S. Kawakishi[3]

[1]Department of Food Science
Tokaigakuen Women's College
Nagoya 468, Japan
[2]Department of Food Science and Technology
University of California
Davis, California 95616
[3]Faculty of Agriculture
Nagoya University
Nagoya 464, Japan

Volatile components formed from the reaction of monosaccharides or disaccharides with β-alanine were investigated in a dry condition as a model system of cookie processing. Maltol is a common compound formed in the Maillard reaction, but it was very difficult to detect it in previous experiments using actual cookie materials. In this work, we investigated the principal compounds and maltol formation from the reaction of monosaccharides or disaccharides with β-alanine at 150 °C for 10 min. Neither the reaction of monosaccharides nor the disaccharides with β-alanine resulted in the formation of maltol. 2,3-Dihydro-3,5-dihydroxy-6-methyl-4(H)-pyran-4-one (DDMP) was detected as a principal product from the reaction of monosaccharides with β-alanine. 5-Hydroxymethyl-2-furfural was also confirmed as being a major product in both reactions.

INTRODUCTION

The Maillard reaction (1912) is a well-known chemical reaction in foods. Hodge (1953a,b) reported that nonenzymatic browning reactions could be demonstrated by using simple model systems instead of actual food systems. Subsequently, many researchers have used simple systems such as a sugar-amino acid mixture to investigate the formation of browning products and flavor. Generally, α-alanine, β-alanine, glycine and proline have been used to supply an amino group, and glucose has been used to provide a carbonyl group. These reactions were also mainly conducted in an aqueous solution. Hodge (1953a,b) considered that maltol is the principal compound produced in the Maillard reaction, and it has a favorable cookie-like aroma. Recently, some researchers have reported that maltol is detected in the thermal degradation mixtures of reducing and nonreducing disaccharides such as maltose, lactose, sucrose, and starch under aqueous conditions. We have already reported that 2,3-dihydro-3,5-dihydroxy-6- methyl-4(H)-pyran-4-one (DDMP) is formed when fructose or glucose is used as a sugar in actual cookie processing (Figure 1). In this contribution, we describe the products formed from the reaction of monosaccharides or disaccharides with added β-alanine in a dry system. Particular attention will be paid to the formation of maltol. Analysis and identification of the volatile compounds were made by HPLC, NMR, GC, FAB-MS and GC-MS.

EXPERIMENTAL

Sample Preparation

For the model system of cookies baking, the reaction mixtures consisted of equimolar amounts of a sugar (either 180 mg of monosaccharides or 360 mg of disaccharides) and β-alanine (89 mg). The components were placed in test tubes, and the mixtures were heated in a heating block at 150 °C for 2–10 min. On cooling, the samples were extracted with methanol-water (3:1 v/v) or dichloromethane, and the extract with dichloromethane was concentrated to 100 mL under a stream of nitrogen.

High Performance Liquid Chromatography (HPLC)

HPLC analysis was performed using a Toyo Soda Model HLC-803D pump. Test parameters were as follows: column, DEVELOSIL ODS (4.6 mm i.d. x 250 mm); mobile

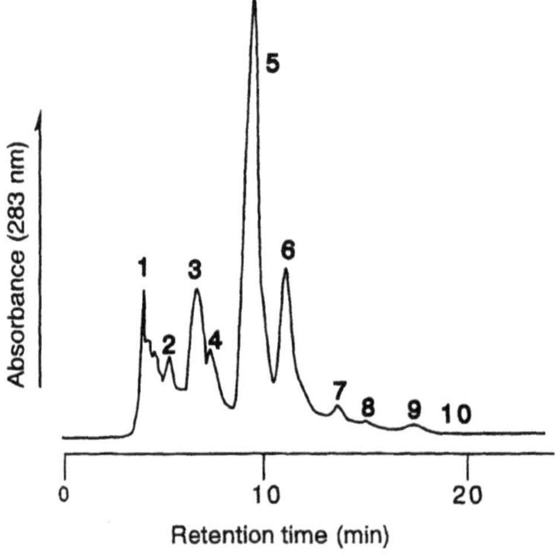

Figure 1. High performance liquid chromatogram of products from cookies backed with fructose at 150°C for 10 min. Chromatographic conditions: column; DEVELOSIL ODS-5 (4.6 mm i.d. x 250 mm), mobile phase; H_2O-MeOH (6:1 v/v), flow rate; 0.8 ml/min, UV detection; 283 nm, chart speed; 5 mm/min. (3); 2,3-dihydro-3,4-dihydroxy-5-acetylfuran, (5); 2,3-dihydro-3,5-dihydroxy-6-methyl-4-(*H*)-pyran-4-one, (6); 5-hydroxymethylfurfural.

phase, water-methanol (6:1 v/v); flow rate, 0.8 mL/min; detector, UV 283 nm. A JASCO model MD-910 photodiode array (PDA) multiwavelength UV/VIS detector was also used for monitoring extraction profiles and chromatograms. A data processing software package DP-910 run on a DELL PC was also used for the post-run data analysis.

Acetylation

Peaks 3 (P3), 5 (P5) and 6 (P6) in Figure 1 were isolated from the preparative HPLC. A mixture of P3 or P5 (each 5 mg), acetic anhydride (0.5 mL) and anhydrous pyridine (0.5 mL) was kept at room temperature for 24 hr. The excess acetic anhydride was then decomposed with methanol, and the mixture containing a small added portion of toluene was concentrated under reduced pressure. Acetylated P3 (Ac. P3) and P5 (Ac. P5) were purified by HPLC fitted with DEVELOSIL SI 60–10, 20 mm i.d. x 250 mm. HPLC conditions were as follows: mobile phase, hexane-ethyl acetate (7:3 v/v); flow rate, 6.0 mL/min; detection, UV 298 nm.

Nuclear Magnetic Resonance (NMR)

NMR spectra were obtained on a JEOL GX-200 (^{1}H, 200 MHz and ^{13}C, 50 MHz). ^{1}H -NMR and ^{13}C-NMR were measured in D_2O and CD_3OD as solvents, respectively. P3, P5 and P6 were sealed in 4 mm i.d. NMR tubes.

Fast Atom Bombardment-Mass Spectrometry (FAB-MS) and Electron Impact-Mass (EI-MS) Spectrometry

Mass spectral analyses were carried out on a JEOL DX-303. Glycerol was used as the matrix for FAB-MS analysis.

Capillary Gas Chromatography (GC)-Mass Spectrometry (MS)

A Hewlett-Packard 5792 gas chromatograph coupled with a VG analytical ZAB-HS-2F mass spectrometer with a VG 11/250 data system was used for mass spectral identification of components. Operating conditions: ionizing voltage, 70 eV; source temperature, 250 °C; accelerating voltage, 8000 eV; filament trap current, 100 mA. The gas chromatography was modified for use with capillary columns. A 30 m x 0.25 mm (film thickness 0.25 mm) fused silica capillary column bonded and cross-linked with DB-1 was temperature-programmed as follows: 50 °C for 8 min, 50- 250 °C at 4 °C /min, isothermal hold at 250 °C. Carrier gas velocity was 30 cm/s He.

RESULTS AND DISCUSSION

Identification of the Compounds in the Reaction of Reducing Sugar and β-Alanine

Equimolar amounts of fructose and β-alanine were heated in a heating block at 150 °C for 4 min. Figure 2 shows the HPLC profiles of the methanol-water extracts from the reaction mixtures of fructose or glucose with β-alanine. The products from the reaction for 2 min gave

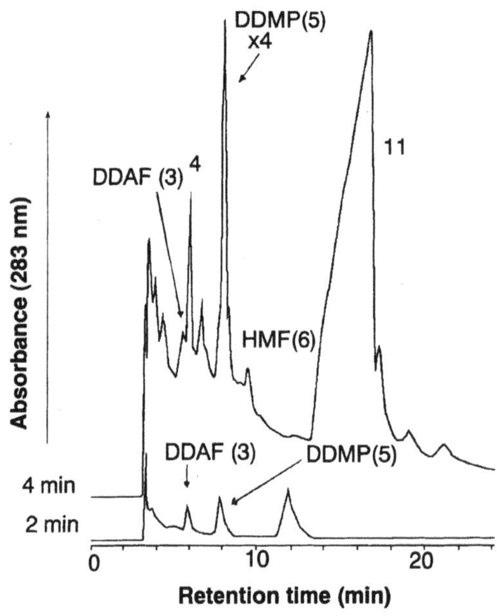

Figure 2. High performance liquid chromatogram of extract from heating of fructose with β-alanine. Chromatographic conditions: column; DEVELOSIL ODS-5 (4.6 mm i.d. x 250 mm), mobile phase; H_2O-MeOH (6:1 v/v), UV detection; 283 nm, chart speed; 5 mm/min.

only two main peaks (Nishibori and Kawakishi, 1991): 2,3-dihydro-3,4-dihydroxy-5-acetylfuran (DDAF) and 2,3-dihydro-3,5-dihydroxy-6- methyl-4(H)-pyran-4-one (DDMP). In the case of 4 min, many compounds were formed from the reaction mixture. Five principal products in the reaction were identified as 3-(N-(2-formyl-5-hydroxymethylpyrrolyl))-propionic acid (3NFHP), DDMP, 3-(N-(2- methyl-3,6-dihydro-4,5,6-trihydroxymethylpyridinyl))-propionic acid (3NMDTP), 5-hydroxymethyl-2-furlfural (HMF) and DDAF (Figure 3). These compounds were isolated and identified by HPLC, UV spectra, FAB-MS, EI-MS and NMR. Spectral data of the compounds were as follows.

2,3-Dihydro-3,5-Dihydroxy-6-Methyl-4H-Pyran-4-One (DDMP). UV spectrum (water): λmax 295 nm. Mass spectrum: molecular weight of 144. ^{13}C-NMR ($CDCl_3$): 15.6 ppm

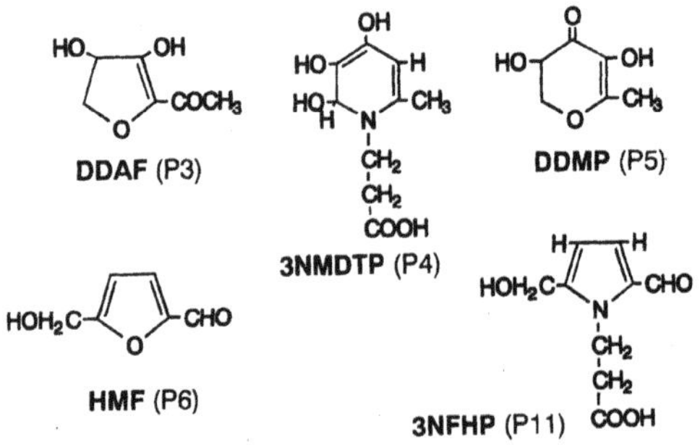

Figure 3. The structures of DDAF, 3NMDTP, DDMP, HMF and 3NFHP. DDAF; 2,3-dihydro-3,4-dihydroxy-5-acetylfuran, 3NMDTP; 3-{N-(2-methyl-3,6-dihydro-4,5,6-trihydroxymethylpyridinyl)}-propionic acid, DDMP; 2,3-dihydro-3,5-dihydroxy-6-methyl-4(H)-pyran-4-one, HMF, 5-hydroxymethylfurfural, 3NFHP; 3-{N-(2-formyl-5-hydroxymethyl-pyrrolyl)}-propionic acid.

(C6), 67.5 ppm (C1), 71.4 ppm (C2), 130.1 ppm (C4), 165.0 ppm (C5) and 188.2 ppm (C3). FAB-MS spectrum: 145 (M+ H) for the molecular ion, m/z 127 for loss of water and m/z 112 for loss of a methyl group. ^{1}H -NMR spectrum: 1.90 ppm (3H) and 4.05–4.25 ppm (5H).

2,3-Dihydro-3,5-Diacetoxy-6-Methyl-4H-Pyran-4-One (Ac DDMP). EI-MS: m/z 186 for addition of a acetyl group and m/z 228 for two acetyl groups. ^{1}H-NMR (CDCl3): a singlet at 1.99 ppm (CH_3), a singlet at 2.11 ppm (AcO-C), 2.2 ppm (AcO-C=C), a quartet at 4.39 ppm (Ha), a quartet at 4.53 ppm (Hb) and a quartet at 5.41 ppm (Hc). The IR spectrum: Vmax 1760, 1700 and 1628 cm^{-1} for 3 carbonyl groups from the configuration of DDMP and the acetylation of DDMP.

2,3-Dihydro-3,4-Dihydroxy-5-Acetylfuran (DDAF). UV spectrum (water): λmax 288 nm. Mass spectrum: 144 (M+, 100), 126 (30), 113 (10), 101 (35), 84 (13), 73 (10), 55 (47), 43 (75), 31 (12), 18 (15). HR-MS: found, m/z 145.0571 [M+H]+; calcd. for $C_6H_9O_4$, 144.0562. ^{13}C-NMR ($CDCl_3$): 13.0 ppm (C6), 60.4 ppm (C1), 83.9 ppm (C2), 135.6 ppm (C4), 174.0 ppm (C5) and 196.5 ppm (C3). ^{1}H-NMR (D_2O): 2.11 ppm (3H, s), 3.71 ppm (1H, q), 3.76 ppm (1H, q), 3.88 (1H, q).

2,3-Dihydro-3,4-Diacetoxy-5-Acetylfuran (Ac DDAF). UV spectrum (MeOH): λmax 274 nm. Mass spectrum: 228 (M+), 186 (M+-$COCH_3$), 144 (M+ -$COCH_3$ x 2), 126, 115, 101, 84, 55, 43. ^{1}H -NMR (D2O): 2.03 ppm (3H, s), 2.16 ppm (3H, s), 2.26 ppm (3H, s), 4.43 (1H, q), 4,52 (1H, q), 5.43 (1H, q).

5-Hydroxymethyl-2-Furfural (HMF). UV spectrum: λmax 283 nm. EI-MS: m/z 126. ^{1}H -NMR: a singlet at 3.30 ppm (OH), a singlet at 4.05 ppm (CH_2), a doublet at 6.58 ppm (H4), a doublet at 7.37 ppm (H3) and a singlet at 9.52 ppm (CHO).

3-{N-(2-Formyl-5-Hydromethylpyrrolyl)}-Propionic Acid (3NFHP). UV spectrum (water): a shoulder at 262 nm (ϵ, 6.2 x 10^3) and a maximum at 298 nm (ϵ, 1.52 x 10^4). ^{1}H-NMR (D_2O): 2.87 (2H, t), 4.65 (2H, t), 4.81(2H, s), 6.28 (1H, d), 7.10 (1H, d), 9.37(1H, s). ^{13}C-NMR ($CDCl_3$): 31.2, 45.6, 55.6, 111.8, 127.1, 132.3, 144.3, 175.5, 181.6 ppm. MS m/z: 197 (M+, 100), 179, 168 (M+ - CHO), 152, 124 (M+ - CH_2-CH_2-COOH), 108. 3NFHP has a strong sweet aroma. The yield was 2.5 mg from 1 m mole of fructose.

Acetylated 3-{N-(2-Formyl-5-Hydroacetylpyrrolyl)}-Propionic Acid. ^{1}H-NMR(D_2O): 2.08 (3H, s), 3.24 (2H, t), 4.75 (2H, t), 6.24 (1H, d), 6.89 (1H, d), 9.55 (1H, s). MS m/z: 239 (M+), 211 (M+ - CHO), 197 (M+ - $COCH_3$), 179, 134, 108, 106. IR: one acetyl carbonyl (max 1740 cm^{-1}), one aldehyde (2650 and 1650 cm^{-1}), one carbonic acid (1725 cm^{-1}).

3-{N-(2-Methyl-3,6-Dihydro-4,5,6-Trihydroxylpyridinyl)}-Propionic Acid (3NMDTP). UV spectrum (water): λmax 290 nm. ^{1}H-NMR (D_2O): 1.35 (3H, d), 2.64 (2H, t), 3.34 (2H, t), 4.06 (1H, s), 4.21(1H, q). MS m/z: 215 (M+), 197 (M+ - H_2O), 180, 156, 144, 126, 115, 93. The yield was 0.4 mg from 1 m mole of fructose.

The volatile compounds from the reaction mixtures of fructose or glucose with β-alanine were extracted with dichloromethane and identified by GC and GC-MS using a fused silica capillary column (Figure 4). Eleven products were detected, and the major component, peak 5, was identified as DDMP (Table 1).

Using GC/MS, Parliment (1992) identified DDMP from the reactions of glucose and anhydrous proline heated in either a microwave or conventional oven. He also detected 2-

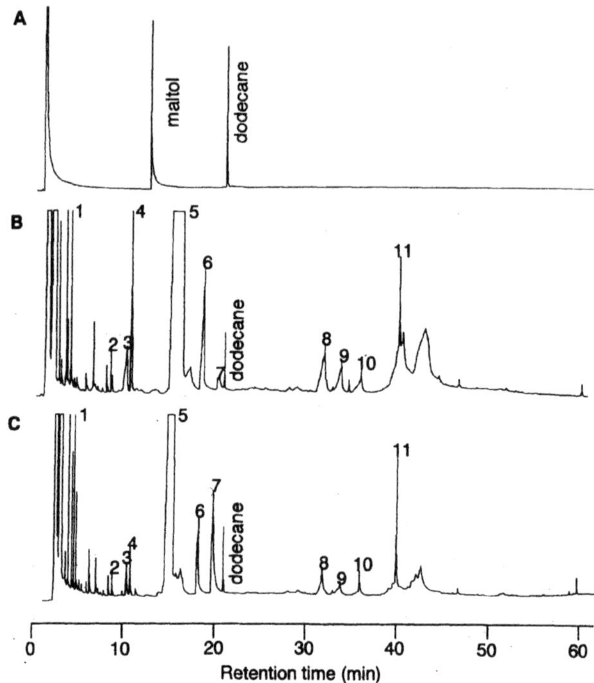

Figure 4. Gas chromatogram of authentic sample and model system extracts. (A) maltol; (B) extracts from the reaction of fructose and β-alanine; (C) the extracts from the reaction of glucose and β-alanine. GC column: 30 m x 0.25 mm (i.d.) fused silica capillary column bonded and cross-linked with DB-1 (film thickness 0.25 μm); temperature program, 50°C for 8 min, 50–250°C/min, isothermal hold at 250°C. Internal standard is dodecane.

Table 1. Compounds of the volatile flavor identified in the reaction of fructose and β-alanine

Products	Peak no.	Kovates index (DB-1)	Mass spectra data, M+ m/e, or ref
Furans			
2-hydroxymethylfuran	1	865	98, Parliment (1992)
2-acetyl-3-hydroxy-4,5-dihydrofuran (isomatol)	2	995	126
Furanones			
2,5-dimethyl-4-hydroxy-3(2H)-furanone (DHF)	3	1030	128, Parliment (1992)
2-acetyl-4-hydroxy-3(SH)-furanone	7	1093	142
Pyrans			
2,3-dihydro-3-hydroxy-6-methyl-4(H)-pyran-4-one	4	1036	126, Parliment (1992)
2,3-dihydro-3,5-dihydroxy-6-methyl-4(H)-pyran-4-one (DDMP)	5	1123	144, Parliment (1929)
2,3-dihydro-3-hydroxy-6-formyl-4(H)-pyran-4-one	6	1123	142, Parliment (1992
Pyrroles			
3-{N-(2-formylpyrrolyl)}-propionic acid	8	1450	167
3-{N-(2-acetylpyrrolyl)}-propionic acid	9	1508	181
3-{N-(2-formyl-5-methylpyrrolyl)}-propionic acid	10	1568	181
3-{N-(2-formyl-5-hydroxymethylpyrrolyl)}-propionic acid (3NFHP)	11	1700	197

hydroxymethylfuran, 2,5- dimethyl-4-hydroxy-3(2H)-furanone (DHF), 2,3-dihydro-3-hydroxy-6-methyl-4(H)-pyran-4-one, and 2,3-dihydro-3-hydroxy-6-formyl- 4(H)-pyran-4-one from the same reaction. Mill (1979) also suggested that DHF was formed from an Amadori compound, such as 1-deoxy-1-sarcosino-D-fructose. Yeo and Shibamoto (1991a,b) reported the formation of DHF and DDMP on the reaction of glucose and cysteine with different moisture contents (0–40%), using microwave irradiation. We and Parliment (1992) confirmed DHF which was produced in the reactions of either fructose or glucose with β-alanine in an anhydrous system. Hodge (1972) described the formation of 2-acetyl-3-hydroxy-4,5-dihydrofuran (isomaltol) from glucose and various amino acids.

In this investigation, we also identified four pyrroles, 3-(N-(2-formylpyrrolyl))-propionic acid, 3-(N-(2-acetylpyrrolyl))- propionic acid, 3-(N-(2-formyl-5-methyl-pyrrolyl))-propionic acid, and 3NFHP from the reactions of both fructose and glucose with β-alanine. There are numerous papers concerning the formation of pyrroles from reactions using glucose and various amino acids (Kato, 1967; Kato et al., 1972; Njoroge and Monnier, 1989). These compounds have somewhat similar structures due to the reducing sugars, but there are slight differences between each compound due to the various amino acids used. In our study, β-alanine was used as the amino acid for the reaction, and pyrroles containing propionic acid were them detected (Nishibori and Kawakishi, 1991, 1995).

From examination of these results, it is apparent that part of the fructose reacted with the amino acids and produced the same compounds as those from glucose via a pathway similar to the reaction of glucose and amino acids.

Confirmation of Maltol in the Reactions of Fructose and Glucose with β-Alanine

Maltol (3-hydroxy-2-methyl-4(H)-pyran-4-one) is a known compound with a sweet aroma and a caramel-like flavor. Maltol is used industrially as a flavor enhancer for improving the quality of food. Patton (1950) reported that maltol was detected when either maltose or lactose was heated with glycine in aqueous solutions. Hodge also described the nonenzymatic formation of maltol from sugars (Hodge and Nelson, 1961) and suggested a reaction mechanism for its formation (Hodge et al., 1972). For many years, maltol was mentioned as a principal product of cookie model systems. Recently, there has been very little mention of the product (Hiebl et al., 1987; Yeo and Shibamoto, 1991). Little maltol was detected in model systems of baking cookies (Nishibori and Kawakishi, 1992; Nishibori and Bernhard, 1993), and we were unable to confirm the identity of compounds at the position of maltol with a Kovats Index of 1065.4 by GC-MS analysis (Figure 4). In addition, Figure 5 shows the absence of maltol in the extract from the reaction of fructose and β-alanine, though authentic maltol was detected at the retention time of 12 min with a λmax of 275 nm.

Formation of Maltol from Disaccharides and Amadori Compounds

Yaylayan and Mandeville (1994) investigated the formation of maltol from disaccharides in aqueous systems under basic and neutral conditions, and reported that maltose was the most efficient source of maltol. In our study, using maltose and lactose, however, maltol could not be detected in the reaction mixture (Figure 5). HMF was confirmed at 7.2 min with lmax 283 nm in each extract from three reactions, and N-substituted compounds were found at 18–19 min with λmax 290–300 nm.

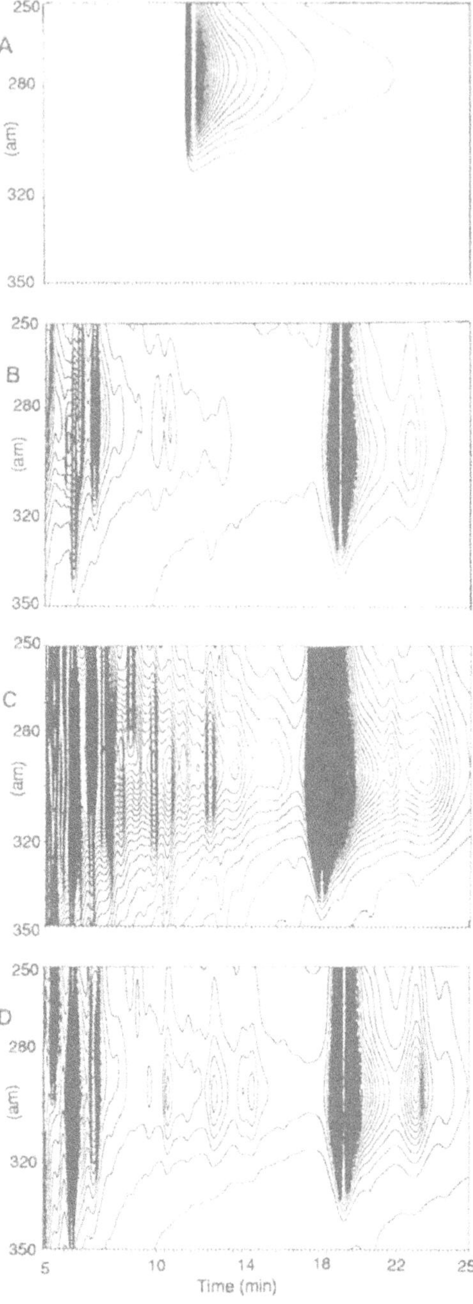

Figure 5. High performance liquid chromatogram of authentic sample and extract from sugar and β-alanine heated at 150°C for 10 min by photodiode array. Chromatographic conditions: column; DEVELOSIL ODS-5 (4.6 mm i.d. x 250 mm), mobile phase; H_2O-MeOH (6:1 v/v), flow rate; 0.8 ml/min. (A) maltol; (B) the extracts from the reaction of maltose and β-alanine; (C) the extracts from the reaction of lactose and β-alanine; (D) the extracts from the reaction of fructose and β-alanine.

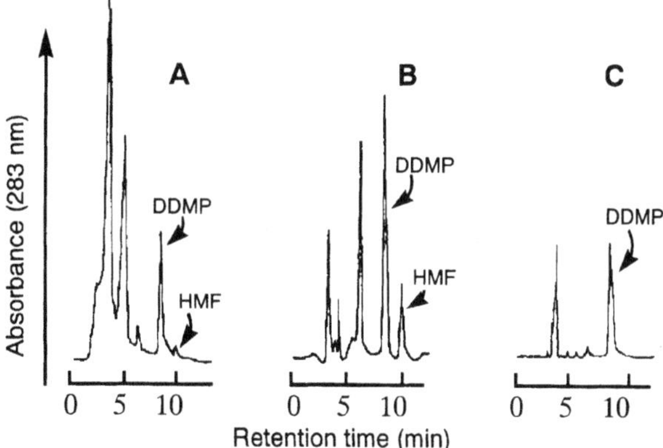

Figure 6. Formation of DDMP and HMF from Amadori compounds heated at 150°C for 10 min. Chromatographic conditions: column; DEVELOSIL ODS-5 (4.6 mm i.d. x 250 mm), mobile phase; H_2O-MeOH (6:1 v/v), flow rate; 0.8 ml/min, UV detection; 283 nm, chart speed; 5 mm/min. A; fructose-β-alanine, B; fructose-ρ-toluidine, C; fructose-Nα-t-Boc-lysine.

On the other hand, Yaylayan and Mandeville (1991) have proposed that the formation of maltol is a consequence of electron impact during mass spectrometry of Amadori products, and maltol is thus an artifact. They also suggested that maltol was formed by the heating of DDMP and amino acids or by the ionization of the ring oxygen in GC analysis. Nishibori and Kawakishi (1994) have reported that Amadori compounds, using fructose-β-alanine, fructose-γ-toluidine, fructose-Na-t-Boc-lysine, formed DDMP and HMF but not maltol (Fig 6). DDMP and HMF from fructose-β-alanine increased according to heating time (Figure7). Therefore, the presence of a pathway to DDMP through Amadori compounds was suggested.

Main Compound Obtained from the Reactions of Glucose or Fructose with β-Alanine

A mixture of 180 mg of either fructose or glucose with 89 mg of β-alanine was heated at 150 °C for 4 - 8 min. Many more volatile compounds were formed from the reaction of fructose and β-alanine than glucose and β-alanine (Nishibori and Bernhard, 1993). However, the principal compounds formed from fructose and glucose with β-alanine were almost the same. In particular, DDMP was formed from fructose and β-alanine during heating periods of 4 and 6 min (Figure 8), and its production was then gradually decreased after heating for over 8 min. In the case of glucose and β-alanine (Nishibori and Bernhard, 1993), the formation of DDMP increased stepwise in proportion to the heating time. This result suggests that the reaction of fructose and β-alanine occurs significantly faster than that of glucose and β-alanine. Figure 9 shows a proposed mechanism for the formation of the main compounds in both reactions; the formation of compounds 4, 5 (DDMP) and 7, each containing a methyl group, was particularly enhanced using fructose (Nishibori and Bernhard, 1993).

Heyns and Meinecke (1953) and Heyns and Breuer (1958) reported that the reaction of fructose proceeds via the Heyns Rearrangement, and that of glucose is via the Amadori Rearrangement. Nishibori and Kawakishi (1994) also investigated the intermediate prod-

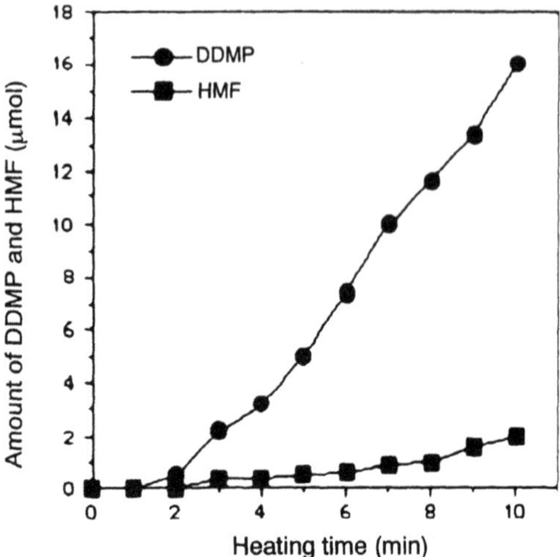

Figure 7. Formation of DDMP and HMF from 1 mmol of fructose-β-alanine heated at 150°C for 10 min.

ucts in the reaction mixture of fructose or glucose with β-alanine by HPLC on a DEVELOSIL NH_2–5 column. An Amadori compound was found to exist in the reaction mixture of fructose with β-alanine similar to that in the reaction mixture of glucose with β-alanine. The amounts of the Amadori compound from fructose and β-alanine was one-fourth that of the glucose and β-alanine. The reaction of fructose and β-alanine was also confirmed to form four kinds of theoretical fructosylamines including the Heyns Rearrangement. Further, it was ascertained that a compound (analogous HR compound) containing a methyl group was present in these fructosylamines according to the molecular weight by FAB-MS and the signals from NMR. This analogous HR compound might contribute to the rapid formation of the products including a methyl group such as DDMP, DDAF and isomaltol more than the Amadori compound does. Maltol also has a methyl group, but it was not present in the reaction mixture of fructose and β-alanine, though it is more available for the formation of compounds containing a methyl group than the reac-

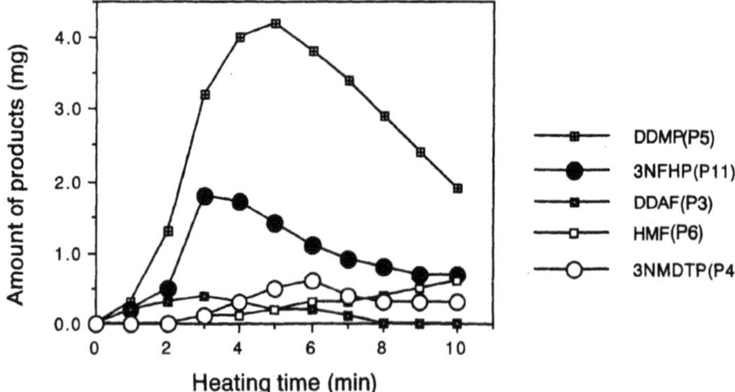

Figure 8. Changes in the products from fructose heated with β-alanine at 150°C for 10 min, detected by HPLC.

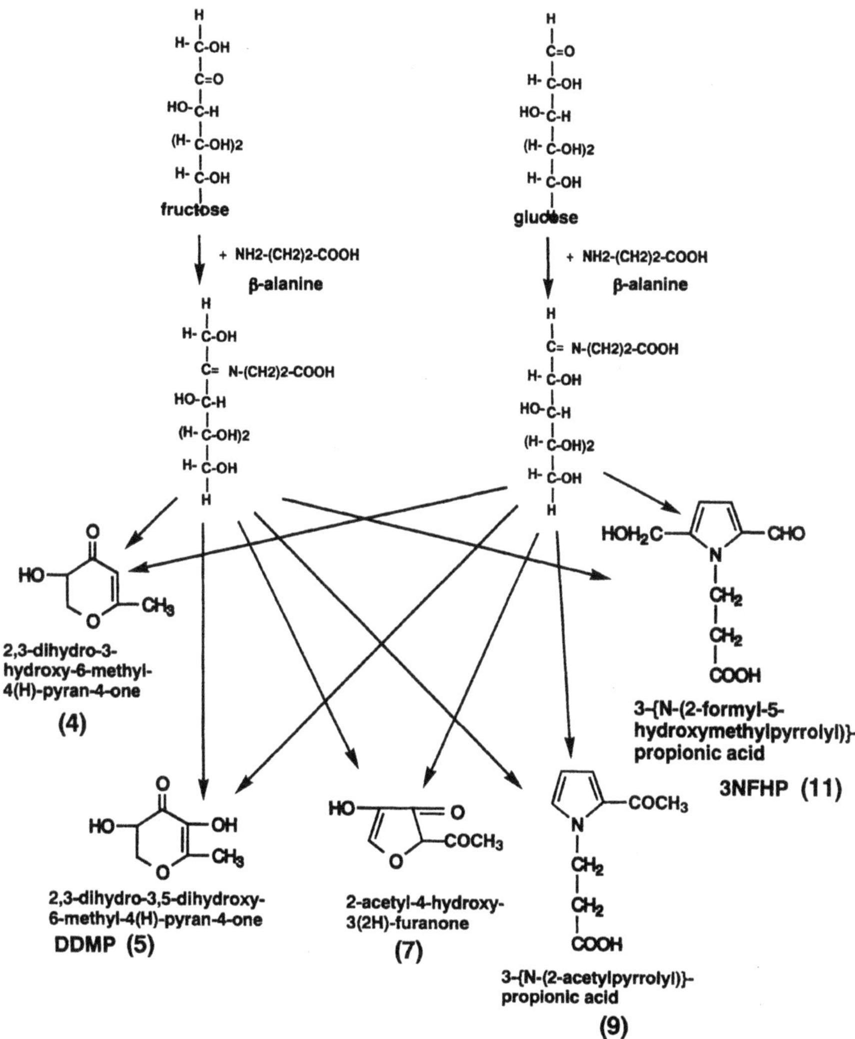

Figure 9. Proposed mechanism of main compounds in the reaction of either fructose or glucose with β-alanine, identified by GC-MS.

tion mixture of glucose and β-alanine. The results indicate that maltol is rarely formed in the reaction of monosaccharides with amino acid.

As mentioned previously, Yaylayan and Mandeville (1994) reported that maltose is the most efficient source of maltol. Maltose and lactose have a structure binding β-D-glucose and α-D-galactose at the position-4 carbon of D-glucose possessing reductive power. In the case of heating with propylamine, Ledl et al. (1989) reported that these disaccharides transform into glycosylated pyridinium betaines, and then the betaines change to small molecular compounds of pyridone rings containing a methyl group at the position-2 and a hydroxy group at the position-3, by the separation of β-D-glucose and α-D-galactose at the position-4 carbon of D-glucose possessing reductive power. If these reactions could occur without N-substitution in the reaction of maltose and lactose with amino acids, maltol might be more easily formed from these reducing disaccharides than from monosaccharides. On the assumption that reducing disaccharides can form maltol, they would be useful in the aroma formation in cookies. However, maltol is still not detected in cookies except in the artificial addition of maltol in our system of cookie processing (Nishibori and Kawakishi, 1988, 1990, 1992).

REFERENCES

Heyns, K.; Meinecke, K.H. Uber bildung und darstellung von D-glucosamin aus fructose und ammoniak. *Chem. Ber.* **1953**, *86*, 1453–1462.

Heyns, K.; Breuer, H. Darstellung und verhalten weiterer N-substituierter 2-amino-2-desoxy-aldosen aus D-fructose und aminosauren. *Chem. Ber.* **1958**, *91*, 2750–2756.

Hiebl, J.; Ledl, F.; Severin, T. Isolation of 4-hydroxy-2-(hydroxymethyl) - 5-methyl-3(2H)- furanone from sugar amino acid reaction mixtures. *J. Agric. Food Chem.* **1987**, *35*, 990–993.

Hodge, J.E. Chemistry of browning reactions in model systems. *J. Agric. Food Chem.* **1953a**, *1*, 928–943.

Hodge, J.E.; Rist, C.E. The Amadori rearrangement under new conditions and its significance for non-enzymatic browning reactions. *J. Amer. Chem. Soc.* **1953b**, **75**, 316–322.

Hogde, J.E.; Mills, F.D.; Fisher, B.E. Compounds of browned flavor derived from sugar-amine reactions. *Cereal Sci. Today* **1972**, *17*, 34.

Hogde, J.E.; Nelson, E. C. Preparation and properties of galactosylisomaltol and isomaltol. *Cereal Chem.* **1961**, *38*, 207–221.

Kato, H. Chemical studies on amino-carbonyl reaction Part III. Formation of substituted pyrrole-2-aldehydes by reaction of aldoses with alkylamines. *Agr. Biol. Chem.* **1967**, *31*, 1086–1090.

Kato, H.; Shigematu, H.; Kurata, T.; Fujimaki, M. Maillard reaction products formed from L-rhamnose and ethylamine. *Agr. Biol. Chem.* **1972**, *36*, 1639–1642.

Ledl, F.; Osiander, H.; Pachmayr, O.; Severin, T. Formation of maltosine, a product of the Maillard reaction with a pyridone structure. *Z. Lebensm Unters Forsch* **1989**, *188*, 207–211.

Maillard, L.C. Action of amino acids on sugars. Formation of melanoidins in a methodical way. *Compt. Rend.* **1912**, *154*, 66–68.

Mill, F.D. Vacuum thermolysis of 1-deoxy-1-sarcosino-D-fructose. *J. Agric. Food Chem.* **1979**, *27*, 1136–1138.

Nishibori, S.; Kawakishi, S. Changes in baking products of cookie dough composed of different materials. *Nippon Shokuhin Kogyo Gakkaishi.* **1988**, *35*, 235–241.

Nishibori, S.; Kawakishi, S. Effects of dough materials on flavor formation in baked cookies. *J. Food Sci.* **1990**, *55*, 409–412.

Nishibori, S.; Kawakishi, S. Formation of 2,3-dihydro-3,4-dihydroxy-5- acetylfuran in the reaction between D-fructose and β-alanine. *Agric. Biol. Chem.* **1991**, *55*, 1993–1998.

Nishibori, S.; Kawakishi, S. Effect of various sugars on the quality of baked cookies. *Cereal Chem.* **1992**, *69*, 160–164.

Nishibori, S.; Bernhard, R. A. Model system for cookies; volatile components formed from the reaction of sugar and β-alanine. *J. Agric. Food Chem.* **1993**, *41*, 2374–2377.

Nishibori, S.; Kawakishi, S. Formation of 2,3-dihydro-3,5-dihydroxy-6-methyl- 4(H)-pyran-4-one from fructose with β-alanine under condition used for baking. *J. Agric. Food Chem.* **1994**, *42*, 1080–1084.

Nishibori, S.; Kawakishi, S. Formation of N-substituted compounds in the reaction between D-fructose and β-alanine. *Biosci. Biotech. Biochem.* **1995**, *59*, 307–308.

Njoroge, F.G.; Monnier, V.M. The chemistry of the Maillard reaction under physiological conditions: A Review. In *The Maillard Reaction in Aging Diabetes, and Nutrition*; Alan R. Liss Inc., 1989; pp. 85–107.

Patton, S. The formation of maltol in certain carbohydrate-glycine systems. *J. Biol. Chem.* **1950**, *184*, 131–135.

Parliment, T.H. Comparison of thermal and microwave mediated proline: glucose reaction. 204th ACS National Meeting Abstract #214, 1992.

Yaylayan, V.A.; Mandeville, S. Structural features of ortho elimination reactions of amadori rearrangement products under electron impact condition. *Spectroscopy* **1991**, *9*, 73–78.

Yaylayan, V.A.; Mandeville, S. Stereochemical control of maltol formation in Maillard reaction. *J. Agric. Food Chem.* **1994**, *42*, 771–775.

Yeo, C.H.H.; Shibamoto, T. Flavor and browning enhancement by electrolytes during microwave irradiation of the Maillard model system. *J. Agric. Food Chem.* **1991a**, *39*, 948–951.

Yeo, C.H.H.; Shibamoto, T. Effects of moisture content on the Maillard browning model system upon microwave irradiation. *J. Agric. Food Chem.* **1991b**, *39*, 1860–1862.

AMINO-REDUCTONES

Formation Mechanisms and Structural Characteristics

T. Kurata[1] and Y. Otsuka[2]

[1]Institute of Environmental Science for Human Life
Ochanomizu University
Tokyo 112, Japan
[2]Faculty of Education
Tottori University
Tottori 680, Japan

Various types of amino-reductones are known to be involved in process-induced chemical changes in foods. Since most amino-reductones, especially enaminol compounds are unstable reaction intermediates and are difficult to isolate, their structural characteristics are still unclear. In order to obtain more precise structural information about amino-reductones, the structures of the enaminol form of fructoseglysine (D-fructoseglycine = FG) and scorbamic acid (L-scorbamic acid = SCA), a relatively stable cyclic enaminol compound were examined by the use of a semi-empirical molecular orbital method. Optimized structures, heat of formations and charge distributions of various enol forms of FG were obtained. Heat of formations of the non-dissociated, mono-anion, and di-anion forms of FG were estimated to be about -302, -347 and -308 kcal/mol, respectively. Optimized structures of non-dissociated, anion, and dipolar ion forms of SCA were also obtained and their heat of formations were estimated to be about -197, -204 and -248 kcal/mol, respectively. The nitrogen atom of the enaminol group in the SCA molecule was found to be positively charged.

INTRODUCTION

There are a number of chemical reactions which can participate in the formation of various types of amino-reductones including enaminols and enediamines. These amino-reductones are known to play important roles in the process-induced chemical changes in foods, especially in browning and in cooked flavor formation, and thus the over-all quality

Process-Induced Chemical Changes in Food
edited by Shahidi *et al.* Plenum Press, New York, 1998

control of foods during processing (Ledl *et al.*, 1989; Njoroge *et al.*, 1989). Most of amino-reductones produced in foods during heat treatment have an enaminol group as its characteristic partial structure. Two major reaction pathways which contribute to the formation of enaminol compounds observed in food processing are given in Figure 1. In the reaction pathway of the sugar-amine reaction (Finot *et al.*, 1990; Fujimaki *et al.*, 1986; Waller *et al.*, 1983), the enaminol form of the Amadori-rearrangement product, also designated as the enol form of Schiff's base, is formed. This enaminol compound is an important reaction intermediate compound in the sugar-amine reaction. On the other hand, the reaction of α-amino acids with α-dicarbonyl compounds, also known as the Strecker degradation, is important for the production of various volatile flavor compounds such as aldehydes and pyrazines which contribute to the aroma of many cooked foods. Furthermore, in the reaction process of decarboxylation and deamination of an α-amino acid, one of two carbonyl groups of an α-dicarbonyl compound usually change to an amino group yielding an enaminol compound.

These amino reductones described above are usually very unstable reaction intermediate compounds and, therefore, isolation and elucidation of their precise chemical structures by ordinary experimental techniques are rather difficult. However, owing to the recent remarkable progress in computational chemistry, various types of molecular orbital methods are now applicable to obtain needed information about their precise structures and chemical reactivities. For instance, the optimized structure of L-ascorbic acid, an important acid-reductone in food and biological systems, was obtained by both semi-empirical and *ab initio* molecular orbital methods (Abe *et al.*, 1987, 1992). Semi-empirical molecular orbital calculations were also used to elucidate the autoxidation mechanism of L-ascorbic acid (Kurata *et al.*, 1996a,b).

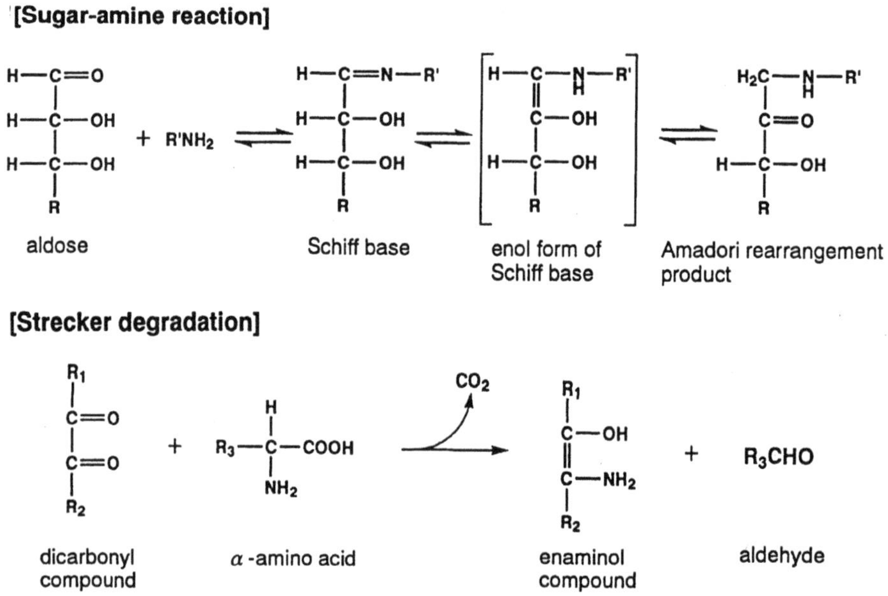

[Sugar-amine reaction]

aldose + $R'NH_2$ ⇌ Schiff base ⇌ [enol form of Schiff base] ⇌ Amadori rearrangement product

[Strecker degradation]

dicarbonyl compound + α-amino acid → ($-CO_2$) enaminol compound + R_3CHO (aldehyde)

Figure 1. Formation of enaminol compounds by sugaramine reaction or strecker degradation.

In this report, some results of calculations made on a simple enaminol compound, 2-amino-3-hydroxy-2-butene are given first. Then the calculation results of the enaminol form of fructoseglycine as an example of the enaminol form of Amadori compound produced in the sugar-amine reaction are described. And last, the results of the calculation made on L-scorbamic acid which is produced by the Strecker degradation of dehydro-L-ascorbic acid (DHA) with an α-amino acid are given.

EXPERIMENTAL

In this study, most of the molecular orbital calculations were carried out using a MOPAC (Stewart, version 6.0), a packaged program of the semi-empirical molecular orbital method. A molgraph (Daikin Co. Ltd.) was used for the visualization of the calculation results.

RESULTS AND DISCUSSION

At first, MO calculations were made on 2-amino-3-hydroxy-2-butene which is one of the most simple enaminol compounds, and can exist in its ketone form, 2-amino-3-oxobutane. In Table 1, the values of the heat of formation, the ionization potential and the dipole moment of keto form of this compound as well as those of the *cis*- and *trans*- forms of this enaminol compound are summarized. The fact that the value of the heat of formation of keto form is larger than that of the enol form, suggests that the former is more stable than the latter. Especially in a polar solvent, the former which has a larger dipolar moment seemed to be more stable than the latter. The value of the ionization potential, which corresponds to the energy level of the highest occupied molecular orbital (HOMO) of the keto form is larger than that of enol form which also seems to indicate that the keto form is more stable than the enol form. On the other hand, though the differences between the values of cis-and trans-isomers are rather small, the trans-isomers seemed to be slightly more stable than the *cis*-isomer.

The values of the heat of formation and the energy levels of HOMO and the lowest unoccupied molecular orbital (LUMO) of various forms of 2-amino-3-hydroxy-2-butene are given in Table 2. The value of the heat of formation of the dipolar ion suggests this form is not stable. However, it should be taken into consideration that all of these calculations were made on isolated molecules, neglecting any possible solvent effects or other

Table 1. An amino-reductone and its characteristics

	CH_3–C(=O)–CH(NH_2)–CH_3	CH_3–C(OH)=C(NH_2)–CH_3	CH_3–C(OH)=C(NH_2)–CH_3
	Keto form	Enol form (enaminol)	
Heat of formation (kcal/mol)	−52.3	−47.6	−49.3
Ionization potential (EV)	9.68	8.30	8.22
Dipole moment (debye)	2.77	1.48	1.60

Table 2. Various forms of an enaminol compound and its characteristics

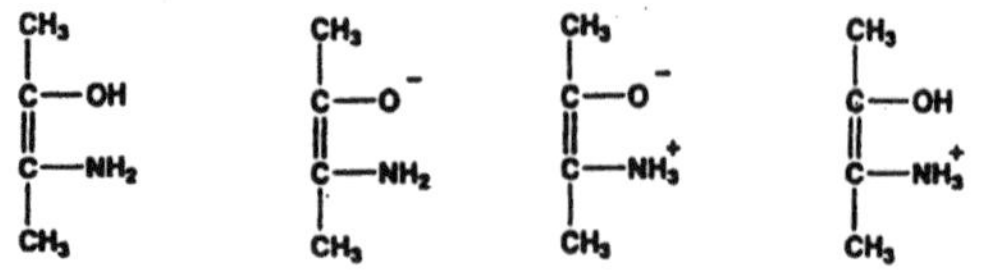

	$CH_3-C(OH)=C(NH_2)-CH_3$ Non-disociated form	$CH_3-C(O^-)=C(NH_2)-CH_3$ Anion form	$CH_3-C(O^-)=C(NH_3^+)-CH_3$ Dipolar ion form	$CH_3-C(OH)=C(NH_3^+)-CH_3$ Cation form
Heat of formation (kcal/mol)	−47.6	−64.9	−40.2	97.2
HOMO (EV)	−8.30	−2.13	−7.97	−14.33
LUMO (debye)	0.99	7.28	1.35	−4.48

kinds of possible inter-molecular interactions. Therefore, the dipolar ion form would be more stable in a polar solvent than is expected from its value of heat of formation. On the other hand, the anion form which seems to be rather stable from its value of the heat of formation has the highest energy value of HOMO, and therefore its electron donating capacity is considered to be the highest among these forms. This form maybe responsible for the characteristic reducing property of this amino-reductone, especially in an alkaline state.

The results of calculations made on the enaminol form of non-dissociated fructoseglycine, the optimized structure, heat of formation, HOMO and LUMO together with their energy levels are given in Figure 3. Heat of formation of this non-dissociated form of fructoseglycine with its enaminol group in trans-form was -302 kcal/mol, while that in *cis*-form was -303.5 kcal/mol, and the *cis*-form seemed to be slightly more stable than the *trans*-form. The HOMO is localized on the enaminol group, which is enaminol double bonded two carbon atoms, nitrogen and oxygen atoms.

The optimized structure and the HOMO and LUMO of the enol form of fructoseglycine in its mono-anion form is given in Figure 4. In this form, the carboxyl group of glycine residue is dissociated, and therefore, the HOMO is partly localized on carboxyl two oxygen atoms, though it is mainly localized on the nitrogen atom of the enaminol group. On the other hand, LUMO is exclusively localized on two doubly bonded carbon atoms of

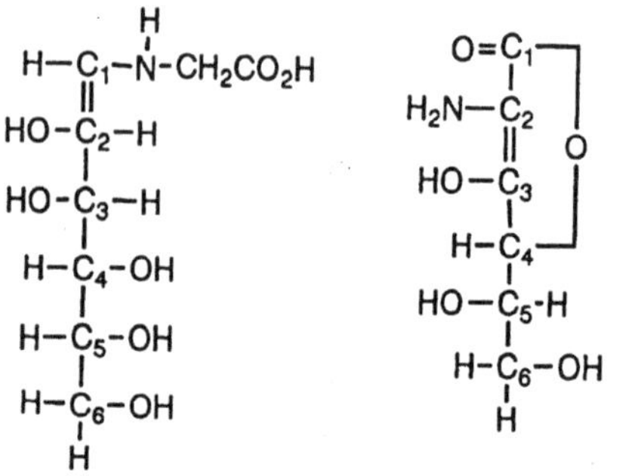

Fructoseglycine (enol, *trans*-form) L-Scorbamic Acid

Figure 2. Chemical structure of enol form of fructoseglycine and L-scorbamic acid.

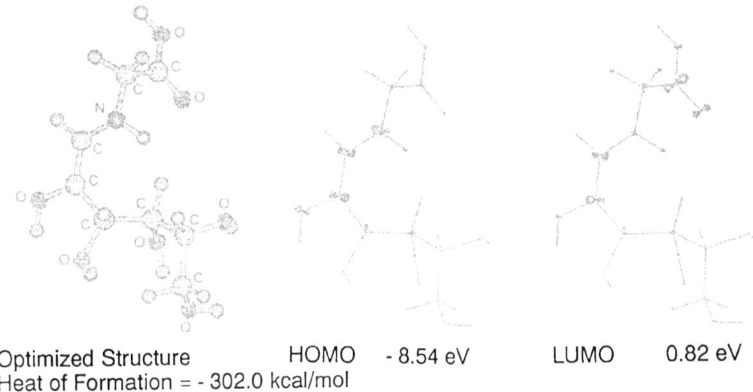

Figure 3. Optimized structure, homo, lumo of enol form of fructoseglycine (non-dissociated form).

the enaminol group. The heat of formation of this enaminol form of fructoseglycine in its *trans*-form was -346.3 kcal/mol, while that in its *cis*-form it was -346.6 kcal/mol, there seemed to be no significant difference between these two stereoisomers in their heat of formation, therefore, it might take either structure almost equally.

The optimized structure, heat of formation, HOMO, LUMO and the energy levels of enol of fructoseglycine in its di-anion form are given in Figure 5. In this form, both the carboxyl group and the hydroxyl group in the enaminol group were dissociated, and the HOMO is largely localized on the carbon atom to which an oxygen atom is attached. The energy level of this HOMO was very high (0.12 kcal/mol), which means that this form has a very strong reducing activity.

The optimized structure of the non-dissociated form of scorbamic acid given in Figure 6 showed that it had a planar γ-lactone ring including an enaminol group conjugated with the lactone carbonyl group. The HOMO is mainly localized on the nitrogen atom,

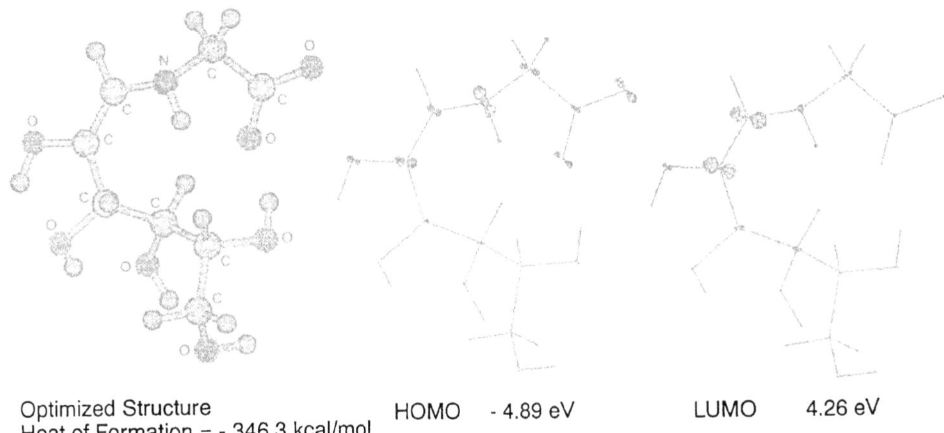

Figure 4. Optimized structure, homo, lumo of enol form of fructuoseglycine (mono-anion).

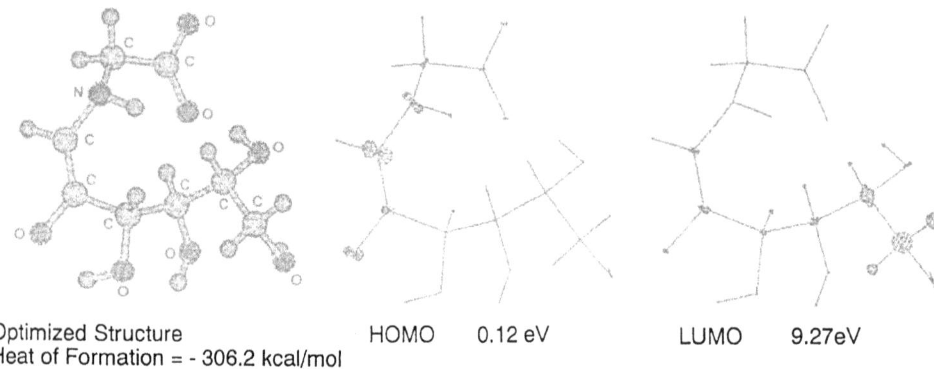

Figure 5. Optimized structure, homo, lumo of enol form of fructoseglycine (di-anion).

and C2, C3 atoms, while the LUMO is largely localized on C3, and the net charge of the nitrogen atom is slightly positive (0.08), and that of C2 is considerably negative (-0.35). The optimized structure of the dipolar ion form of scorbamic acid is given in Figure 7. The HOMO is localized on the C2 atom while the LUMO is on the nitrogen atom. The bond length of C1-O and C3-O were both 2.2 A, suggesting that these bonds have a similar double bond character. The net charge of the nitrogen atom was 1.17, strongly positive as expected, and that of the oxygen atom (C3-O) is -0.38, while that of the C2 atom is more strongly negative (-0.987). The heat of formation of the dipolar ion form of scorbamic acid was -203.5 kcal/mol, an intermediate value of the other two forms, the non-dissociated form (-196.9 kcal/mol) and the anion form (-247.7 kcal/mol). However, in a polar solvent like water, this dipolar form is considered to be more stabilized than as expected from this calculated value. Recently, an X-ray crystal analysis of scorbamic acid was made (Ohashi *et al.*, unpublished results), and the structure was clarified to be this

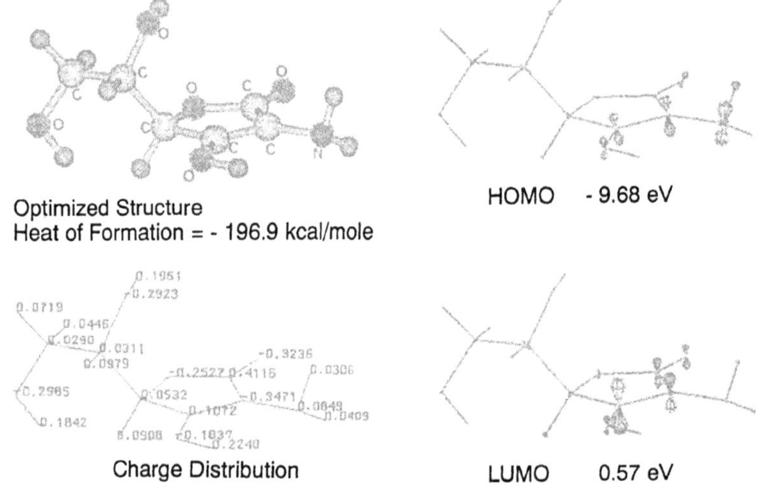

Figure 6. Optimized structure, homo, lumo and charge distribution of L-scorbamic acid (non-dissociated form).

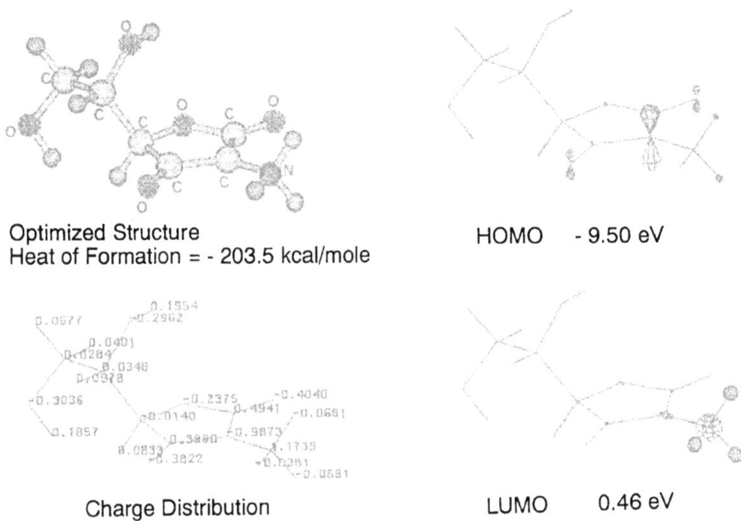

Figure 7. Optimized structure, homo, lumo, and charge distribution of L-Scorbamic Acid (di polar ion form).

dipolar ion form. In an aqueous solution, scorbamic acid will usually take the dipolar ion form, though it can exist in its dipolar ion form, cation form, and anion form depending on the pH value of the solution, as is the case of most α-amino acids in an aqueous solution.

The optimized structure of anion form of scorbamic acid is given in Figure 8. Its heat of formation, -247.7 kcal/mol was larger than those of the dipolar ion and non-dissociated forms. HOMO is localized on C2, and its energy level is the highest among these three forms. Therefore, this anion form is considered to be most important for the reducing activity of this amino-reductone. The length of the C3-O bond is 2.3A and the same with

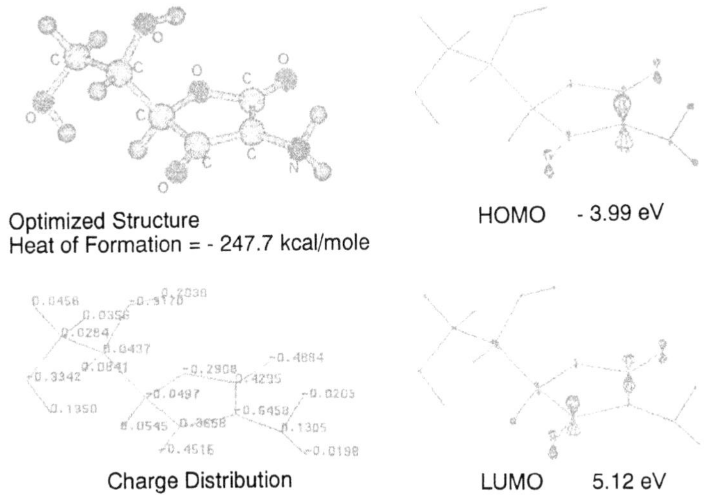

Figure 8. Optimized structure, homo, lumo, and charge distribution of L-scorbamic acid (anion form).

that of C1-O bond. Therefore, the C3-O bond should be regarded as C=O double bond. The nitrogen atom is weakly positively charged (0.13), while C2 is strongly negatively charged (-0.65), and a similar tendency in charge distribution was also observed in the case of scorbamic acid dipolar ion form as described above.

REFERENCES

Abe, Y.; Okada, S.; Horii, H.; Taniguchi, S.; Yamabe, S. A theoretical study on the mechanism oxidation of L-ascorbic acid. *J. Chem. Soc. Perkin Trans. II*, **1987**, 715–720.

Abe, Y.; Okada, S.; Nakao, R.; Horii, T.; Inoue, H.; Taniguchi, S.; Yamabe, S. A molecular orbital study on the reactivity of L-ascorbic acid towards OH radical. *J. Chem. Soc. Perkins Trans. II*, **1992**, 2221–2232.

Finot, P.A.; Aeschbacher, H.U.; Hurrell, R.F.; Liardon, R. *The Maillard Reaction in Food Processing, Human Nutrition and Physiology*, Birkhauser Verlagm Basel, 1990.

Fujimaki, M.; Namiki, M.; Kato, H. *Amino-Carbonyl Reactions in Food and Biological Systems*, Kodansha, Ltd., Tokyo, 1986.

Kurata, T.; Fujimaki, M.; Sakurai, Y. Red pigment produced by the oxidation of L-scorbamic acid. *J. Agric. Food Chem.* **1973**, *21*, 676–680.

Kurata, T.; Fujimaki, M.; Sakurai, Y. Red pigment produced by the reaction of dehydro-L-ascorbic acid with L-amino acid. *Agric. Biol. Chem.* **1973**, *37*, 1471–1477.

Kurata, T.; Miyake N.; Suzuki, E.; Otsuka, Y. Autoxidation of L-ascorbic acid, and its significance in food processing. In *Chemical Markers for the Quality of Processed and Stored Foods*, Lee, T.C.; Kim, H.J. (Eds.); American Chemical Society: Washington, DC, 1996a; pp. 137–145.

Kurata, T.; Miyake, N.; Otsuka, Y. Formation of L-threonolactone and oxalic acid in the autoxidation reaction of L-ascorbic acid - Possible involvement of singlet oxygen. *Biosci. Biotech. Biochem.* **1996b**, *60*, 1212–1214.

Ledl, F.; Bech, J.; Sengl, M.; Osiander, H.; Estendorfer, S.; Severin, T.; Huber, B. Chemical pathways of the Maillard reaction in aging, diabetes and nutrition. In *The Maillard Reaction in Aging, Diabetes, and Nutrition*, Alan R. Liss, Inc. 1989; pp. 23–42.

Njoroge, F.G.; Monnier, V.M. The chemistry of the Maillard reaction under physiological conditions: A review, In *The Maillard Reaction in Aging, Diabetes, and Nutrition*, Alan R. Liss, Inc., 1989; pp. 85–107.

Stewart, J.J.P. MOPAC version 6.0, Frank Seiler, Research Laboratory, U.S. Air Force Academy, Colorado, U.S.A., 1996.

Waller, G.W.; Feather, M.S. The Maillard reaction in foods and Nutrition, ACS Symp. Ser. No. 215, American Chemical Society: Washington DC, 1983.

EFFECTS OF GAMMA IRRADIATION ON THE FLAVOR COMPOSITION OF FOOD COMMODITIES

Jui-Sen Yang

Institute of Marine Biology
National Taiwan Ocean University
Keelung, Taiwan, R.O.C.

Fifteen food products including potato, sweet potato, shallot, onion, garlic, ginger, papaya, mango, rice, tobacco, small red bean, mungbean, soybean, wheat, flour and spices have been approved for irradiation by the National Health Administration in Taiwan. Market tests (Wu *et al.*, 1996) provided strong proof that Taiwanese consumers would accept irradiated foods. However, researchers in the food industry are concerned about the possibility of chemical changes, especially in volatile composition, during irradiation processing. This study considers several food commodities, including garlic, ginger, shiitake, onion, potato, day-lily, tilapia, silver carp and shrimp. Food samples were irradiated with optimum doses and then studied for possible occurrance of chemical changes and effects on compositional characteristics of foods.

GARLIC IRRADIATION

Garlic bulbs (*Allium sativum* L.) were harvested from March to May in Taiwan. The bulbs sprouted 4–6 months after harvesting and storing at ambient temperatures. Early studies in our laboratory (Yang *et al.*, 1979) has indicated that the treatment of garlic bulbs with 0.15 kGy of ^{60}Co gamma rays could inhibit sprouting and reduce weight loss during storage. The use of gamma irradiation to inhibit sprouting extended the storage life of garlic from September to February of the following year. Market tests provided a strong proof that consumers would accept irradiated garlic (Wu *et al.*, 1996a).

Volatile compounds of garlic were collected using a Liken-Nickerson steam distillation and solvent extraction apparatus (Romer and Renner, 1974). The volatile compounds were determined using gas chromatography and then further examined by GC-MS. The

Process-Induced Chemical Changes in Food
edited by Shahidi *et al.* Plenum Press, New York, 1998

major volatile compounds of garlic were diallyl disulfide, methyl 2-propenyl disulfide, *trans*-1,2-dimethoxycyclohexane and di-2-propenyl trisulfide. The content of diallyl disulfide in irradiated garlic was decreased immediately after irradiation. After being stored for 4 months unirradiated garlic also showed a decrease in its content of diallyl disulfide. However, after 8 months storage diallyl disulfide underwent an increase in both irradiated and unirradiated garlic. Gamma irradiation with 0.15 kGy had little influence on the amount of volatile components, except diallyl disulfide (Wu *et al.*, 1996b). However, Kwon and Yoon (1985) reported that the content of diallyl disulfide in garlic during storage was apparently reduced by irradiation of 0.5–3.0 kGy.

On the other hand, Ceci *et al.* (1991) did not find any difference in the contents of allicin and diallyl disulfide upon comparison of 0.5 kGy irradiated with unirradiated bulbs during storage.

GINGER IRRADIATION

The ginger was harvested during the month of December in Taiwan. Nutrition and weight loss occurred in stored ginger rhizome due primarily to sprouting. Cold storage (13°C) could delay sprouting and weight loss of ginger (Liu *et al.*, 1982). However, gamma irradiation (0.05 kGy) could inhibit ginger sprouting very well. After irradiation, ginger could be economically stored at ambient temperature.

Volatile compounds of ginger were collected and determined by gas chromatography and GC-MS. Slices of ginger were boiled for making a traditional Chinese ginger soup for panel triangle test.

The volatile compounds of ginger were α-terpinene, α-bergamoteme, geranial, 1,8-cineole, neral, farnesene, β-sesquiphellandrene, β-phellandrene, α-curcumene, α-pinene, myrcene, nerol and α-zingiberene. The volatile compounds of irradiated gingers sampled immediately or one month after 0.05 kGy irradiation had the same GC profile as unirradiated gingers. However, the amount of the volatile compounds of irradiated gingers, such as α-zingiberene, β-bergamotene, neral, geranial and α-curcumene were significantly decreased 3 months after irradiation. It suggested that irradiation affected the stability of volatile compounds, especially neral, during storage. In the panel triangle test, the flavor, taste, pungency and color of ginger soup were not different from those of unirradiated ginger one month after irradiation, but after 5 months of storage the ginger soup showed a marked difference in its flavor, taste and color when unirradiated and irradiated samples were used. Irradiation (0.05 kGy) did not offer a benefit for flavor, taste and color after long-term storage but instead induced the loss of flavor, taste and color, although the treatment could inhibit sprouting. The ginger provided a low flavor quality 5 months after irradiation. The data from the chemical analysis showed that volatile components decreased after irradiation (Wu and Yang, 1994a).

The cell viability was determined by using 2,3,5-triphenyltetrazolium chloride (TTC) to measure dehydrogenase activity. No difference in cell viability was found between irradiated and unirradiated gingers (Wu and Yang, 1994b).

SHIITAKE IRRADIATION

Shiitake mushroom (*Lentinus edodes* Sing) is an important ingredient in the Oriental cuisine because it contributes a special aroma to foods. There were two classes of volatiles in shiitake, including eight-carbon and sulfur-containing compounds (Chen and Wu, 1984;

Chen *et al.*, 1984). The eight-carbon compounds are the major flavor components of the fresh mushroom, while the sulfur-containing compounds are the important characteristic compounds in the dried product (Chyau and Wu, 1990; Chang *et al.*, 1991). The eight-carbon compounds included 3-octanone, 3-octanol, 1-octen-3-ol, n-octanol and (*E*)-2-octen-1-ol. The sulfur-containing compounds were dimethyl disulfide, dimethyl trisulfide, methyl (methylthio)methyl disulfide and 1,2,4-trithiolane.

During storage dried shiitakes were easily contaminated with microorganisms or infestation by insects. Irradiation of 5–10 kGy reduced the microbial count of dry shiitakes to near sterility (Lai *et al.*, 1992). The amount of total volatile compounds of unirradiated shiitakes (113 mg/kg) was much higher than that of irradiated (10 kGy) samples. The eight-carbon compounds and sulfur-containing compounds of dry shiitake were decreased after irradiation. The changes in the proportions of volatile compounds were shown in the irradiated shiitakes. The eight-carbon compounds made up 72% of the total volatiles in the unirradiated shiitake, but only 47% in 5 kGy and 21% in 10 kGy irradiated samples. It was determined that the difference in the content of the sulfur-containing compounds proportionate to the total volatiles was not significantly changed before and after irradiation. However, Nishimura *et al.* (1971) reported that irradiation up to 0.7 kGy had no effect on the volatile components of onion. The results suggested that an optimal dose of irradiation increased the levels of volatile compounds, while high doses decreased them. Irradiation doses of both 5 and 10 kGy seemed high enough to decrease the levels of volatiles and to change the proportions of flavor components in shiitakes. The characteristic flavor compounds, including 3-octanone, 3-octanol and 1-octen-3-ol, decreased in proportion to the increased doses from 5 to 10 kGy, although the unidentified compounds were increased quite dramatically by irradiation (Lai *et al.*, 1994).

The irradiation with 2 kGy, however, increased the eight-carbon volatile components which are the flavor compounds of fresh shiitake. The treatment of 1 kGy irradiation in fresh shiitake produced some new volatile compounds in the dry product, such as methyl ethyl disulfide, sulfinylbismethane, methyl (methylthio)ethyl disulfide and N-(3-methylbutyl) acetamide. The eight-carbon compounds had almost disappeared after the drying process. The content of sulfur-containing volatile compounds in the dried shiitake became low during irradiation. Conclusively, the amount of the sulfur-containing compounds in dry shiitake without irradiation was considerably higher than that in irradiated with doses from 1 to 10 kGy in either fresh or dry shiitakes. With irradiation from 1 to 10 kGy, the higher the dose used, the lower the content of the sulfur-containing volatile compounds detected in fresh or dry shiitakes. In the present work irradiation did not offer any benefit in the production of volatile compounds in dry shiitake. The panel evaluation by triangle testing did not show any significant difference between irradiated and unirradiated samples. However, irradiation could effectively control microbial contamination and insect infestation in shiitake (Lai *et al.*, 1994; Yang *et al.*, 1996).

DAY-LILY FLOWER

Day-lily flower (*Hemerocallis spp.*) is an important vegetable in the Orient in either fresh or dried style. The fresh day-lily flower is always harvested before reaching full-bloom, as fully bloomed flowers lose their market value. Therefore, harvested flowers must be sold within two days otherwise they bloom. The dried flowers are also available in the oriental market. The problems associated with the dried product include microbial contamination and insect infestation.

Irradiation with a dose above 2 kGy inhibits the blooming of day-lily flowers. After irradiation above 2 kGy damage was detected with scanning microscope examination in the cells of the petal, sepal and receptacle of the lily flowers. With 5 kGy irradiation the appearance of the flowers was seriously injured and the flowers became pale and then dropped. Irradiation at a dose of 0.5 or 1 kGy did not injure the flower appearance and could not inhibit the blooming of the flowers (Yang, 1994).

The dried day-lily flowers contained volatile components including ethyl formate, ethyl acetate, nonanol, 3-methylbutanol, isoamyl alcohol, methylpyrazine, 2,6-dimethylpyrazine, furfural, acetic acid, terpinolene, furfuryl formate, furfuryl acetate, phenylacetaldehyde and furfuryl alcohol. A low dose (0.5 kGy) of irradiation could control mold contamination and insect infestation in dried day-lily flowers. No significant difference in the volatile compositions between the irradiated (0.5, 1.0, 2.0 and 5.0 kGy) and unirradiated dried day-lily flowers were found. However, the quality of dried day-lily flowers was acceptable after irradiation (at 0.5 kGy) (Yang, 1994).

POTATO IRRADIATION

Potatoes were gamma irradiated with a dose of 0.125 kGy to inhibit tuber sprouting. The amount of amino nitrogen in irradiated potatoes was higher than that in unirradiated potatoes. The chips of irradiated potatoes appeared to be a darker brown after frying in 160°C oil than that of unirradiated potatoes (Yang *et al.*,1979). When the filter papers were soaked in a glucose solution, they showed a darker brown color after frying in 160°C oil, however, those soaked in sucrose and other sugars did not darken. The chips of irradiated potatoes may contain a higher amount of glucose than their unirradiated counterpart.

With a scanning electron microscope examination, the stomata in the irradiated potatoes showed an opening feature. The openning of the stomata might be due to the inhibition effect of gamma irradiation on the turgor of the guard cell. On the other hand, the hole size in the lenticles of the unirradiated potatoes increased due to a swelling reaction of the lentical tissues during storage. However, the swelling in the lentical tissues was not found in the irradiated tubers. The lenticels of irradiated potatoes were "pluged" by suberous filling cells. In the starch grains of unirradiated potatoes, splits appeared which radiated from the helium, but not in the irradiated potatoes. The splits may be the result of dehydration of starch in grains, suggesting that irradiation inhibited starch dehydration as well as sprouting (Tzeng *et al.*,1987). When the fresh chips of irradiated potatoes were exposed to a 366 nm UV light, blue fluorescence was visible on the chips, but no fluorescence could be found in the chips prepared from unirradiated potatoes (Tzeng *et al.*,1988). The substrate which excited UV fluorescence in the irradiated potatoes needs to be further identified.

TIGER PRAWN IRRADIATION

Gamma irradiation with a dose of 5 kGy was used in treating tiger prawn (*Penaeus monodon* Fabricius, also named grass shrimp in Taiwan) (Perng and Yang, 1990) to increase their refrigerated self-life (Rothbart, 1986). With an electron microscope examination, no significant change was found in the muscle ultrastructure of shrimp meat irradiated at doses of 2 or 5 kGy and then stored at 4°C for 0 and 8 days. However, 10 kGy irradiation at ambient temperatures induced myosin depolymerization in the myofibrils of shrimp muscle and damaged sarcoplasmic membranes. Shrimp frozen at -18°C and then irradiated with 10 kGy did not exhibit any myofibril damage. It has been suggested that freezing offers protection to shrimp meat against irradiation injury (Perng and Yang, 1990). Using an immunocytochemical technique with electron microscopy the *in vivo* sites of melanogenesis in shrimps, unirradiated and irradiated (5 kGy), were investigated by locating polyphenoloxidase in the heptopancreas, muscle and epidermis. The results suggested that polyphenoloxidase was synthesized in the R-cells of the heptopancreas or chromatophores in the epidermis and transferred to the muscle. There was no difference in the location of polyphenoloxidase in shrimp tissues between unirradiated and irradiated samples (Yang *et al.*, 1993). Medium-dose (5 kGy) irradiation could inhibit the dispersion of erythrin in chromatophores in epidermis but not the dispersion of melanins (Yang *et al.*, 1993). By using an electron microprobe (EDX) and an electron microscope the calcium ion distribution in shrimp muscle, unirradiated and 5 kGy irradiated, was investigated. The calcium ions in the sarcoplasmic reticulua and nuclei in irradiated shrimp muscle cells were not released to the myofibrils, and remained *in situ* during storage (8 days) at 4°C after irradiation. However, without irradiation the concentration of calcium ions in the myofibrils of muscle cells increased after storage. Calcium could play the role of an activator of proteinase, which induced the softening of shrimp meat. Therefore, shrimp meat without irradiation treatment could be broken down more extensively by proteinases both from microorganisms and shrimp hepatopancreas, and became soft faster than irradiated shrimp meat (Yang and Perng, 1995).

No significant difference in the composition of volatile compounds was found between 5 kGy irradiated and unirradiated shrimps. After irradiation no off-odor or irradiation volatile compounds was detected by gas chromatography. This suggested that medium-dose irradiation did not produce off-odor compound, and did not affect volatile composition in shrimp (Yang *et al.*, 1993).

FRESHWATER FISH IRRADIATION

A great benefit of low dose radiation is killing or suppression of parasites while leaving the fish meat closer to its unprocessed state (CAST, 1989) for use as sashimi or for cooking. Gamma irradiation at 0.5 kGy was effective in inactivating the parasites *Clonorchis sinensis* and *Hyplorchis spp.* in freshwater fish and *Hemiculter kneri* and *Angiostrongylus cantonensis* in African snails (Pai *et al.*, 1988, 1989). It is suggested that a dose range from 0.5 to 1 kGy would initiate the effect.

Fresh fillets from tilapia (*Oreochromis mossambicus* Peters) and silver carp (*Hypophthalmichythys molitrix* Cuvier and Valenciennes) were irradiated with 1 kGy at 2–4°C to inactivate the parasites. After irradiation, the thiamin content of silver carp fillets was reduced by 34%, but the thiamin content of tilapia was not affected by irradiation. The other constituents, such as moisture, ash, crude fat, crude protein, and riboflavin, of tilapia and silver carp were not affected by the treatment of irradiation. Irradiation (1 kGy) did not increase the acid value of fat in tilapia and silver carp. It suggested that 1 kGy irradiation did not show a significant effect on lipid cleavage in the two freshwater fishes. Irradiation at 1 kGy had little effect on the concentrations of nucleotide catabolites, such as inosine monophosphate (IMP), inosine (Ino) and hypoxanthin (Hx), in both fishes. The K value was calculated using the following fomula for expressing the extents of IMP breakdown.

$$K = (Ino + Hx) / (ATP + ADP + Ino + Hx)$$

Table 1. Effects of gamma irradiation on the flavor composition of food commodities

Food commodities	Irradiation dose (kGy)	Irradiation purpose	Change after irradiation	References
Garlic	0.15	sprouting inhibition	1. dially disulfide decreased immediately after irradidtion but increased during storage	Wu et al. 1996
Ginger	0.05	sprouting inhibition	1. α-curcumene, α-zingiberene, α-bergamotene, neral and geranial decreased 3 months after irradiation	Wu and Yang, 1994a 1994b
			2. panel test changed 5 month after irradiation	
			3. no difference in cell viability	
Shiitake	10	mold and insect	1. total volatile decreased	Lai et al.
(dry)	5	control	2. eight-carbon volatile compounds decreased	1992
			3. sulfur-containing volatile compounds decreased	1984
			4. 3-octanone, 3-octanol and 1-octen-3-ol decreased	Yang et al. 1996
(fresh)	2	opening	1. eight-carbon compounds in fresh shiitake increased	
(fresh)	1	inhibition, mold and insect control	1. new compounds as methyl ethyl disulfide, sulfinylbis methane, methyl (methylthio) ethyl disulfide N-(3-methyl butyl) acetamide appeared	
			2. sulfur-containing volatile compounds decreased	
Day-lily flower	5	blooming	1. blooming inhibited	Yang, 1995
(fresh)	2	inhibition	2. damaged the cells in petal, sepal and receptacle	
	0.5-1		1. no effect detected	
Day-lily flower (dry)	0.5	mold and insect control	no significant effect detected	
Potato	0.125	sprouting inhibition	1. amino nitrogen increased	Tzeng *et al.*
			2. darker browning after oil frying	1987
			3. opening stomata	1988
			4. "pluged" lentical	
			5. no split helium in the starch grain	
			6. blue fluorescence in fresh potato chips under 366 nm U.V. light	
Tiger prawn (grass shrimp)	10 (ambient)	microorganism control	1. myosin depolymerization	Perng and Yang. 1990
			2. sarcoplasmic membrane damaged	Yang *et al.* 1993 Yang and Perng, 1995
	10(-18°C)	microorganism control	1. no effect on microstructure	
	5	microorganism control	1. erythrin dispersion inhibited	
			2. calcium in sarcoplasmic reticulum and nuclei remained	
			3. meat softing delayed	
			4. no effect on volatile compounds	
Fresh-water fish - tilapia and silver carp	0.5-1.0	parasite control	1. thiamin in silver carp reduced	Pai *et al.* 1988, 1989
		microorganism control	2. slightly increased reddish color in raw fillets of silver carp	Liu *et al.* 1991

There were no significant differences in the concentrations of the nucleotide catabolites and the K value between irradiated and unirradiated fishes (Liu *et al.*, 1991).

Irradiation at a medium dose of 1 kGy reduced the bacterial load of tilapia and silver carp in one order. A reduction in the level of bacteria in fish flesh was another beneficial effect of gamma irradiation in addition to its control of parasites. The effect of 1 kGy irradiation on the meat color was minimal. Hunter "a" value of silver carp and tilapia increased with the irradiation dose. The slightly reddish color of raw fillets of irradiated silver carp was detected by a panel examination. However, no significant differences in raw color, cooked color and cooked flavor were observed between irradiated and unirradiated tilapia. In silver carp the cooked flavor and color of irradiated samples did not exhibit any significant difference as compared to the unirradiated fish species (Liu *et al.*, 1991).

CONCLUSION

Irradiation provides several uses (Table 1) in agriculture and food processing operations, including sprouting inhibition (garlic, ginger and potato), mold and insect control (shiitake and dried day-lily flower), blooming inhibition (fresh day-lily flower) and microorganism and parasite control (tiger prawn and fresh water fishes). Different irradiation doses were used for different purposes, as shown in Table 1. Although high dose irradiation, as well as cooking, canning or other processes, induced some changes in flavor composition, tissue ultrastructure and panel testing, an optimum dose of irradiation did not significantly affect the quality of food commodities. Irradiation may have benefits for agriculture, the food industry and consumers as an alternate technique to provide a good quality food supply.

REFERENCES

CAST. Ionizing energy in food processing and pest control: II. Application. Task Force Report No 115. Council for Agricultural Science and Technology, Ames, Iowa., 1989.

Chang, C. H.; Chyau, C.C.; Wu, C.M. Volatile components of various shiitake products. *Food Sci.* **1991**, *18*, 199–204 (in Chinese).

Ceci, L.N.; Curzio, O.A.; Pomilio, A.B. Effect of irradiation and storage on the flavor of garlic bulbs cv 'red'. *J. Food Sci.* **1991**, *56*, 44–46.

Chen, C.C.; Wu, C.M. Studies on the volatiles of shiitake (Lentinus edodes). Reserch Rep. No. 327, Food Industry Research and Development Institute: Hsinchu, Taiwan, ROC (in Chinese), 1984.

Chen, C.C.; Chen, S.D.; Chen, J.J.; Wu, C.M. Effects of pH value on the formation of volatiles of shiitake (*Lentinus edodes*), an edible mashroom. *J. Agric Food Chem.* **1984**, *32*, 999–1001.

Chyau, C.C.; Wu, C.M. Study of generation of volatile compounds from shiitake. Research Rep. No. 617–1. Food Industry Research and Development Institute, Hsinchu, Taiwan, ROC, 1990.

Kwon, J.H.; Yoon, H.S. Changes in flavor components of garlic resulting from gamma irradiation. *J. Food Sci.* **1985**, *50*, 1193–1195.

Lai, C.L.; Yang, J.S.; Liu, M.S. Effect of γ-irradiation on the flavour of dry shiitake (*Lentinus edodes* Sing). *J. Sci. Food Agric.* **1994**, *64*, 19–22.

Lai, C.L.; Tsai, M.J.; Yang, J.S.; Liu, M.S. Effects of gamma irradiation on the quality and flavor of shiitake mushroom. Research Rep. No. 682, Food Industry Research and Development Institute, Hsinchu, Taiwan, ROC (in Chinese), 1992.

Liu, M.S.; Chen, R.Y.; Tsai, M.J.; Yang, J.S. Effect of gamma irradiation on the keeping quality and nutrients of tilapia (*Oreochromis mossambicus*) and silver carp (*Hypophthalmichthys molitrix*) stored at 1°C. *J. Sci. Food Agric.* **1991**, *57*, 555–563.

Liu, M.S.; Chang, T.C.; Liao, M.L.; Yang, J.S. Inhibition of sprouting and decay of ginger, shallot and garlic during storage. Research Rep. 258. Food Industry Research and Development Institute, Hsinchu, Taiwan, ROC (in Chinese), 1982.

Nishimura, H.; Asahi, A.; Fujiwara, K.; Mizutani, J.; Obata, Y. Changes in flavor components of onion by r-irradiation. *Agric. Biol. Chem.* **1971**, *35,* 1831–1835.

Pai, H.H.; Yang, M.G.; Yang, J.S.; Chen, E.R.; Liu, M.S. Parasite (*Colnorchis sinensis*) control of fresh water fish by gamma irradiation. Research Rep. No. 551. Food Industry Research and Development Institute, Hsinchu, Taiwan, ROC (in Chinese), 1988.

Pai, H.H.; Yang, M.G.; Yang, J.S.; Chen, E.R.; Liu, M.S. Low dose irradiation of snail and fish for parasite control. Research Rep. No. 561. Food Industry Research and Development Institute, Hsinchu, Taiwan, ROC (in Chinese), 1989.

Perng, F.S.; Yang, J.S. Ultrastructural effect of gamma radiation on grass shrimps (*Penaeus monodon* Fabricius). *Radiat. Phys. Chem.* **1990**, *35*, 258–262.

Romer, G.; Renner, E. Sample methods for isolation and concentration of flavor compounds from foods. *Z. Lebensm. Unters. Forsch.* **1994**, *156,* 329–335.

Rothbart, H.C. Potential application of food irradiation in the United States. USA-ROC and ROC-USA Economic Councils, Tenth Anniversary Joint Business Conference, Taipei, Taiwan, ROC, 1986.

Tzeng, S.S.; Yang, J.S.; Liu, M.S. Sprouting inhibition in irradiated tubers of paddy field potatoes. Research Rep. 480, Food Industry Research and Development Institute, Hsinchu, Taiwan, ROC (in Chinese), 1987.

Tzeng, S.S.; Yang, J.S.; Liu, M.S. Sprouting inhibition in irradiated turbers of paddy field potatoes (II). Research Rep. 520, Food Industry Research and Development Institute, Hsinchu, Taiwan, ROC (in Chinese), 1988.

Wu, J.J.; Yang, J.S. Effects of γ-irradiation on the volatile compounds of ginger rhizome (*Zingiber officinale* Roscoe). *J. Agric. Food Chem.* **1994a**, *42*, 2574–2577.

Wu, J.J.; Yang, J.S. Ginger irradiation. *Food Sci. (Taiwan).* **1994b**, *21*, 485–494 (in Chinese).

Wu, J.J.; Yang, J.S.; Liu, M.S.; Kuo, J.D. Shelf-life extension and promotion of irradiated garlic blubs. *Nuclear Sci. J. (Taiwan)* **1996a**, *33*, 52–57 (in Chinese).

Wu, J.J.; Yang, J.S.; Liu, M.S. Effect of irradiation on the volatile compounds of garlic (*Allium sativum* L.). *J. Sci. Food Agric.* **1996b**, *70*, 506–508.

Yang, J.S.; Perng, F.S.; Liou, S.E.; Wu, J.J. Effect of gamma irradidtion on chromatophores and volatile components of grass shrimp muscle. Radiat. *Phys. Chem.* **1993**, *42*, 319–322.

Yang, J.S.; Perng, F.S. Effect of gamma irradidtion on the distribution of calcium ions in grass shrimp (*Penaeus monodon* F.) muscle. *Meat Sci.* **1995**, *39*, 1–7.

Yang, J.S., Perng, F.S.; Marshall, M.R. Immunohistochemical localization of phenoloxidase in heptopancreas, muscle and epidermis of grass shrimp (*Penaeus monodon* F.). *J. Food Biochem.* **1993**, *17*, 115–124.

Yang, J.S.; Fu, Y.H.; Liu, T.Y. Effects of irradiation treatment on nutritive constituents of garlics, gingers, onions and potatoes. Research Rep. 144, Food Industry Research and Development Institute, Hsinchu, Taiwan, ROC, 1979

Yang, M.S.. Effect of gamma irradiation on the quality of shiitake (*Lentinula edodes* Sing) and day-lily (*Hemerocallis fulva* L.). M. S. Thesis, Dept. of Honticulture, National Chung-Hsing University, Taichung, Taiwan, ROC, 1995

Yang, M.S.; Chyau, C.C.; Horng, D.T.; Yang, J.S. Effect of irradiation and drying process on the flavor of fresh shiitake (*Lentinus edodes* Sing). *J. Sci. Food Agric.* **1997**, in press.

FLAVOR DETERIORATION IN YOGURT

Naomi Harasawa, Hideki Tateba, Nobuko Ishizuka, Toshiyuki Wakayama, Katsumi Kishino, and Mitio Ono

Ogawa & Co., Ltd.
6-32-9 Akabanenishi, Kita-ku
Tokyo 115, Japan

Volatile components of flavored yogurt preserved at 5°C in the dark for 0 day, 3 days and 10 days were recovered by simultaneous distillation extraction (SDE) and headspace (HS) procedures. Gas chromatography (GC) and GC-mass spectrometry (GC-MS) analyses of those samples showed remarkable changes in some compounds. Aldehydes which contribute to expressing citrusy notes were reduced to alcohols during fermentation process and storage. As a result, the strength of flavors which expressed well-balanced citrusy notes in yogurt were weakened, and fatty or oily notes mainly caused from alcohols were strengthened reversely. Hydrocarbons were also digested by bacteria during a fermentation process. A small amount of other compounds, such as esters and terpene alcohols changed. Fewer effects of sorption into a package material and chemical reactions, such as hydrolysis esters, hydration or oxidation of hydrocarbons, were observed.

INTRODUCTION

Yogurt is widely consumed in Japan as a healthy food. Improving flavor quality of yogurt is important for its palatability. Flavor added to yogurt is well known to deteriorate during fermentation and storage, resulting in off-flavor. The deterioration of typical flavor components of non-flavored yogurt, such as acetaldehyde, diacetyl, acetoin, acetone, 2-butanone, and acetic acid was investigated (Gaafar, 1992). The content of most of them decreased with age, except acetic acid. The deterioration of strawberry flavor was also studied on firm yogurt (Marion and Chardon-Fayard, 1990). These authors reported that some esters and acetals were unstable. However, those data are not enough to understand the behavior of aroma components in flavored yogurt. Flavors expressing citrus notes changed remarkably. The quantitative study of aroma components may help creation of

Process-Induced Chemical Changes in Food
edited by Shahidi *et al.* Plenum Press, New York, 1998

more suitable and stable flavors for yogurt. Selection of a model flavor containing various compounds is deemed necessary.

The objective of this research was to shed light in how the aroma components deteriorate during processing and storage of flavored yogurt.

EXPERIMENTAL

Preparation of Yogurt

Experimental yogurt base was produced by inoculating a culture consisting of *Lactobacillus bulgaricus* and *Streptococcus Thermophilus* prepared from a commercial yogurt product into a heat-treated milk mixture. The above milk mixture, which was adjusted to 2% fat content using vegetable oil, was composed of milk and non-fat dry milk. A glass bottle and a paper cup coated with polyethylene film were used for packaging. A model flavor expressing citrus and fruity notes composed of aldehydes, terpene-hydrocarbons, esters and alcohols was prepared. The analysis on 0 day was carried out at once after the model flavor was added to the experimental yogurt base in which the fermentation was already finished. For other analyses, the experimental yogurt base with the model flavor was fermented and then stored at 5°C in the dark for 3 days and 10 days.

ISOLATION OF VOLATILE COMPOUNDS

Simultaneous Distillation Extraction (SDE)

Yogurt (280 g) was poured into a 2 L conical flask containing hot water (900 mL, 70°C). The suspension of yogurt was distilled and continuously extracted for 30 min with 80 mL of dichloromethane using a Likens-Nickerson apparatus. After drying over sodium sulfate, an internal standard (IS: amyl hexanoate) was added and the extract was concentrated to 1 mL on a Vigreux column and to 100 μL under nitrogen stream. The concentrate was analyzed by gas chromatography (GC) and GC-mass spectrometry (GC-MS). Identification of volatile components was carried out by comparing their mass spectral data and retention indices on the polar column with those of the authentic samples. The amount of each flavor component obtained was estimated by comparison with the internal standard (IS). The response factor in flame ionizing detector (FID) of each flavor component to the IS was measured.

Dynamic Headspace Extraction (HS)

Yogurt (280 g) was poured into a 1 L conical flask equipped with Porapak Q (Millipore Co., Ltd., MA) column and a magnetic stirrer bar. During the collection of the volatile components, the yogurt was stirred and maintained at 30°C in a water bath. The volatiles were swept from the flask to the adsorbent in the trap using a flow of purified nitrogen (35 mL/min), and the collection was continued for 30 min. The concentrate was injected into GC-MS directly through the thermal desorption cold trap injector (TCT; Chrompack, The Netherlands) to analyze the components. The TCT was operated under the following conditions in all runs: pre-cooling, the cold trap was cooled to -100°C; thermal desorption, 180°C, 5min, cold trap was maintained at -100°C; and for injection, the cold trap was heated to 220°C. The quantitative data were calculated by correcting the SDE data using allyl hexanoate.

Vanillin used in the model flavor was recovered with methanol from yogurt and analyzed using high-performance liquid chromatography (HPLC, Nakazawa *et al.*, 1982).

GAS CHROMATOGRAPHY (GC)

GC was carried out using a Hewlett-Packard 5890 Series II gas chromatograph equipped with FID. Separation was achieved on a fused silica capillary column, 30 m x 0.25 mm i.d.; 0.25 μm film thickness (DB-WAX, J&W Inc.). The oven temperature was programmed from 80°C to 200°C at 2°C /min. The temperature of injection and that of the detector were 250°C. The nitrogen gas flow rate was 30 mL/s using a split ratio of 200:1.

GAS CHROMATOGRAPHY-MASS SPECTROMETRY (GC-MS)

Electron impact mass spectrometric data were collected using a Hewlett-Packard 5971A mass spectrometer interfaced to a Hewlett-Packard 5890 Series II gas chromatograph. Separation was achieved on a fused silica capillary column, 60 m x 0.25 mm i.d.; 0.25 μm film thickness (Supelcowax 10, Supelco Inc.). The oven and injection temperatures were the same as those described for the GC analysis. The helium gas flow rate was 1mL/s using a split ratio of 50:1.

HIGH-PERFORMANCE LIQUID CHROMATOGRAPHY (HPLC)

HPLC analyses of vanillin were performed using a HP1090M instrument equipped with a photodiode array detector. Vanillin was monitored by reversed-phase using a Capcell Pak C18 SG120 (S-5 μm) column (Shishedo, 4.6 mm x 250 mm) with a constant elution gradient from 10% (V/V) acetonitrile in water to pure acetonitrile. The flow rate was 1 mL/min and the detection wavelength was 280 nm.

RESAZURIN TEST

The test was used to confirm the increase of reducing power of yogurt (Pharmaceutical Society of Japan, 1984). Yogurt was centrifuged for 20 min at 3000 xg and the supernatant was filtrated through 0.45 μm pore size membrane filters. The filtrate was diluted 5 times with 0.1 M phosphate buffer solution (pH 6.0). A 0.5 mL resazurin solution, prepared by dissolving 10 mg of resazurin in 100 mL of water, was added to a 5 mL of yogurt solution and then placed at 37°C for 30 min. UV spectra of solutions were measured using a UV/VIS spectrophotometer (JASCO, Ubest 30).

RESULTS AND DISCUSSION

Although a dynamic headspace extraction (HS) method was often applied for the analysis of yogurt flavor (Imhof and Bosset, 1994), simultaneous distillation extraction

Table 1. Variations on recoveries of typical components obtained by SDE

Component	Storage time, days		
	0	3	10
Ethyl acetate	[1]19.6 (15.4)	21.7 (7.6)	17.1 (5.9)
Butyl acetate	759.8 (0.1)	621.5 (3.5)[2]	534.8 (9.0)
Myrcene	43.8 (12.8)	9.8 (17.1)[2]	10.8 (30.1)
Limonene	1922.4 (10.2)	495.1 (2.0)[2]	428.0 (10.7)
p-Cymene	48.4 (5.9)	22.4 (18.3)[2]	15.2 (18.7)
3-Hydroxy-2-pentanone	68.6 (2.8)	97.2 (9.9)	72.7 (13.0)
2-Hydroxy-3-pentanone	31.0 (4.4)	43.6 (9.9)	29.1 (11.9)
Allyl hexanoate	703.0 (0.6)	456.4 (7.8)[2]	389.6 (4.3)
Nonanal	46.9 (6.1)	tr. (–)[2]	tr. (–)
Furfural	20.7 (34.0)	15.0 (4.7)	23.4 (7.6)[2]
Decanal	264.2 (1.2)	tr. (–)[2]	tr. (–)
Linalool	463.4 (1.5)	332.4 (2.6)	301.1 (0.2)
Octanol	64.6 (2.0)	659.1 (9.8)[2]	568.5 (3.4)
2-Undecanone	tr. (–)	16.4 (18.4)[2]	13.6 (7.4)
Nonanol	n.d. (–)	43.8 (11.5)[2]	38.9 (6.3)
Neral	782.4 (3.9)	293.2 (3.3)[2]	102.9 (1.7)[2]
4-Terpineol	48.8 (0.2)	35.3 (0.4)[2]	41.6 (5.4)
γ-Terpineol132.9 (1.5)	109.9 (7.1)	105.5 (0.9)	105.5 (0.9)
Neryl acetate	44.9 (14.9)	19.1 (4.3)[2]	18.4 (6.8)
Geranial	1010.9 (6.1)	379.7 (1.8)[2]	135.2 (3.8)[2]
Geranyl acetate	45.3 (8.6)	22.7 (6.7)[2]	23.5 (15.3)
Decanol	n.d. (–)	151.6 (10.4)[2]	145.3 (1.0)
Nerol	33.4 (8.4)	37.2 (3.2)	40.7 (14.6)
Hexanoic acid	372.4 (5.9)	493.0 (11.8)	460.8 (8.1)
Geraniol	24.2 (6.4)	85.1 (2.1)[2]	98.2 (0.2)[2]
5-Decanolide	24.9 (15.9)	31.6 (3.9)	26.0 (8.2)

[1]μg/kg

[2]Statistically significant ($p<0.05$).

Values in parentheses are the standard deviation of three experiments.

Underlines indicate compounds from a model flavor and their deterioration products.

(SDE) procedure for preparing the aroma concentrates was used in addition to HS method to recover the volatile components comprising a wide range of boiling points in the study to afford reproducible data with relative standard deviation (RSD) for amounts and the recovery of desirable odor. The recoveries of the added flavor compounds by SDE were 40–100%. Variation on recoveries of some components by SDE was shown in Table 1. The components underlined are the main compounds used for model flavor. Vanillin was not recovered in these extractions. The response factor in flame ionizing detector (FID) of each volatile component to the IS was measured. The collection of volatile compounds using SDE and subsequent analysis were repeated three times, which gave 0.1 to 34.0% of RSD for volatile components, and statistically significant ($P<0.05$) data were calculated. Production of furfural obtained as an artifact after SDE was minimized by the brief distillation. During processing and storage, the initial well-balanced citrus and fruity aroma in yogurt changed to a more fatty, oily and sweeter odor, and this corresponded with generation of a sour and bitter taste.

Figure 1 shows the change of aldehydes and alcohols on SDE. Aldehydes, such as octanal, nonanal, decanal, neral and geranial, decreased drastically during the initial three days and their corresponding alcohols appeared, thus causing flavor change of the yogurt. The rate of disappearance of octanal was less than those of other alhehydes, because oc-

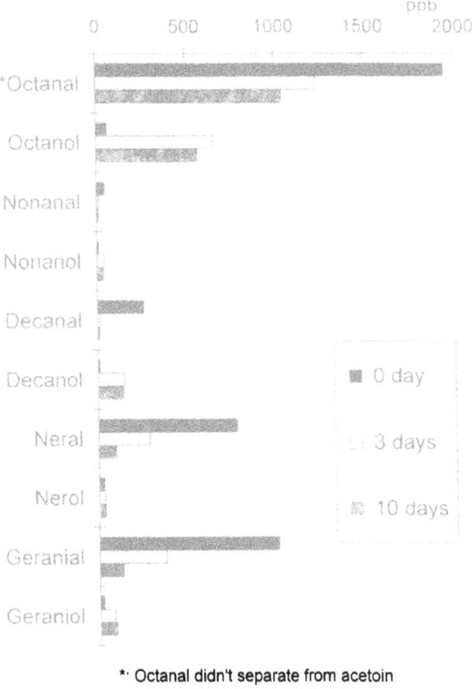

Figure 1. Aldehydes and alcohols in fresh and stored yogurt (* Octanal didn't separate from acetoin in this condition [see HS data]).

tanal could not be separated from acetoin, which increased during fermentation under GC conditions employed. However HS analysis indicated a decrease in the content of octanal (Figure 2). Meanwhile, HPLC analysis showed no decrease in the amount of vanillin under this condition. *Lactobacillus bulgaricus* has been reported to be sluggish in digesting vanillin (Nakazawa etal, 1982); similarly *Streptococcus thermophilus* also might be ineffective in its digestion. Figure 3 shows the change of terpenes; compounds such as limonene, γ-terpinene, *p*-cymene and myrcene decreased markedly during initial 3 days, but terpene alcohols changed less, except for linalool. Figure 4 shows the change of esters, such as butyl acetate and allyl hexanoate all of which exhibited a decrease during the initial 3 days.

The changes of original compounds in yogurt base are shown in Figures 5 and 6. Acids generated from yogurt base showed a maximum level on the 3rd day after manufacturing and then decreased slightly over the next 7 days (Figure 5). Large increases of some methylketones, such as 2-undecanone and 2-tridecanone were observed during initial 3 days; others changed marginally (Figure 6).

In a resazurin test, the decrease of absorbance of yogurt in 500–600 nm in comparison with that of milk suggested that the reducing activity of alcohol dehydrogenase increased in yogurt by fermentation (Figure 7), thus decreasing the concentration of aldehydes. Alcohol dehydrogenase activity has been reported for *Streptococcus thermophilus* but not for *Lactobacillus bulgaricus* (Lees and Jago, 1978). Therefore, reduction of

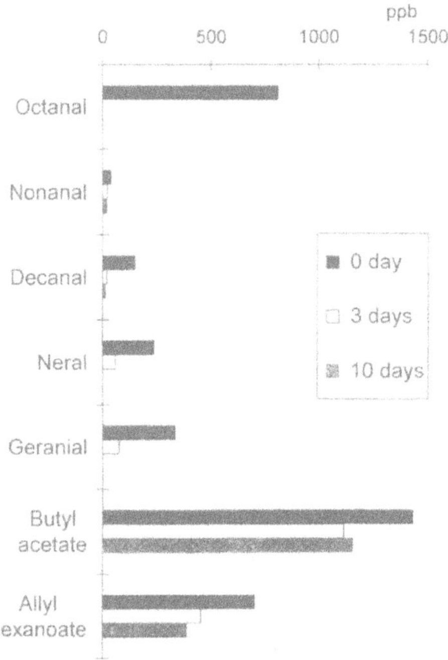

Figure 2. Aldehydes and esters (HS) in fresh and stored yogurt.

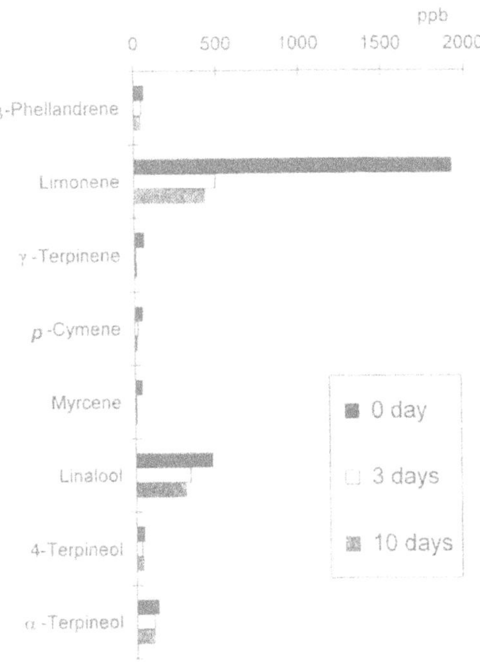

Figure 3. Terpenes in fresh and stored yogurt.

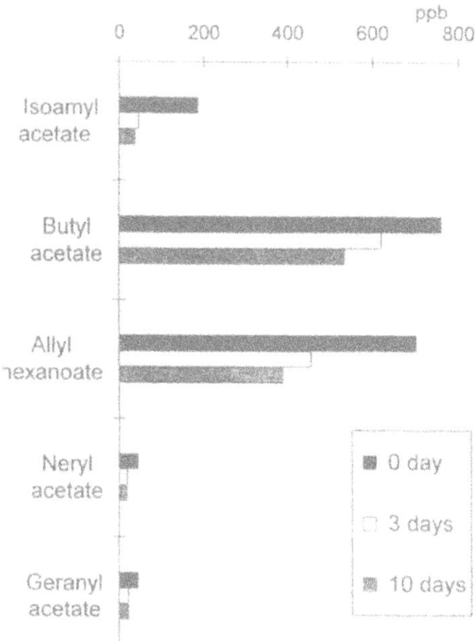

Figure 4. Esters in fresh and stored yogurt.

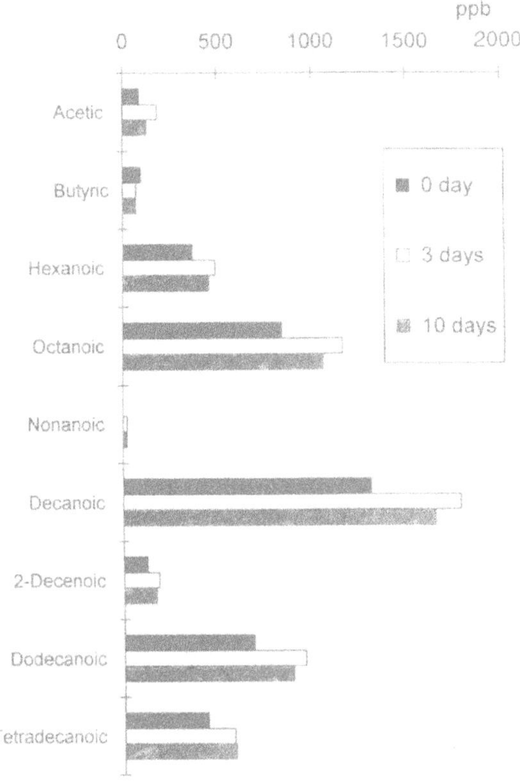

Figure 5. Acids in yogurt.

aldehydes to alcohols may be brought about by the metabolism of *Streptococcus thermophilus*.

Effects of sorption into packaging material and chemical reactions are assembled in Table 2. The degree of change seen in a glass bottle was compared with that in a paper cup(coated with film). The 0 day's amount of each component was expressed as 100. The 3rd day's and the 10th day's data were calculated from those for the 0 day. On the one hand, defferences after comparing 0 day's data with 3rd day's data mainly suggested the effects of fermentation and sorption into the packaging material. The degree of change in a glass bottle was similar to that in a paper cup except for a few hyrocarbons and alcohols. On the other hand, differences of results for the 3 and 10 days data mainly showed the effects of chemical reactions. Therefore, most components, except aldehydes and alcohols, changed marginally over the latter seven days in both packaging material. Thus, the effect of sorption into the packaging material (Matsui, 1993) of alcohols, esters and hydrocarbons, and the effect of chemical reactions, such as hydrolysis of esters, hydration or oxidation of hydrocarbons (Clark and Chamblee, 1992; LaGrange and Hammond, 1993) were considered to have little effect on yogurt flavor.

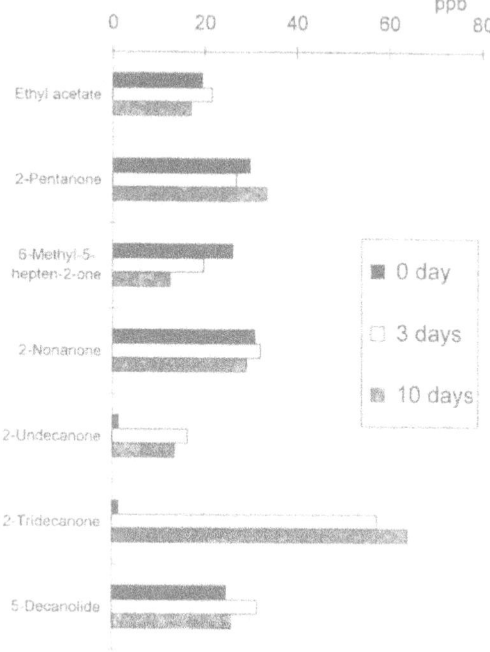

Figure 6. Presence of other compounds in fresh and stored yogurt.

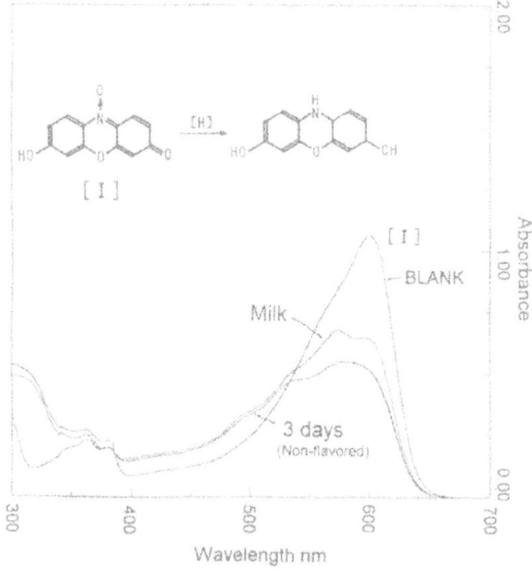

Figure 7. UV/VIS spectrum (300–700 nm) of resazurin [I] solution containing yogurt.

Table 2. Effects of a) sorption into packaging material (top) and b) chemical reactions (bottom)

		3 days		10 days	
Component	0 day	Paper	Glass	Paper	Glass
Neral	100	37[1]	37	13	29[2]
Nerol	100	111	94	122	103
Geranial	100	38	38	13	30[2]
Geraniol	100	352	258[2]	406	296[2]
β-Phellandrene + 1,8-Cineole	100	78	53	71	57
Limonene	100	26	39	22	54[2]
γ-Terpinene	100	17	42[2]	26	54[2]
p-Cymeme	100	46	41	31	56
Butyl acetate	100	82	60	70	73
Allyl hexanoate	100	65	48	55	57
Neryl acetate	100	43	52	41	47
Geranyl acetate	100	50	57	52	65
Linalool	100	72	66	65	70
4-Terpineol	100	72	79	85	85
α-Terpineol	100	83	81	79	88

		Paper		Glass	
		3 days	10 days	3 days	10 days
Neral	100	37	13[2]	37	29[2]
Nerol	100	111	122	94	103
Geranial	100	38	13[2]	38	30[2]
Geraniol	100	352	406[2]	258	296[2]
β-Phellandrene + 1,8-Cineole	100	78	71	53	57
Limonene	100	26	22	39	54
γ-Terpinene	100	17	26	42	54
Butyl acetate	100	82	70	60	73
Allyl hexanoate	100	65	55	48	57
Neryl acetate	100	43	41	52	47
Geranyl acetate	100	50	52	57	65
Linalool	100	72	65	66	70
4-Terpineol	100	72	85	79	85
α-Terpineol	100	83	79	81	88

[1]Recovery (%) after fermentation or storage.
[2]Statisticslly significant ($p<0.05$).

CONCLUSIONS

Three main factors are involved in relation to flavor deterioration of yogurt: bacterial digestion, sorption into packaging material and chemical reactions. Bacterial digestion was clearly observed. Aldehydes which contribute to expressing citrus notes were especially reduced to alcohols during fermentation process and storage. As a result, the well-balanced flavor which expressed citrus notes were altered and weakened, and fatty or oily notes mainly caused by an increase in the content of alcohols were strengthened. However, effects of sorption into packaging material and chemical reactions were less pronounced. Therefore, the bacterial digestion of flavor components should be firstly controlled in manufacturing flavored yogurt.

REFERENCES

Clark, Jr. B.C.; Chamblee, T.S. Acid-catalyzed reactions of citrus oils and other terpene-containing flavors. In *Off-Flavors in Foods and Beverages*; Charalambous, G., Ed.; Elsevier Science Publishers, Amsterdam, 1992; pp 229–285.

Gaafar, A.M. Volatile flavour compounds of yoghurt. *Int. J. Food Sci. Technol.* **1992**, *27*, 87–91.

Imhof, R.; Bosset, J.O. Qantitative GC-MS analysis of volatile flavour compounds in pasteurized milk and fermented milk products applying a standard addition method. *Lebensm. -Wiss. u. Technol.* **1994**, *27*, 265–269.

LaGrange, W.S.; Hammond, E.G. The shelf life of daily products. In *Shelf Life Studies of Foods and Beverages*; Charalambous, G., Ed.; Elsevier Science Publishers B.V.: Amsterdam, 1993; pp 1–20.

Lees, G.J.; Jago, G.R. Role of acetaldehyde in metabolism: A review 2. The metabolism of acetaldehyde in cultured dairy products. *J. Dairy Sci.* **1978**, *61*, 1216–1224.

Marion, J.P.; Chardon-Fayard, S. Incorporation of flavours in foodstuffs. In *Flavors and Off-flavors*; Charalambous, G., Ed.; Elsevier Science Publishers, Amsterdam, 1990; pp 687–699.

Matsui, T. Sorption of flavors into packaging film and its depression. *Nippon Shokuhin Kogyo Gakkaishi.* **1993**, *40*, 895–904.

Nakazawa, Y.; Wada, R.; Izumitani, M. Studies on the vanilla flavours for food processing (V). *Jap. J. Dairy Food Sci.* **1982**, *31* (2), A-51 - A-56.

Pharmaceutical Society of Japan. Standard methods of analysis of milk and milk products with commentary. Kanehara and Co., Ltd.: Tokyo, 1984; p 137.

FLAVOR GENERATION DURING EXTRUSION COOKING

William E. Riha, III[1] and Chi -Tang Ho[1,2]

[1]Department of Food Science and
[2]The Center for Advanced Food Technology
Cook College
New Jersey Agricultural Experiment Station
Rutgers, The State University of New Jersey
New Brunswick, New Jersey 08903

Extrusion cooking is a high temperature-short time process which is ideal for the production of flavor volatiles. However, due to the nature of this process there are many opportunities for these volatiles to be lost. This has led many researchers to study the formation and retention of volatiles during extrusion. This review will focus on the flavor related reactions which may occur during extrusion and their occurrence in various flours and due to the addition of reactive precursors.

INTRODUCTION

Extrusion cooking is currently being used extensively in the food industry for fast and economical production of a number of food items. Food extrusion began in the 1930's with the use of single-single screw extruders to form and shape macaroni and ready-to-eat cereals. Today, the number of food products produced by extrusion consist of expanded snacks, a multitude of breakfast cereals, soup and drink bases, textured soy protein, and many others (Harper, 1992). Extrusion processing of foods involves the exposure of the raw material to high temperature, pressure, and shear force to mix and cause chemical changes which constitute cooking. These chemical changes, among many others, involve the production of flavors which give the final product its characteristics. These flavor compounds may be formed from a number of compounds in the starting material and through a number of different reactions including lipid oxidation, breakdown of carotenoids, deamidation of glutamine and asparagine and the formation of Maillard reaction products.

Although extrusion technology has been used in food processing for quite sometime there is still much which remains ambiguous about this technology. As more research is done in the area of food extrusion, a better understanding of the physical and chemical changes which take place is gained. Much of the recent extrusion research focuses on the transport phenomenon, rheology, scale-up, and control associated with extrusion processing. However, the area of flavor production during extrusion has not been well researched. The addition of precursors and processing conditions will not only aid in the development of final extruded products with better flavor, but will also provides an opportunity to enhance our understanding of the chemical reactions which occur during extrusion.

EXTRUSION COOKING

Extrusion is a processing technique that involves the addition of thermal and mechanical energy to an uncooked mass, such as wheat flour. This is accomplished by feeding the raw material into a hopper which introduces the raw material between a rotating screw and a stationary barrel. At this stage, heat may be added at various zones along the barrel. After the material is passed through the barrel, it is forced through a die of a specific shape. The result is a cooked, shaped product. The high shear and temperature conditions inside the screw channel result in the mixing of the material and leads to the chemical reactions that constitute the cooking process. Some of the chemical changes that occur include gelatinization of starch molecules, crosslinking of proteins, and the generation of flavors.

The texture of the final product is established as the material leaves the die of the extruder. If the product temperature is over 100°C, then as the hot material leaves the high pressure of the extruder, water is flashed off. This flashing off of water results in an expanded product. This volatilization of water may have a negative implication on the flavor of the extruded product. This may result in the steam stripping of the volatiles and loss of overall flavor. Volatiles may also become bound to proteins or starch during extrusion or thermally degraded due to high temperatures. In order to compensate for these losses, manufacturers may add volatiles near the die of the extruder or post-extrusion (Camire, 1991). Another method to enhance flavor, which will be discussed in a later section, includes the addition of reactive flavor precursors to the flour or water supply prior to extrusion in order to increase production of flavor volatiles or change the flavor profile.

FLAVOR RETENTION DURING EXTRUSION PROCESSING

Research in the area of flavor retention during processing has many practical applications. Due to the loss of flavors by volatilization at the die, thermal degradation of the flavorants, and binding to the starch/protein matrix, processors must look to other means to flavor extruded products. One such method is the addition of flavorants directly to the raw feed material. From this concept, a number of research articles have been published dealing with the flavoring and retention of flavorants during extrusion. Maga (1989) has reviewed most of the research which was available at that time. The study of the effectiveness of natural and artificial extrusion stable volatiles (Lazarus and Renz, 1985) or the effect of high or low sugar concentration on the retention of orange volatiles (Mariani *et al.*, 1985) were reported. Use of surface response methods (Lane, 1983) or calculated vapor pressures (Chen *et al.*, 1986) to optimize or predict volatile retention during extrusion was also demonstrated.

A recent study on flavor retention during extrusion was conducted by Kim and Maga (1994). These researchers added C6, C8, and C10 acids, alcohols and aldehydes to high amylose corn starch and extruded the mixture at 115, 125, or 135°C. After extrusion, the "free" and "total" volatiles were measured and the difference was calculated as the "bound" volatiles. These researchers found that the retention of longer chain length compounds was superior, perhaps due to their lower volatility. Acids, at all temperatures, and aldehydes, at higher temperatures, were retained in the free form, however, alcohols were found to bind more to the starch during extrusion. This binding may have allowed alcohols to achieve the highest total retention of 85–98%.

ADDITION OF PRECURSORS

There has been limited research dealing with the addition of reactive precursors to flour prior to extrusion to enhance the flavor of the final product. However, because extrusion conditions favor a number of chemical reactions, specifically the Maillard reaction, it would seem that exploiting these reactions would be beneficial to the production of flavored extrudates. In this contribution, a reactive precursor can be defined as any substance added to the flour which can react to produce flavor compounds. For example, in terms of the Maillard reaction, this would include substances with amino or carbonyl groups. Of course, the flour itself usually contains a large number of these flavor precursors, as will be discussed in the following section. All studies dealing with reactive flavor precursors have focused on the formation of volatiles through Maillard reaction and therefore have dealt with the addition of a reactive nitrogen source and, in some cases, sugars or sugar fragments. These reactive nitrogen sources have taken the form of a protein, an amino acids or an ammonium salt.

Proteins

Maga and Kim (1989) studied the effects of a number of proteins including defatted soy flour, soy protein concentrate, sodium caseinate, whey protein concentrate, and gluten on the levels of volatiles produced during extrusion of corn starch. In addition to protein type, these researchers also examined the effect of protein concentration (0–50%), feed moisture (12 or 25%), and product temperature (120 or 150°C). The samples were extruded on a single-screw extruder at 120 rpm, ground and distilled with water. The volatiles were extracted from the distillate with diethyl ether and separated using gas chromatograpic analysis, however, resulting peaks were not identified. A comparison of the chromatograms obtained for each protein source showed that some peaks were unique to the protein source being used, for example, some peaks only appeared for the milk protein sources, others for soy, and still others for gluten. In terms of extrusion conditions, the more mild condition of low temperature/high moisture had the most peaks, however, they were in lower concentrations than the peaks which were retained in the high temperature/low moisture sample. This indicates that perhaps at the more severe condition greater concentrations of volatiles are being produced, but more of the volatiles are being lost. These researchers also found that upon increase in protein concentration some of the volatiles remained at constant levels while others increased.

Another study on the effect of added protein on flavor produced during extrusion cooking was conducted by Bailey *et al.* (1994) who added whey protein concentrate to corn meal at concentrations ranging from 0 to 20%. This mixture was extruded on a twin-screw extruder at various screw speeds, feed loads, and moisture contents. Headspace analysis was used to extract volatiles and volatile separation and identification was accomplished using GC and GC/MS. These researchers identified 71 volatiles produced from lipid oxidation and Maillard reaction. They found that the production of pyrazines, furans, and other heterocyclics did increase as the concentration of whey protein concentrate, an excellent source of lysine and sulfur containing amino acids, was increased.

Amino Acids

Apart from proteins, addition of amino acids as a source of nitrogen for the production of flavor volatiles may be practiced. If proteolysis must occur during extrusion in order to supply a suitable nitrogen source for the Maillard reaction, then the addition of amino acids may speed up the formation of flavors. The addition of amino acids with different functional groups may also influence the types of flavors produced and therefore, the flavor of the extrudate may be controlled by the processor. There has been limited work done on the addition of amino acids during extrusion and is an area which may require further attention.

Cysteine

The addition of the amino acid cysteine to wheat flour during extrusion has recently been studied by Riha *et al.* (1996). Due to the presence of both a thiol group and an amino group on the amino acid cysteine, one would expect a large number of different volatiles to be produced. These researchers added 0, 0.25, 0.5,0.75, or 1.0% cysteine to high gluten wheat flour and extruded the mixture using a twin-screw extruder at 185°C, 500 rpm, and 16% moisture. The volatiles were extracted from the ground extrudate using a thermal desorption sample-collection system and were analyzed using GC and GC/MS.

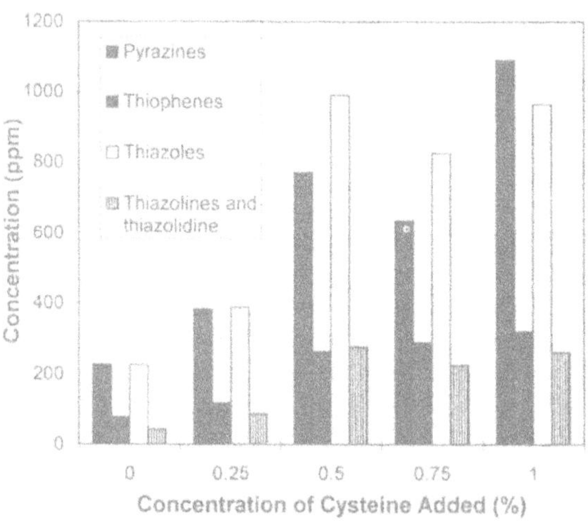

Figure 1. Classes of compounds identified from the extrusion of wheat flour with various levels of cysteine added. Adapted from Riha *et al.* (1996).

Figure 1 presents the concentration of the major classes of volatile compounds tentatively identified in the solid extrudate for all five concentrations of cysteine added. In general, the concentration of most classes of volatiles increased for up to 0.5% added cysteine, above this concentration there was very little change in the volatile concentration. The authors propose that this may be due to the fact that cysteine had a marked effect on physical properties of the extrudate, such as expansion ratio and oil/water binding capacity, and perhaps the extrudate with its limited ability to retain volatiles became saturated with flavor compounds. Although the majority of compounds identified were sulfur containing, the addition of cysteine did increase the production of pyrazines. Of the sulfur-containing volatiles produced, there were much higher concentrations of thiazoles than thiophenes. Thiazoles have been found to be formed mainly from the thermal degradation of cysteine, whereas thiophenes are formed through the Maillard reaction. The high concentration of thiazoles may indicate that thermal degradation of cysteine occurs instead of the Maillard reaction. This may be due to the fact that starch does not have enough time in the extruder to break down into sugar fragments which are suitable to participate in the Maillard reaction. Another possible explanation offered by the authors is that the thiazole, a nitrogen and sulfur-containing volatile, may be more efficiently retained by the extrudate as shown by Palkert and Fagerson (1980).

Other Precursors

In order to investigate the effect of ammonia and a sugar fragment, pyruvaldehyde, on the flavor produced during the extrusion of wheat flour, Izzo *et al.* (1994) added ammonium bicarbonate or ammonium bicarbonate and pyruvaldehyde to the water source during the extrusion of wheat flour. The twin-screw extruder used for these experiments was equipped with a pump which could control the addition of water directly into the barrel of the extruder. The moisture content was controlled at 16%, the melt temperature was 180°C and the extruder operated at 450 rpm. The volatiles were extracted from the extrudate using a thermal desorption sample-collection system and analyzed using GC and GC/MS. Table 1 lists the volatiles identified and quantified under various conditions. It is interesting to note the volatiles present in unextruded wheat flour consist mainly of lipid oxidation products, such as aldehydes, alcohols, and ketones. The authors propose that these compounds are a result of the milling process and, as Table 1 shows, they are quickly lost during extrusion. It was interesting to note that the addition of ammonia bicarbonate resulted only in an increase in the concentration of unsubstituted pyrazine, whereas, the addition of both ammonium bicarbonate and pyruvaldehyde resulted in further production of unsubstituted pyrazine and alkylpyrazines. These authors proposed that during extrusion with the addition of ammonium bicarbonate the breakdown of starch led to the formation of two and three-carbon unit fragments which can condense to form pyrazine and methylpyrazine. With the addition of pyruvaldehyde, a three-carbon sugar fragment, the kinetics of the reaction shifted to allow two three-carbon fragments to condense and form dimethylpyrazines. These researchers also suggested that the residence time in the extruder was too short to facilitate the breakdown of proteins and starch and the reactions which lead to the formation of flavors.

Another study by Izzo and Ho (1992) involved the addition of glucose and/or ammonium bicarbonate to autolyzed yeast extract prior to extrusion. These researchers measured the pyrazine formation and brown color development when the yeast extract mixtures were extruded on a single screw extruder at 115°C, and 140 rpm screw speed. Liquid-solid extraction was used to extract the volatiles; pyrazines were identified and quantified using

Table 1. Volatiles generated in extruded wheat flour systems with added precursors

	Volatile concentration (ppb)			
Compounds	UNEX[1]	Ex[2]	EX + Ab[3]	EX+AB+PA[r]
Aldehydes				
3-Methylbutanal	–	0.0208	0.0059	0.0198
Pentanal	0.3114	0.0423	0.0221	0.0212
Hexanal	4.6571	0.8382	0.5559	0.3591
Heptanal	0.1334	0.0169	0.0171	0.0094
Benzaldehyde	–	0.1382	0.0735	0.0789
Ethylpentanal	0.1139	0.0472	0.0242	0.0103
Nonanal	0.3233	0.0178	0.0609	0.1218
2-Nonenal	–	–	–	0.0286
Subtotal	5.5391	1.1214	0.7596	0.6205
Alcohols				
Pentanol	0.5949	–	–	–
2-Furanmethanol	–	–	–	0.0174
Hexanol	7.1221	0.0345	0.0430	0.0248
1-Octen-3-ol	0.2846	0.0312	0.0191	0.0144
Subtotal	8.0016	0.0657	0.0621	0.0566
Ketones				
5-Methyl-2-hexanone	–	0.0102	0.0107	–
2-Heptanone	0.1549	0.0172	0.0166	0.0077
3-Octen-2-one	0.1948	0.1140	0.0568	0.0191
Subtotal	0.3497	0.1414	0.0841	0.0268
Furans				
2-Propylfuran	0.0603	0.0148	0.0060	0.0202
2-Pentylfuran	0.4249	0.0666	0.0870	0.0640
Subtotal	0.4852	0.0814	0.0930	0.0842
Pyrazines				
Pyrazine	–	0.0119	0.1000	0.3231
Methylpyrazine	–	0.0261	0.0210	0.8595
2,6-Dimethylpyrazine	–	–	–	0.2582
2,5-Dimethylpyrazine	–	–	–	0.0205
	–	–	–	0.0107
2-Vinyl-5-methylpyrazine				
Subtotal	0.0000	0.0380	0.1210	1.4720
Sulfur-Containing				
Dimethyl trisulfide	–	–	–	0.0096

Adapted from Izzo *et al.*, 1994.

[1]UNEX=Unextruded wheat flour, [2]EX= Extruded Wheat Flour, [3]AB= Ammonium Bicarbonate added, [4]PA= Pyruvaldehyde added.

GC/MS. These researchers identified 24 pyrazines in the extruded autolyzed yeast extract systems. The addition of 2% ammonia bicarbonate to the yeast extract reduced the concentration of pyrazines by about 60%. These authors felt that ammonia was too reactive of a nitrogen source and contributed more to brown color formation than to flavor formation. The addition of 5% glucose to the yeast extract greatly increased pyrazine production, thus indicating that reducing sugar is the limiting reagent in this extrusion system.

EXTRUDED MATERIAL

As mentioned in the introduction, the flour itself contains protein, starch, and lipid which may be broken down as a result of the high temperature and shear forces within the extrusion to form flavor compounds. Because of the different composition of flours from different sources, the resulting flavor profiles may take on different characteristics. The following is a review of some of the flavor research that has been done on different types of flours.

Corn

The volatiles produced from an extruded corn-based model system has been studied by Ho *et al.* (1989). In this model system, zein (a corn protein), corn amylopectin, and corn oil were mixed and extruded using a single-screw extruder at 120 or 165°C. The extruded samples were ground and extracted with ethyl ether, the volatiles were analyzed by GC and GC/MS. For comparison, the mixture was also baked in an oven at 120 or 180°C for 30 min. The results so obtained are presented in Table 2. In all cases examined, the higher temperature resulted in the formation of higher amounts of all volatiles. It is also interesting to note how different ingredients interacted to produce different aroma volatiles.

Another interesting study on flavor production during extrusion of corn flour was carried out by Nair *et al.* (1994). These researchers used a twin-screw extruder to extrude corn flour at 178°C and a cold trap collection apparatus to collect volatiles as they were released at the die. Gas chromatography/mass spectrometry was used to identify, but not quantify, volatiles from both the extrudate and condensate which was collected at the die. Figure 2 presents the classes of compounds and the number of each compound found in both the extrudate and condensate.

Potato

There have been a number of studies conducted on the flavor of extruded potato products. Dehydrated potatoes are high in starch content, which allows excellent expansion properties, and are also high in free amino acids, which can be utilized in the Maillard reaction to produce aroma compounds. Raw potato is particularly high in the amino acids glutamine and asparagine. These amide containing amino acids may undergo a deamidation reaction which results in the release of ammonia. As demonstrated by Hwang *et al.* (1993), the ammonia released can also participate in the Maillard reaction to produce flavors. Extruded potato product flavor has been previously reviewed by Maga (1994). Furthermore, Maga and Sizer (1979a, 1979b) found that during extrusion there was a large loss of free amino acids while production of pyrazine compounds increased at higher temperatures (160°C) and lower moisture contents (25%).

Table 2. Concentration of volatile compounds in extruded and baked zein/corn amylopectin/corn oil samples

	Concentration (ppb)					
	Extruded				Baked	
	Z		Z + O + A[1]		Z + O + A	
Compounds	120°C	165°C	120°C	165°C	120°C	180°C
From Lipids						
Hexanal	t	t	11	42	86	709
2-Heptanone	2	6	4	16	8	91
Benzaldehyde	5	18	17	51	20	351
3-Octanone	t	t	1	6	–	34
3,5-Octadien-2-one	1	3	3	7	–	66
2,4-Decadienal	t	2	6	16	–	185
2-Methyl-3-phenyl-2-propenal	t	t	6	17	–	103
From Carotenoids						
Toluene	t	t	7	29	10	115
6-Methyl-5-heptene-3-one	t	t	1	4	–	26
Isophorone	8	28	33	94	24	798
α-Ionone	1	3	6	19	–	28
β-Ionone	t	t	3	9	–	18
6,10-Dimethyl-5,9-undecadien-2-one	t	t	1	4	–	
From Proteins and Carbohydrates						
2,5-Dimethylpyrazine	12	40	33	113	18	255
2-Methyl-5-ethylpyrazine	2	6	6	18	–	59
2,5-Dimethyl-3-ethylpyrazine	t	2	4	10	–	31
From Lipids, Proteins and Carbohydrates						
2-Methyl-3(or 6)-pentylpyrazine	t	1	7	26	–	29
2-Methyl-3(or 6)-hexylpyrazine	t	t	1	4	–	–
2,5-Dimethyl-3-penylpyrazine	t	t	1	4	–	6

Adapted from Ho et al., 1989.
[1]A=Amylopectin; O=Oil; Z=Zein.

Others

A thorough study on the pyrazines produced during the extrusion of partially dried, green malt was conducted by Fors and Eriksson (1986). In this work, milled or whole kernel malt was extruded on a twin-screw extruder using various temperatures, moisture contents and screw speeds. Headspace sampling was used to collect volatiles which were then analyzed and identified by gas chromatography and mass spectrometry. These researchers found that temperature played the dominant role in pyrazine formation and that the concentration of pyrazines increased as temperature was increased from 130 to 160°C, but dropped as the temperature increased to 190°C. In addition, if the malt kernels were milled, making the particle size smaller and increasing surface area, the production of volatiles increased. Other conditions, such as moisture content and residence time in the extruder also affected, not only the pyrazine production, but color, pH and soluble α-amino nitrogen.

The flavor properties of other extruded materials have also been studied. Maga and Kim (1994) extruded dried, ground taro corns, which are reported to have a bland cooked flavor similar to potatoes, at 120°C and 15% moisture on a single screw extruder. Volatiles were extracted and identified using GC/MS. Unlike potatoes, the heterocyclics identified from ex-

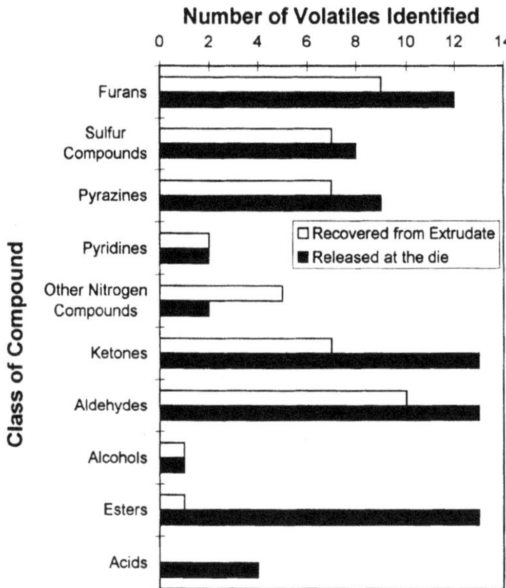

Figure 2. Number of compounds trapped in the extrudate or released at the die during the extrusion of corn meal. Adapted from Nair *et al.* (1994).

truded taro were more sulfur-containing, such as thiophenes and thiazoles, and no pyrazines were identified. Another study by Lui and Maga (1994) investigated the sensory properties of mechanically deboned pork extruded with soy protein concentrate, rice flour and spices. In this study the flavor, texture, and overall acceptability were evaluated for different processing conditions. Using sensory evaluation, these researchers optimized the product quality based on the level of flavors, soy protein concentrate and non-glutinous rice flour.

CONCLUSIONS

Extrusion cooking has been shown to be an efficient method for the production of many types of food products. One disadvantage of this processing technique is the possible loss of flavor due to steam distillation of volatiles at the die, flavor binding to starch or protein, or the thermal degradation of flavors. In order to compensate for these losses, processors may add flavors during extrusion. In response to this, a number of researchers have studied how flavors are retained in an extruded material and what can be done to help retain these compounds.

Another approach is to use the extruder as a reactor in order to optimize the production of flavor volatiles during extrusion. There are a number of chemical reactions which occur during extrusion which lead to the production of flavors, the most important being the Maillard reaction. Recently, more researchers are attempting to enhance or change the flavor of extruded materials by adding precursors to the flour which will react in the extruder and provide the flavor desired. This may be a sensible approach to enhancing flavors because many times these precursors can take the form of inexpensive by-products such as whey protein concentrate. The addition of precursor may also provide insight into the chemistry which is occurring during extrusion and answer some questions about a processing technique which in many ways remains a mystery.

REFERENCES

Bailey, M.E.; Gutheil, R.A.; Hsieh, F.H.; Cheng, C.W.; Gerhardt, K.O. Maillard reaction volatile compounds and color quality of a whey protein concentrate - corn meal extruded product. In *Thermally Generated Flavors: Maillard, Microwave, and Extrusion Processes*; Parliment, T.H., Morello, M.J., McGorrin, R.J., Eds.; American Chemical Society: Washington, D.C., 1994; pp. 315–327.

Camire, M.E. Protein functionality modification by extrusion cooking. *J. Am. Oil Chem. Soc.* **1991**, *68*, 200–205.

Chen, J.; Reineccius, G.A.; Labuza, T.P. Prediction and measurement of volatile retention during extrusion processing. *J. Food Technol.* **1986**, *21*, 365–383.

Fors, S.M.; Eriksson, C.E. Pyrazines in extruded malt. *J. Sci. Food Agric.***1986**, *37*, 991–1000.

Harper, J.M. Comparative analysis of single- and twin-screw extruder. In *Food Extrusion and Technology*; Kokini, J.L., Ho, C.-T., Karwe, M.V., Eds.; Marcel Dekker, Inc.: New York, NY, 1992, pp.139–148.

Ho, C.-T.; Bruechert, L.J.; Kuo, M.C.; Izzo, M.T. Formation of volatile compounds from extruded corn-based model systems. In *Thermal Generation of Aromas*; Parliment, T.H., McGorrin, R.J., Ho, C.-T., Eds.; American Chemical Society: Washington, D.C., 1989; pp.504–511.

Hwang, H.I.; Hartman, T.G.; Rosen, R.T.; Ho, C.-T. Formation of pyrazines from the Maillard reaction of glucose and glutamine-amide-^{15}N. *J. Agric. Food Chem.* **1993,** *41*, 2112–2115.

Izzo, H.V.; Ho, C.-T. Ammonia affects Maillard chemistry of an extruded autolyzed yeast extract: Pyrazine aroma generation and brown color formation. *J. Food Sci.* **1992**, *57*, 657–659 & 674.

Izzo, H.V.; Hartman, T.G.; Ho, C.-T. Ammonium bicarbonate and pyruvaldehyde as flavor precursors in extruded food systems. In *Thermally Generated Flavors: Maillard, Microwave, and Extrusion Processes*; Parliment, T.H., Morello, M.J., McGorrin, R.J., Eds.; American Chemical Society: Washington, D.C., 1994; pp. 328–333.

Kim, C.H.; Maga, J.A. Chain length and functional group impact on flavor retention during extrusion. In *Thermally Generated Flavors: Maillard, Microwave, and Extrusion Pocesses*; Parliment, T.H., Morello, M.J., McGorrin, R.J., Eds.; American Chemical Society: Washington, D.C., 1994; pp. 380–384.

Lane, R.P. Formulation variables affecting the flavor of extruded snacks and crackers. *Cereal Foods World.* **1983**, *28*, 181–183.

Lazarus, C.R.; Renz, K.H. The influence of cereal flours on the taste perception of extrusion-stable flavors. *Cereal Foods World.* **1985**, *30*, 319–320.

Liu, M.B.; Maga, J.A. Flavor properties of extrusion cooked mechanically deboned pork. In *Thermally Generated Flavors: Maillard, Microwave, and Extrusion Processes,* Parliment, T.H., Morello, M.J., McGorrin, R.J., Eds.; American Chemical Society: Washington, D.C., 1994; pp. 356–364.

Maga, J.A. Flavor formation and retention during extrusion. In "Extrusion Cooking"; Mercier, C., Linko, P., Harper, J.M., Eds.; American Association of Cereal Chemists: St. Paul, MN., 1989; pp. 387–398.

Maga, J.A. Potato flavor. *Food Rev. Int.* **1994,** *10*, 1–48.

Maga, J.A.; Kim, C.H. Protein-generated extrusion flavors. In *Thermal Generation of Aromas*; Parliment, T.H., McGorrin, R.J., Ho, C.-T., Eds.; American Chemical Society: Washington, D.C., 1989; pp. 494–503.

Maga, J.A.; Liu, M.B. Extruded taro (*Colocasia esculenta*) volatiles. In *Thermally Generated Flavors: Maillard, Microwave, and Extrusion Processes*; Parliment, T.H., Morello, M.J., McGorrin, R.J., Eds.; American Chemical Society: Washington, D.C., 1994; pp. 365–369.

Maga, J.A.; Sizer, C.E. Pyrazine formation during the extrusion of potato flakes. *Lebensm. Wiss. -u. Technol.* **1979,** *12*, 15–16.

Maga, J.A.; Sizer, C.E. Influence of water hardness on the sensory and physical properties of extruded potato flakes. *Lebensm. Wiss. -u. Technol.* **1979,** *12*, 17–18.

Mariani, M.; Scotti, A.; Colombo, E. Flavours for foodstuffs obtained by extrusion-cooking: analytical and application aspects. In "Progress in Flavour Research 1984"; Adda, J., Ed.; Elsevier: Amsterdam, 1985; pp. 549–562.

Nair, M.; Shi, Z.; Karwe, M.V.; Ho, C.-T.; Daun, H. Collection and characterization of volatile compounds released at the die during twin screw extrusion of corn flour. In *Thermally Generated Flavors: Maillard, Microwave, and Extrusion Processes*; Parliment, T.H., Morello, M.J., McGorrin, R.J., Eds.; American Chemical Society: Washington, D.C., 1994; pp. 332–347.

Palkert. P.E.; Fagerson, I.S. Determination of flavor retention in pre-extrusion flavored textured soy protein. *J. Food Sci.* **1980,** *45*, 526–533.

Riha, W.E.; Hwang, C.F.; Karwe, M.V.; Hartman, T.G.; Ho, C.-T. Effect of cysteine addition on the volatile of extruded wheat flour. *J. Agric. Food Chem.* **1996,** *44*, 1847–1850.

PROCESS-INDUCED COMPOSITIONAL CHANGES OF FLAXSEED

P. K. J. P. D. Wanasundara and F. Shahidi

Department of Biochemistry
Memorial University of Newfoundland
St. John's, NF, Canada, A1B 3X9

Flaxseed has been used as an edible grain in different parts of the world since ancient times. However, use of flaxseed oil has been limited due to its high content of polyunsaturated fatty acids. Nonetheless, α-linolenic acid, dietary fiber and lignans of flaxseed have regained attention. New varieties of flaxseeds containing low levels of α-linolenic acid are available for edible oil extraction. Use of whole flaxseed in foods provides a means to utilise all of its nutrients and require minimum processing steps. However, the presence of cyanogenic glucosides and diglucosides in the seeds is a concern as they may release cyanide upon hydrolysis. In addition, the polyunsaturated fatty acids may undergo thermal or autooxidation when exposed to air or high temperatures that are used in food preparation. Studies todate on oxidation products of intact flaxseed lipids have not shown any harmful effects when flaxseed is included, up to 28%, in the baked products. Furthermore, cyanide levels produced as a result of autolysis are below the harmful limits to humans. However, the meals left after oil extraction require detoxification but, by solvent extraction, to reduce the harmful effects of cyanide when used in animal rations. Flaxseed meal is a good source of proteins; these could be isolated by complexation with sodium hexametaphosphate without changing their nutritional value or composition. In addition, the effect of germination on proteins, lipids, cyanogenic glycosides, and other minor constituents of flaxseed is discussed.

INTRODUCTION

Flax or linseed (*Linum usitatissimum* L.) is one of the oldest arable crops of the world and the seed has a long tradition of use for its oil, and a grain for edible and medici-

Process-Induced Chemical Changes in Food
edited by Shahidi *et al.* Plenum Press, New York, 1998

nal purposes. Recent nutritional research on flaxseed has shown its potential for incorporation into different food products. Other new applications of flaxseed relevant to human consumption are enriching the meat (Romans, 1995), eggs (Jiang *et al.*, 1991) and milk (Kennelly and Khorasani, 1992) of the animals with omega-3 polyunsaturated fatty acids which are present in flax; their use in fish feed has also been accepted (Dick and Yang, 1995). The Food and Drug Administration of the United States of America allows inclusion of up to 12% (by weight) flaxseed in foods but flaxseed and cold-pressed flaxseed oil have not attained the GRAS (generally recognized as safe) status (Carter, 1993). The interest in consumption of flaxseed is related to its high content of α-linolenic acid (ALA, 18:3 ω3), dietary fiber known as mucilage, lignans and phenolic compounds which are considered beneficial in reducing the risk factors for coronary heart diseases (CHD) and cancers.

Flaxseed is composed of a seed coat (testa, true hull or spermoderm), embryo or the germ, a thin endosperm and two cotyledons. Two large flattened cotyledons constitute the bulk of the seed (57%) while the seed coat and endosperm account for 38% and the embryo comprises 5% of the total seed weight (Dorrel, 1970). Cotyledons are the major storage tissues for the oil and contain 75% of the seed oil. The seed coat and endosperm contain 22% of the seed oil while the embryo contains only 3% oil (Dorrel, 1970). Proteins are the next largest component of flaxseed which are mainly stored in cotyledons. Polysaccharides or mucilage of flaxseed are present in the outermost cell layer of the seed coat. In addition to that, all other micronutrients that are needed for seed germination and seedling development such as macro- and microelements, vitamins and enzyme precursors are also stored, mainly in the cotyledons. Flaxseed is known to contain cyanogenic glycosides, a vitamin B_6 antagonist, phytic acid and phenolic acids, all of which are considered to have antinutritive properties at high levels.

The Canadian grown flaxseed contains, approximately 41% oil (on a moisture-free basis), 26% protein (%N × 6.25), 4% ash, 5% acid detergent fiber and 24% nitrogen-free extract (Oomah and Mazza, 1993). Proximate composition of the whole flaxseed varies considerably depending on the cultivar, growing conditions, seed processing and also the methods employed for analyses. There are other minor constituents of flaxseed such as vitamins, minerals, cyanogenic glycosides and phenolic compounds or their precursor lignans.

PROCESSING AND FOOD USES OF FLAXSEED

The type of processing of flaxseed varies depending on the end use of the product. The primary commercial product of flaxseed is its oil as it is the major seed component and a rich source of polyunsaturated fatty acids. Cold-pressing or prepressing followed by solvent extraction is normally practised to obtain flaxseed oil. It has been found that cold pressing generally produces a lighter color oil than that obtained via solvent extraction. Kolodziejczyk and Fedec (1995) have suggested that solvent extraction process used for other commercial oilseeds is suitable for edible oil preparation from flaxseed. Traditionally, flaxseed (linseed) oil has been used primarily in industrial applications such as drying oil in paints, varnishes and leather tanning. Cold pressed flaxseed oil has also been used as a source of ALA in pharmaceutical products or as an ingredient for salad dressing. Oil from genetically modified flaxseed contains lower amounts of ALA (1.8–2.0%) than traditional varieties (45–63%) and can be used as a salad oil or other edible products. Possible application of a two-phase solvent extraction system consisting of alkanol-ammonia-water/hexane for processing of flaxseed has been examined in order to obtain oil and meal with improved quality (Wanasundara and Shahidi, 1994a,b). Flaxseed meal is the main co-product of oil extraction and its major traditional uses have been for animal feed or fertilizer.

Preparation of whole or ground flaxseed as a food ingredient requires minimum processing. Flaxseed as such or in the ground form in foods provides a complete spectrum of nutrients. Whole or ground flaxseed could be incorporated into almost any baked product (breads, muffins, popovers, bagels, buns, waffles, pancakes, cakes, crackers and biscuits, etc.), ready-to-eat breakfast cereals, breakfast drinks, salad toppings, meat extenders, soups and fiber bars (Carter, 1993). The high temperatures involved in preparation of such products may chemically alter their heat sensitive macro- and micronutrients (*e.g.* lipids, proteins, heat labile vitamins). In addition, sprouted or germinated flaxseed may be used directly as an alternative food commodity (Wanasundara, 1995).

Preparation of protein ingredients from seed meal is also possible since it contains about 26% of crude protein. In addition to the traditional uses of seed meal, recovery of its protein fraction, in the form of isolate or concentrate, may provide value-added products for use in food and feed formulations. The flaxseed polysaccharide (mucilage) fraction, which is located on the outer seed coat, once isolated has potential use as a food gum or dietary fiber (Mazza and Oomah, 1995).

FLAXSEED OIL

Oil is the major constituent of flaxseed varying from 35 to 45% of the seed weight, depending on genetic and environmental conditions of growth. Triacylglycerols compose a major portion (approximately 95%) of flaxseed oil. Flaxseed oil is unique in that it contains a very high level (45–63%) of α-linolenic acid (ALA, C18:3 ω3), very low levels of saturated fatty acids and oleic acid (C18:1 ω9; Table 1). However, new varieties of flaxseed have been developed and these contain low levels of linolenic acid. Since 1994 new varieties of flaxseed, namely Linola™ or Solin™ which contain elevated levels (65–76%) of linoleic acid (C18:2 ω6) (Table 1), have become available and may serve as a suitable source of vegetable oil (Green and Dribnenki, 1994).

Solvent Extraction

Wanasundara and Shahidi (1994a) have compared extraction efficiency of oil from flaxseed using a monophasic hexane and bi-phasic alkanol-ammonia-water/hexane solvent

Table 1. Fatty acid (%) composition of flaxseed oil[1]

Fatty acid	Traditional[2]	Low-linolenic[3]
Palmitic (16:0)	3.8-9.2	5.6-5.8
Stearic (18:0)	1.3-6.2	3.4-4.0
Oleic (18:1 ω9)	13.3-25.2	14.5-15.9
Linoleic (18:2 ω6)	10.4-20.9	71.9-73.9
Linolenic (18:3 ω3)	45.5-63.1	1.8-2.0

[1]Both samples contained 35-45% oil.
[2]Adapted from Green and Marshall (1981)
[3]Adapted from Green and Dribnenki (1994); Kolodziejczyk and Fedec (1995).

Table 2. The effect of different two-phase solvent extraction systems on mass balance of processed of flaxseed[1]

Solvent system	Yield on a dry weight basis (%) Meal	Oil	Gums	Loss
Hexane	48.9±1.0	49.2±1.5	–	1.9
Methanol/hexane	46.7±1.0	45.9±1.9	5.2±0.1	2.0
Methanol-ammonia-water/hexane (95:10:5, v/w/v)	46.4±2.0	47.1±0.1	5.7±0.1	0.9
Ethanol-ammonia-water/hexane (95:10:5, v/w/v)	48.1±1.0	46.8±1.8	4.2±0.1	0.9
Isopropyl alcohol-ammonia-water/hexane (95:10:5, v/w/v)	50.0±3.0	48.8±0.5	No phase separation	1.2

[1]Adapted from Wanasundara and Shahidi (1994a).

systems. Table 2 presents the yields of the oil, meal and gums (polar lipids) from different solvent extraction systems used. Hexane alone extracted the most amount of oil from flaxseed (49.2%), however, when alkanol was present in the primary extraction system the yield of oil was reduced slightly (Table 2). Presence of ammonia or water in the alkanol resulted in the removal of a considerable amount of polar matter from flaxseed oil and meal (Table 2) which may be considered as an advantage when using the two-phase solvent extraction process.

According to Oomah *et al.* (1996), extraction of residual oil from commercially available flaxseed meal with hexane alone resulted in a higher oil recovery than methanol/hexane or methanol-ammonia-water/hexane. However, it should be noted that their starting material was meal and not seeds of flax. Analysis of the extracted lipids from the meal showed (Table 3) that hexane preferentially extracted the neutral lipids primarily composed of triacylglycerols (TAG). Neutral lipid fraction of oil extracted with methanol/hexane solvent system contained more monoacylglycerols (MAG), diacylglycerols (DAG) and free fatty acids (FFA) than the lipids extracted by other solvent systems (Table 3). Methanol/hexane solvent system extracted more polar lipids than methanol-water-ammonia/hexane system and this lipid fraction was composed mainly of phosphotidylcholine (PC). Polar lipids of methanol-ammonia-water/hexane extracted lipids contained also phosphotidylethanolamine (PE), lysophosphatidylcholine (LPC) and some unidentified compounds.

Table 3. Lipid composition of flaxseed meal extracted with different solvents[1]

		Lipid fractions/100g oil								
		Neutral lipids[2]					Polar lipids[3]			
Solvent system	Total lipids (g/100g)	TAG	FFA	MAG	DAG	Others	PC	PE	LPC	Others
Hexane	11.49	95.5	0.3	1.1	1.9	1.2	–	–	–	–
Methanol/hexane	8.53	5.7	4.1	7.3	5.0	–	72.9	5.0	–	–
Methanol-ammonia-water/hexane (90:5:5, v/v/v)	2.76	26.8	4.6	2.3	3.1	2.7	27.0	6.7	13.5	13.3

[1]Adapted from Oomah *et al.* (1996).
[2]TAG: triacylglycerol, FFA: free fatty acids, MAG: monoacylglycerols, DAG: diacylglycerols.
[3]PC: phosphotidylcholine, PE: phosphotidylethanolamine, LPC: lysophosphatidylethanolamine.

Fatty acid composition of total lipids extracted from crushed flaxseeds with hexane or methanol-ammonia-water/hexane (95:10:5, v/w/v) solvent systems did not show any considerable difference (Wanasundara and Shahidi, 1994a) and retained the same proportion of polyunsaturated, monounsaturated and saturated fatty acids. However, when oil-extracted flaxseed meal was further extracted with methanol-ammonia-water/hexane (90:5:5, v/w/v) system the amount of ALA in the residual oil so extracted was almost 50% less than that extracted with hexane (Oomah *et al.*, 1996).

High Temperature Processing

Due to high content of polyunsaturated fatty acids, mainly ALA, flaxseed lipids undergo rapid autooxidation. Chen *et al.* (1994) have studied the oxidative stability of flaxseed lipids during baking of whole or ground seeds or when they were used as an ingredient in muffin mix. The oxidative stability of flaxseed ingredients was measured as oxygen consumption and changes in ALA content under different process conditions. When ground flaxseed was heated to 178°C in a sealed tube, headspace oxygen decreased from 21 to 2% within 30 min. However, for whole flaxseeds the decrease of headspace oxygen was only slight, even when heated to 178°C for up to 90 min. Under the same temperature conditions, the oxygen consumption of lipids extracted from an equivalent amount of flaxseed was in between that for whole and ground flaxseeds. The ALA content of lipids decreased from 55.1 to 51.3% in ground flaxseed and to 51.7% in lipid extracts of the meal after heating to 178°C for 90 min; however, this amount remained nearly unchanged in the intact seeds. The muffin mix containing 28.5% (w/w) flaxseed flour, consumed oxygen more rapidly than the control muffin mix with no flaxseed flour at a baking temperature of 178°C for 120 min, but the ALA content remained unchanged in both muffin mixes. This indicated that minimal loss of ALA from flaxseed could be expected under typical baking conditions even if flax flour was used. According to the same study, polymers derived from TAG oxidation and process-induced formation of *trans*-isomers of ALA were not present in heat processed flaxseed or oil.

Cunnane *et al.* (1993) have studied the generation of secondary lipid oxidation products such as malonaldehyde (MA) in muffin mixes containing whole flaxseed, flour or oil (15g/kg) and observed that the MA levels were similar to the muffin mixes of wheat flour with no flax ingredients. Storage of flaxseed flour at -20°C for up to two years did not affect the MA levels of muffins when compared with those containing freshly ground flaxseed. According to Ratnayake *et al.* (1992), feeding of rats with meals containing up to 40% (by weight) ground flaxseed did not increase the oxidative stress in the animals.

Ground flaxseed having large (<20 mesh, >950 μm) or small (>35 mesh, <500 μm) particle size absorbed more oxygen than samples with medium particle size (20–35 mesh, 500–950 μm) when treated at 122°C. Therefore, the extent of size reduction should be optimized in order to minimize oxidation of PUFA. Whole or ground flaxseed or its lipid extracts showed long term storability (up to 280d) with 12h dark/light cycles at room temperature (Chen *et al.,* 1994).

Germination

Germination changes the lipid constituents of flaxseed both qualitatively and quantitatively because they serve as major seed reserve which are utilized in the development of

Table 4. Changes (g/100g) in drymatter, total lipid content and major lipid classes of flaxseed during an 8-day germination period[1]

Germination period (days)	Drymatter	Total lipid	Major lipid classes		
			Neutral lipids	Glycolipids	Phospholipids
0 (ungerminated)	73.5±3.5	32.0±1.0	30.7±1.9	0.73±0.15	0.45±0.09
2	67.6±2.7	28.3±0.9	26.8±2.8	0.76±0.17	0.65±0.11
4	55.4±1.2	22.5±1.1	21.2±2.9	0.94±0.13	0.34±0.10
6	53.8±2.4	14.7±1.5	13.3±2.5	1.27±0.15	0.16±0.10
8	47.5±2.0	8.6±1.1	7.2±3.0	1.17±0.21	0.03±0.10

[1]All values are given on a fresh weight basis; adapted from Wanasundara (1995).

seedlings (Wanasundara, 1995). The drymatter content of flaxseeds (variety Somme) decreased from 73.5 to 47.5 g/100g as the germination proceeded for 8 days (Table 4). At the same time total lipid content of the seeds was reduced from 31.9 to 8.6 g/100g indicating its participation in the germination and seedling development. Table 4 also shows changes in the content of neutral, glyco- and phospholipids of flaxseed as germination progressed for 8 days. The individual components of neutral lipids also showed a considerable change during germination (Table 5). The content of triacylglycerols, which comprised a major portion of neutral lipids, decreased while that of free fatty acids increased over a 6-day germination period. The content of mono- and diacylglycerols also increased up to the 2nd day of germination and then started to decrease. The amount of sterols in the oil was trace. Phospholipid composition of flaxseed lipids also changed quantitatively (Table 5). The fatty acid composition of total lipids of flaxseed showed an increase in the total content of saturated and monounsaturated fatty acids with a concurrent decrease in that of polyunsaturated fatty acids. The neutral and phospholipid fractions followed different patterns of change as illustrated in Figure 1.

Table 5. Changes in the composition of neutral and phospholipids of flaxseed during germination[1]

Lipid class	Germination period (days)				
	0	2	4	6	8
Neutral lipids					
Monoacylglycerols	0.338	0.641	0.292	0.833	0.341
Diacylglycerols	0.639	1.033	0.888	0.691	0.390
Triacylglycerols	29.62	23.77	18.23	10.39	5.720
Free fatty acids	0.421	1.373	1.325	1.35	0.708
Sterols	trace	ND	trace	trace	trace
Phospholipids					
Phosphatidic acid	ND[2]	0.132	0.130	0.065	0.013
Phosphotidylcholine	0.155	0.348	0.050	0.027	0.002
Phosphatidylinositol	0.002	0.005	0.044	0.009	0.001
Phosphatidylserine + Phosphatidylethanolamine	0.220	0.073	0.059	0.023	0.002
Lysophosphatidylcholine	0.067	0.094	0.052	0.021	0.007
Lysophosphatidylethanolamine	ND	ND	0.002	0.015	0.001

[1]Values are in g/100g on a fresh weight basis; adapted from Wanasundara (1995).
[2]Not detected.

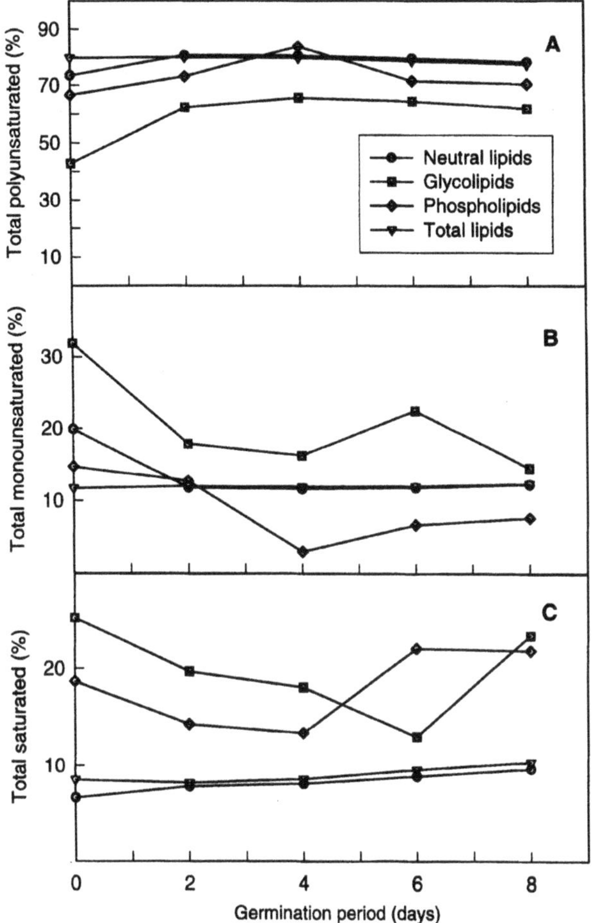

Figure 1. Changes in the total contents of (A) polyunsaturated, (B) monounsaturated and (C) saturated fatty acids of flaxseed during germination.

PROTEINS AND AMINO ACIDS

The total protein content of the Canadian grown flaxseed cultivars is 36–38% on an oil-free, moisture-free, basis with differences attributed to both environmental and genetic factors. Solubility of flaxseed nitrogenous compounds is dependent on pH, composition of the solvent, ionic strength, meal-to-solvent ratio and temperature (Madhusudhan and Singh, 1983; Dev and Quensel, 1986). The broad pattern of nitrogen extractability of flaxseed at varying pH and ionic strength is comparable to other oilseed meals. About 20 to 24% of total nitrogen of flaxseed meal is soluble at the minimum solubility pH (3.5 to 3.8) due to the presence of high levels of non-protein nitrogenous compounds (Bhatty *et al.*, 1973; Sosulski and Bakal, 1969). Approximately 25% of flax protein are water-soluble, 34–47% soluble in 5% (w/v) NaCl, 1–2% soluble in 70% (v/v) ethanol and 3–3.5% soluble in 0.2% (w/v) NaOH (Sosulski and Bakal, 1969). Madhusudhan and Singh (1983) reported that 70–85% of flaxseed proteins are globulins, of which two thirds are high-molecular-weight and the remainder are low-molecular-weight proteins.

Linin is the major protein and conlinin is the low-molecular-weight protein of flaxseed (Vassel and Nesbitt, 1945). The purified linin was shown to be homogeneous with an isoelectric pH of 4.75, and contained 17% nitrogen, 0.6% sulfur and 0.5% carbohydrate (Madhusudhan and Singh, 1985a). The major protein has a sedimentation coefficient ($S_{20,w}$) of 12 in the presence of 1 M NaCl, however, Dev and Sienkiewicz (1987) have reported a sedimentation coefficient of 11.5 for flaxseed globulins. The molecular weight of purified flax globulins varied between 252 and 294 kDa depending on the method of determination. The 12S globulin had a largely nonhelical secondary structure. Sodium dodecylsulphate-polyacrylamide gel electrophoresis (SDS-PAGE) showed that the protein contained at least five nonidentical subunits with molecular weights ranging from 11 to 61 kDa. Urea-polyacrylamide gel electrophoresis at acidic and alkaline pHs showed six bands. From the mobility of protein bands, it was assumed that it contained one acidic, two neutral and three basic subunits (Madhusudhan and Singh, 1985a). The low molecular weight protein (albumin) may constitute up to 42% of the total seed protein (Youle and Huang, 1981). Madhusudhan and Singh (1985d) isolated the major low molecular weight protein (conlinin) using carboxymethylcellulose-sephadex C-50 chromatography; its sedimentation coefficient was 1.6 and it had a molecular weight of 16 kDa.

Solvent Extraction

The content of crude protein of the flaxseed meal extracted with different solvents is shown in Table 6. The two-phase solvent extraction system consisting of an alkanol and a hexane phase, increased the crude protein content of the meal, perhaps due to a concentration effect brought about by the removal of some polar compounds into the alkanol phase (Wanasundara and Shahidi, 1994a). The content of non-protein nitrogen compounds of the meal did not change considerably as a result of solvent extraction.

The pH for minimum extractability of nitrogen from oil-free flaxseed meal was between 3.5 and 3.8 (Smith *et al.,* 1946; Painter and Nesbitt, 1946; Sosulski and Bakal, 1969; Wanasundara and Shahidi, 1994b), however, a broader range of pH (3.0–6.0) was reported for demucilaged, defatted and dehulled flaxseed meal (Madhusudhan and Singh, 1983). Solvent extraction did not change the minimum solubility pH of flaxseed meal (Wanasundara and Shahidi, 1994b).

Table 6. Crude protein and non-protein nitrogen content of flaxseed meal extracted with different solvent systems[1]

Solvent system	Crude protein (%N × 6.25)	Non-protein nitrogen (% of total nitrogen)
Hexane	42.9 ± 0.3	11.0 ± 0.5
Methanol-ammonia-water/ hexane (95:10:5, v/w/v)	48.6 ± 0.3	11.7 ± 0.3
Ethanol-ammonia-water/hexane (95:10:5, v/w/v)	46.3 ± 0.2	11.8 ± 0.2
Isopropanol-ammonia-water/hexane (95:10:5, v/w/v)	46.0 ± 0.4	9.5 ± 0.2

[1]Adapted from Wanasundara and Shahidi (1994a).

Table 7. Amino acid composition of flaxseed meal

Amino acid	Petroleum ether extracted[1]	Hexane extracted[2]	Methanol-ammonia-water/hexane extracted[b]
Histidine	2.9 ± 0.2	2.69 ± 0.24	2.46 ± 0.05
Isoleucine	5.2 ± 0.2	4.78 ± 0.54	4.54 ± 0.15
Leucine	6.8 ± 0.3	6.70 ± 0.62	6.39 ± 0.15
Lysine	4.1 ± 0.1	4.38 ± 0.37	4.14 ± 0.08
Methionine	2.2 ± 0.1	1.45 ± 0.09	1.41 ± 0.05
Cysteine	3.8 ± 0.3	3.29 ± 0.56	3.39 ± 0.21
Phenylalanine	5.3 ± 0.2	5.13 ± 0.05	4.91 ± 0.08
Tyrosine	2.9 ± 0.1	2.21 ± 0.19	2.12 ± 0.03
Threonine	4.9 ± 0.2	3.40 ± 0.30	3.33 ± 0.04
Tryptophan	1.8 ± 0.1	0.46 ± 0.10	0.46 ± 0.05
Valine	5.6 ± 0.2	5.75 ± 0.07	5.64 ± 0.17
Alanine	5.4 ± 0.2	4.81 ± 0.50	4.64 ± 0.10
Aspartic acid	12.5 ± 0.5	9.18 ± 0.60	9.16 ± 0.60
Arginine	11.8 ± 0.6	11.50 ± 0.33	11.20 ± 0.10
Glycine	7.0 ± 0.3	6.44 ± 0.35	6.26 ± 0.25
Glutamic acid	26.3 ± 1.0	16.70 ± 0.43	16.36 ± 0.33
Proline	5.2 ± 0.4	3.64 ± 0.24	3.65 ± 0.10
Serine	5.8 ± 0.2	4.94 ± 0.03	4.99 ± 0.24

[1]Adapted from Bhatty and Cherdkiatgumchai (1990).
[2]Adapted from Wanasundara and Shahidi (1994b).

The amino acid composition of flaxseed meal has been reported by Bhatty and Cherdkiatgumchai (1990) and Wanasundara and Shahidi (1994b) and it shows that it is deficient in methionine, lysine and tryptophan (Table 7). The amino acid composition of hexane and methanol-ammonia-water/hexane extracted seed meals was almost similar.

Protein Product Preparation

Isolation of protein from defatted flaxseed meal using alkali extraction, acid precipitation, and separation and drying of the curd has been reported (Smith *et al.*, 1946). However, procedures suitable for isolating soybean protein are not effective for flaxseed, mainly due to the interference of flaxseed hull polysaccharides with the isolation and settling of proteins (Oomah and Mazza, 1993; Wanasundara and Shahidi, 1997a). Therefore, removal of seed mucilage by chemical or enzymatic treatment prior to protein extraction is required (Wanasundara and Shahidi, 1997a). Soaking of the whole seeds in a bicarbonate solution or treating with commercial carbohydrases reduced the content of seed coat polysaccharides and enhanced nitrogen solubility and protein recovery from aqueous meal extracts (Wanasundara and Shahidi, 1997a). The meal obtained from mucilage-reduced flaxseed was used for preparation of protein isolates via sodium hexametaphosphate (SHMP) complexation, and under optimum pH, SHMP concentration and meal-to-solvent ratio, 78% of the total meal nitrogen was present in the resultant protein isolate (Wanasundara and Shahidi, 1996). The amino acid composition of the protein isolate was similar to the amino acid composition of total seed proteins. The percentage ratios of the essential to total amino acids for the seed and protein were well above the 36% value that is reported for ideal proteins as recommended by FAO/WHO (1973). According to the electrophoretic analysis (SDS-PAGE), the protein isolate contained major seed protein and also low molecular weight proteins (Wanasundara, 1995).

Flaxseed protein products containing different levels of polysaccharides were prepared by alkali extraction followed by isoelectric precipitation and subsequently tested in food systems (Dev and Quensel, 1986; 1988 and 1989). The protein product containing high levels of polysaccharides exhibited better water absorption, emulsifying properties and foaming capacity but had a lower nitrogen solubility, oil absorption and foam stability (Dev and Quensel, 1988). A good emulsion stabilizing effect was observed in canned fish sauce with a high polysaccharide content flax protein concentrate. The product was a creamy and smooth fish sauce devoid of any undesirable flavor and exhibited a marked reduction of the red color of the sauce. These protein products when used as meat extenders resulted in a reduced fat, texture and meaty flavor loss during cooking. In ice cream, both low- and high-mucilage protein products gave a stabilizing effect similar to gelatin and increased product viscosity, specific gravity and overrun but a reduced melt-down time (Dev and Quensel, 1989).

Chemical Modification

Flaxseed protein isolate obtained from the SHMP complexation process was acetylated and succinylated in order to improve its functional properties (Wanasundara and Shahidi, 1997b). Similar to other oilseed proteins, a higher degree of modification of free amino groups was observed due to the acetylation as compared to the succinylation process. The color of the acylated proteins became lighter as the degree of acylation was increased. Emulsification property of protein isolates was improved due to acylation, especially with succinylation. Foaming properties of flax protein isolates were not improved by acylation, however, solubility was markedly improved. Low degrees of acetylation improved fat binding capacity of flax protein isolates, but succinylation did not.

CYANOGENIC GLYCOSIDES

Flaxseed contains cyanogenic glycosides which are nitrogenous secondary plant metabolites capable of liberating HCN upon hydrolysis. Linamarin (2-[(6-O-β-D-glucopyranosyl)-oxy]-2-methylpropanenitrile) and lotaustralin ([(2R)-[(6-O-β-D-glucopyranosyl)-oxy]-2-methylbutanenitrile]) are the monosaccharide cyanogenic glycosides which may be present in flaxseed (Butler, 1965; Conn, 1981). Disaccharide cyanogenic glycosides namely linustatin (2-[(6-O-β-D-glucopyranosyl-β-D-glucopyranosyl)-oxy]-2-methylpropanenitrile) and neolinustatin ([(2R)-[(6-O-β-D-glucopyranosyl-β-D-glucopyranosyl)-oxy]-2-methylbutanenitrile]) have also been isolated from flaxseed (Smith *et al.*, 1980). Recent studies by Oomah *et al.* (1992) and Wanasundara *et al.* (1993) have shown the presence of linustatin and neolinustatin, and possibly low amounts of linamarin, in the Canadian grown flaxseed. The content of these three glycosides depended on the cultivar, location and year of production; with cultivar having the most important effect. The predominant cyanogenic glycoside of the Canadian cultivars was linustatin (213 to 352 mg/100 g seed) which accounted for 54 to 76% of the total amount. The content of neolinustatin ranged from 91 to 203 mg/100 g seed and linamarin was present in less than 32 mg/100 g seeds (Oomah *et al.*, 1992).

The aliphatic cyanogenic glycosides of flaxseed are referred to as bound cyanide. The molecular HCN is regarded as free non-glycosidic cyanide. Under normal physiological conditions, tissues of cyanophoric plants contain little or no detectable HCN. When plant tissues are disrupted, HCN may be released rapidly from cyanogenic glycosides upon enzymatic hydrolysis. The catabolism of cyanogenic glycosides is initiated by cleavage of their carbohydrate moiety by one or more β-glucosidases, thus yielding the corresponding α-hydroxynitrile. This intermediate may decompose either spontaneously or enzymatically in the presence of α-hydroxynitrile lyase to yield HCN and an aldehyde or a ketone (Poulton, 1989).

Crude flaxseed extracts contain two distinct β-glucosidases which cooperate in stepwise removal of glucose residues. Linustatinase catalyses the hydrolysis of β-(*bis*-1,6)- and β-(*bis*-1,3)-glucosides but is inactive toward linamarin and cyanogenic disaccharides having terminal xylose or arabinose moieties. Thus, linustatin and neolinustatin are hydrolysed to linamarin and lotaustralin, respectively, by linustatinase. These monosaccharides are further degraded to their corresponding α-hydroxynitriles by linamarase (Fan and Conn, 1985). Flaxseed linamarase shows only a moderate degree of substrate specificity with respect to the aglycone moiety. It catalyses the hydrolysis of both aliphatic and aromatic cyanogenic monosaccharides but is virtually inactive towards cyanogenic disaccharides (Butler *et al.*, 1965). Therefore, presence of cyanogenic glycosides is of concern and may limit the use of whole seeds and meals in large quantities in food and livestock feed formulations. Thus, all processing methods employed for flaxseed are intended to reduce the content of cyanogenic glycosides of flaxseed and meal.

Solvent Extraction

Hexane extraction of flaxseed may not reduce the cyanogenic glycoside content of the meal, however, the two-phase solvent system composed of methanol-ammonia-water (95:10:5, v/w/v)/hexane simultaneously extracted oil and cyanogenic glycosides of flaxseed (Wanasundara *et al.*, 1993), together with other polar components. The two-phase solvent system was able to reduce the content of both linustatin and neolinustatin by >50% in a single extraction step. Addition of higher amounts of water (10 or 15%, v/v) in the methanol phase enhanced the removal of cyanogenic glycosides (Figure 2A). The content of residual glycosides was reduced further as the contact time (quiescent period) during extraction or the amount of extraction solvent (solvent-to-seed ratio) was increased (Figure 2B). Increasing the volume and the contact time of the extraction solvent enhanced the removal of the original linustatin and neolinustatin present in the meal by over 80% (Figure 2B). The extraction of the meal for a second or third time with methanol-ammonia-water (95:10:5, v/w/v) and then hexane enhanced the removal of cyanogenic glycosides from 56 to 80% and over 90%, respectively (Figure 2C). Therefore, solvent-to-seed ratio and contact period of the meal with solvent were the main factors affecting the degree of detoxification of the meal. Thus, a multistage extraction process involving a two-phase solvent system (methanol-ammonia-water/hexane, 95:10:5, v/w/v) may reduce the content of cyanogenic glycosides of flaxseed meal to below 10% of their original amount (Wanasundara *et al.*, 1993). Varga and Diosady (1994) have also confirmed that two-phase solvent extraction effectively reduced total cyanogenic glycosides of flaxseed, however, these authors have mistakenly considered linamarin as being the main cynogenic compound of flaxseed.

High Temperature Treatments

High temperature treatments such as boiling in water, dry and wet autoclaving and acid treatment followed by autoclaving have been used to reduce the content of cyanogenic glycosides of flaxseed (Madhusudahan and Singh, 1985b,c). Chadha *et al.* (1995)

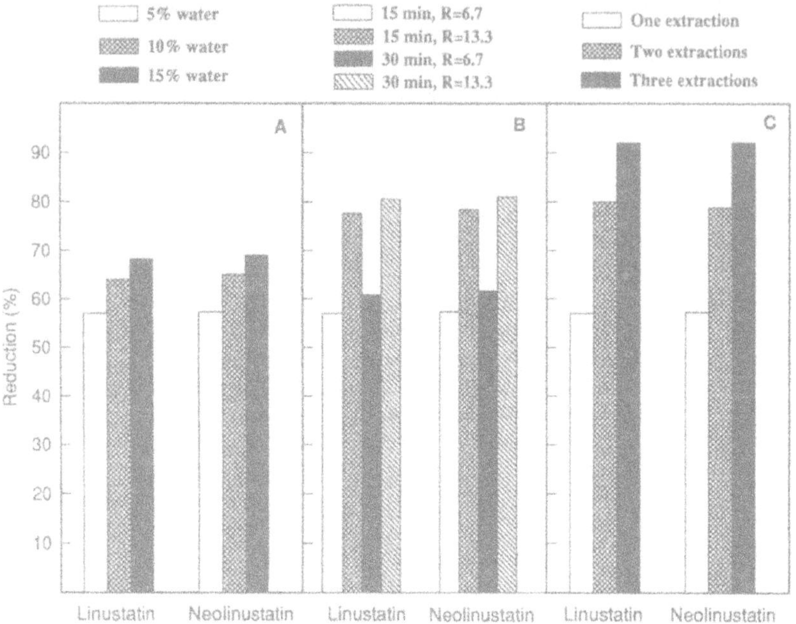

Figure 2. Removal of cyanogenic glycosides of flaxseed due to two-phase solvent extraction with methanol-ammonia-water/hexane (95:10:5, v/w/v); (A) change in water content, (B) change in contact time and meal-to-solvent ratio and (C) change in the number of extraction cycles.

have studied the release of HCN due to enzymic breakdown of flaxseed cyanogenic glycosides in a distilled water homogenate after subjecting the intact seeds to different temperature treatments. Heating of the seeds to 177°C for 1h yielded products with 80% lower cyanogens compared to the original seeds as determined by autolysis over a 4.5 h period. However, under the same conditions, corresponding reductions in cyanogens of ground flaxseed was only 20%. According to Chadha *et al.* (1995), this difference in cyanide production is whole and ground flaxseeds may be due to the existing differences in moisture content resulting from heat treatment. The water loss from whole flaxseed is much slower than that from ground flaxseed during heating and this may hasten the reaction between the glycoside and glycosidase enzyme, thus facilitating the release of more cyanide. Boiling of flaxseed for 5 min in water produced little cyanide due to autohydrolysis. This may be attributed to the efficient inactivation of glycosidase enzyme by moist heat, thus producing a lesser amount of cyanide. A similar observation was noted when an untreated flaxseed homogenate was boiled. A commercial cereal mix containing 5% (w/w) whole flaxseed did not produce any detectable amount of cyanide upon boiling as instructed on the package for cereal preparation. A bread sample that contained flaxseed as an ingredient also did not produce any detectable amount of cyanide, perhaps due to the inactivation of enzymes responsible for degradation of cyanogeic glycosides during the baking process. According to Cunnane *et al.* (1993), the content of linustatin and neolinustatin in intact flaxseed was decreased to undetectable levels when flaxseed flour was included in muffin mixes and then baked.

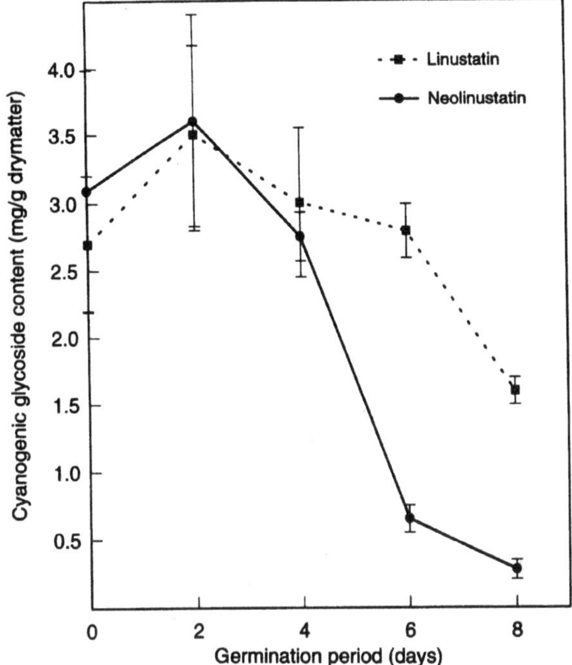

Figure 3. Changes of linustatin and neolinustatin content of flaxseed during an 8-day germination period.

Germination

Germination process is known to utilize cyanogenic glycosides of seeds as a nitrogen source. Selmar *et al.* (1988) have described that *Hevea brasiliensis* seed which contains similar types of cyanogenic mono- and diglucosides of flaxseed, exhibited a decrease in its content of monoglucoside during the onset of germination. The diglucoside content started to decline once the monoglucosides were depleted. Germination of flaxseeds reduced the content of linustatin and neolinustatin after a 2-days period (Figure 3; Wanasundara, 1995).

Other Processing Methods

Mazza and Oomah (1995) reported that dry fractionation of flaxseed meals, depending on their particle size, influences the profile and content of cyanogenic glycosides. The lowest concentration of cyanogenic glycoside was observed for particles smaller than 850 μm but larger than 450 μm. Mechanically separated flaxseed hulls (mostly particles >850 μm) contained the highest amount of neolinustatin, regardless of the seed color, whereas the endosperm contained the highest level of linustatin.

PHENOLIC COMPOUNDS

Major phenolics of flaxseed include phenolic acids, tannins and lignans. Flaxseed, as compared to other oilseeds, contains low levels of phenolic acids (Kozlowska *et al.*,

1983; Dabrowski and Sosulski, 1984; Shahidi and Naczk, 1989; Wanasundara and Shahidi, 1994a). Kozlowska *et al.* (1983) reported the presence of free phenolic acids (22.4 mg/100 g) in the methanolic extracts of defatted flour, but Dabrowski and Sosulski (1984) did not detect any free phenolic acids in the meal, perhaps due to the use of tetrahydrofuran for extraction. Phenolic acids released from soluble esters constituted the largest fraction (55–89%) of the total content of phenolic acids. Furthermore, Dabrowski and Sosulski (1984) reported that ferulic acid was the predominant phenolic acid in the soluble esters and insoluble residues of flaxseed.

Amarowicz *et al.* (1994) studied the phenolic compounds of flaxseed meal extracted into 80% ethanol (v/v) by chromatographic techniques. Both hydrophilic and hydrophobic phenolic compounds were present. The separated hydrophilic fractions had a UV maximum absorption between 270 and 290 nm which was different from that observed for phenolic acids. Meanwhile, the tannin content of flaxseed meal was very low (125 to 137 mg/100 g of defatted meal) when compared to that of high glucosinolate rapeseed and canola (Shahidi and Naczk 1988; Shahidi *et al.*, 1988; Wanasundara and Shahidi, 1994a).

Flaxseed is a source of lignan precursor secoisolariciresinol in the form of its diglucoside (Bekke and Klosterman, 1956; Axelson *et al.*, 1982; Amarowicz *et al.*, 1994). Secoisolariciresinol diglucoside (SD) is metabolized by intestinal bacteria to the mammalian lignan enterodiol, which is then oxidized to enterolactone. Enterolactone may also arise from metabolism of the plant lignan, matairesinol (Setchell *et al.*, 1981). Harris and Haggerty (1993) were able to quantify SD at levels of 0.7–0.9 μg/g from a methanolic extract of flaxseed meal using an HPLC methodology. Thompson *et al.* (1991) screened various plant foods for their production of mammalian lignans by simulated *in vitro* colonic fermentation. Flaxseed flour and its defatted meal were able to produce the highest amount of mammalian lignans ranging from 52.7 to 67.5 mg/100 g sample. However, changes of flaxseed lignans due to processing have not been reported.

Solvent Extraction

Wanasundara and Shahidi (1994a) reported that the total content of phenolic acids of flaxseed meal was about 220 mg/100g, on a dry weight basis. Methanol-ammonia-water/hexane extraction lowered the total phenolic acid content of the meals by approximately 48%. The content of condensed tannins of flaxseed meal was about 136 mg/100g, determined as (+)catechin equivalents, and this must have originated from the seed coat. The two-phase solvent extraction system (methanol-ammonia-water/hexane) reduced the content of tannins in the meal by 74% (Wanasundara and Shahidi, 1994a).

Furthermore, extraction of flaxseed with methanol-ammonia/hexane reduced both soluble phenolic acid esters and insoluble (bound) phenolic acids by 20 and 29%, respectively, but free phenolic acids remained unchanged (Varga and Diosady, 1994). The total content of phenolic acids was 442 and 355 mg/100g for hexane extracted and methanol-ammonia treated flaxseed meals, respectively. Esterified or soluble phenolic acid esters constituted 50–54% of the total amount and insoluble bound phenolics comprised 26–29% of the total amount.

SUGARS AND DIETARY FIBER

The sugar composition of flaxseed meal has been reported by Bhatty and Cherdkiatgumchai (1990). Glucose was the major sugar followed by xylose, galactose, arabinose, rhamnose and fucose. Total sugar content of the seed meal was 28.5%. According to Wanasundara and Shahidi (1994a), total soluble sugars of meal was 7.8% as sucrose equivalents and their content was reduced upon treatment with methanol-ammonia-water/hexane. In another study, it was observed that during germination, the content of soluble sugars of flaxseed tripled. The soluble sugars were mainly composed of sucrose, raffinose, glucose and fructose. Sucrose was the most abundant sugar and its content during germination was reduced to one third of the original content. Meanwhile, the content of glucose and fructose increased by a factor of 40 and 98 during germination, but all the seed raffinose was depleted after four days of germination (Wanasundara, 1995).

Flaxseed meal is high in crude, acid detergent (cellulose and lignin), neutral detergent and total fiber (cellulose, lignin and hemicellulose). The total dietary fiber content of flaxseed meal was 39–45% (Bhatty and Cherdkiatgumchai, 1990). This is somewhat more than the amount present in barley and oat bran (Bhatty, 1993). The yellow-seeded flaxmeal contained a lesser amount of crude (8.7 versus 11.7%) and neutral detergent fiber (24 versus 29%) that from its brown-seeded counterpart (Bell and Keith, 1993). Most of the total dietary fiber in flaxseed meal is in the insoluble form and the ratio of soluble to total dietary fiber is 1:7 as compared to 1:13 for that of ground full-fat seeds (Ranhotra *et al.*, 1993).

Flaxseed polysaccharides (gum or mucilage; 5–8% of the seed weight) may be isolated via aqueous extraction of the whole seed or meal (BeMiller, 1973, Mazza and Biliaderis, 1989; Oomah *et al.*, 1995). The soluble fiber fraction of flaxseed is mainly composed of polysaccharides. Flaxseed gum is a mixture of acidic and neutral polysaccharides (Erskine and Jones, 1957; Hunt and Jones, 1962). The neutral fraction is composed of a β- (1→4) linked D-xylose backbone to which arabinose and galactose side chains are attached at the 2 and/or 3 position(s) (Muralakrishna *et al.*, 1987; Cui *et al.*, 1994). The acidic fraction contains mainly (1→2) linked D-galactopyranosyluronic acid backbone with fucose and galactose side chains (Muralakrishna *et al.*, 1987). Changes in dietary fiber or mucilage of flaxseeds due to solvent extraction, heat treatment or germination have not been reported.

PHYTIC ACID

Phytic acid or *myo*-inositol hexaphosphate (1,2,3,4,5,6-hexakis-dihydrogen phosphate) is the major storage form of phosphorus (60–90% of total phosphorus) in seeds and is produced as a secondary metabolism product of carbohydrates (Loewus and Loewus, 1980). Phytic acid exists typically as salts of calcium, magnesium or potassium (Mills and Chong, 1977; Yiu *et al.*, 1983). In cottonseed, peanut (Saio *et al.*, 1977), rapeseed (Yiu *et al.*, 1983) and soybean (Maga, 1982), phytic acid is found in a complexed form in globoids inside the protein bodies of cotyledon cells. Therefore, its removal by traditional processing is not possible due to its strong association with proteins and may result in its concentration with proteins during preparation of protein concentrates and isolates (Thompson, 1989).

The phytic acid molecule is highly reactive towards positively charged groups such as metal ions and proteins (Erdman, 1979; Thompson, 1990) as it possesses 12 replaceable protons and is negatively charged at pH conditions generally encountered in food and feedstuff. In general, one or two phosphate groups of phytic acid may bind with cations (Gosselin and Coughlan, 1953). The mixed salt of phytic acid is formed when several cations complex within the same phytic acid molecule. The binding of phytic acid with minerals is pH dependent, and complexes with varying solubilities are formed (Cheryan, 1980). Most polyvalent metal ions, especially calcium (Reinhold *et al.,* 1973), magnesium (Nolan *et al.*, 1987), zinc (Erdman, 1979; Nosworthy and Cladwell, 1988; Champagne and Phillipy, 1989) and iron (Davis and Nightingale, 1975) bind to phytic acid and form insoluble complexes which makes them unavailable for metabolism.

Solvent Extraction

The content of phytic acid in flaxseed meals is not reduced as a result of solvent extraction. Due to the association of phytic acid with proteins, in fact, its content tends to increase due to the removal of non-protein matter of seeds (Wanasundara and Shahidi, 1994a).

Germination

Germination is effective in lowering the content of phytic acid in flaxseed. After 8 days of germination, flax seedlings contained 27% lower phytic acid than that present in ungerminated seeds (Wanasundara, 1995). Degradation of phytic acid for the release of phosphorus for metabolic requirements may account for this observation.

MINERALS AND VITAMINS

Bhatty and Cherdkiatgumchai (1990) have reported the mineral and vitamin composition of Canadian grown flaxseeds (Table 8). Flaxseed (100g) may provide 100% of the

Table 8. Mineral and vitamin contents of flaxseed meal[1]

Mineral/vitamin (unit)	Content
Minerals	
Sodium (mg/g)	0.6 ± 0.2
Potassium (mg/g)	12.1 ± 2.4
Calcium (mg/g)	4.5 ± 0.6
Magnesium (mg/g)	6.1 ± 0.5
Phosphorus (mg/g)	9.9 ± 2.6
Sulfur (mg/g)	4.0 ± 0.1
Zinc (μg/g)	123.2 ± 30.7
Iron (μg/g)	207.6 ± 29.0
Copper (μg/g)	20.0 ± 2.5
Manganese (μg/g)	58.5 ± 9.7
Vitamins	
A (IU/100g)	18.8
E (IU/100g)	0.6
B_1 (mg/100g)	0.5
B_2 (mg/100g)	0.2
B_3 (mg/100g)	9.1
B_6 (mg/100g)	0.8
B_{12} (mg/100g)	0.5

[1]Adapted from Bhatty and Cherdkiatgumchai (1990).

recommended daily allowance (RDA) of potassium and manganese, 87% of magnesium, 57–65% of iron and phosphorus and 13–35% of zinc, calcium and copper (Stitt, 1986, Bhatty and Cherdkiatgumchai, 1990). Flaxseed is deficient in zinc, copper and calcium. Mineral composition of flaxseed does not change considerably as a result of 8-day germination (Wanasundara, 1995).

REFERENCES

Amarowicz, R.; Wanasundara, P.K.J.P.D.; Shahidi, F. Chromatographic separation of flaxseed phenolics. *Die Nahrung* **1994**, *38*, 20–26.

Axelson, M.; Sjovall, J.; Gustafsson, B.; Setchell, K.D.R. Origin of lignans in mammals and identification of precursor from plants. *Nature* **1982**, *298*, 659–660.

Bemiller, J.N. Quince seed, psyllium seed, flaxseed and okra gums. *Industrial Gums*. 2nd ed.; Whistler, R.L. and Bemiller, J.N., Eds. Academic Press: New York, NY, 1973; pp 331- 337.

Bell, J.M.; Keith, M.O. Nutritional evaluation of linseed meals from flax with yellow or brown hulls, using mice and pigs. *Anim. Feed Sci. Technol.* **1993**, *43*, 1–18.

Bhatty, R.S. Physicochemical properties of roller-milled barley bran and flour. *Cereal Chem.* **1993**, *70*, 397–402.

Bhatty, R.S.; Cherdkiatgumchai, P. Compositional analysis of laboratory-prepared and commercial samples of linseed meal and of hulls isolated from flax. *J. Am. Oil Chem. Soc.* **1990**, *67*, 79–84.

Bhatty, R.S.; Sosulski, F.W.; Wu, K.K. Protein and non-protein nitrogen contents of some oilseeds and peas. *Can. J. Plant Sci.* **1973**, *53*, 651–657.

Bakke, J.E. and Klosterman, H.J. A new diglucoside from flaxseed. *Proc. North Dakota Acad. Sci.* **1956**, *10*, 18–22.

Butler, G.W. The distribution of cyanogenic glucosides, linamarin and lotustralin in higher plants. *Phytochem.* **1965**, *4*, 127–131.

Butler, G.W.; Bailey, T.W.; Kennedy, L.D. Studies on the glucosidase "linamarase". *Phytochem.* **1965**, *4*, 369–381.

Carter, J.F. Potential of flaxseed and flaxseed oil in baked goods and other products in human nutrition. *Cereal Foods World* **1993**, *38*, 753–759.

Chadha, R.K.; Lawrence, J.F.; Ratnayake, W.M.N. Ion chromatographic determination of cyanide released from flax under autohydrolysis conditions. *Food Addit. Contam.* **1995**, *12*, 527–533.

Champagne, E.T.; Phillippy, B.Q. Effects of pH on Ca, Zn, and phytate solubilities and complexes following in vitro digestion of soy protein isolates. *J. Food Sci.* **1989**, *54*, 587–590.

Chen, Z-Y.; Ratnayake, W.M.N.; Cunnane, S.C. Oxidative stability of flaxseed lipids during baking. *J. Am. Oil Chem. Soc.* **1994**, *71*, 629–632.

Cheryan, M. Phytic acid interactions in food systems. *CRC Crit. Rev. Food Sci. Nutr.* **1980**, *13*, 297–334.

Conn, E.E. Cyanogenic glycosides. *The Biochemistry of Plants*, Vol. 7, *Secondary Plant Products*; Conn, E.E., Ed.; Academic Press, New York, NY, 1981; pp 479–500.

Cui, W.; Mazza, G.; Biliaderis, L.C. Chemical structure, molecular size distribution and rheological properties of flaxseed gum. *J. Agric. Food Chem.* **1994**, *42*, 1891–1895.

Cunnane, S.C.; Ganguli, S.; Menard, C.; Liede, A.C.; Hamadeh, M.J.; Chen, Z-Y.; Wolever, T.M.S.; Jenkins, D.A. High-α-linolenic acid flaxseed (*Linum ussitatissimum* L.); some nutritional properties in humans. *Brit. J. Nutr.* **1993**, *69*, 443–453.

Dabrowski, K.J.; Sosulski, F.W. Comparison of free and hydrolysable phenolic acids in defatted flours of ten oilseeds. *J. Agric. Food Chem.* **1984**, *32*, 128–130.

Davis, N.T.; Nightingale, R. The Effects of phytate on intestinal absorption and secretion of Zn, and whole body retention of Zn, Cu, Fe and Mn in rats. *Brit. J. Nutr.* **1975**, *34*, 8A.

Dev, D.K.; Quensel, E. Functional properties and microstructural characteristics of linseed flour and protein isolate. *Lebensm. Wiss. u. Technol.* **1986**, *19*, 331–337.

Dev, D.K.; Quensel, E. Preparation and functional properties of linseed protein products containing differing levels of mucilage. *J. Food Sci.* **1988**, *53*, 1834–1837,1857.

Dev, D.K.; Quensel, E. Functional properties of linseed protein products containing different levels of mucilage in selected food systems. *J. Food Sci.* **1989**, *54*, 183–186.

Dev, D.K.; Sienkienicz, T. Isolation and subunit composition of 11S globulin of linseed (*Linum ussitatissimum* L.). *Die Nahrung* **1987**, *31*, 767–769.

Dick, T.A.; Yang, X. Flaxseed in arctic char and rainbow trout nutrition. *Flaxseed in Human Nutrition*, Cunnane, S.C; Thompson, L.U., Eds.; American Oil Chemists' Society, Champaign, IL, 1995, pp 295–314.

Dorrel, D.G. Distribution of fatty acid within the seed of flax. *Can. J. Plant Sci.* **1970**, *50*, 71–75.

Erdman, J.W. Oilseed phytates, nutritional implications. *J. Am. Oil Chem Soc.* **1979**, *56*, 736–741.

Erskine, A.J.; Jones, J.K.N. The structure of linseed mucilage. Part I. *Can. J. Chem.* **1957**, *35*, 1174–1182.

Fan, T.W.M.; Conn, E.E. Isolation and characterization of two cyanogenic β-glucosidases from flax seeds. *Arch. Biochem. Biophys.* **1985**, *243*, 361–373.

FAO/WHO. Energy and protein requirements. Report of a Joint FAO/WHO Adhoc Expert Committee. World Health Organization Technical Report Series 522. WHO, Geneva, 1973.

Gosselin, R.E.; Coughlan, E.R. The stability of complexes between calcium and orthophosphate, polymeric phosphate and phytate. *Arch. Biochem. Biophys.* **1953**, *45*, 301–305.

Green, A.G.; Marshell, D.R. Variation for oil quantity and quality in linseed. *Aust. J. Agric. Res.* **1981**, *32*, 599–607.

Green, A.G.; Dribinenki, J.C.P. Linola - a new premium polyunsaturated oil. *Lipid Technol.* **1994**, *6*, 29–33.

Harris, R.K.; Haggerty, W.J. Assay for potentially anti-carcinogenic phytochemicals in flaxseed. *Cereal Foods World* **1993**, *38*, 147–151.

Hunt, K.; Jones, J.K.N. The structure of linseed mucilage. Part II. *Can. J. Chem.* **1962**, *40*, 1266–1279.

Jiang, Z.; Ahn, D.U.; Sim, J.S. Effects of feeding flax and two types of sunflower seeds on fatty acid composition of yolk lipid classes. *Poult. Sci.* **1991**, *70*, 2467–2475.

Kennelly, J.J.; Khorasani, R.G. Influence of flaxseed feeding on fatty acid composition of cow's milk. *Proc. Flax Inst. US.* **1992**, *54*, 99–105.

Kolodziejczyk, P.P.; Fedec P. Processing flaxseed for human consumption. *Flaxseed and Human Nutrition*, Cunnane, S.C.; Thompson, L.U., Eds.; American Oil Chemists' Society, Champaign, IL, 1995, pp 261–280.

Kozlowska, H.; Zadernowski, R.; Sosulski, F.W. Phenolic acids in oilseed flours. *Die Nahrung.* **1983**, *27*, 449–453.

Loewus, F.A.; Loewus, M.N. Myo-inositol biosynthesis and metabolism. *Biochemistry of Plants 3. Carbohydrates: structure and functions*, Preiss, J., Ed.; Academic press, London, 1980, pp. 43–100.

Madhusudhan, K.T.; Singh, N. Studies on linseed protein. *J. Agric. Food Chem.* **1983**, *31*, 959–963.

Madhusudhan, K.T.; Singh, N. Isolation and characterization of major fraction (12S) of linseed protein. *J. Agric. Food Chem.* **1985a**, *33*, 673–677.

Madhusudhan, K.T. and Singh, N. Effect of detoxification treatment on the physicochemical properties of linseed. *J. Agric. Food Chem.* **1985b**, *33*, 1219–1222.

Madhusudhan, K.T. and Singh, N. Effect of heat treatment on the functional properties of linseed meal. *J. Agric. Food Chem.* **1985c**, *33*, 1222–1226.

Madhusudhan, K.T. and Singh, N. Isolation and characterization of small molecular weight protein of linseed meal. *Phytochem.* **1985d**, *24*, 2507–2509.

Maga, J.A. Phytate, its chemistry, occurrence, food interactions, nutritional significance and methods of analysis-a review. *J. Agric. Food Chem.* **1982**, *30*, 1–9.

Mazza, G.; Biliaderis, C.G. Functional properties of flaxseed mucilage. *J. Food Sci.* **1989**, *98*, 237–238.

Mazza, G.; Oomah, D.B. Flaxseed dietary fibre and cyanogens. *Flaxseed in Human Nutrition*, Cunnane, S.C.; Thompson, L.U., Eds.; American Oil Chemists' Society, Champaign, IL, 1995, pp 56–81.

Mills, J.T.; Chong, J. Ultrastructure and mineral distribution in heat damaged rapeseed. *Can. J. Plant. Sci.* **1977**, *57*, 21–34.

Muralikrishna, G.; Salimanth, P.V.; Tharanathan, R.N. Structural features of an arabinoxylan and rhamnogalacturonan derivative from linseed mucilage. *Carbohyd. Res.* **1987**, *161*, 265–271.

Nolan, K.B.; Duffin, P.A.; McWeeny, D.J. Effects of phytate on mineral bioavailability: *In vitro* studies on Mg, Ca, Fe, Cu, and Zn (also Cd) solubilities in the presence of phytate. *J. Food Sci.* **1987**, *40*, 79–85.

Nosworthy, N.; Cladwell, R.A. The interaction of Zn and phytic acid with soybean glycinin. *J. Sci. Food Agric.* **1988**, *44*, 143–150.

Oomah, B.D.; Mazza, G. Flaxseed proteins-a review. *Food Chem.* **1993**, *48*, 109–114.

Oomah, B.D.; Mazza, G.; Kenaschuk, E. Cyanogenic compounds in flaxseeds. *J. Agric. Food. Chem.* **1992**, *40*, 1346–1348.

Oomah, B.D.; Kenaschuk E.O.; Cui, W.; Mazza, G. Variation in the composition of water soluble polysaccharides in flax. *J. Agric. Food Chem.* **1995**, *43*, 1484–1488.

Oomah, B.D.; Mazza, G.; Przybylski, R. Comparison of flaxseed meal lipids extracted with different solvents. *Food Sci. Technol.* **1996**, *29*, 654–658.

Painter, A.; Nesbitt, G. Nitrogenous constituents of flaxseed. *Indust. Eng. Chem.* **1946**, *38*, 95–98.

Poulton, J.E. Toxic compounds in plant foodstuffs: cyanogens. *Food Proteins*, Kinsella, J.E.; Soucie, W.G., Ed.; American Oil Chemists' Society, Champaign. IL. 1989, pp 381–401.

Ranhotra, G.S.; Gelroth, J.A.; Glaser, B.K.; Potnis, P.S. Lipidemic response in rats fed flaxseed, oil and meal. *Cereal Chem.* **1993**; *70*, 364–366.

Ratnayake, W.M.N.; Behrens, W.A.; Fischer, W.F.; L'Abbé, M.R.; Mongeau, R.; Beare-Rogers, J.L. Chemical and nutritional studies of flaxseed (variety- Linott) in rats. *J. Nutr. Biochem.* **1992**, *3*, 232–240.

Reinhold, J.G.; Nasr, K.; Lahimqarzadeh, A.; Hedayati, H. Effects of purified phytate and phytate rich bread upon metabolism of Zn, Ca, P and N in man. *Lancet* **1973**, *1*, 283–288.

Romans, J.R. Flaxseed and the composition and quality of pork. *Flax in Human Nutrition*, Cunnane, S.C.; Thompson, L.U., Eds.; American Oil Chemists' Society, Champaign, IL, 1995, pp 348–362,.

Saio, K.; Gallant, D.; Petit, L. Electron microscope research in sunflower protein bodies. *Cereal Chem.* **1977**, *54*, 1171–1181.

Selmar, D.; Lieberei, R.; Bichl, B. Metabolization and utilization of cyanogenic glycosides. The linustatin pathway. *Plant Physiol.* **1988**, *86*, 711–716.

Setchell, K.D.R.; Lawson, A.M.; Boriello, S.P.; Harkness, R.; Gordon, H.; Morgan, D.M.L.; Kirk, D.N.; Anderson, L.C.; Adlercreutz, H; Axelson, M. Lignan formation in man-microbial involvement and possible roles in relation to cancer. *Lancet*, **1981**, *2*, 4–7.

Shahidi, F.; Naczk, M. Effect of processing on the phenolic constituents of canola. *Proc. of XIVth Inernational Conference of Groupe Polyphenols*, Vol. 14, St. Catharines, ON, 1988, pp 89–92.

Shahidi, F.; Naczk, M. 1989. Effect of processing on the content of condensed tannins in rapeseed meals. J. Food Sci. 54:1082–1083.

Shahidi, F.; Naczk, M.; Rubin, L.J.; Diosady, L.L. A novel processing approach for rapeseed and mustard seed - Removal of undesirable constituents by methanol-ammonia. *J. Food Protect.* **1988**, *51*, 743–749.

Smith, A.K.; Johnsen, V.L.; Beckel, A.C. Linseed proteins, alkali dispersion and peptisation. *Ind. Eng. Chem.* **1946**, *38*, 353–356.

Smith, C.R. Jr.; Weisidler, D.; Miller, R.W. Linustin and neolinustatin; cyanogenic glycosides of linseed meal that protects animals against selenium toxicity. *J. Org. Chem.* **1980**, *45*, 507–510.

Sosulski, F.W.; Bakal, A. Isolated protein from rapeseed, flax and sunflower meals. *Can. Inst. Foods Sci. Technol. J.* **1969**, *2*, 28–32.

Stitt, P. Nutritional importance of flax. *Proc. Flax Inst. US* **1986**, *51*, 23.

Thompson, L.U. Nutritional and physiological effects of phytic acid. *Food Proteins*, Kinsella, J.E.; Soucie, E. Eds.; American Oil Chemists' Society, Champaign, IL, 1989, pp. 410–431.

Thompson, L.U. Phytates in canola and rapeseed. *Canola and Rapeseed - Production, Chemistry, Nutrition and Processing Technology*, Shahidi, F.; Ed.; Van Nostrand Reinhold, New York, NY, 1990, pp. 173–192,.

Thompson, L.U.; Robb, P.; Serraino, M.; Cheung, F. Mammalian lignan production from various foods. *Nutr. Cancer* **1991**, *16*, 43–52.

Varga, T.K.; Diosady, L.L. Simultaneous extraction of oil and antinutritional compounds from flaxseed. *J. Am. Oil Chem. Soc.* **1994**, *71*, 603–607.

Vassel, B.; Nesbitt, L.L. The nitrogenous constituents of flaxseed II. The isolation of a purified protein fraction. *J. Biol. Chem.* **1946**, *159*, 571–584.

Wanasundara, P.K.J.P.D. Protein products and sprouts from flaxseed. 1995. Ph.D. Thesis, Memorial University of Newfoundland, St.John's Canada,

Wanasundara, P.K.J.P.D.; Shahidi, F. Removal of flaxseed mucilage by chemical and enzymatic treatments. *Food Chem.* **1997a**, *59*, 47–55.

Wanasundara, P.K.J.P.D.; Shahidi, F. Functional properties of acylated flaxseed protein isolates. *J. Agric. Food Chem.* **1997b**, In press.

Wanasundara, P.K.J.P.D.; Shahidi, F. Optimization of hexametaphosphate-assisted extraction of flaxseed proteins using response surface methodology. *J. Food Sci.* **1996**, *61*, 604–607.

Wanasundara, P.K.J.P.D.; Shahidi, F. Alkanol-ammonia-water/hexane extraction of flaxseed. *Food Chem.* **1994a**, *49*, 39–44.

Wanasundara, P.K.J.P.D.; Shahidi, F. Functional properties and amino acid composition of solvent extracted flaxseed meal. *Food Chem.* **1994b**, *49*, 45–51.

Wanasundara, P.K.J.P.D.; Amarowicz, R.; Kara, M.T.; Shahidi, F. Removal of cyanogenic glycosides of flaxseed. *Food Chem.* **1993**, *48*, 263–266.

Yiu, S.H.; Altosaar, I.; Fulcher, R.G. The effects of commercial processing on the structure and microchemical organisation of rapeseed. *Food Micro. Struc.* **1983**, *2*, 165–173.

Youle, R.J.; Huang, A.H.C. Occurrence of low-molecular weight and high cysteine-containing albumin storage protein in oilseeds of diverse species. *Am. J. Bot.* **1981**, *68*, 44–48.

EFFECT OF PROCESSING ON PHENOLICS OF WINES

V. Z. Blanco, J. M. Auw, C. A. Sims, and S. F. O'Keefe

Food Science and Human Nutrition Department
Institute of Food and Agricultural Sciences
University of Florida
Gainesville Florida, 32611-0370

Phenolics of grapes are the main compounds responsible for color, taste, mouth feel, oxidation and other chemical reactions in wine and juice. Phenolic levels in wine and juice are affected by numerous processing conditions (crushing, pressing, sulfite addition, skin contact, oak aging). Studies were conducted to better understand the effect of processing on phenolic composition of three varieties of grapes. Three different processing steps: immediate press, hot press, and hull treatment (skin contact) for 7 and 14 days were applied to three different grape varieties, *Vitis rotundifolia* cv. Noble, *Vitis vinifera* cv. Cabernet Sauvignon, and the French-American hybrid Chambourcin. Phenolic compounds were identified and quantified by High Performance Liquid Chromatography (HPLC) and bitterness/astringency were assessed using a trained sensory panel. V. rotundifolia wines had higher levels of epicatechin and gallic acid but lower caftaric acid and procyanidins compared to the other varieties and were more astringent and bitter. Processing treatment affected phenolics and color differently among the three varieties.

INTRODUCTION

Phenolics are important to wine and grape juice because they contribute to color, flavor, oxidation and other reactions. Phenolics are also important because of their desirable health benefits as *in vivo* antioxidants. Three of the principal factors affecting the phenolic content of wines are: the phenolic composition of the grape, the procedure used to make the wine, and reactions that take place during aging. This chapter will discuss processing effects on phenolic composition in wines and juices and present information on effect of processing on phenolics in three different grape species.

Process-Induced Chemical Changes in Food
edited by Shahidi *et al.* Plenum Press, New York, 1998

PHENOLICS AND THEIR IMPORTANCE

Phenolic compounds contribute to the color, taste, mouth feel, oxidation/antioxidant properties and related characteristics of juices and wines (Wulf and Nagel, 1976; Oszmianski and Lee, 1990; Ricardo da Silva *et al.*, 1993; Zoecklein *et al.*, 1990; Ramos *et al.*, 1993). The main sources of phenolic compounds in grapes are skin and seeds. The major classes of phenolics that have been identified in grapes and wines include flavanoids such as procyanidins (catechins, flavonols), anthocyanins (malvidin, cyanidin, delphinidin, peonidin, petunidin and their glucosides) and polymers, as well as nonflavanoid compounds (cinnamic, gallic, caftaric and other phenolic acids) (Figure 1). Hydroxycinnamate derivatives comprise the majority of the non-flavonoid phenolics in both white and red wines, with caftaric and coutaric acids as the major cinnamics (Lee and Jaworski, 1987; Roggero *et al.*, 1990; Singleton *et al.*, 1978). The neutral phenolics are composed of flavan-3-ols, with catechin and epicatechin as the two major compounds (Singleton and Noble, 1976). The procyanidins or condensed tannins arise from flavan-3-ols. In grapes, these phenolic polymers range mostly from dimers through decamers (Kantz and Singleton, 1991).

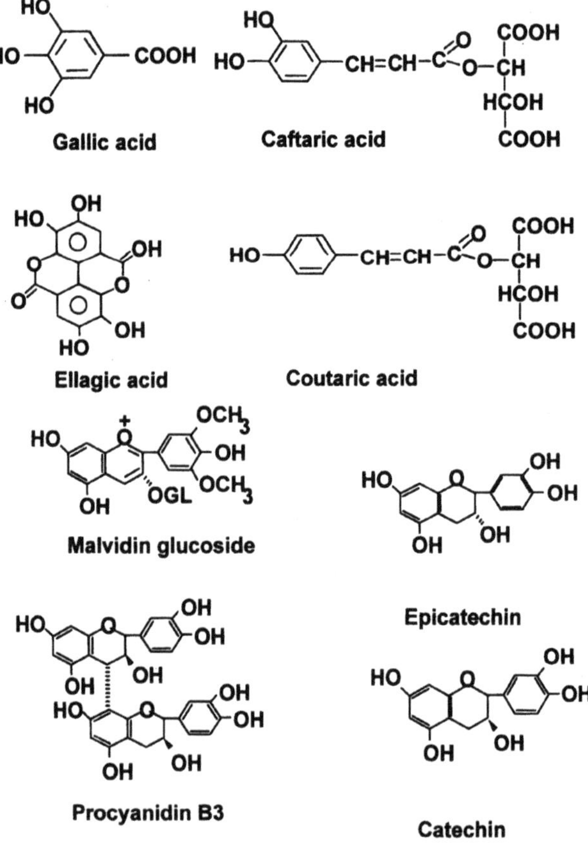

Figure 1. Structures of some important phenolics in wines.

Due to the heterogeneity of phenolics of wines, phenolics are commonly reported as gallic acid equivalents (GAE). The total phenolic content in white wines with no skin contact varies from 100 to 250 mg/ L GAE (Fischer and Noble, 1994) and total phenolics in red wines range from 1000 to 4000 mg/ L GAE (Shahidi and Naczk, 1995). Variations are due to differences in raw materials and wine making techniques. Some of the factors affecting phenolic content will be discussed in this chapter.

The normal light yellow color as well as undesirable brown color of white wine are due to phenolic compounds (Zoecklein *et al.*, 1990). The color of red wine is largely due to anthocyanins that are extracted from the skins during fermentation. Color stability closely corresponds to the degree of polymerization between phenolics and anthocyanins (Sims and Morris, 1985). Astringency and bitterness in wines are primarily due to flavonoid phenolics, most of which are extracted from the skins and seeds of grapes during fermentation (Robichaur and Noble, 1990). It has been reported that bitterness increases with polymerization of flavonoids up to tetramers, and decreases with further polymerization (Robichaur and Noble, 1990).

The amount and types of phenolics present in wines and juices play an important role in controlling oxidation (Lee and Jaworski, 1988). Phenolics are among the most easily oxidized compounds in grape juice and wines (Main and Moris, 1991). Monomeric catechins and dimeric procyanidins brown more intensely than do other phenolics (Main and Morris, 1991). Oxidation reactions involving phenolics can lead to undesirable changes such as loss of flavor and color, development of off-flavors and brown colors. However, oxidation reactions can also lead to improved color, color stability and sensory characteristics (Nagel and Wulf, 1979). Browning susceptibility of grape musts was related to the initial concentration of hydroxycinnamic acids (i.e., caftaric and coutaric) (Nagel and Graber, 1998). In the presence of oxygen, they are both oxidized by grape polyphenoloxidase (PO) and form grape reaction products, which are resistant to further attack by PO and are not involved in subsequent browning. Oxidation of musts prior to fermentation produced wine that had increased resistance to browning and had similar sensory properties to control wine (when sulfite is added at crush) (Nagel and Graber, 1988; Cheynier *et al.*, 1990).

Phenolic compounds, primarily flavonoids, have antioxidant properties which may contribute to health benefits of wine consumption (Kanner *et al.*, 1994; Frankel, 1994). Wine consumption has a possible cardioprotective effect and may prevent thrombosis (Frankel, 1994). Epicatechin and quercetin are more effective in preventing LDL (low density lipoprotein) oxidation than a-tocopherol (vitamin E) (Frankel, 1994). Quercetin is found at an average concentration of 25 mg/L in red wine. Catechin and epicatechins are among the most abundant phenolics in wine and are present at about 150 mg/ L in red wine and about 15 mg/ L in white wine. Resveratrol is thought to be of major importance as a dietary antioxidant in red wines, but levels are much lower than catechin/epicatechin (Frankel, 1994).

PROCESSING EFFECTS ON THE PHENOLICS COMPOSITION OF WINES AND JUICES

A. Grape Cultivar or Species

Grape cultivar reportedly affects anthocyanin content more than wine making procedures (Etievant *et al.*, 1988; Conzales San Jose *et al.*, 1990). The effects of season and cultivar on amounts and types of phenolics in grape are significantly greater at early stages of ripening and decrease toward harvest (Lee and Jaworski, 1989).

B. Crushing

Crushing and pressing are the first processing protocols that can alter phenolic composition of wines (Zoecklein *et al.*, 1990; Lamuela-Raventos and Waterhouse, 1994). In general, gentle crushing and handling of the crushed grapes can significantly lower phenol extraction from the skins. The process of tearing or shredding the skins can cause an increase in the extraction of phenols that may be undesirable for many wines (Zoecklein *et al.*, 1990).

C. Immediate and Hot Press Treatments

The objective of heat treatment is to increase color intensity in grape juices and wines. Heat treatment of crushed grapes also increases extraction of other phenols and destroys oxidative enzymes (Garrido *et al.*, 1993). The temperature and time used during the heat extraction will influence the color intensity, acidity, pH, and extraction of phenolics (Wagener, 1981).

D. Skin Fermentation

Of all the processing factors, the skin fermentation regime probably has the largest impact on phenol levels, sensory characteristics, and wine style (Ricardo da Silva *et al.*, 1993; Lin and Vine, 1990; Sims and Bates, 1994). The practice of extending pomace contact beyond dryness, which is known as extended skin fermentation, may produce a wine lower in color, higher in soft tannins and with increased body (Zoecklin *et al.*, 1990). The optimum skin fermentation time to achieve both the proper level and composition of phenols depends on the wine style that is to be produced, the cultivar and other factors (Sims and Bates, 1994). Total phenol levels increase with skin fermentation time (Kantz and Singleton, 1991; Lin and Vine, 1990; Scudamore-Smith *et al.*, 1990). Both the length of maceration time (Sims and Bates, 1994; Scudamore-Smith *et al.*, 1990) and the presence of high quantity of pomace, seeds, and stems in contact with the must during fermentation (Ricardo da Silva, 1993; Scudamore-Smith *et al.*, 1990) led to wines with a higher content of catechin and procyanidins. However, the effects of skin-contact time on total and polymeric phenol concentration in wines was significantly different between varieties (Kantz and Singleton, 1991).

E. Sulfite Addition

The proper time of sulfite addition can significantly affect phenolic levels. Delaying the initial sulfite addition until after crushing and pressing or until the end of fermentation reduces phenol levels in the wine, which can improve the color stability of white wines by eliminating some of the phenolics that are substrates for browning. The reduced phenol levels may also improve the sensory characteristics of some wines by reducing astringency, bitterness, and overall harshness (Panagiotakopoulou and Morris, 1991; Schmidt and Noble, 1983). Delaying the addition of sulfite does not improve the color stability of muscadine wine due to the lack of polyphenoloxidase (PO) and the type of phenolics present (i.e. lack of readily oxidizable phenolics), but significantly affects the sensory characteristic of wine (less harshness) compared to the addition of sulfite at crush (Schmidt and Noble, 1983). Instead of just delaying sulfite, hyperoxidation or aeration of the juice and crushed grape is an alternative technique which may result in even greater reduction of phenols in the wine (Ricardo da Silva *et al.*, 1993).

F. Fining Agents

Another common way of altering phenolic composition of juices and wines is through the use of fining agents, such as polyvinylpolypyrrolidinone (PVPP), gelatin, egg albumin, casein, and carbons (Lamuela-Raventos and Waterhouse, 1994; Boyle and Hsu, 1990). These compounds bind to various phenols, and the fining agent-phenol complex either precipitates or is removed by filtration (Lamuela-Raventos and Waterhouse, 1994). The removal of phenolic compounds has been shown to be directly proportional to the level of fining agents used in juice treatments (Boyle and Hsu, 1990).

PVPP reportedly removes mainly smaller molecular weight phenols such as single flavonoid units (i.e. catechin, leucoanthocyanidins, etc.) which can result in improved color stability and reduced bitterness or harshness, with little alteration of flavor (Zoecklein *et al.*, 1990). Gelatin is composed of collagen, a large molecular weight protein that reportedly binds to larger molecular weight phenols (i.e. some of the tannins) which may reduce astringency as well as improve color or color stability (Sims *et al.*, 1992). Egg albumin is used to reduce astringency in red wines by binding to and removing tannins, but is not frequently used in white wines due to over-fining problems (Zoecklein *et al.*, 1990). Casein has been reported to bind to leucoanthocyanidins (and probably other flavonoids) which may reduce browning and harshness in some wines (Zoecklein *et al.*, 1990). Carbons are usually used to remove brown phenolic pigments from wine or juices, but carbons may also strip flavor from wines, and may not be appropriate for many wines (Zoecklein *et al.*, 1990).

Utilization of gelatin, egg albumin and PVPP can also reduce ellagic acid sediment that form in muscadine juice (Lamuel-Raventos and Waterhouse, 1994). Treatment with 1.08 g PVPP/L juice was more effective in lowering ellagic acid content than other treatments examined (Boyle and Hsu, 1990).

G. Aging

In general, the flavonoid content is initially very low during fermentation of wine, but increases with continued pomace contact (Singleton, 1967). Catechin and epicatechin increase during the first part of the fermentation, reaching a maximum by the fourth day and decreasing slowly with age (Nagel and Wulf, 1979).

In young wines, procyanidins are found mainly in dimeric and trimeric form, and in aged wines, the relative degree of polymerization increases to 8–10 (Amerine *et al.*, 1980). Since procyanidins arise from flavans-3-ols, catechin and epicatechin diminish as the dimeric procyanidin content increases (Amerine *et al.*, 1980). The procyanidins have the ability to complex strongly with polysaccharides and proteins. In red wines, anthocyanins and condensed tannins react to form polymeric pigments during aging of red wines. These polymers are more complex than the condensed tannin polymers found in grape tissues, less reactive toward proteins and are less astringent than tannin polymers of similar molecular weight present in young wines (Kantz and Singleton, 1991). A young red wine is often bitter, astringent and/or harsh, but during aging, the wines become more palatable and acceptable (Kantz and Singleton, 1991).

The color of red wines is initially from the extraction of anthocyanins from the skins (Thorngate, 1993). Anthocyanin concentration increases to a maximum by the third day of skin contact, when the level of ethanol is about 3 to 6 %, and decreases thereafter (Nagel and Wulf, 1979; Lin and Vine, 1990). During aging, the concentration of the original anthocyanins decrease steadily with the subsequent formation of more stable oligomeric pigments, which maintain wine color (Berg and Akiyoshi, 1956; Thorngate, 1993). Anthocyanins condense with catechins or procyanidins with substitution at C-4 (Thorngate, 1993; Singleton and Trousdale, 1992).

During aging, the color of the wine changes from a bright red to a reddish-brown color. This corresponds to a decrease in the concentration of monomeric anthocyanins and other phenolic compounds and an increase in colored polymeric compounds. If aging continues for long periods, those polymers may precipitate (Thorngate, 1993; Liao *et al.*, 1992). The intensity of wine color declines with aging, with 40 to 80% color loss (Sims and Bates, 1994). Anthocyanins are important to color for at least three years after fermentation and they continue to be present for up to five years (Baranowski and Nagel, 1983). They suggest that anthocyanin loss rates are determined by the type of anthocyanins present and to the extent that they polymerize with tannin.

EXPERIMENTAL

To better understand the role of grape variety and winemakeing treatments on phenolics in red wines, we conducted several studies on three varieties of wine grapes. The varieties were: *Vitis vinifera* cv. Cabernet Sauvignon, a French-American hybrid cv. Chambourcin and *Vitis rotundifolia* cv. Noble (muscadine). The Chambourcin and Cabernet Sauvignon grapes were obtained from Stoneypile vineyard, Clarksville, GA and Noble grapes were obtained from Crevasse vineyard, Archer, FL.

The three treatments were: immediate press, hot press (60 °C for 15 min), or skin fermentation (7 and 13–14 days). Total phenolics were measured as gallic acid equivalents (GAE) /L and were separated and measured by HPLC (Lamuela-Raventos and Waterhouse, 1994). A Perkin Elmer HPLC quaternary pump (Model 400) was fitted with a Rheodyne 7125 injector and Waters Novapack C18 (30 cm x 0.39 cm., i.d., 4 μm particle size) column. A diode array detector (Perkin Elmer model LC 135) was connected to a Perkin Elmer Model 1022 integrator. The solvent system used was described earlier (Lamuela-Raventos and Waterhouse, 1994).

Titratible acidity was determined by diluting 5 mL wine with 125 mL distilled water and titrating to pH 8.2 using 0.1N NaOH. Titratible acidity was expressed as g tartaric acid/100mL. pH was determined using a Corning pH meter (Model 7). The hue and color intensity were determined using a Beckman spectrophotometer (Model DU 640). The absorbance was determined in 0.1 cm cells at 420 nm and 520 nm using distilled water blanks. Color intensity was calculated by the sum of the absorbances and the hue as the ratio of absorbances at 420 nm / 520 nm. Chemical age was measured using a previously described procedure (Zoecklein *et al.*, 1990).

Trained panelists were used to determine bitterness and astringency in wines for all treatments after 3 months of storage at 55 °C using quantitative descriptive analysis. Two training sessions were held to train four experienced panelists with standards for bitterness and astringency. The astringency standard was natural grapeseed tannin (Presque Isle Wine Cellars, PA) and the bitterness standard was epicatechin (Sigma, MO). The analysis was conducting with 15-point scales. The standard training anchor points for the sensory astringency standards were: 3 = 0.05%, 9 = 0.10% and 15 = 0.25% grapeseed tannin. For sensory bitterness the training anchor points were were 2 = 1440 ppm, 9 = 2440 and 15 = 3600 ppm epicatechin. Wines were coded with 3-digit random numbers and were presented randomly in black wine glasses (20 mL) at 25 °C. The sensory tasting sessions were replicated.

Results were analysed by Analysis of Variance using a completely random experimental design. Means were separated by LSD test when the ANOVA was significant.

RESULTS AND DISCUSSION

The total phenolics were higher in hot press compared to immediate press for wines from all varieties (Figure 2). Skin contact increased the phenolics relative to the hot press for Cabernet Sauvignon and Noble, but for Chambourcin there was slight decrease ($p>0.05$) in phenolics for skin contact compared to hot press wines. The total phenolics in 14 day wines were highest for Noble (~1200 mg GAE/L compared to ~800 mg GAE/L for the others).

The HPLC analysis worked well for Chambourcin and Cabernet Sauvignon wines, but there was considerable interference by anthocyanins in Noble (Figure 3). Muscadine wines contain diglucoside anthocyanins which elute and interfere with the HPLC analysis of other phenolics (Scudamore-Smith *et al.*, 1990). We found that fractionation using C-18 SepPaks was required before HPLC analysis (Oszmianski and Lee, 1990) for these wines.

Cabernet Sauvignon wine had the highest content of procyanidins and flavan-3-ols (Table 1). Noble and Chambourcin seemed to be similar in several aspects. They have a very similar color intensity and low contents of procyanidins and catechin. Gallic acid and epicatechin were the main phenolics for all three cultivars. However, Noble wines con-

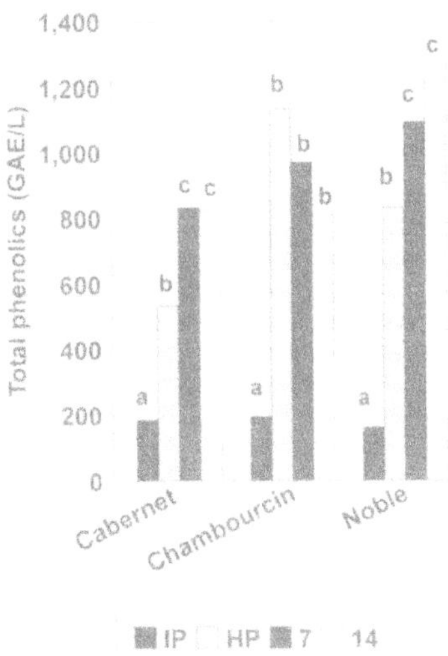

Figure 2. Total phenolics (GAE/L) in wines from three different grape cultivars and 4 different processing conditions.

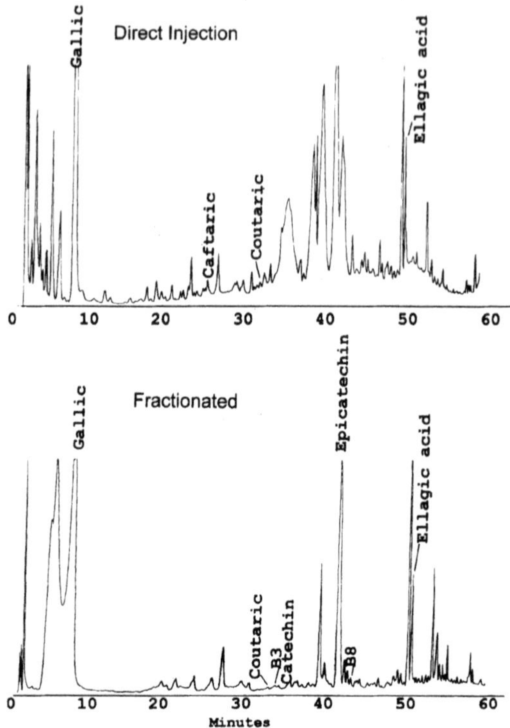

Figure 3. HPLC chromatograms of directly injected and fractionated Noble wines.

Table 1. Levels of phenolics and other characteristics of wines of three different grape cultivars under various processing conditions[1]

Phenolics/	Cabernet				Chambourcin				Noble			
Characteristics	IP	HP	7D	14D	IP	HP	7D	14D	IP	HP	7D	14D
Gallic acid	1.8	2.9	7.2	15.9	2.9	6.6	14.7	15.0	3.9	13.5	148	200
Epicatechin	1.5	2.1	28.8	40.3	2.6	6.9	11.8	14.8	2.3	22.3	118	205
Catechin	6.8	10.5	30.5	32.6	4.1	9.3	5.7	5.6	0.3	4.6	2.8	2.6
Caftaric acid	7.7	9.0	5.6	4.6	0.7	15.3	13.8	9.5	0.5	0.7	1.1	0.9
Coutaric acid	UD[2]	UD	UD	UD	0.9	9.3	4.3	UD	ND	0.3	0.3	0.3
Ellagic acid	ND[3]	ND	ND	ND	ND	ND	ND	ND	0.5	8.2	6.1	4.1
Pro B3	ND	23.5	30.2	30.2	2.5	4.3	9.6	UD	0.3	0.3	0.7	0.9
Pro B4	ND	ND	6.1	5.2	UD	UD	UD	UD	UD	UD	UD	UD
Color intensity ($A_{520+420}$)	0.20	3.7	3.5	3.3	0.44	11.7	8.8	6.5	0.21	4.7	7.1	7.7
Hue ($A_{420/520}$)	1.8	0.78	0.80	0.76	1.1	0.44	0.44	0.48	1.6	0.48	0.50	0.52
Chemical age	0.19	0.14	0.12	0.14	0.08	0.06	0.03	0.03	0.03	0.10	0.03	0.03
TA (% w/v)	0.66	0.73	0.57	0.52	0.81	0.73	0.72	0.71	0.43	0.53	0.57	0.52
pH	3.35	3.63	3.78	3.75	3.38	3.33	3.33	3.33	3.43	3.33	3.31	3.41
Sensory astringency	6.2	7.4	8.0	9.1	5.3	6.3	6.0	5.6	8.0	6.8	10.3	11.1
Sensory bitterness	4.3	4.5	6.5	5.8	6.5	9.5	8.6	7.9	6.6	5.4	7.3	8.0

[1]Concentration is in ppm; UD, undetectable due to interference; ND, not detected; TA, titratible acidity as % w/v tartaric acid; IP, immediate press; HP, hot press; 7D, 7 days skin contact; and 14D, 13-14 days skin contact.

tained extremely high amounts of these two phenolics, which may be the major factors in contributing to the bitterness and harshness of Noble wines.

The results of the present studies showed that wines made from immediate press juices of the three varieties had the lowest total phenolics among all the treatments (Table 1). Heat treatment of the crushed grapes extracted phenols from the skins, but to a lesser extent than any of the hull treatments. In all varieties, hot press wines and juices were much higher in flavonoids and total phenolic content than immediate press wines and juices. The total phenolic concentrations of hot press wines and juices from Cabernet Sauvignon and Noble were shown to be lower than all the skin contact wines. However, for Chambourcin, hot press wine had the highest total phenol concentration among all treatments, and hot press juice had a similar phenol concentration to skin contact wines.

The amounts of gallic acid and epicatechin were low in immediate press wines and juices from Cabernet Sauvignon, Chambourcin and Noble. Hot press wines and juices had slightly higher levels of gallic acid and epicatechin, but less than all skin fermentation wines. The immediate and hot press wines of Cabernet Sauvignon had significantly higher amounts of caftaric acid than all skin contact wines. Caftaric acid was lower in both immediate and hot press juices than the corresponding wines. Hot press Chambourcin wines had the highest level of caftaric acid among all the treatments. Caftaric acid in immediate and hot press juices of Chambourcin was low. Caftaric acid was present in very low levels in both wines and juices from Noble. Coutaric acid could not be quantified in Cabernet Sauvignon due to poor HPLC separation. In Chambourcin, the juice and wine from immediate press contained the lowest amounts of coutaric acid. A significantly ($p<0.05$) higher amount of coutaric acid was found in hot press wine compared to other treatments. The hot press juice had the next highest coutaric acid concentration. Coutaric acid was found in very low levels in hot press Noble wine and juice, and could not be quantified in immediate press wine and juice.

Ellagic acid has not been reported in grape species other than *Vitis rotundifolia* (Amerine and Ough, 1974). The formation of ellagic acid precipitates in *Vitis rotundifolia* wine and especially juices has been reported (Lamuela-Raventos and Waterhouse, 1994; Quinn and Singleton, 1985; Boyle and Hsu, 1990). However, few data are available on levels of soluble ellagic acid in wines and juices. In this study, it was found that hot press wine extracted the highest amount of ellagic acid among all treatments. Immediate press wine and hot press wine had higher amounts of ellagic acid than their juices.

For Cabernet Sauvignon, an interesting observation was that procyanidin B8 was present in immediate press wine at similar concentrations to those for skin contact wines, but low in immediate press juice. Procyanidin B3 and B4 were not detected in immediate press wine and juice. In hot press wine and juice, procyanidin B3 was present at lower amounts compared to all skin contact wines. However, hot press wine contained the highest amount of procyanidin B8 among all treatments. The amounts of catechin were low in both hot press wine and juice.

In Chambourcin, immediate press wine and juice had low levels of catechin and procyanidin B3. Procyanidin B4 could not be quantified in both immediate and hot press wines and juices. The hot press wine had lower amounts of procyanidin B3 than skin contact wines, but the hot press juice had the highest amounts of B3 among all treatments. Catechin levels were the highest in hot press wine and juice compared to all other treatments.

In Noble, immediate press wine and juice contained the lowest amounts of catechin, while both hot press wine and juice had the highest levels. A similar trend was observed for procyanidin B8, in which hot press wine also had the highest level. Procyanidin B3 was found in very low level in immediate and hot press wines and juices.

The immediate press wines and juices of all three varieties had very light red color, and hot press wine and juice had much darker color as a result of increased anthocyanin extraction. The color intensities of all Chambourcin wines and juices were higher than those for Cabernet Sauvignon and Noble. All immediate press wines and juices had browned to some extent as indicated by high hue values (Table 1). All hot press wines and juices had much lower hue values than immediate press wines and juices, which indicated less browning and better red color. In all three varieties, immediate press wines had the highest chemical age (Table 1), which agrees previous work (Lin and Vine, 1990). Among these three varieties, all Cabernet Sauvignon wines and juices had the highest chemical age, which indicated more anthocyanin-phenol polymerization.

The astringency of hot press wines from Cabernet Sauvignon was rated lower than all the skin contact wines, while the immediate press wines was rated as the lowest in astringency. The bitterness in hot press and immediate press wines was similar, and lower than skin contact wines (Table 1). These results can be explained because of their lower content of phenols than all skin contact wines. In Chambourcin, the hot press wine was rated as the most astringent and bitter, which was probably due to the highest phenol level. There were no significant difference in bitterness between Chambourcin wines (Table 1). For Noble, hot press wine had the lowest bitterness and astringency among all treatments. Immediate press wine was rated slightly higher in both astringency and bitterness than hot press wine, but lower than all skin fermentation wines.

The results obtained in our studies showed that the concentration of total phenols of both *Vitis vinifera* cv. Cabernet Sauvignon and *Vitis rotundifolia* cv. Noble wines increased with increasing skin fermentation time. These concentrations varied significantly from 831 to 821 mg/L GAE (from 7 to 14 days) for Cabernet Sauvignon and from 1096 to 1269 mg/L GAE (from 7 to 14 days) for Noble. However, in Chambourcin wines, the concentration of total phenols decreased significantly with increasing skin fermentation time (from 971 to 699 mg/L GAE).

Gallic acid and epicatechin of all three varieties increased with increasing skin fermentation time. Similar results were found previously (Kovac *et al.*, 1992), and the increase in gallic acid and epicatechin could be due to an extraction from the seeds and/or hydrolysis from epicatechin gallate. The presence of combined forms of gallic acid (gallocatechin and epicatechin gallate) has already been reported in other wines and grape seeds (Singleton and Trousdale, 1983). In Cabernet Sauvignon, the amounts of gallic acid increased significantly between 7 to 13 days of skin fermentation. Epicatechin increased between 7 to 14 days skin fermentation. In Chambourcin, the increase in gallic acid was not significant from 7 to 13 days of skin contact. In Noble wines, gallic acid and epicatechin were extremely high (at least 10 higher compared to Cabernet Sauvignon and Chambourcin) and significantly increased between 3 to 14 days.

Some researchers have reported that there is no increase of non-flavonoid phenolics with increasing skin contact time (Sims and Bates, 1994), while others have reported the opposite (Kovac *et al.*, 1992). In this research, Cabernet Sauvignon showed no significantly increases in caftaric acid concentration with increasing skin contact time. However, in the Chambourcin wines, there was a significant decrease of caftaric acid with increasing skin contact time. This could be due to tyrosinase or polyphenoloxidase (PO) activity during maceration (Ricardo da Silva *et al.*, 1993). In Noble, caftaric acid was present in very low levels compared to the other two varieties.

The content of coutaric acid was low compared with caftaric acid for all three varieties, which agrees with previous reports (Nagel and Wulf, 1979). In Cabernet Sauvignon, it was not possible to quantify coutaric acid in all the treatments due to poor separation. In

Chambourcin, coutaric acid was able to be quantified in all treatments except in 14 day skin fermentation wine. Increasing the skin fermentation time significantly decreased the coutaric acid concentration. In Noble, coutaric acid was present in very low levels in all skin fermentation wines.

Ellagic acid in Noble wines did not increase with skin fermentation time. Previous observations (Oszmianski *et al.*, 1986) indicated that extended skin fermentation time increased the precipitation of ellagic acid sediment. Therefore, skin fermentation could have increased the extraction of ellagic acid, but most of this ellagic acid may have precipitated.

Increasing the skin contact time increased the extraction of catechin and procyanidins in Cabernet Sauvignon, with 14 day skin contact wine having the highest content of catechin and procyanidins. Catechin and procyanidins remained stable from 7 to 14 day skin fermentation time. Procyanidin B8 remained almost stable from 7 to 14 days skin contact.

In Chambourcin, the contents of catechin and procyanidins B3 were low compared to Cabernet Sauvignon. In general, the amount of catechin did not increase with increasing skin fermentation time. The transfer of phenols and pigments into wine depends on the concentration of phenols in skins and seeds (Kantz and Singleton, 1991). It could be that concentration of anthocyanins in this cultivar are higher than that of other phenols. Procyandin B3 decreased with skin contact time. Due to the size and presence of more peaks eluting near or with procyanidin B4, it was not possible to quantify it.

In Noble, skin contact increased the extraction of procyanidin B3, but catechin and procyandin B8 decreased with increasing skin fermentation time. One explanation could be that catechin and procyanidin B8 were subject to oxidation or further polymerization. Overall, the levels of procyanidins and catechin in Noble were very low compared to Chambourcin and Cabernet Sauvignon.

The color density of wine generally reaches a maximum before the completion of fermentation (38). Extended pomace contact did not lead to increased color extraction in Cabernet Sauvignon wines (Sims and Bates, 1994). The percentage of red monomers tends to decrease during maceration, but red and brown polymers increase (Scudamore-Smith *et al.*, 1990). Some researchers have shown that extended pomace contact led to greater polymerization of pigments following fermentation, imparting greater color stability in the early stages of red wine, especially at low pH (Sims and Morris, 1985; Sims and Bates, 1994). Pomace contact wines had higher absorbances at 420 nm compared to a wine processed without skin contact (Ricardo da Silva *et al.*, 1993). The effects of skin fermentation time on color and anthocyanin extraction varied between fruit maturities, but resulted in *Vitis rotundifolia* wines which were more astringent and with less aroma than shorter skin fermentation wines (Scudamore-Smith *et al.*, 1990).

In our studies, Chambourcin and Noble wines had higher color intensities than Cabernet Sauvignon wines. In Cabernet Sauvignon and Chambourcin wines, the color intensity decreased slightly with time of skin contact. Hence, the wines fermented with skin for 14 days had lower color intensities than wines fermented with skin for 7 days. In Cabernet Sauvignon, the difference was not significant. However, for Chambourcin, the color intensity decreased significantly with skin contact time. For Noble wines, the color intensity did not increase significantly with increasing skin fermentation time.

Cabernet Sauvignon wines had more browning than Chambourcin and Noble wines as indicated by visualization and higher hue values (Table 1). There was no significant increase in browning with increasing skin fermentation time among these three cultivars. It is known that color intensity and hue are dependent upon pigment composition and the particular state of pigment equilibrium. pH is one of the major influences upon anthocyanin color. The high pH (3.8–3.9) of the Cabernet Sauvignon wines could be in part responsible for the poor color.

The chemical age index refers to the extent of polymerization of anthocyanins, and is related to good color stability during storage (Berg and Akiyoshi, 1956). The chemical age (degree of anthocyanin polymerization) in Cabernet Sauvignon was almost 4 to 6 times higher than that of Chambourcin and Noble respectively (Table 1). For Cabernet Sauvignon, the chemical age increased from 7 to 14 days of skin fermentation. In Chambourcin, the chemical age in skin fermentation wines was almost constant. Noble wines had by far the lowest chemical age of the three cultivars, and chemical age slightly increased from 3 to 14 days of skin contact. This low degree of anthocyanin polymerization is believed to be a major contributing factor of color instability of red muscadine wines (Sims and Morris, 1985).

Previous research has shown that both total aroma and astringency increase with ex/tended pomace contact (Lin and Vine, 1990; Somers, 1978). In our study, astringency tended to increase and bitterness tended to decrease from 7 to 21 days skin fermentation Cabernet Sauvignon wines (Table 1). In Chambourcin, there were no significant differences in astringency and bitterness among skin fermentation wines. For Noble, both astringency and bitterness increased with increasing skin fermentation time. For all three cultivars, the increased astringency in skin fermentation wines was probably due to the higher total phenol levels and the type of phenols extracted.

When comparing cultivars (7 days skin fermentation wines), it can be seen that Chambourcin is not as astringent as Cabernet Sauvignon, which was probably related to the lower content of procyanidins in Chambourcin compared to Cabernet Sauvignon. However, Noble wine, which had the lowest content of procyanidins, was rated as the most astringent and bitter. This evidence was probably due to the presence of extremely high amounts of gallic acid and epicatechin.

Extended skin contact treatment did not result in tannic wines and could produce wines with good color stability (Sims and Bates, 1994). However, some species or cultivars (such as *Vitis rotundifolia*) may not be suitable for extended skin fermentation, since they may not possess the type of phenols and/or anthocyanins that will polymerize, and consequently soften and provide color stability (Sims and Morris, 1985).

Skin fermentation time had no significant effect on pH, but titratable acidity increased slightly (by 0.05–0.1 g/100 mL) with increasing skin fermentation time for red muscadine wines (Lin and Vine, 1990). However, total acidity and tartaric acid content decreased and pH increased with skin contact on *Vitis vinifera* grapes (Ricardo da Silva *et al.*, 1993). All of the Cabernet Sauvignon wines had higher pH than Chambourcin and Noble wines.

In Cabernet Sauvignon with extended skin contact time, pH increased significantly from 3.8 to 3.9, while titratable acidity decreased significantly from 5.7 to 4.2 g/L, in agreement with previous work (Ricardo da Silva *et al.*, 1993). In Noble, there was also a significant increase of pH from 3.3 to 3.4 (from 3 to 14 days) and a non significant decrease of titratable acidity from 5.5 g/L to 5.3 g/L with increasing skin fermentation time. The pH of all the skin contact Chambourcin wines was similar, but titratable acidity increased slightly between 7 to 14 days of skin fermentation.

There are many factors that affect the composition and content of phenolic compounds in wines. The color, stability, possible nutritive benefit and sensory characteristics of the wine will be strongly affected by these changes. Increased understanding will allow winemakers to produce more pleasant and beneficial wines.

REFERENCES

Amerine, M.A.; Ough, C.S. *Wine and Must Analysis.* John Wiley and Sons, Inc., New York, 1974.

Amerine, M.A.; Berg, H.W.; Kunkee, R.E.; Ough, C.S.; Singleton, V.L.; Webb, A.D. Technology of wine making. AVI Publishing Company, Inc., Westport, Connecticut, 1980.

Baranowski, E.S.; Nagel, C.W. Kinetics of malvidin-3-glucoside condensation in wine model system. *J. Food Sci.* 1983, *48,* 419–429.

Berg, H.W.; Akiyoshi, M. The effect of contact time of juice with pomace on the color and tannin content of red wines. *Am. J. Enol. Vitic.* **1956,** *7,* 84–90.

Boyle, J.A.; Hsu, L.M. Identification and quantification of ellagic acid in muscadine grape juice. *Am. J. Enol. Vitic.* **1990,** *41,* 43–47.

Cheynier, V.; Rigaud, J.; Souquet, J.M.; Duprat, F.; Moutounet, M. Must browning in relation to the behavior of phenolic compounds during oxidation. *Am. J. Enol. Vitic.* **1990,** *41,* 346–349.

Etievant, P.; Schlich, P.; Bertrand, A.; Symonds, P.; Bouvier, J.C. Varietal and geographic classification of French red wines in terms of pigments and flavonoid compounds. *J. Sci. Food Agric.* **1988,** *42,* 39–54.

Fischer, U.; Noble, A.C. The effect of ethanol, catechin concentration, and pH on sourness and bitterness of wine. *Am. J. Enol. Vitic.* **1994,** *45,* 6–10.

Frankel, E.N. Red wine antioxidants and potential health benefits. Presentation at One Hundred and Eighty-Seventh Dinner Meeting of the Society of Medical Friends of Wine, Mark Hopkins Inter-Continental Hotel, San Francisco, 1994.

Garrido, V.M.; Sims, C.A.; Marshall, M.R.; Bates, R.P. Factors influencing ellagic acid precipitation in muscadine grape juice during storage. J. Food Sci. **1993,** *58,* 193–196.

Gonzales San Jose, M.L.; Santa-Maria, G.; Diez, C. Anthocyanins as parameters for differentiating wines by grape variety, wine-growing region, and wine-making methods. *J. Food Composition Anal.* **1990,** 54–66.

Kanner, J.; Frankel, E.; Granit, R.; German, B.; Kinsella, J.E. Natural antioxidants in grapes and wines. *J. Agric. Food Chem.* **1994,** *42,* 64–69.

Kantz, K.; Singleton, V.L. Isolation and determination of polymeric polyphenols in wines using sephadex LH-20. *Am. J. Enol. Vitic.* **1991,** *42,* 309–316.

Kovac, V.; Alonso, E.; Bourzeix, M.; Revilla, E. Effect of several enological practices on the content of catechins and proanthocyanidins of red wines. *J. Agric. Food Chem.* **1992,** *40,* 1953–1957.

Lamuela-Raventos, R.M.; Waterhouse, A.L. A direct HPLC separation of wine phenolics. *Am. J. Enol. Vitic.* **1994,** *45,* 1–5.

Lee, C.Y.; Jaworski, A. Phenolic compounds in white grapes grown in New York. *Am. J. Enol. Vitic.* **1987,** *38,* 277–281.

Lee, C.Y.; Jaworski, A.W. Phenolics and browning potential of white grapes grown in New York. *Am. J. Enol. Vitic.* **1988,** *39,* 337–340.

Lee, C.Y.; Jaworski, A.W. Major phenolic compounds in ripening white grapes. *Am. J. Enol. Vitic.* **1989**, *40,* 43–46.

Liao, H.; Cai, Y.; Haslam, E. Polyphenol interactions. Anthocyanins: Co-pigmentation and color changes in red wines. *J. Sci. Food Agric.* **1992,** *59,* 299–305.

Lin, T.Y. and Vine, R. P. Identification and reduction of ellagic acid in muscadine grape juice. *J. Food Sci.* **1990,** *55,* 1607–1609, 1613.

Main, G.L.; Morris, J.R. Color of Riesling and Vidal wines as affected by bentonite, cufex, and sulfur dioxide juice treatments. *Am. J. Enol. Vitic.* **1991**, *42*, 354–357.

McCloskey, L.P.; Yengoyan, L.S. Analysis of anthocyanins in *Vitis vinifera* wines and red color versus aging by HPLC and spectrophotometry. *Am. J. Enol. Vitic.* **1981,** *32,* 257–261.

Nagel, C.W.; Wulf, L.W. Changes in the anthocyanins, flavonoids and hydroxycinnamic acid esters during fermentation and aging of Merlot and Cabernet Sauvignon. *Am. J. Enol. Vitic.* **1979,** *30,* 111–116.

Nagel, C.W.; Graber, W.R. Effect of must oxidation on quality of white wines. *Am. J. Enol. Vitic.* **1988,** *39,* 1–4.

Oszmianski, J.; Romeyer, F.M.; Sapis, J.C.; Macheix, J.J. Grape seed phenolics: Extraction as affected by some conditions occuring during wine processing. *Am. J. Enol. Vitic.* **1986,** *37,* 7–12.

Oszmianski, J.; Lee, C.Y. Isolation and HPLC determination of phenolic compounds in red grapes. *Am. J. Enol. Vitic.* **1990,** *41*, 204–206.

Panagiotakopoulou, V.; Morris, J.R. Chemical additives to reduce browning in white wines. *Am. J. Enol. Vitic.* **1991**, *42*, 255–260.

Quinn, M.K.; Singleton, V.L. Isolation and identification of ellagitannins from white oak wood and an estimation of their roles in wine. *Am. J. Enol. Vitic.* **1985,** *36,* 148–155.

Ramos, T.; Fleuriet, A.; Rascalou, M.; Macheix, J.J. The effect of anaerobic metabolism of grape berry skins on phenolic compounds. *Am. J. Enol. Vitic.* **1993,** *44,* 13–16.

Ricardo da Silva, J.M.; Cheynier, V.; Samson, A.; Bourzeix, M. Effect of pomace contact, carbonic maceration, and hyperoxidation on the procyanidin composition of Grenache Blanc wines. *Am. J. Enol. Vitic.* **1993,** *44,* 168–172.

Robichaud, J.L.; Noble, A.C. Astringency and bitterness of selected phenolics in wines. *J. Sci. Food Agric.* **1990,** *53,* 343–353.

Roggero, J.P.; Coen, S.; Archier, P. Wine phenolics: Optimization of HPLC analysis. *J. Liq. Chromatogr.* **1990,** *13,* 2593–2603.

Schmidt, J.O.; Noble, A.C. Investigation of the effects of skin contact time on wine flavor. *Am. J. Enol. Vitic.* **1983,** *34,* 135–138.

Scudamore-Smith, P.D.; Hooper, R.L.; McLaran, E.D. Color and phenolic changes of Cabernet Sauvignon wine made by simultaneous yeast/bacterial fermentation and extended pomace contact. *Am. J. Enol. Vitic.* **1990, 41,** 57–67.

Shahidi, F.; Naczk, M. *Food Phenolics. Sources, Chemistry, Effects, and Applications.* Technomic Publishing Company, Inc., Lancaster, PA, USA, 1995.

Sims, C.A.; Morris, J.R. A comparison of the color components and color stability of red wine from Noble and Cabernet Sauvignon at various pH levels. *Am. J. Enol. Vitic.* **1985,** *36,* 181–184.

Sims, C.A.; Bates, R.P. Effects of skin fermentation time on the phenols, anthocyanins, ellagic acid sediment, and sensory characteristics of a red *Vitis rotundifolia* wine. *Am. J. Enol. Vitic.* **1994,** *45,* 56–62.

Sims, C.A.; Bates, R.P.; Mortensen, J.A. Effects of must polyphenoloxidase activity and timing of sulfite addition on the color and quality of *Vitis rotundifolia* and *Euvitis* hybrid white wines. *Am. J. Enol. Vitic.* **1991,** *42,* 128–132.

Sims, C.A.; Garrido, V.M.; Bates, R.P. Methods to limit ellagic acid precipitation in muscadine juice and wine. *Proc. Fla. State Hort. Soc.* **1992,** *105,* 135–139.

Singleton, V.L. Adsorption of natural phenols from beer and wine. Tech. Quart. Master brewers Assoc. *America* **1967,** *4,* 245.

Singleton, V.L.; Noble, A.C. Wine flavor and phenolic substances. In *Phenolic, Sulfur, and Nitrogen Compounds in Food Flavors*; Charalambous, G.; Katz, I., Eds.; ACS Symposium Series; ACS: Washington, DC, 1976; Vol. 26, pp 47–70.

Singleton, V.L.; Trousdale, E. White wine phenolics: Varietal and processing differences as shown by HPLC. *Am. J. Enol. Vitic.* **1983,** *34,* 27–34.

Singleton, V.L.; Trousdale, E. Anthocyanin-tannin interactions explaining differences in polymeric phenols between white and red wines. *Am. J. Enol. Vitic.* **1992,** *43,* 63–70.

Singleton, V.L.; Timberlake, C.F.; Lea, A.G.H. The phenolic cinnamates of white grapes and wine. *J. Sci. Food Agric.* **1978,** *29,* 403–410.

Somers, T.C. Interpretations of colour composition in young red wines. *Vitis* **1978,** 161–167.

Thorngate, J.H. Flavan-3-ols and their polymers: Analytical techniques and sensory considerations. In *Beer and Wine Production*. Gump, B.H., Ed.; American Chemical Society, Washington, DC, 1993.

Wagener, G.W.W. The effect of different thermovinification systems on red wine quality. *Am J. Enol. Vitic.* **1981,** *32,* 179–184.

Wulf, L.W.; Nagel, C. W.. Analysis of phenolic acids and flavonoids by high pressure liquid chromatography. *J. Chromatography* **1976** *116,* 271–279.

Zoecklein, B.W.; Fugelsang, K.C.; Gump, B.H.; Nury, F.S. Production wine analysis. Van Nostrand Reinhold, New York, 1990.

28

PHOTOCHEMICAL REACTIONS OF FLAVOR COMPOUNDS

Chung-Wen Chen and Chi-Tang Ho

Department of Food Science
Cook College
New Jersey Agricultural Experiment Station
Rutgers, The State University of New Jersey
New Brunswick, New Jersey 08903

Photochemical reaction is a chemical reaction which is initiated by light. In addition to light, photosensitizer and oxygen are the two important factors which contribute to the formation of photochemical products. In this contribution, photochemical reactions of flavor compounds are classified into four categories according to the factors of photosensitizer and oxygen. Photochemical reaction with or without sensitizer in the absence of oxygen and unsensitized photochemical reaction in the presence of oxygen usually involve free radical reactions; while in the presence of oxygen and sensitizer, the singlet oxygen can be generated that then reacts with flavor compounds which contain double bonds to give the oxygenated products.

INTRODUCTION

Photochemical reaction is a chemical reaction initiated by light. The photochemical reaction has been used to synthesize new chemicals, such as rose oxide, which is an important component of Bulgarian rose oil (Shibamoto, 1983). On the other hand, some reactive oxygen species such as singlet oxygen and/or superoxide radical, which are believed to cause cell damaging or mutagenic effects, may be generated when the photosensitized reaction occurs. In addition, the photochemical reaction may lead to formation of off-flavor or even toxic compounds in foods (Rosenthal, 1985).

Flavor compounds are widely used in foods, cosmetics, and other household products. It is unavoidable that the products which contain those flavor materials are exposed to light because of the psychological factor in modern retail marketing of "visual appealing". Therefore, it is necessary to more intensively study the photochemical reactions on flavor chemicals.

Process-Induced Chemical Changes in Food
edited by Shahidi *et al.* Plenum Press, New York, 1998

The photochemical reactions may proceed under light exposure with or without sensitizer in the presence or absence oxygen. According to the factors of sensitizer and oxygen, the whole photochemical reaction can be categorized into four reactions: photoreaction without sensitizer and oxygen, photosensitized reaction without oxygen, photooxidation without sensitizer, and photosensitized oxidation. Most of the photochemical reactions involve free radicals which are formed from homolytic breakdown of covalent bonds when molecules absorb certain quantities of energy from light. Whereas, the photosensitized oxidation reaction promote the generation of singlet oxygen, which is able to react with substrates to afford oxidized products. The details of the photochemical reactions of organic compounds have been reported by Shibamoto and Mihara (1983), Shibamoto (1983), Coxon and Halton (1974), Horspool (1976), and Buchardt (1976). The present review focuses on the photochemical reactions of selected flavor compounds.

PHOTOREACTION WITHOUT SENSITIZER AND OXYGEN

1. Fragmentation Reaction

There are two types of fragmentation reaction, Norrish type I and Norrish Type II reactions, which were reported to occur during the photochemical reaction of either cyclic or acyclic ketones (Shibamoto and Mihara, 1983; Horspool, 1976). Norrish Type I reaction is an α-cleavage reaction. The carbon-carbon bond adjacent to the carbonyl group is cleaved to produce a radical pair. Cyclic ketones such as cyclopentanone, in contrast to the acyclic ketones, show a greater tendency to undergo α-cleavage to produce acyl-alkyl biradicals, which then form pent-4-enal on excitation either at 313 or 254 nm as shown in Scheme 1 (Horspool, 1976; Dunion and Trumbore, 1965; Simonaitis *et al.*, 1967; Wagner and Spoerke, 1969; Dalton *et al.*, 1971).

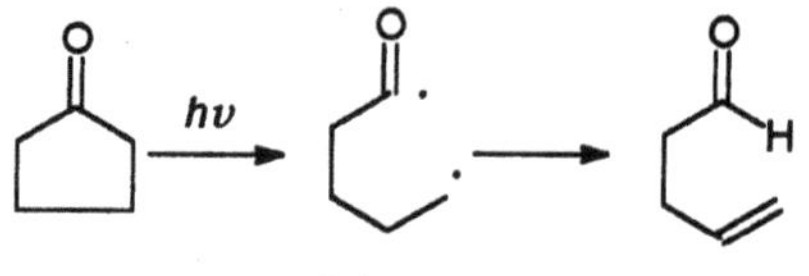

O hν O H

Scheme 1

Alpha-cleavage may occur at both sides of carbonyl group resulting in the loss of a carbon monoxide molecule. This reaction is demonstrated in Scheme 2. Thujone, which possesses a minty-camphoraceous odor, decomposes into 2-isopropyl-1,4-hexadiene after losing a carbonyl group during irradiation (Eastman *et al.*, 1963).

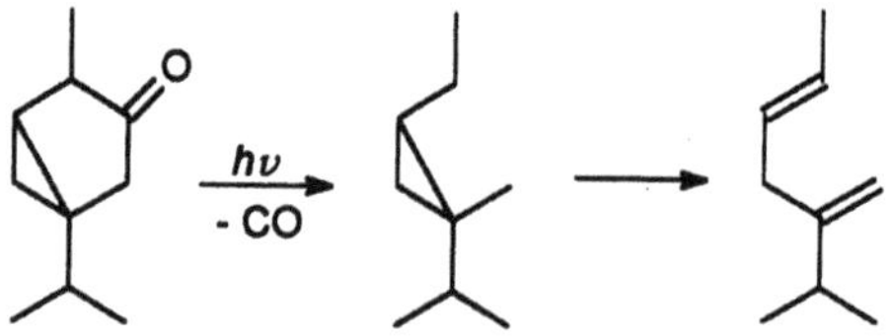

O hν - CO

Scheme 2

Norrish Type II fragmentation reaction is the excitation of the carbonyl group of cycloketones by irradiation followed by intermolecular hydrogen abstraction usually from the γ-site with respect to the carbonyl group leading to a 1,4-biradical intermediate formation. The 1,4-biradical can be cleaved to an olefin and an enol to form a cyclobutanol in solution. It may also back-transfer the hydrogen atom to reform starting material (Horspool, 1976). Scheme 3 shows 1-isopulgone, which has a powerful minty-woody mild green odor, undergoing Norrish Type II fragmentation to give a cycloalcohol derivative (Matsui *et al.*, 1967).

2. Addition Reaction

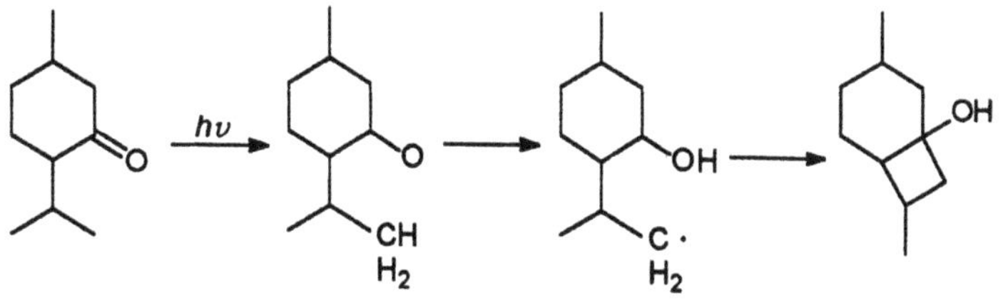

Scheme 3

Oxetane Formation. Reactions of carbonyl compounds such as ketones and aldehydes with electron-rich olefins results in the formation of oxetanes. The oxetane formation involves the addition of the carbonyl oxygen to the olefinic π-system to produce a biradical intermediate, which then undergoes spin inversion to produce the oxetane (Horspool, 1976). Typical examples of oxetane formation include the photoreaction of furan compounds with aldehyde or alkene compounds as shown in Scheme 4 (Cantrell, 1977; Whipple and Evanega, 1968).

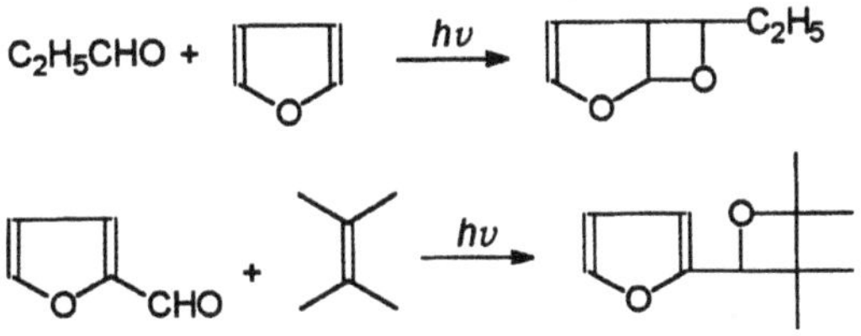

Scheme 4

Dimerization Reaction. A typical example of dimerization reaction is the photoreaction of diacetyl (Scheme 5) which contributes the butter flavor. A cyclic dimer 1,4-dihydroxyl-1,4-dimethylcyclohexane-2,5-dione is obtained from photoreaction of diacetyl in hexane or ether (Bowen and Horton, 1934). However, when this photoreaction is conducted in an alcoholic solution, dimerization reaction of diacetyl is replaced by reduction reaction (Yamada and Hanai, 1951). This indicates that the photoreaction products of diacetyl in a solution depend on the solvent systems.

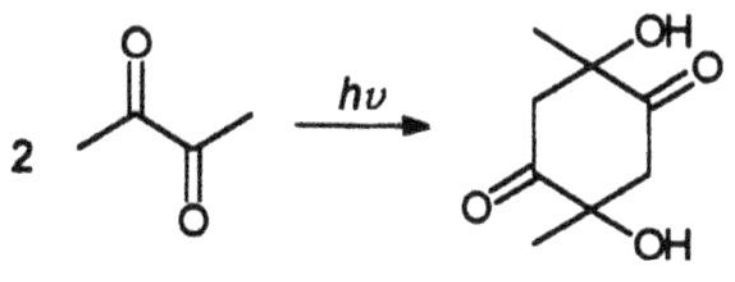

Scheme 5

3. Isomerization Reaction

Isomerization reaction may occur by both direct and sensitized irradiation. The direct irradiation is difficult to achieve in the case of the simple olefins. However, with more substituted olefins the direct excitation can be more readily achieved. When the *cis-trans*-isomerization occurs, it usually produces more *cis*-form. The reason is that the *trans*-form absorbs more light at the excitation wavelength than does the *cis*-form and consequently the *trans*-isomer is converted more readily into the *cis*-form (Horspool, 1976).

The examples of the *cis-trans*-isomerization are the photoreaction of α-ionone *1* and jasmone *2* as shown in Scheme 6. α-Ionone has a fresh violet-like odor and a warm-woody balsamic-floral odor of deep sweetness and moderate tenacity. It is used widely in all types of perfume compositions (Shibamoto and Mihara, 1983). *cis*-α-Ionone is yielded from *trans*-α-ionone by photo irradiation (Büchi and Yang, 1955). Jasmone, which has a jasmine-like odor, is the principal aroma component in jasmine oil (Nofal *et al.*, 1982; Toda *et al.*, 1983). On irradiation, jasmone gives the final products *trans*- and *cis*-2-(2'-ethylcyclopropyl)-3-methyl-2-cyclopenten-1-one (Tateba and Mihara, 1986). These two products are described as having a waxy, oily, and *cis*-jasmone-like odor for *trans*-form and a *cis*-jasmone-like odor with a mild, minty, and woody undertone for *cis*-form.

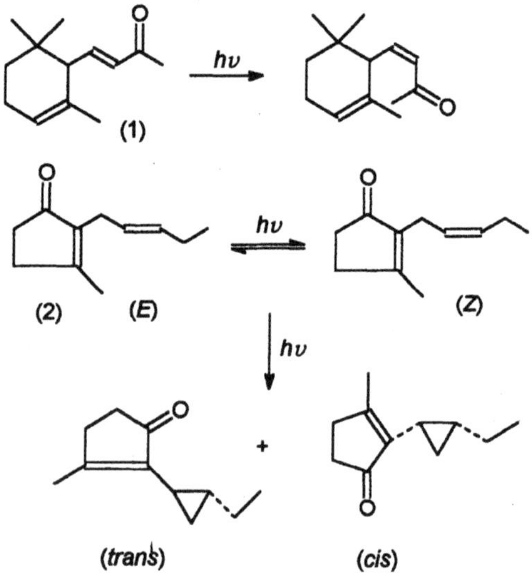

(1) hυ

(2) (E) hυ (Z)

hυ

(trans) + (cis)

Scheme 6

The isomerization can occur on the photo-irradiation of aromatic compounds, in which the rearrangement was shown to involve the skeletal reorganization of the molecular. The pyrazine skeletally rearranges as shown in Scheme 7 probably via intermediate *3* to form a pyrimidine after a photochemical isomerization reaction (Shibamoto, 1983; Horspool, 1976).

PHOTOSENSITIZED REACTION WITHOUT OXYGEN

Photosensitized reaction occurs in the presence of light and photosensitizers with or without oxygen. A sensitizer, such as chlorophyll, rose bengal, or methylene blue, can absorb light energy and becomes an excited singlet state sensitizer. This singlet state sensitizer can be converted to the ground state by emitting fluorescence, or it may undergo electronic rearrangement via an intersystem crossing mechanism resulting in the formation of an excited triplet state sensitizer. In the absence of oxygen, the excited triplet state sensitizer can activate a substrate to form free radical species (Foote, 1976).

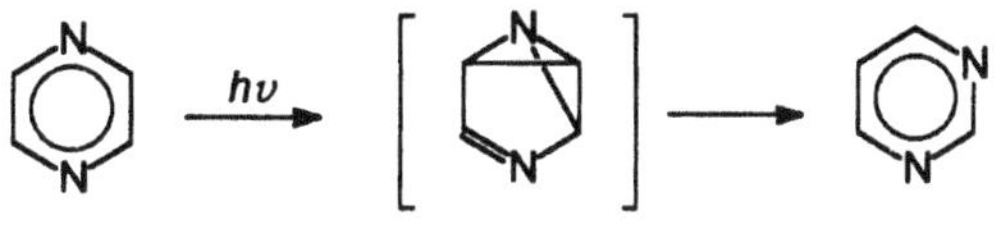

(3)

Scheme 7

1. Addition Reaction

When menthene is irradiated in a methanol solution in the presence of alkylbenzene as a sensitizer, its isomeric compound and methanol addition products are obtained as shown in Scheme 8 (Marshall and Carroll, 1966; Kropp and Drauss, 1967). The mechanism of this reaction could be a radical reaction, which is activated by the excited sensitizer, followed by hydrogen abstraction from solvent or methene itself.

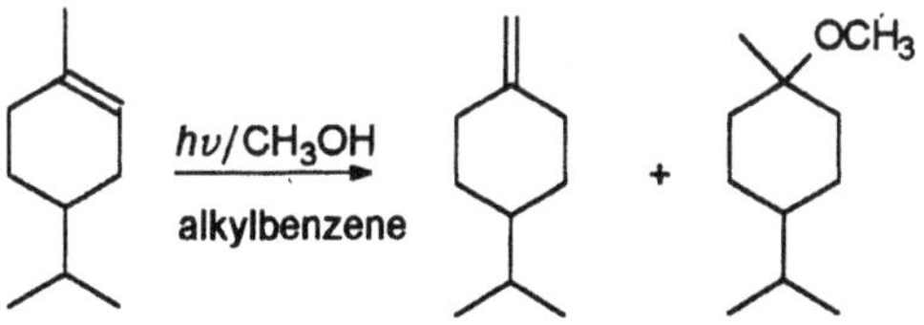

Scheme 8

2. Isomerization Reaction

Photosensitized irradiation of myrcene, Which possesses a sweet balsamic-resinous odor of poor tenacity, gives a bicyclic isomer as shown in Scheme 9 (Liu and Hammond, 1967; White and Grpta, 1969). The biradical mechanism could be used to elucidate this reaction.

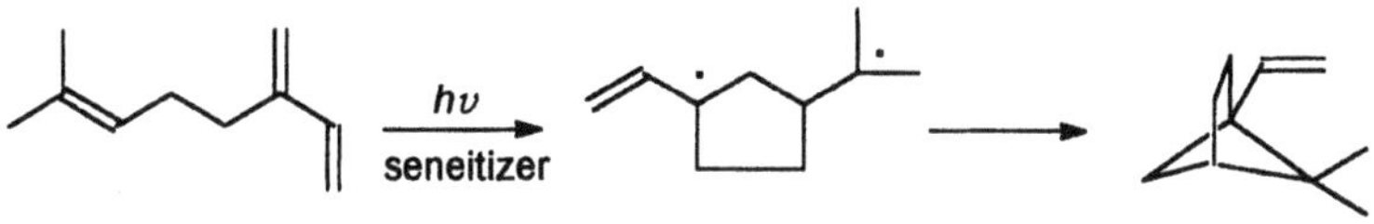

Scheme 9

3. Dimerization Reaction

The naphthalene-photosensitized irradiation of α-phellandrene (Scheme 10), which has a peppery odor, gives the dimmers (Baldwin and Nelson, 1966). The sensitized irradiation is more complex than direct irradiation since the dimer can be converted photochemically into the other isomers. Dimerization reaction is readily achieved when the conjugation is increased in diene system which results in more accessible energy levels (Horspool, 1976).

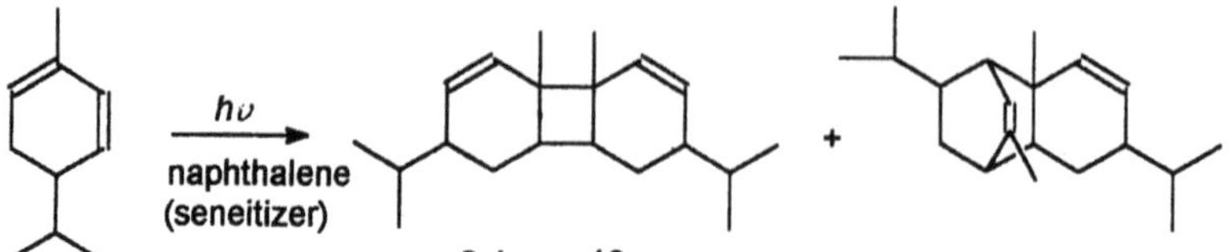

hυ
naphthalene
(seneitizer)

Scheme 10

UNSENSITIZED PHOTOOXIDATION

The mechanism of the unsensitized photoreaction with oxygen is similar to that of photoreaction without oxygen. Both of them involve free radical or biradical formation. Photooxidation without sensitizer generally gives a more complex result than sensitizer photooxidation, probably because of co-occurrence of the photoreaction of the substrate itself, oxidation of the initial photoproduct, or further photochemical transformation of the photooxdation product (Matsuura and Saito, 1976).

1. Hydroperoxide Formation

Irradiation of cyclic ethers such as tetrahydrofuran in the presence of oxygen yields mainly the α-hydroperoxides (Scheme 11)(Stenburg *et al.*, 1970). A mechanism involving charge transfer excitation followed by free radical formation can be used to explain this reaction.

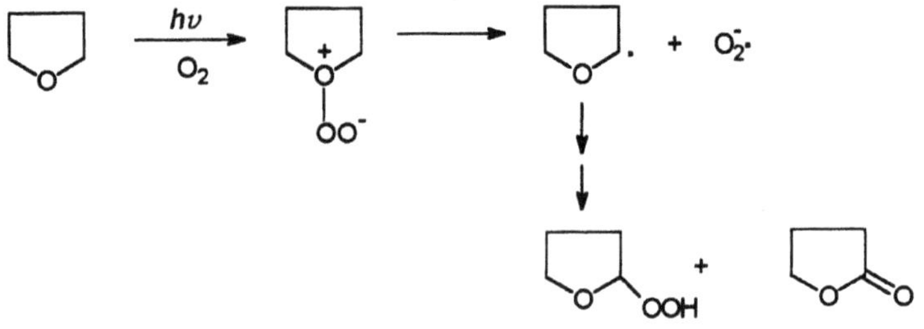

Scheme 11

Photooxidation of pulegone (Scheme 12), which has been widely used as a flavoring agent (Opdyke, 1978; Hall and Oser, 1965), leads to the formation of 8-hydroxy-$^{4(5)}$-p-menthen-3-one (Fröhlich and Shibamoto, 1990). The mechanism of this reaction could involve biradical formation followed by the far formation of a hydroperoxide.

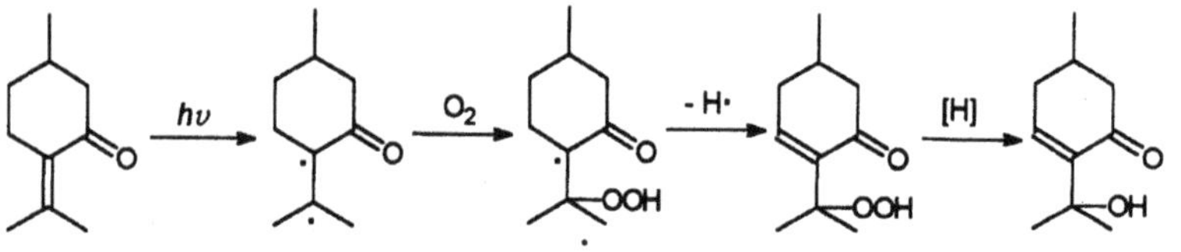

Scheme 12

2. Dehydrogenation Reaction

Some dihydroheterocyclic compounds such as pyrroline *4* and imidazoline *5* are known to undergo dehydrogenation reaction in the presence of oxygen to give the corresponding heterocyclic compounds (scheme 13)(Kawana and Emoto, 1969; Matsuura *et al.*,

1973). The mechanism of a free radical formation followed by hydrogen abstraction has been proposed (Matsuura *et al.*, 1973).

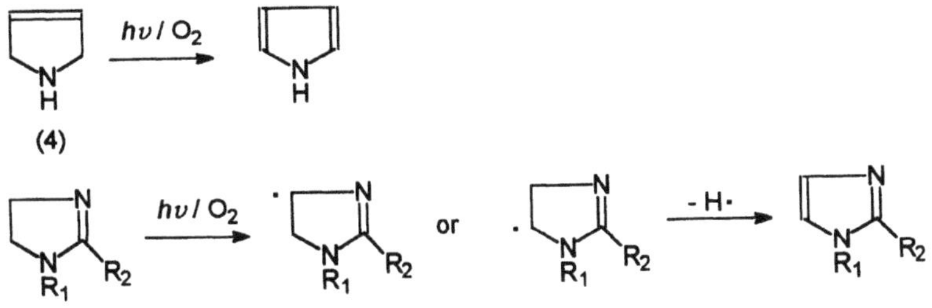

Scheme 13

3. Addition Reaction

Irradiation of perillaldehyde, which has a powerful fatty-spicy, oily-herbaceous odor, and sweet-herbaceous taste, in a methanol solution in the presence of oxygen gives a methanol addition product as shown in Scheme 14 (Tateba *et al.*, 1992a). This mechanism follows the Norrish Type II fragmentation reaction to produce a biradical intermediate, which then undergoes hydrogen abstraction from methanol to give an alcoholic intermediate with a free radical. Methanol as a solvent participates in this reaction followed by oxygenation to give a methoxyhydroperoxide intermediate. The final product, a methoxy ketone, is yielded after elimination of a formic acid molecule.

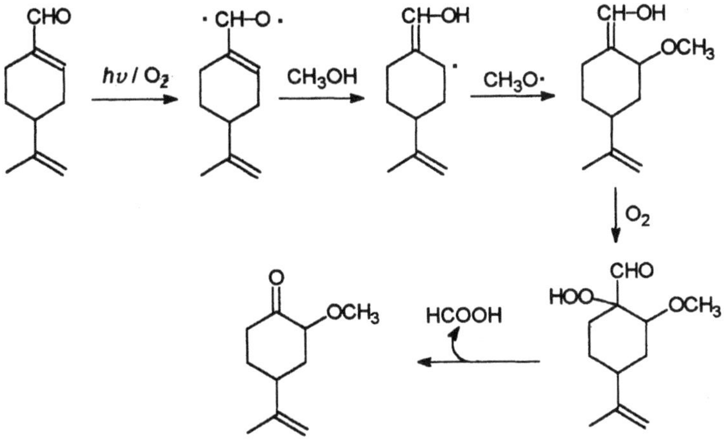

Scheme 14

4. Epoxide Formation

The formation of nootkatone epoxide from photooxidation of nootkatone (Scheme 15), which is the principal flavoring constituent in grapefruit oil (MacLeod and Buigues, 1964), has been observed (Tateba *et al.*, 1992b). The mechanism of this reaction involves the biradical formation, oxygenation, dimerization, and fragmentation leading to the formation of two molecules of epoxide.

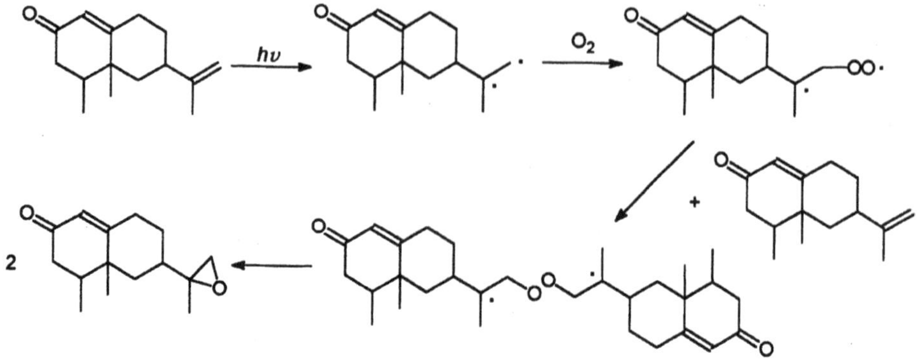

Scheme 15

PHOTOSENSITIZED OXIDATION

Two types of mechanisms are involved: Type I and Type II reactions, both of which have been reported to occur during photosensitized oxidation. In Type I reaction, a substrate is activated by the excited triplet state sensitizer resulting in the formation of free radical species; whereas, the Type II reaction involves the activation of oxygen molecule to generate singlet oxygen. Singlet oxygen easily initiates the photooxidation of flavor compounds because of its low energy of only 22.4 kCal above the ground state; its relatively long lifetime, and its highly electrophilic nature allows easy attack on moieties of high electron density (Bradley and Min, 1992; Rawls and Van Santen, 1970). The photosensitized oxidation of flavor compounds follows these two mechanisms, but the Type II mechanism outweighs the Type I mechanism.

The reactions of singlet oxygen with organic compounds can be classified into five categories (Matsuura and Saito, 1976). However, three categories can be classified for the reactions of singlet oxygen with flavor compounds including "ene" reaction, 1,4-cycloaddition to conjugated dienes, and 1,2-addition reactions.

1. "Ene" Reaction

When singlet oxygen reacts with ene system of flavor compounds, the oxygen is added to one carbon atom of the double bond to give allylic hydroperoxides with a shift of the double bond. The examples of ene reaction are the photosensitized oxidation of nootkatone *6* (MacLeod and buigues, 1964), perillaldehyde *7* (Matsuura *et al.*, 1973), jasmone *8* (Tateba *et al.*, 1993), and limonene *9* (Foote *et al.*, 1965) as shown in Scheme 16. It is worthwhile to note that singlet oxygen directly attacks the double bond on the ring structure of limonene; whereas, in the case of nootkatone, perillaldehyde, and jasmone, singlet oxygen attacks the double bonds on the alkyl substituents. Maybe the electron withdrawing effect of carbonyl group in the nootkatone, perillaldehyde, and jasmone renders the double bonds in these ring structures less reactive towards singlet oxygen. Limonene does not contain the carbonyl conjugated double bond in the ring structure which is easily attached by singlet oxygen because the oxygen is transferred much faster to tri- and tetra-substituted ethylene than to di-substituted ethylenes (Gollnick and Schenck, 1964).

Another interesting example is the photosensitized oxidation of 2,5-dimethyl-4-hydroxy-3(2H)-furanone (DMHF), which possesses a caramel-like, sweet, fruity, and burnt pineapple-like flavor (Chen *et al.*, 1996). When DMHF was irradiated in absolute alcohol containing chlorophyll as a sensitizer, ten photooxidation products, including ethyl pyruvate *10*, ethyl lactate *11*, acetic acid *12*, ethyl 2-acetoxypropionate *13*, acetoxyacetone *14*, acetoxy-2,3-butanedione *15*, 1,2-ethanediol *16*, 2-oxopropyl 2-acetoxypropionate *17*, lactic acid *18*, and 2-acetoxypropionic acid *19*, were generated (Scheme 17). The possible mechanism for the formation of these products during the photosensitized oxidation reaction involves the reaction of singlet oxygen with DMHF via an ene reaction to form a hydroperoxide, which then decomposes to primary products or intermediates. The acyclic esters are formed after primary products or intermediates react further with alcohols or acids (Chen *et al.*, 1996).

2. 1,4-Cycloaddition to Conjugated Dienes

This reaction mostly occurs in the photosensitized oxidation of heterocyclic five-membered ring compounds. Heterocyclic, five-membered ring compounds including furans, pyrroles, oxazoles, thiazoles, imidazoles and thiophenes, which contain conjugated diene systems, are the volatile constituents of cooked meats (Shahidi, 1989). Singlet oxygen reaction of these compounds often leads to complicated results. This complexity is due to the multiplicity of pathways for the decomposition of the initially formed peroxide or hydroperoxide intermediates. Since the singlet oxygen reacts with conjugated heterocyclic five-membered ring compounds via a 1,4-cycloaddition reaction, the electron density at C-1 and C-4 decides the photooxidative reactivity on these five-membered ring compounds. Generally, the more electronegative heteroatoms such as nitrogen and oxygen facilitate these reaction more significantly than do those heteroatoms such as sulfur, and the presence of electron-donating groups at C-1 or/and C-4 tends to increase the reactivity, while electron-withdrawing substituents tend to decrease the reactivity (Chen and Ho, 1996).

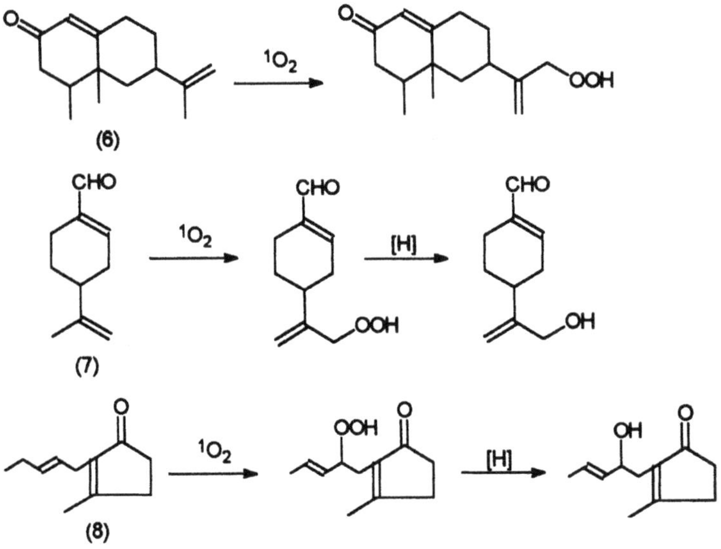

Scheme 16

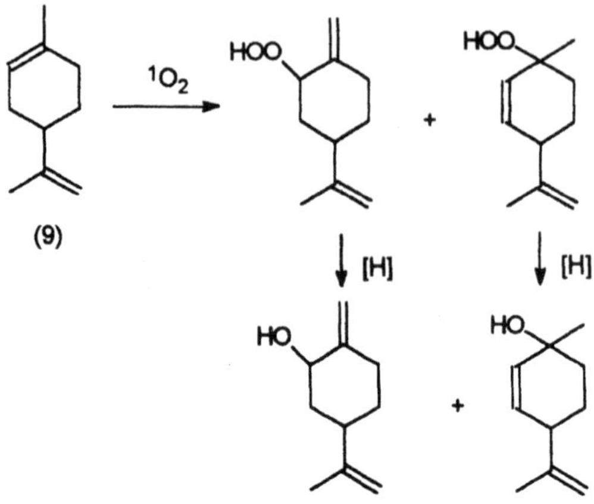

(9)

Scheme 16 (continued)

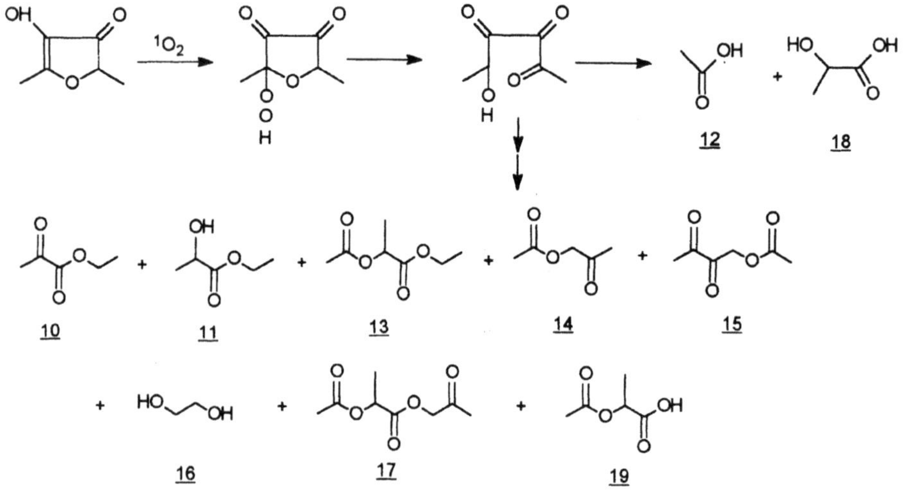

Scheme 17

Furans. Our study shows that on sensitized photooxidation of 2-ethylfuran, singlet oxygen undergoes 1,4-cycloaddition reaction to give a six-membered ring intermediate (Scheme 18). This intermediate is then decomposed to yield a variety of products depending on the solvent, temperature, and other conditions. In case the reaction is carried out in alcoholic solution at room temperature, the alcohol participates in this reaction to form the α,β-unsaturated γ-lactones as demonstrated in Scheme 18. The photosensitized oxidation of furfural has been reported to yield not only an α,β-unsaturated γ-lactone but a δ-lactone (Scheme 19) (Mensah *et al.*, 1986).

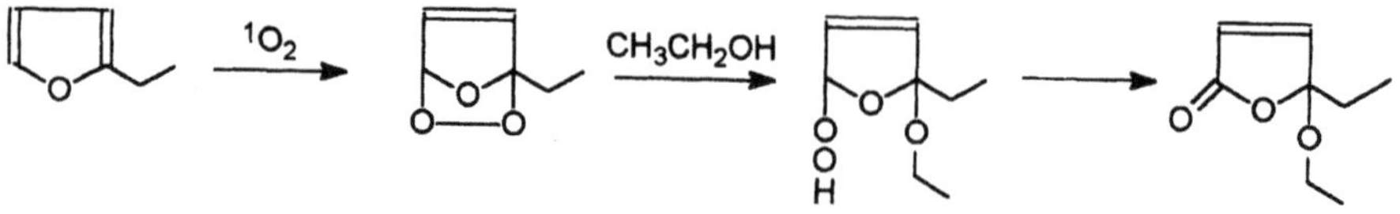

Scheme 18

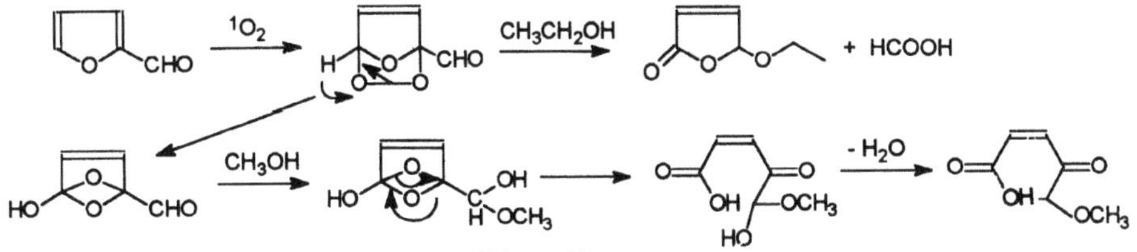

Scheme 19

Pyrroles. It has been known that pyrroles are very sensitive to photosensitized oxidation and give a complex mixture of products. The pathway of pyrrole reaction is very similar to that of furans. In our study, the photosensitized oxidation of 2-ethylpyrrole gave a 5-ethyl-5-ethoxy-1H-pyrrole-2-one and a 1H-pyrrole-2,5-dione in absolute alcohol as shown in Scheme 20.

Thiophenes. It has been reported that thiophenes, unlike the furans and pyrroles, do not react by sensitized photooxidation; however, when thiophene is substituted with alkyl group, it does undergo reaction with singlet oxygen (Matsuura and Saito, 1976). Photosensitized oxidation of 2,5-dimethylthiophene in either chloroform or methanol gives a diketone and a sulfine as shown in Scheme 21 (Skold and Schlessinger, 1970). Three possible pathways which lead to sulfine formation have been proposed (Scheme 21).

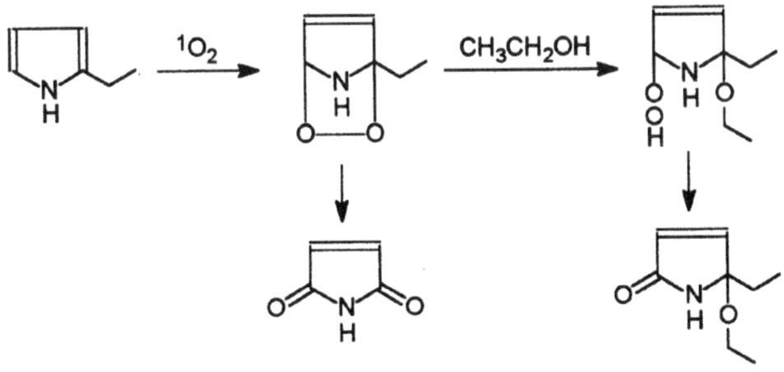

Scheme 20

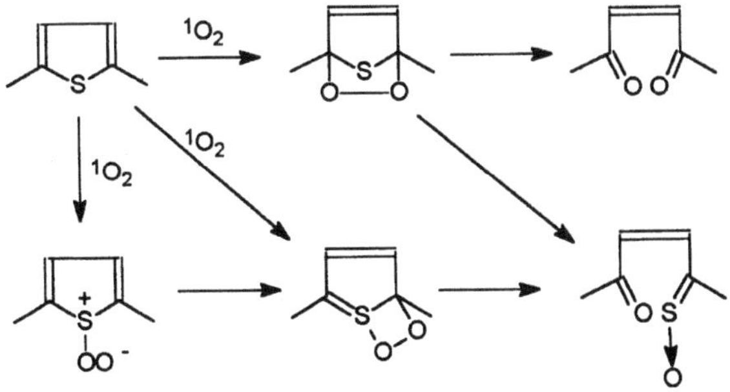

Scheme 21

Imidazoles. Imidazoles behave in many respects like furans and pyrroles on photosensitized oxidation, but the α,β-unsaturated γ-lactones are not the final products. These α,β-unsaturated γ-lactones, which usually are the photosensitized oxidation products of furans and pyrroles, are further cleaved at their enamine double bond to yield a dimethoxy product. Scheme 22 shows the formation of the photosensitized oxidation product dimethoxyhydantoin from imidazole in methanol (Sasserman *et al.*, 1968).

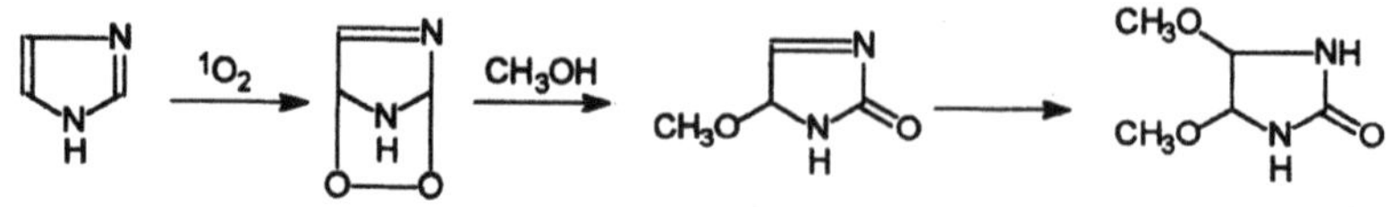

Scheme 22

Oxazoles. Chlorophyll-sensitized photooxidation of 2,4,5-trimethyloxazole in absolute alcohol yields a triacetamide according to our study. The possible pathway for the formation of this product is shown in Scheme 23. Trimethyloxazole reacts with singlet oxygen via a 1,4-cycloaddition mechanism to give a 1,4-endoperoxide, which then proceeds a Baeyer-Villiger rearrangement followed by an O-acyl to N-acyl migration to form a triacetamide.

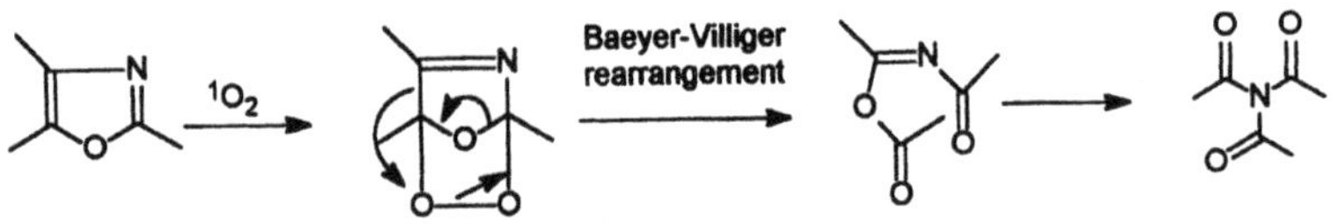

Scheme 23

Thiazoles. The reaction of thiazoles with singlet oxygen is similar to that of oxazoles in many respects, but it gives more complicated results. In our study, the photosensitized oxidation of 2,4,5-trimethylthiazole yielded five compounds including diacetylthioacetamide *20*, 2,4,5-trimethy-loxazole *21*, N-acetyl-2-imino-3-butanone *22*, acetamide *23*, and acetylthioacetamide *24* as shown in Scheme 24. The possible mechanism for the photosensitized oxidation of 2,4,5-trimethylthiazole has been proposed. Singlet oxygen undergoes 1,4-cycloaddition reaction with 2,4,5-trimethylthiazole to give the six-membered endoperoxide ring, which then undergoes a Baeyer-Villiger rearrangement followed by an O-acyl to N-acyl migration to form diacetylthioacetamide. The decomposition of diacetylthioacetamide results in the formation of acetylthioacetamide and acetamide. There is a possible pathway that endoperoxide loses a sulfur atom generating a N-acetyl-2-imino-3-butanone, which is then reduced to afford a 2,4,5-trimethyloxazole.

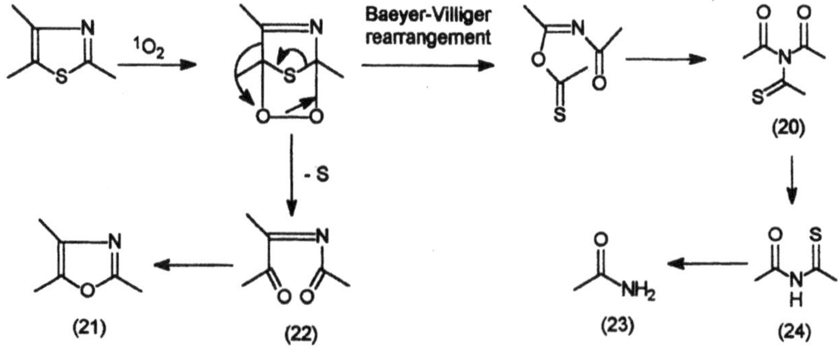

Scheme 24

Other Compounds. In addition to the five-membered heterocyclic ring flavor compounds, the other flavor compounds which contain a conjugated diene system can also react with singlet oxygen by 1,4-cycloaddition reaction. The photosensitized oxidation of β-ionone, which has a violetlike odor, in methanol solution yields the final products *25*, *26* and *27* as shown in Scheme 25 (Shibamoto, 1983).

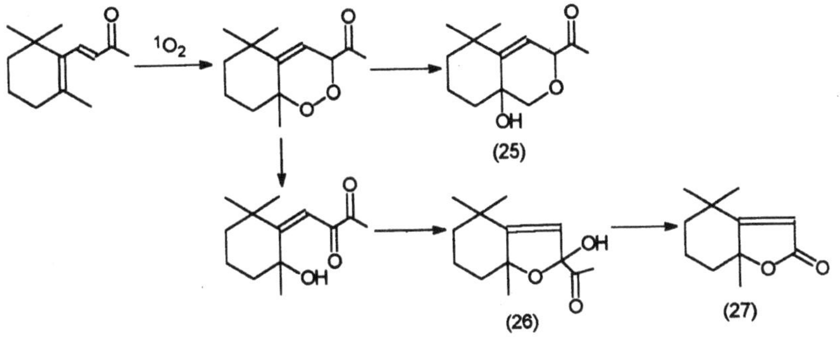

Scheme 25

3. 1,2-Addition Reaction

The typical example of 1,2-addition reaction with singlet oxygen is the photosensitized oxidation of isoeugenol (Eskin, 1979). On sensitized photo-irradiation in sodium a solution of hydroxide, the isoeugenol molecule is attacked by singlet oxygen to form a dioxetane intermediate *28*, which is then converted to a vanillin (Scheme 26). Methylene blue is the most effective sensitizer in terms of vanillin yield.

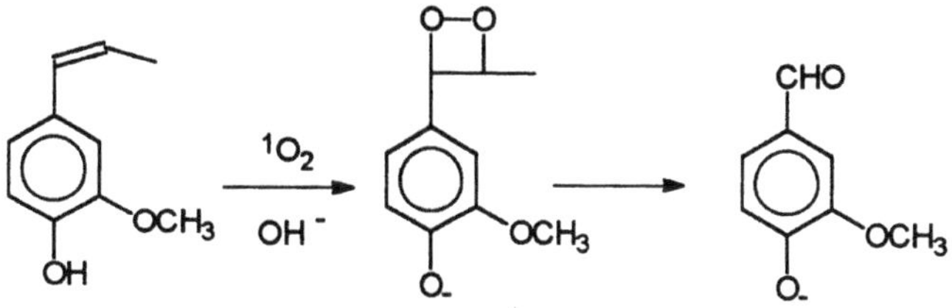

Scheme 26

4. Other Reactions

Type I mechanism, which does not involve the singlet oxygen reaction, is the another mechanism for the photosensitized oxidation of flavor compounds. Benzophenone-sensitized photooxidation of tetrahydrofuran leads to formation of the α-hydroperoxide (Scheme 27) (Schench *et al.*, 1963), which is the same product formed in the unsensitized photooxidation of tetrahydrofuran as demonstrated in Scheme 11.

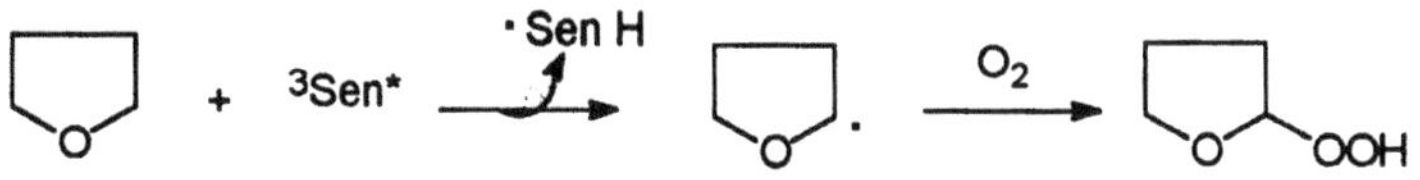

Scheme 27

REFERENCES

Baldwin, J.E.; Nelson, J.P. Cycloadditions. VI. Photosensitized dimerization of -phellandrene. *J. Org. Chem.* **1966,** *31*, 336–338.

Bowen, E.J.; Horton, A.T. Photoreactions of liquid and dissolved ketones. *J. Chem. Soc.* **1934,** 1503–1504.

Bradley, D.G.; Min, D.B. Singlet oxygen oxidation of foods. *CRC Crit. Rev. Food. Sci. Nutr.* **1992,** *31,* 211–236.

Buchardt. O.*Photochemistry of heterocyclic compounds.* John Wiley & Sons Inc., New York, 1976.

Büchi, G.; Yang, N.C. Light-catalyzed organic reactions. III. cis- -Ionone. Helv. Chim. Acta. **1955,** *38,* 1338–1341.

Cantrell, T.S. Reactivity of photochemically excited 3-acylthiophenes, 3-acylfurans, and the formythiophenes and furans. *J. Org. Chem.* **1977,** *4,* 3774–3776.

Chen, C.-W.; Ho, C.-T. Reactivity on photooxidation of selected five-membered heterocyclic flavor compounds. *J. Agric. Food Chem.* **1996,** *44,* 2078–2080.

Chen, C.-W.; Shu, C.-K.; Ho, C.-T. Photosensitized oxidative reaction of 2,5-dimethyl-4-hydroxy-3(2H)-furanone. *J. Agric. Food Chem.* **1996,** *44,* 2361–2365.

Coxon, J.M.; Halton, B. *Organic photochemistry.* Cambridge University Press, London, 1974.

Dalton, J.C.; Dawes, K.; Turro, N.J.; Weiss, D.S.; Barltrop, J.A.; Coyle, J.D. Type I and type II photochemical reactions of some five- and six-membered cycloalkanones. *J. Am. Chem. Soc.* **1971,** 93, 7213–7221.

Dunion, P.; Trumbore, C.N. The role of the triplet excited state in the photolysis and radiolysis of liquid cyclopentanone. *J. Am. Chem. Soc.* **1965,** *87,* 4211–4212.

Eastman, R.H.; Starr, J.E.; Martin, R.S.; Sakata, M.K. The photolysis of thujone. *J. Org. Chem.* **1963,** *28,* 2162–2163.

Eskin, K. Sensitized photooxidation of isoeugenol to vanillin. *Photochem. Photobiol.* **1979,** *29,* 609–610.

Foote, C.S. Phtosensitized oxidation and singlet oxygen: consequences in biological systems. In *Free Radicals in Biology;* Pryor, W.A., Ed.; Academic Press, New York, 1976; Vol. 2, pp. 85–133.

Foote, C.S.; Wexler, S.; Ando, W. Chemistry of singlet oxygen. III. Product selectivity. *Tetrahedron Lett.* **1965,** *46,* 4111–4118.

Fröhlich, O.; Shibamoto, T. Stability of pulegone and thujone in ethanolic solution. *J. Agric. Food Chem.* **1990,** *38,* 2057–2060.

Gollnick, K.; Schenck, G.O. Mechanisms and stereoselectivity of phototsensitized oxygen transfer reaction. *Pure Appl. Chem.* **1964,** *9,* 507–525.

Hall, R.A.; Oser, B.L. Recent progress in the consideration of flavoring ingredients under the food additives amendment III GRAS substances. *Food Technol.* **1965,** *19,* 253–271.

Horspool, W.M.*Aspects of Organic Photochemistry.* Academic Press, New York, 1976; pp. 124–186.

Kawana, M.; Emoto, E. Synthesis of pyrrole nucleoxide by the photo-dehydrogenation of ^{3}-pyrroline derivatives. *Bull. Chem. Soc. Jpn.* **1969,** *42,* 3539–3546.

Kropp, P.J.; Drauss, H.J.Photochemistry of cycloalkenes. III. Ionic behavior in protic media and isomerization in aromatic hydrocarbon media. *J. Am. Chem. Soc.* **1967,** *89,* 5199–5208.

Liu, R.S.H.; Hammond, G.S. Photosensitized internal addition of dienes to olefins. *J. Am. Chem. Soc.* **1967,** *89,* 4936–4944.

MacLeod, W.D.; Buigues, N.M. Sesquiterpenes I. Nootkatone, a new grapefruit flavor constituent. *J. Food Sci.* **1964,** *29,* 565–568.

Marshall, J.A.; Carroll, R.D. The photochemically initiated addition of alcohols to 1-methene. A new type of photochemical addition to olefins. *J. Am. Chem. Soc.* **1966,** *88,* 4092–4093.

Matsui, T.A.; Komatsu, A.; Moroe, T. Photochemistry of isopulegone. *Bull. Chem. Soc. Jpn.* **1967,** *40,* 220?.

Matsuura, T.; Ito, Y.; Saito, I. Photoinduced reactions. LXVIII. Photochemical dehydrogenation of imidazolines to imidazoles. *Bull. Chem. Soc. Jpn.* **1973,** *46,* 3805–3809.

Matsuura, T.; Saito, I. Photooxidation of heterocyclic compounds. In *Photochemistry of Heterocyclic Compounds,* Buchardt, O., Ed.; John Wiley & Sons Inc., New York, 1976; pp. 456–523.

Mensah, T.A.; Ho, C-.T.; Chang, S.S. Products identified from photosensitized oxidation of selected furanoid flavor compounds. *J. Agric. Food Chem.* **1986,** *34,* 336–338.

Nofal, M.A.; Ho, C-.T.; Chang, S.S. New constituents in Egyptian jasmine absolute. *Perfum. Flavor. 1982,* **6,** 24–34.

Opdyke, D.L. Monographs on fragrance raw materials - plegon. *Food Cosmet. Toxicol.* **1978,** *16,* 867–868.

Rawls, H.R.; Van Santen, P.J. A possible role for singlet oxygen in the initiation of fatty acid autoxidation. *J. Am. Oil Chem. Soc.* **1970,** *47,* 121–125.

Rosenthal, I. Photooxidation of food. In *Singlet O_2; Vol IV Polymers and Biomolecules*, Frimer, A.A., Ed.; CRC Press, Florida, 1985; pp. 145–163.

Schenck, G.O.; Becker, H.-D.; Schulte-Elte, K.-H.; Krauch, C.H. The benzophenone-photosensitized autoxidation of sencondary alcohols and ethers. Preparation of -hydroperoxides. *Chem. Ber.* **1963,** *96,* 509–516.

Shahidi, F. Flavor of cooked meats. In *Flavor Chemistry-Trends and Developments*, Teranishi, R.; Buttery, R.G.; Shahidi, F., Ed.; American Chemical Society, Washington, DC, 1989; pp. 189–201.

Shibamoto, T. Photochemistry of fragrance materials. II. Aromatic compounds and phototoxicity. *J. Toxicol.-cut. & Ocular Toxicol.* **1983,** *2,* 267–375.

Shibamoto, T.; Mihara, S. Photochemistry of fragrance materials. I. unsaturated compounds. *J. Toxicol. -cut. & Ocular Toxicol.* **1983,** *2,* 153–192.

Simonaitis, R.; Cowell, G.W.; Pittis, Jr. J.N. Photoreduction of cyclopentanone and cyclohexanone. *Tetrahedron Lett.* **1967,** 3751–3754.

Skold, C.N.; Schlessinger, R.H. The reaction of singlet oxygen with a simple thiophene. *Tetrahedron Lett.* **1970,** *10,* 791–794.

Stenburg, V.I.; Wang, C.T.; Kulevsky, N. Photochemical oxidations. III. Photochemical and thermal behavior of α-hydroperoxytetrahydrofuran and its implications concerning the mechanism of photooxidation of ethers. *J. Org. Chem.* **1970,** *35,* 1774–1777.

Tateba, H.; Mihara, S.Photochemical products obtained from jasmone. *Agric. Biol. Chem.* **1986,** *50,* 2681–2683.

Tateba, H.; Morita, K.; Kameda, W.; Tada, M. Photochemical reaction of perillaldehyde under various conditions. *Biosci. Biotech. Biochem.* **1992a,** *56,* 614–619.

Tateba, H.; Morita, K.; Tada, M. Photochemical products obtained from jasmone. *J Chem. Res.* **1992b,** 140–141.

Tateba, H.; Morita, K.; Kameda, W.; Tada, M. Photochemical reactions of (Z)-jasmone under various conditions. *Biosci. Biotech. Biochem.* **1993,** *57,* 220–226.

Toda, H.; Mihara, S.; Umano, K.; Shibamoto, T. Photochemical studies on jasmin oil. *J. Agric. Food Chem.* **1983,** *31,* 554–558.

Wagner, P.J.; Spoerke, R.W. Triplet lifetime of cyclic ketones. *J. Am. Chem. Soc.* **1969,** *91,* 4437–4440.

Wasserman, H.H.; Stiller, K.; Floyd, M.B. The reactions of heterocyclic systems with singlet oxygen. Photosensitized oxygenation of imidazoles. *Tetrahedron Lett.* **1968,** 3277–3280.

Whipple, E.B.; Evanega, G.R. The assignment of configuration to the photoaddition products of unsymmetrical carbonyls to furan using pseudocontact shifts. *Tetrahedron.* **1968,** *24,* 1299–1310.

White, J.D.; Gupta, D.N.Photochemical cyclization of -farnesene. *Tetrahedron.* **1969,** *25,* 3331–3339.

Yamada, M.; Hanai, S. *Biacetyl., J. Soc. Brewing (Jpn).* **1951,** *46,* 47–49.

INDEX

Acids, in yogurt, 293
Actin, 28
Activation energies, 94–95
Active oxygen method, 194
β-Alanine, 263–265
Albumin, egg, 101–107
Alcohol dehydrogenase, 38
Aldehydes, in yogurt, 290
Aldol condensation, 240
Alkylpyrazines, 217
Alkylresorcinols, 118
Allium sativum L. 277–278
Amadori compounds, 216–217, 261–263
Amadori rearrangement product, 202, 207–209, 214–215
Amino acids, 300
  flaxseed, 313
Amino-reductones, 269–276
Ammonia, 301
  flavor from, 301
Ammonium bicarbonate, 301
Amylopectin, 304
Amylose-lipid complex, 113
1-Anilinonaphthalene-8-sulfonic acid, 10–11, 37
Anthocyanins, 328
Antioxidants, 153
Antioxidative properties, 201–211, 221–225
  phenolic compounds in wines, 329
Antioxidizing potentials, 181–187
Aseptic processing, 91–98
  conventional, 97
  particulate foods, 91–98
Asparagine, 303
Autolysis assays, 61–62

Bacillus stearothermophilus, 92, 95
Beer, black, 226
BHA, 185–187
BHT, 185–187
Bluefish, 75
Brix, 131

Caramel, 227–228
Carotene, 116, 138
Casein, 205
Cathepsins, 27
Chemical marker compounds, 92–93
Chemiluminescence intensity, 205
Chemometric, 91–98
Chemometric model, 96
Chlorogenic acid, 119
Chlorophyll, 138, 155, 173
Chromatography
  high performance liquid, 128
  ion-exchange, 128
Chymotrypsin, bovine, 71
Circular dichroism, 16
  far-UV, 16
  vibrational, 19
cis-Parinaric acid, 10–11
Clostridium botulinum, 96
Clonorchis sinensis, 281
Coagulation, 105
Cod, 75
Color analysis, 62
Color compound, polyhydroxy phenolic, 124
Color formation, 128–132
Conlinin, 314
Curing, peanut, 38–39
Cystamine, 220
Cysteine, 300
  flavor from, 300

Day-lily flower, 279–280
Degradation, proteolytic, 26–29
Denaturation, 115
  extrusion-induced protein, 115
  pressure-induced protein, 51
Diacylglycerols, 162
Die geometry, 110
Dietary fiber, 114
  flaxseed, 320
Difructose dianhydrides, 124, 132

2,3-Dihydro-3,5-dihydroxy-6-methyl-4(H)-pyran-4-one, 91
Diketopiperazine, 217
2,5-Dimethyl-4-hydroxy-3(2H)-furanone, 348–349
Dipole moment, 271
Disaccharides, 261
Duncan's multiple range test, 163
D-value, 95

Egg white, dried, 229
Electron microscopy, high-resolution transmission, 8
Ellagic acid, 335
Enaminols, 269
Enediamines, 269
ESR spectra, 211
Esters, in yogurt, 292
2-Ethylfuran, 350
Extraction
  dynamic headspace, 286–288
  simultaneous distillation, 286–288
Extrusion, 109–119, 297–305
  corn, 303
  potato, 303

Fatty acid, 162
  free, 146, 186, 310
  omega-3, 181
  polyunsaturated, 138
  trans, 141
Feed moisture, 110
Feed rate, 110
Fiber
  dietary, 114
  soluble, 114
Fish product, 181–187
Flavonoids, 328–329
Flavor compounds, 215–221
  peptide generated, 217–218
  photochemical reaction, 341–353
Flavor deterioration, in yogurt, 285–295
Flavors
  binding of, 36
  precursors, 299
  yogurt, 285–295
Flavor composition, effect of gamma irradiation on, 277–283
Flavor generation, during extrusion cooking, 297–305
Flavor retention, during extrusion cooking, 298–299
Flax, 307
Flaxseed, 307–320
Formaldehyde, 50
Fluorescent probe, 37
Free radical interceptors, 192
Freezing, 50
Fructose, 125–126, 261
Fructosylglycine, 206–207, 274
Frying, 145–149
Frying oils, 168–170
Furfural, 350

Galactomannan, 229
Gallic acid, 335
Gamma irradiation, 277–283
Garlic, 277–278
Gels
  breaking strength of, 103
  elasticity of, 78
  firmness of, 78
  protein, 26
  shear strain, 27, 47
  shear stress of, 27, 47
Gel strength, 60
Gelatinization, 112
Genistein, 119
Germination, flaxseed, 311
Glass transition temperature, 52
Glucose, 125, 261
Glucosinolates, rapeseed, 118
Glutamine, 303
Glycoalkaloids, potato, 118
Glycolipids, 135
Glycosidase, 318
Glycosides
  cyanogenic, 316–319
  flaxseed, 316–319
Glycolysis, 68
Green tea, 198
Guava, 81–89
Guava juice, 81–89
  cloud content of, 87

Heating
  roasting, 152–153
  microwave, 152–153
Hemerocallis spp., 279–280
High hydrostatic pressure, 57–64, 69–70, 81–89
High performance liquid chromatography, 128, 130, 256–257, 262, 287
High pressure processing, 45–53, 67–79
Horse mackerel, 182
Hydroperoxides, 175
Hydrophobic-hydrophilic balance, 105
Hydrophobicity values, 11
Hydroxycinnamate, 328
Hydroxyl radical, 206
4-Hydroxy-5-methyl-3(2H)-furanone, 91
5-(Hydroxymethyl)-2-furaldehyde, 125
Hyplorchis spp., 281
Hypophthalmichythys molitrix Cuvier, 281
Hypophthalmichythys molitrix Valenciennes, 281

Ionization potential, 271
Imidazoles, 351

Juice, guava, 81–89

Kestoses, 124

Lard, 144
Lactobacillus bulgaricus, 286
Lentinus edodes Sing, 278
Lethality, 96
Light scattering, 7–8
  dynamic, 8
  quasi-elastic, 8
Lignans
Linin, 313
Linoleic acid, 161
Linolenic acid, 161
Linseed, 307
Linum usitatissimum L., 307
Linustatin, 316
Lipases, 143
Lipids, 112, 135–157
  modification of , 140–152
  thermal oxidation, 146–149
Lysine, 115
Lysophosphatidylcholine, 186
Lysozyme, 208–209

Mackerel, 75
Macroglobulin, 76
Maillard reactions, 115, 213–231, 237–244, 297
Maillard reaction products, 201
  antinutritional effects of, 231
  antioxidative effect of, 221–225, 245–253
  desmutagenicity of, 226–228
  metal chelating activity, 246–247, 249
Malonaldehyde, 311
Maltol, 261
Mass spectrometry
  capillary gas chromatography-, 257
  elctrospray ionization, 8
  fast atom bombardment-, 257
  gas chromatography, 287, 301, 303
Maturity, peanut, 38
Meal, flaxseed, 310
Melanoidins, 202, 207–208, 226, 229–231
Merluccius productus, 25
Metal chelating activity, 245
Metal ions, 191
Microbial analysis, 62
Minerals, 117
  flaxseed, 322–23
Molasses, 227
Molecular orbital
  highest occupied, 271–276
  lowest unoccupied, 271–276
Monoacylglycerols, 162–164
Monosaccharides, acid degradation of, 125–126
Mushroom, Shiitake, 278–280
Myosin, 28
  muscle, 45
Myrcene, 345

Natural toxins, 117–118
Neolinustatin, 316
Neutron diffraction, 8
Nitrosamines, 227
NMR, 9, 12–14, 124, 163

Ohmic heating, 25–33, 101–107
Oil
  alkali refining of, 137
  bleaching of, 138
  corn, 304
  degumming of, 136
  deodorization of, 138
  edible, 135–157
  extraction of, 136
  flaxseed, 153, 309
  fractionation, 145
  hydrogenation, 140
  neutralization of, 137
  olive, 139
  refining of, 136–140
  sesame, 139, 155
  soybean, 161–175
  transesterification, 142
  winterization of, 138
Oreochromis mossambicus Peters, 281
Oxazoles, 352
Oxetane, 343
Oxidation, lipid, 112
Oxidative stability, 161–179, 189–199
Oxygen, 189
Oxygen absorber, 186
Oxygen consumption measurement, 249–250

Parasies, 281
Pectin, 81, 86
  molecular characteristics, 87
Pectin esterification, degree of, 86
Penaeus monodon Fabricius, 280–281
Perilaldehyde, 347
Peroxidants, 190
Peroxides, 175
Peroxide value, 192, 203
pH analysis, 63–64
Phase change, 107
α-Phellandrene, 345
Phenolic compounds, 119
  flaxseed, 319–320
  grape juice, 327
  wines, 327–338
Phosphatidylcholine, 164
Phosphatidylethanolamine, 169
Phosphatidyl inositol, 164
Phospholipids, 135–137, 164–168
  flaxseed, 312
Photochemical reactions, 341–353
Photosensitized reaction, 341–342
Phytic acid, flaxseed, 321
Phytoestrogens, 119
Polarimetry, 128
Pollock, Alaska, 26, 47, 76

Polyphenols, 138
Polysaccharides, flaxseed, 321
Potato, 280
Procyanidins, 328
Proline, 219, 242
Protease, 36, 68
  fish meat, 52
Proteins, 5–23
  fish, 25–33, 45–53
    effects of high pressure on, 45–53
    gelation of, 29–32
  fish muscle, 46, 68
  flaxseed, 313
  myofibrillar, 27, 49
  molten globule, 9
  peanut, 35–41
  quaternary structure of, 6–8
  soy, 35
  surface hydrophobicity of, 35
  tertiary structural changes of, 8–16
Protein surface hydrophobicity, 35, 37–41
Pulegone, 346
Pyrazines, 218–219, 301302
Pyrolysis-GC/MS, 238–239
Pyrroles, 351
Pyrrolidines, 219
Pyrrolines, 219
Pyruvaldehyde, 240, 301

Quadratic model, 95
Quenchers, 192

Rate constants, 94–95
Reducing sugars, 113, 125, 257
Resazurin test, 287, 289
Rosemary, 198

Salmon, 195–196
Scavenging activity, 205–206
Schiff base, 214
Scobamic acid, 274
Screw configuration, 110
Screw speed, 110
Seafoods, 67–79
Secoisolariciresinol, 320
Sedimentation, 7
  equilibrium, 7
  velocity, 7
Sensitizer, 342
Sesame oil, 139
Sesame seed, 140
Sesamin, 153
Sesaminol, 140
Sesamol, 140
Sesamoline, 140, 153, 156
Silver carp, 281
Singlet oxygen, 341
Solubility, protein, 115
Soybean paste, 204
Soy isolate, 115
Soy sauce, 226
Specific mechanical energy, 110
Spectrophotometry, derivative, 9
Spectra
  ESR, 211
  infrared, 163
  mass, 163
Spectroscopy
  fluorescence, 10
  FTIR, 17
  near-UV circular dichroism
  nuclear magnetic resonance, 12–14
  optical mixing, 8
  photon correlation, 8
  Raman, 14–15
    visible laser, 14
  spectrum analyzer, 8
  vibrational, 17
Spores, 92
Squid mantle, 76
Sterilization, 92
Strecker aldehydes, 217–218
Strecker degradation, 240
Streptococcus thermophilus, 286
Sucrose, 123–132
  acid hydrolysis of, 124
Sugar, 123–132
Sulfite, 330
Superoxide radical, 341
Surfimi, 47
  Alaska pollock, 26, 58, 76
  gelation of, 49
  hoki, 26
  Pacific whiting, 25–29, 57–64
  sardine, 26
  Southern blue whiting, 26

TBHQ, 185–187
Tea, green, 198
Temperature
  barrel, 110
  mass, 110
  product, 110
Terpenes, 291
Texturization, extrusion, 115
Theragra chalcogramma, 47
Thiamine, 117
Thiazoles, 352
Thiazolidines, 220
Thiazolines, 220
2-Thiobarbituic acid, 155, 205, 251
Thiophenes, 351
Tiger prawn, 280–281
Tilapia, 281
Time, residence, 110
Titratible acidity, 332
Tocopherols, 170–173, 184–187
  alpha-, 171
  oxidized alpha-, 172

Trachurus japonicus, 182
Transglutaminase, 47
Transglycosidation, 113
Triglycerides, 135
  hydrolysis of, 112, 151
  thermal decomposition of, 150
  thermally oxidized, 168
  pyrolysis of, 151
Tripeptides, 217
Trypsins, fish, 71
Trypsin inhibitors, 118
Turbidity, 84

Ultracentrifugation, 7
Unsaponifiable matter, 176

Viscosity, 84
Vitamins, 116–117
  flaxseed, 322–323
Vitis vinifera cv. Cabernet sauvignon, 332–333
Vitis rotundifolia cv. Noble, 332–333
Volatile compounds, 255–266
  day-lily flower, 279–280
  garlic, 277–278
  ginger, 278
  Shiitake mushroom, 278–279
  tiger prawn, 280–281

Whiting, Pacific, 25, 57–64

Xanthophyll, 138
X-ray scattering, small angle, 104
X-ray diffraction, 8

Yeast extract, autolyzed, 301
Yogurt, 285–295

Zein, 303
Z-value, 95

www.ingramcontent.com/pod-product-compliance
Ingram Content Group UK Ltd.
Pitfield, Milton Keynes, MK11 3LW, UK
UKHW020059200726
13856UKWH00002B/291

* 9 7 8 1 4 8 9 9 1 9 2 6 7 *